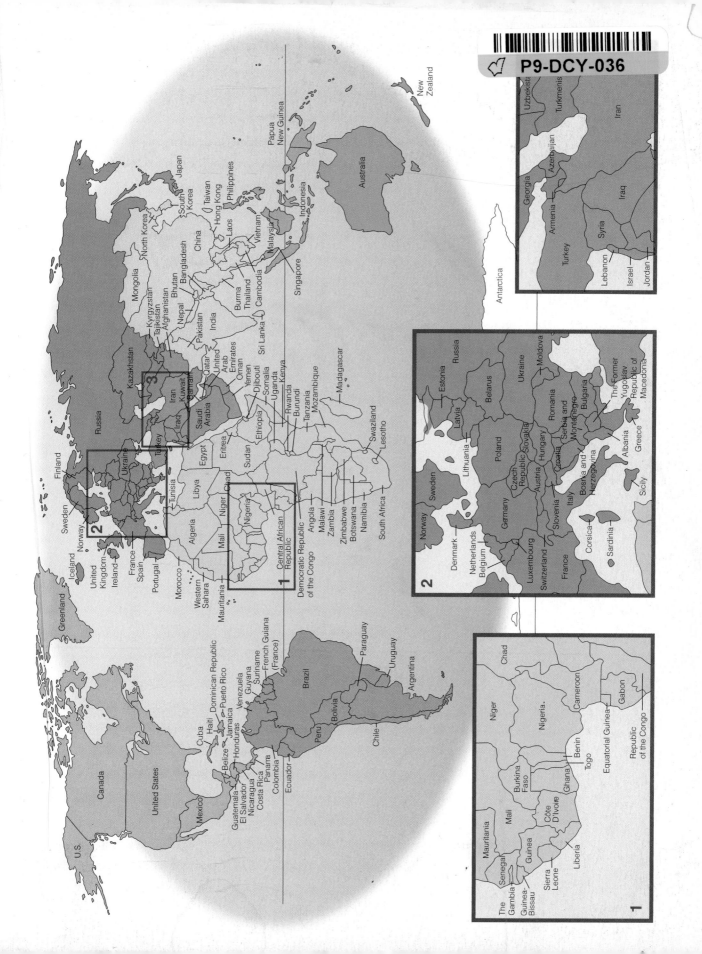

EIGHTH EDITION

Essentials of Sociology

David B. Brinkerhoff
University of Nebraska, Lincoln

Suzanne T. Ortega
University of New Mexico

Lynn K. White
University of Nebraska, Lincoln

Rose Weitz
Arizona State University

WADSWORTH
CENGAGE Learning™

Australia • Brazil • Japan • Korea • Mexico • Singapore • Spain • United Kingdom • United States

Essentials of Sociology, Eighth Edition
David B. Brinkerhoff, Lynn K. White,
Suzanne T. Ortega, Rose Weitz

Senior Publisher: Linda Schreiber-
　Ganster

Sociology Editor: Erin Mitchell

Developmental Editor:
　Kristin Makarewycz

Assistant Editor: Erin Parkins

Editorial Assistant: Rachael Krapf

Media Editor: Melanie Cregger

Marketing Manager: Andrew Keay

Marketing Assistant: Jillian Myers

Marketing Communications Manager:
　Laura Localio

Content Project Manager:
　Cheri Palmer

Creative Director: Rob Hugel

Art Director: Caryl Gorska

Print Buyer: Judy Inouye

Rights Acquisitions Account Manager,
　Text: Roberta Broyer

Rights Acquisitions Account Manager,
　Image: Leitha Etheridge-Sims

Production Service:
　Macmillan Publishing Solutions

Copy Editor: Gary Morris

Text Designer: tani hasegawa

Photo Researcher:
　Tim Herzog, Bill Smith Group

Illustrator: MPS Limited, A Macmillan
　Company

Cover Designer: RHDG

Cover Image: Brian Stauffer/The iSpot

Compositor: MPS Limited, A Macmillan
　Company

Library of Congress Control Number: 2009939348
ISBN-13: 978-0-495-81295-1
ISBN-10: 0-495-81295-1

Wadsworth
20 Davis Drive
Belmont, CA 94002-3098
USA

Cengage Learning is a leading provider of customized learning solutions with office locations around the globe, including Singapore, the United Kingdom, Australia, Mexico, Brazil, and Japan. Locate your local office at: **www.cengage.com/global**

Cengage Learning products are represented in Canada by Nelson Education, Ltd.

To learn more about Wadsworth, visit **www.cengage.com/Wadsworth**

Purchase any of our products at your local college store or at our preferred online store www.**CengageBrain.com**

Printed in Canada
1　2　3　4　5　6　7　13　12　11　10　09

Brief Contents

Contents

CHAPTER 2

Culture   31

CHAPTER 3

Socialization   55

CHAPTER 4

Social Structure and Social Interaction   76

CHAPTER 5

Groups, Networks, and Organizations   100

CHAPTER 6

Deviance, Crime, and Social Control   126

CHAPTER 10

Health and Health Care   232

CHAPTER 11

Family 257

CHAPTER 12

Education and Religion   283

Pedagogical Features

Focus on American Diversity

Focus on A Global Perspective

Focus on Media and Culture

Maps

Preface

Like other sociologists, the writers of this textbook can all remember when they first recognized the power of a sociological perspective. Rose Weitz, for example, remembers the moment she first realized that her graduate Research Methods professor never called on any female student if a male student had his hand raised. Before that point, she had thought she could become more involved in classroom discussions if she only gave better answers or looked more alert or sat in the front of the classroom or . . . . Afterwards, she realized that the professor's behavior reflected his attitudes toward women rather than anything Rose or the other female students were doing. This started Rose on a career exploring (among other things) the sources of stereotypical attitudes toward girls and women, the power structure that allowed discriminatory behavior and attitudes to continue, the consequences of those attitudes and behaviors, and the means through which destructive behaviors and attitudes might be changed.

As this example suggests, sociology offers a set of tools for looking at the world that can help all of us to better understand both individual behaviors and the broader context in which those behaviors occur. Sociology deals with all the crucial issues—both micro and macro—that confront our lives, our nation, and our planet. At the micro level, sociology explores the substance of ordinary life—getting a job, getting married (and maybe divorced), caring for children, and having fun. At the macro level, sociology grapples with the critical national and international problems of our times, including homelessness, health-care reform, environmental degradation, poverty, and war. This textbook is designed both to instill a sociological perspective in students and to increase their knowledge about and passion for all these issues.

Organizing Principles

This book is organized around two sets of principles. First, we have structured the book to emphasize concision, balance, critical thinking, accessibility, and up-to-date research. Second, we have chosen material to emphasize race, class, and gender; global issues; and the everyday life of young Americans.

- **Brevity:** *Essentials of Sociology* is designed to meet the needs of professors who prefer a more concise textbook. Our goal is to provide enough material to spark students' curiosity about and interest in sociology while keeping the book concise enough to be used by those who teach on a quarter system, prefer a less expensive text, or prefer a less detailed text to use on its own or with supplementary readings.
- **Balance:** A central goal of this textbook is to provide a balanced, unbiased approach to the three central theoretical perspectives in sociology. Each of these perspectives is described in detail in Chapter 1 and addressed in each subsequent chapter. Faculty should feel comfortable teaching from this textbook regardless of their personal approach to the field.

- **Critical thinking:** Throughout the textbook, professors and students will find critical thinking questions (described below), engaging features (also described below), intellectual arguments, and other materials designed to encourage students to engage in critical thinking.
- **Accessibility:** As in earlier editions, we have paid careful attention to writing this book in a style that students will find interesting and accessible. In addition, this edition includes an added emphasis on visual aids (such as Concept Summaries and maps) designed to capture the attention of those students who are more attuned to visual materials.
- **Up-to-Date Research:** The textbook has been revised to take into account recent developments in the field as well as "hot" new topics such as the rise in atheism and the current economic crisis. We have reviewed all the major journals and many of the specialty journals to provide students with the most current findings and have revised or added tables and figures to incorporate the very latest data available.
- **Race, Class, and Gender Focus:** Race, class, and gender (and the ways that they interweave) serve as central concerns throughout *Essentials of Sociology*. Each of these topics is discussed in its own chapter, in the "Focus on American Diversity" boxes (discussed below), and in numerous other points.
- **Global Focus:** An emphasis on global issues continues to characterize this edition of *Essentials of Sociology*. This emphasis is evidenced in the "Focus on a Global Perspective" boxes (discussed below), in maps on topics such as world religions and maternal mortality, in the sections on global stratification and on health care around the world, and in examples, tables, and case studies in various chapters.
- **Focus on Students' Everyday Lives:** A central goal of this book is to engage students' interest in sociological concepts and perspectives. One way we have done this is by using examples from the everyday lives of contemporary young people. For example, we use Twitter as an example of social networks and "hooking up" as an example of changing sexual scripts.

Retained Features

- **Maps, Figures, and Tables:** The use of visual materials has increased in this edition. Each chapter now includes at least one map, on topics such as poverty rates (Chapter 1), the percentage of first-graders internationally who reach fifth grade (Chapter 3), and the rate of home foreclosure (Chapter 13). Tables and figures are also distributed throughout the textbook.
- **Focus Boxes:** Throughout the textbook, boxed inserts introduce provocative and engaging issues. "Focus on American Diversity" boxes analyze the diversity of American lives and society, covering topics from American Muslims to methamphetamine use in rural America. "Focus on a Global Perspective" boxes introduce students to a comparative approach to social issues. Topics include water inequality in developing nations and the impact of cell phones on Indian culture. "Focus on Media and Culture" boxes offer sociological insights into a wide range of topics that students will find especially interesting, including the use of extreme body modification and the heavy-drinking culture of spring break.
- **Concept Summaries:** When several related concepts are introduced (for example, pluralist, power elite, and state autonomy models of American government), a "Concept Summary" table is included to summarize the definitions, give examples, and clarify differences.

- **Margin Glossary:** In addition to the full glossary at the end of the book, whenever new terms and concepts first appear in the text, they are printed in boldface type, with concise definitions set out clearly in the margin.
- **"Where This Leaves Us" sections:** Each chapter concludes with a section that ties together the concepts in the chapter, discusses the chapter's implications, and links it to larger themes.
- **Chapter Summary:** Each chapter includes a short point-by-point summary of its chief points. These summaries will aid beginning students as they study the text and help them distinguish the central concepts from the supporting points.
- **"Thinking Critically" Questions:** At the end of each chapter, students will find several critical thinking questions that challenge them to apply sociological concepts and theory to problems relevant to their lives. These questions also can be used for group discussion or individual writing assignments.

New Features

- **"Sociology and You" boxes:** Each chapter includes two small boxes that pull students into the material by having them apply sociological concepts to their own lives. For example, a box in Chapter 7 invites students to think about their own ascribed and achieved statuses, and a box in Chapter 11 explores how growing up in a nuclear, extended, or blended family may affect students' life chances.
- **"Decoding the Data" boxes:** These boxes introduce students to the process of thinking about sociological data. Each box provides students with provocative data about critical social issues, such as racial and ethnic self-identification, demographic differences in strong ties, and the impact of single motherhood in different nations. These boxes also provide students with questions designed to help them both *understand* the implications of the data and *critique* the data. These boxes should work well for class or small group discussions.

New Topics and Coverage in the Eighth Edition

In addition to thorough research updates throughout the chapters, significant chapter-by-chapter changes include:

Chapter 1: The Study of Society

Using a bachelor's degree in sociology
Using existing statistics
Content analysis
Focus on American Diversity: Studying Life in "The Projects"
Concept Summary: Understanding Spurious Relationships
Decoding the Data: Alcohol Use
Map: States with Low, Medium, and High Percentages of Residents in Poverty
Figure: College Grades and Frequency of Alcohol Use

Chapter 2: Culture

Focus on Media and Culture: The Media and Self-Esteem
Decoding the Data: International Disapproval of Aspects of Globalization
Map: Percent of U.S. Residents 5 Years and Over Who Speak a Language Other Than English at Home
Figure: Ethnocentrism around the World
Figure: Debt versus Savings in U.S. Households with Savings

Chapter 3: Socialization

Case Study: Learning social class at the toy store
Case Study: Resocializing young offenders
Concept Summary: Types of Socialization
Concept Summary: The Looking-Glass Self
Decoding the Data: Attitudes Toward Spanking
Map: Percentage of First-Graders Who Continue through Fifth Grade
Table: Daily Hours and Minutes of Media Usage among 8–18 Year Olds

Chapter 4: Social Structure and Social Interaction

Postindustrial societies
Impression management
Focus on Media and Culture: Alcohol and Spring Break
Focus on American Diversity: Becoming Goth
Concept Summary: Using Disclaimers and Accounts
Decoding the Data: American Diversity
Map: Mixed-Race People in the United States

Chapter 5: Groups, Networks, and Organizations

Milgram experiments
Reference groups and relative deprivation
Focus on A Global Perspective: Talking About AIDS in Mozambique
Focus on Media and Culture: Gaming and Social Life
Decoding the Data: Strong Ties
Map: Number of Persons Infected with HIV per 1,000 Residents, Ages 15 to 49

Chapter 6: Deviance, Crime, and Social Control

Focus on Media and Culture: Extreme Body Modification
Decoding the Data: Legalizing Marijuana
Map: Violent Crime Rate by State

Chapter 7: Stratification

Concentrated poverty
The near poor
Global inequality and war
Global inequality and terrorism
Focus on A Global Perspective: Water and Global Inequality
Map: Most- to Least-Developed Nations
Figure: Income Inequality and *Lack* of Social Mobility

Chapter 8: Racial and Ethnic Inequality

The stages of genocide
Multiracial Americans
White privilege
Focus on American Diversity: The Election of Barack Hussein Obama
Decoding the Data: Race and Job Interviews
Map: Genocide and Genocide Risk Internationally

Chapter 9: Sex, Gender, and Sexuality

Sexual scripts
Compulsive heterosexuality
Focus on A Global Perspective: Pregnancy and Death in Less-Developed Nations
Map: Lifetime Risk of Dying from Pregnancy or Childbirth

Chapter 10: Health and Health Care

Health care in other countries
Single-payer health care
Focus on Media and Culture: The Internet and Health
Map: Percent Sometimes Unable to Afford Needed Medical Care or Drugs

Chapter 11: Family

Blended families
Commodification of Children
Focus on Media and Culture: Understanding Bachelorette Parties
Decoding the Data: Poverty and Single Motherhood
Map: Nuclear Families as a Percentage of Households
Figure: Intermarriage among Persons Born in the United States
Figure: Sources of Recent International Adoptions by U.S. Residents
Figure: Trends in Divorce Rates per 1,000 People

Chapter 12: Education and Religion

Denominations
New religious movements
Trends in U.S. religious membership
The rise of "no religion"
Focus on American Diversity: American Muslims
Figure: Percentages of College Grads, Incomes over $60,000, and Liberals among Different Religions

Chapter 13: Politics and the Economy

The economy in crisis
Unemployment and underemployment
Decoding the Data: Attitudes toward Government Responsibilities
Map: Number of Foreclosed Homes per 10,000 Homes on Market
Figure: Percentage Increase or Decrease in Voting Democrats, 2004–2008 Presidential Elections
Figure: The Loss of Jobs for U.S. Workers

Chapter 14: Population and Urbanization

Focus on American Diversity: Methamphetamine in Rural America
Figure: The Demographic Transition in the West
Figure: Population Pyramids for Ghana and Italy

Chapter 15: Social Change

Figure: Environmental versus Economic Concerns
Focus on A Global Perspective: India Meets the Cell Phone

Supplements for the Eighth Edition

For the Instructor

Instructor's Edition

The Instructor's Edition contains a visual walkthrough of the book's themes and features and available supplements. Also, this annotated text contains supplementary info (often in a different color) throughout the body of the text.

Instructor's Resource Manual with Test Bank

Written by Margaret Weinberger of Bowling Green State University, this revised and updated Instructor's Resource Manual contains learning objectives, detailed chapter outlines, discussion and lecture topics, in-class projects and activities, video suggestions, and Internet and InfoTrac® College Edition exercises, all designed to help guide your lecture. The test bank provides you with multiple choice, true-false, short answer, and essay questions to build tests and quizzes from. The multiple-choice and true-false questions are linked to a learning objective from the IRM, and all questions are marked as new or modified if they are new to this edition or have been revised from the previous edition.

PowerLecture® with JoinIN™ and ExamView®

PowerLecture instructor resources are a collection of book-specific lecture and class tools on either CD or DVD. The fastest and easiest way to build powerful, customized media-rich lectures, PowerLecture assets include chapter-specific Microsoft® PowerPoint® presentations (written by Carla Norris-Raynbird of Bemidji State University), images, animations and video, instructor manuals, test banks, useful web links, and more. PowerLecture media-teaching tools are an effective way to enhance the educational experience.

Online Activities for Introductory Sociology Courses

Made up of contributions from introductory sociology instructors, this new online supplement will be offered free to adopters of our Intro books and features new classroom activities for professors to use.

Classroom Activities—Introductory Sociology Courses

Made up of contributions from introductory sociology instructors, this new supplement will be offered free to adopters of our Intro books and features new classroom activities for professors to use.

Film Book: *Spicing Up Sociology*

Written by Marisol Clark-Ibanez and Richelle Swan of California State University San Marcos, *Spicing Up Sociology* is designed to address the growing interest in using film in the classroom. The authors start the book with the rationale for using film in the classroom, methods for incorporating film into the classroom, and learning outcomes. The authors give a synopsis of the film and a description of what concept in that chapter it gets across. Accompanying each feature film is an activity for students to complete.

Tips for Teaching Sociology

Written by veteran instructor Jerry M. Lewis of Kent State University, this booklet contains tips on course goals and syllabi, lecture preparation, exams, class exercises, research projects, course evaluations, and more. It is an invaluable tool for first-time instructors of the introductory course and for veteran instructors in search of new ideas.

Introduction to Sociology Group Activities Workbook

This supplement by Lori Ann Fowler of Tarrant County College contains both in- and out-of-class group activities (utilizing resources such as MicroCase® Online Data exercises from Wadsworth's Online Sociology Resource Center) that students can tear out and turn in to the instructor once complete. Also included are ideas for video clips to anchor group discussions, maps, case studies, group quizzes, ethical debates, group questions, group project topics, and ideas for outside readings for students to base group discussions on. Both a workbook for students and a repository of ideas, instructors can use this guide to get ideas for any introductory sociology class.

Extension: Wadsworth's Sociology Reader Database Sampler

Create your own customized reader for your introductory class, drawing from dozens of classic and contemporary articles found on the exclusive Wadsworth Cengage Learning TextChoice2 database. Create a customized reader just for your class containing as few as two or three seminal articles or more than a dozen edited selections. With Extension, you can preview articles online, make selections, and add original material of your own to create your printed reader for your class.

ABC® Videos—Introductory Sociology, Volumes I–IV

ABC Videos feature short, high-interest clips from current news events as well as historic raw footage going back 40 years. Perfect for discussion starters or to enrich your lectures and spark interest in the material in the text, these brief videos provide students with a new lens through which to view the past and present, one that will greatly enhance their knowledge and understanding of significant events and open up to them new dimensions in learning. Clips are drawn from such programs as *World News Tonight, Good Morning America, This Week, PrimeTime Live, 20/20*, and *Nightline*, as well as numerous ABC News specials and material from the Associated Press Television News and British Movietone News collections.

AIDS in Africa DVD

Expand your students' global perspective of HIV/AIDS with this award-winning documentary series focused on controlling HIV/AIDS in Southern Africa. Films focus on caregivers in the faith community; how young people share messages of hope through song and dance; the relationship of HIV/AIDS to gender, poverty, stigma, education, and justice; and the story of two HIV-positive women helping others.

Sociology: Core Concepts Videos

An exclusive offering jointly created by Wadsworth Cengage Learning and Dallas Tele-Learning, this video contains a collection of video highlights taken from "Exploring Society: An Introduction to Sociology Telecourse" (formerly The Sociological Imagination). Each 15- to 20-minute video segment will enhance student learning of the essential concepts in the introductory course and can be used to initiate class lectures, discussion, and review. The video covers topics such as the sociological imagination, stratification, race and ethnic relations, social change, and more.

Wadsworth Lecture Launchers DVD for Introductory Sociology

An exclusive offering jointly created by Wadsworth Cengage Learning and Dallas Tele-Learning, this DVD contains a collection of video highlights taken from "Exploring Society: An Introduction to Sociology Telecourse" (formerly The Sociological Imagination). Each 3- to 6-minute video segment has been specially chosen to enhance and enliven class lectures and discussions of 20 key topics covered in the introductory sociology course. Accompanying the DVD is a brief written description of each clip, along with suggested discussion questions to help effectively incorporate the material into the classroom.

For the Student

Study Card for Intro Sociology

Prepared by Matisa Wilbon of Bellarmine University, this handy card provides all the important sociological concepts covered in introductory sociology courses, broken down into sections. Providing a large amount of information at a glance, this study card is an invaluable tool for a quick review.

Careers in Sociology Module

Written by Joan Ferrante of Northern Kentucky University, the *Careers in Sociology* module offers the most extensive and useful information on careers available. This module provides six career tracks, each of which has a "featured employer," a job description, and a letter of recommendation (written by a professor for a sociology student) or application (written by a sociology student). The module also includes resume-building tips on how to make the most out of being a sociology major and offers specific course suggestions along with the transferable skills gained by taking them.

Sociology of Sports Module

Written by Jerry Lewis of Kent State University, the *Sociology of Sports* module examines why sociologists are interested in sports and discusses the links between sports and the mass media, popular culture, religion, drugs, and violence.

WebTutor™ ToolBox on WebCT® and Blackboard®

WebTutor ToolBox offers a full array of text-specific online study tools, including learning objectives, glossary flash cards, practice quizzes, web links, and a daily news feed from NewsEdge, an authoritative source for late-breaking news to keep you and your students on the cutting edge.

Companion Website

When you adopt *Essentials of Sociology*, Eighth Edition, you and your students will have access to a rich array of teaching and learning resources that you won't find anywhere else. This outstanding site features chapter-by-chapter online tutorial quizzes, a final exam, chapter outlines, chapter review, chapter-by-chapter web links, flash cards, and more!

Acknowledgments

As is true with any intellectual project, we could not have written this book without the help of many others. We are especially grateful to Carolyn Gilstrap, who proved a tremendous help with our research, and to Helen Triller, whose commitment and attention to detail was a real boon. Jill Traut and Cheri Palmer made our lives much easier by keeping the production process running smoothly, and Kristin Makarewycz did a great job overseeing the project as a whole. We also thank Andrew Keay, Erin Parkins, Rachael Krapf, Caryl Gorska, Melanie Cregger, and Tim Herzog for their contributions to this edition.

We also would like to take advantage of this opportunity to express our gratitude to those who offered suggestions and comments on how to improve the textbook for this edition:

James Ballard, *California State University–Northridge*
Tom Barry, *Central Oregon Community College*
Robert L. Boyd, *Mississippi State University*
Silvio Dobry, *Hostos Community College*
Carla Norris-Raynbird, *Bemidji State University*
Jennifer Worley, *Rasmussen College*

Finally, we wish once again to thank those who offered suggestions and comments on previous editions of this book:

Margaret Abraham, *Hofstra University, New York*; Paul J. Baker, *Illinois State University*; G. Thomas Behler, *Ferris State University, Michigan*; Robert Benford, *University of Nebraska*; Susan Blackwell, *Delgado Community College, Louisiana*; Tim Brezina, *Tulane University, Louisiana*; Marie Butler, *Oxnard College, California*; John K. Cochran, *Wichita State University, Kansas*; Carolie Coffey, *Cabrillo College, California*; Paul Colomy, *University of Akron, Ohio*; Ed Crenshaw, *University of Oklahoma*; Raymonda P. Dennis, *Delgado Community College, Louisiana*; Stanley DeViney, *University of Maryland*; Lynda Dodgen, *North Harris County College, Texas*; David A. Edwards, *San Antonio College, Texas*; Laura Eells, *Wichita State University, Kansas*; Thomas P. Egan, *Western Kentucky University*; William Egelman, *Iona College, New York*; Constance Elsberg, *Northern Virginia Community College*; Christopher Ezell, *Vincennes University, Indiana*; Joseph Faltmeier, *South Dakota State University*; Daniel E. Ferritor, *University of Arkansas*; Charles E. Garrison, *East Carolina University, North Carolina*; James R. George, *Kutztown State College, Pennsylvania*; Harold C. Guy, *Prince George Community College, Maryland*; Rose Hall, *Diablo Valley College, California*; Sharon E. Hogan, *Blue River Community College, Missouri*; Michael G. Horton, *Pensacola Junior College, Florida*; Cornelius G. Hughes, *University of Southern Colorado*; Jon Ianitti, *State University of New York, Morrisville*; Carol Jenkins, *Glendale Community College, California*; William C. Jenné, *Oregon State University*; Dennis L. Kalob, *Loyola University, Louisiana*; Sidney J. Kaplan, *University of Toledo, Ohio*; Florence Karlstrom, *Northern Arizona University*; Diane Kayongo-Male, *South Dakota State University*; William Kelly, *University of Texas*; James A. Kithens, *North Texas State University*; Phillip R. Kunz, *Brigham Young University, Utah*; Billie J. Laney, *Central Texas College*; Charles Langford, *Oregon State University*; Mary N. Legg, *Valencia Community College, Florida*; John Leib, *Georgia State University*; Joseph J. Leon, *California State Polytechnic University, Pomona*; J. Robert Lilly, *Northern Kentucky University*; Jan Lin, *University of Houston*; Lisa Linares, *Madison Area Technical College, Wisconsin*; James Lindberg, *Montgomery College, Maryland*; M. D. Litonjua, *College of Mount St. Joseph, Ohio*; Richard L. Loper, *Seminole Community College, Florida*; Ronald Matson, *Wichita State University*; Carol May, *Illinois Central College*; Rodney C. Metzger, *Lane Community College, Oregon*; Vera L. Milam, *Northeastern Illinois University*; Purna C. Mohanty, *Paine College,*

Georgia; Mel Moore, *University of Northern Colorado*; James S. Munro, *Macomb College, Michigan*; Lynn D. Nelson, *Virginia Commonwealth University*; J. Christopher O'Brien, *Northern Virginia Community College*; Charles O'Connor, *Bemidji State University, Minnesota*; Jane Ollenberger, *University of Minnesota-Duluth*; Robert L. Petty, *San Diego Mesa College, California*; Ruth A. Pigott, *Kearney State College, Nebraska*; John W. Prehn, *Gustavus Adolphus College, Minnesota*; Adrian Rapp, *North Harris County College, Texas*; Lisa Rashotte, *University of North Carolina at Charlotte*; Mike Robinson, *Elizabethtown Community College, Kentucky*; Joe Rogers, *Big Bend Community College, Washington*; Will Rushton, *Del Mar College, Texas*; Rita P. Sakitt, *Suffolk Community College, New York*; Martin Scheffer, *Boise State University, Idaho*; Richard Scott, *University of Central Arkansas*; Ida Harper Simpson, *Duke University, North Carolina*; James B. Skellenger, *Kent State University, Ohio*; Ricky L. Slavings, *Radford University, Virginia*; John M. Smith, Jr., *Augusta College, Georgia*; Evelyn Spiers, *College of the Canyons, California*; James Steele, *James Madison University, Virginia*; Michael Stein, *University of Missouri–St. Louis*; Barbara Stenross, *University of North Carolina*; Jack Stirton, *San Joaquin Delta College, California*; David L. Strickland, *East Georgia College*; Deidre Tyler, *Salt Lake Community College, Utah*; Emil Vajda, *Northern Michigan University*; Henry Vandenburgh, *State University of New York, Oswego*; Steven L. Vassar, *Mankato State University, Minnesota*; Peter Venturelli, *Valparaiso University, Indiana*; Allison Vetter, *University of Central Arkansas*; Allison L. Vetter, *University of Central Arkansas*; Leslie T. C. Wang, *The University of Toledo, Ohio*; Jane B. Wedemeyer, *Sante Fe Community College, Florida*; Dorether M. Welch, *Penn Valley Community College, Missouri*; Thomas J. Yacovone, *Los Angeles Valley College, California*; David L. Zierath, *University of Wisconsin*.

About the Authors

David Brinkerhoff is Professor Emeritus of Sociology at the University of Nebraska-Lincoln. He holds a Ph.D. in sociology from the University of Washington in Seattle, with B.S. and M.S. degrees from Brigham Young University. He has been at the University of Nebraska since 1978, having served as Associate Vice Chancellor since 1991. His research covered topics such as children's work in the family and the effect of economic marginality on the family. He is concerned with issues such as introducing technology in the classroom.

Lynn White is Professor Emerita of Sociology at the University of Nebraska, Lincoln. She holds a Ph.D. in sociology from the University of Washington in Seattle. She has been at the University of Nebraska since 1974, having served as Chair of the Department of Sociology and Director of the Bureau of Sociological Research. She taught social demography, family, and research methods. Her research focused on relationships between parents and their adult children over the life course, covering such topics as the empty nest, co-residence, the link between marital quality and parenting experiences, and intergenerational exchange. Her work appeared in *American Sociological Review* and *Social Forces* as well as in family journals.

Suzanne Ortega, who holds a Ph.D. in sociology from Vanderbilt University, was named Provost and Executive Vice President for Academic Affairs at the University of New Mexico in the fall of 2008. Before starting at the University of New Mexico, she was Graduate Dean and Professor of Sociology at the University of Washington. Her teaching interests have been primarily in introductory sociology and social problems, while her research interests are primarily in the areas of health services research and the social psychological consequences of inequality. Her work has appeared in *American Journal of Sociology, Rural Sociology, Criminology* and other journals. In addition, she has served as Chair of the Boards of GRE and the Council of Graduate Schools and several American Sociological Association committees.

Rose Weitz, Professor of Sociology and of Women and Gender Studies at Arizona State University, received her doctoral degree in Sociology from Yale University. Her research focuses on gender, health, sexuality, and the body. She is co-author of the textbook *Essentials of Sociology* and is the author of numerous scholarly publications, including the books *Life with AIDS, The Politics of Women's Bodies,* and *Rapunzel's Daughters: What Women's Hair Tells Us about Women's Lives.* Professor Weitz has won several teaching awards (including the Pacific Sociological Association Distinguished Contributions to Teaching Award, the ASU Last Lecture Award, and the ASU College of Liberal Arts and Sciences Alumni Association Outstanding Teaching Award) and has served in past years as Director of ASU's graduate and undergraduate sociology programs. In addition, she has served as President of Sociologists for Women in Society, Chair of the Sociologists AIDS Network, and Chair of the Medical Sociology Section of the American Sociological Association.

CHAPTER 1

The Study of Society

© Angela Hampton/Bubbles Photolibrary/Alamy

What Is Sociology?

Each of us starts the study of society with the study of individuals. We wonder why Theresa keeps getting involved with men who treat her badly, why Mike never learns to stop drinking before he gets sick, why our aunt puts up with our uncle, or why anybody ever liked the Spice Girls. We wonder why people we've known for years seem to change drastically when they get married or change jobs.

If Theresa were the only woman with bad taste in men or Mike the only man who drank too much, then we might try to understand their behavior by peering into their personalities. We know, however, that there are millions of men and women who have disappointing romances and who drink too much. We also know that women are more likely than men to sacrifice their needs to keep a romance alive, and that men are more likely than women to drown their troubles in drink. To understand Mike and Theresa, then, we must place them in a larger context and examine the forces that lead some *groups* of people to behave so differently from other groups.

Sociology is the systematic study of human society, social groups, and social interactions. It emphasizes the larger context in which Mike, Theresa, and the rest of us live.

Sociologists tend to view common human interactions as if they were plays. They might, for example, title a common human drama *Boy Meets Girl*. Just as *Hamlet* has been performed around the world for more than 400 years with different actors and different interpretations, *Boy Meets Girl* has also been performed countless times. Of course, people act out this drama a little differently each time, depending on the scenery, the people in the lead roles, and the century, but the essentials are the same. Thus, we can read nineteenth- or even sixteenth-century love stories and still understand why those people did what they did. They were playing roles in a play that is still performed daily.

More formal definitions will be introduced later, but the metaphor of the theater can be used now to introduce two of the most basic concepts in sociology: role and social structure. By **role**, we mean the expected performance of someone who occupies a specific position. Mothers, teachers, students, and lovers all have roles. Each position has an established script that suggests appropriate gestures, things to say, and ways to interact with others. Discovering what each society offers as a stock set of roles is one of the major themes in sociology. Sociologists try to find the common roles that appear in society and to determine why some people play one role rather than another.

The second major sociological concept is **social structure**, the larger structure of the play in which the roles appear. What is the whole set of roles that appears in this play? How are the roles interrelated? Do some actors and roles have more power than others? And how does this affect the outcome of the play? Thus, we understand the role of student in the context of the social structure we call *education*, a context in which teachers have more power than students, and administrators more power than teachers. By examining roles and social structure, sociologists try to understand the human drama.

The Sociological Imagination

The **sociological imagination** refers to the ability to recognize how apparently personal issues at least partly reflect broader social structures (Mills 1959, 15). According to C. Wright Mills, the sociological imagination is what we use when we realize that

---

Sociology is the systematic study of human society, social groups, and social interactions.

A **role** is a set of norms specifying the rights and obligations associated with a status.

A **social structure** is a recurrent pattern of relationships among groups.

The **sociological imagination** is the ability to recognize how apparently personal issues at least partly reflect broader social structures.

MAP 1.1: States with Low, Medium, and High Percentages of Residents in Poverty
Poverty is more common in rural areas, in the south and southwest, on isolated Native American reservations, and in states with many less-educated, Hispanic, and African American residents.
SOURCE: U.S. Bureau of the Census (2009b).

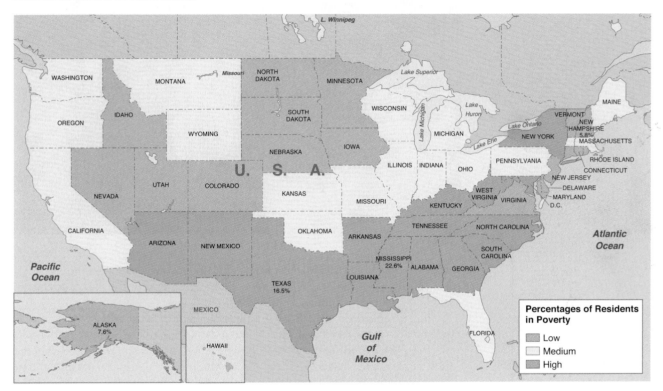

some personal troubles (such as poverty, divorce, or loss of faith) are actually common public issues that reflect a larger social context. Mills suggests that many of the things we experience as individuals are really beyond our control. Instead, they reflect the way society as a whole is organized. For example, Mills writes:

> When, in a city of 100,000, only one man is unemployed, that is his personal trouble, and for its relief we properly look to the character of the man, his skills, and his immediate opportunities. But when in a nation of 50 million employees, 15 million men are unemployed, that is [a public] issue, and we may not hope to find its solution within the range of opportunities open to any one individual. The very structure of opportunities has collapsed. Both the correct statement of the problem and the range of possible solutions require us to consider the economic and political institutions of the society, and not merely the personal situation and character of a scatter of individuals. (Mills 1959, 9)

Map 1.1 illustrates this issue. As it shows, the percentage of people living in poverty varies from 6 percent in New Hampshire to 23 percent in Mississippi. These data suggest that poverty does not result simply from personal characteristics but instead reflects something about where we live—most likely, the number of jobs and the number of people chasing those jobs.

In everyday life, we rarely consider the impact of history, economic patterns, and social structures on our own experiences. If a child becomes a drug addict, parents tend to blame themselves; if spouses divorce, each tends to blame the other; if a

Unemployment is so high in some areas that hundreds of people now show up at job fairs, such as at this one in San Mateo, California.

student does poorly in school, most blame only the student. To develop the sociological imagination is to understand how outcomes such as these are, in part, a product of society and not fully within the control of the individual.

Some people do poorly in school, for example, not because they are stupid or lazy but because they are faced with conflicting roles and role expectations. The "this is the best time of your life" play calls for very different roles and behaviors from the "education is the key to success" play. Those who adopt the student role in the "best time of your life" play will likely earn lower grades than those in the "education is the key to success" play. Other people may do poorly because they come from a family that does not give them the financial or psychological support they need. In fact, their family may need them to earn an income to help support their younger brothers and sisters. These students may be working 25 hours a week in addition to going to school; they may be going to school despite their family's lack of understanding of why college is important, or why college students need quiet and privacy for studying. In contrast, other students may find it difficult to fail: Their parents provide tuition, living expenses, and emotional support, as well as a laptop, iPhone, and new car. As we will discuss in more detail in Chapter 12, parents' social class is one of the best predictors of who will fail and who will graduate. Success or failure depends to a large extent on social factors.

The sociological imagination—the ability to see our own lives and those of others as part of a larger social structure—is central to sociology. Once we develop this imagination, we will be less likely to believe that individual behavior results solely from individual personalities. Instead, we will also consider how roles and social structures affect behavior. Similarly, we will recognize that to solve social problems, we will likely have to change social structures and roles, not just change individuals. Although poverty, divorce, and racism are experienced as intensely personal hardships, we can't eliminate or alleviate them by giving everyone personal therapy. To solve these and many other social problems, we need to change social structures; we need to rewrite the play and rebuild the theater. The sociological imagination offers a new way to look at—and a new way to solve—common troubles and dilemmas that individuals face.

The sociological imagination does *not*, however, imply that individuals have no options or bear no responsibility for their choices. Even slaves can choose to work more slowly, to ridicule their owners in private, or to commit suicide. The sociological imagination does, however, suggest the benefits of considering the impact both of social forces and of the personal choices that we more often notice.

Sociology as a Social Science

Sociology focuses on how people (and groups) interact, as well as on the rules of behavior that structure those interactions. Its emphasis is on patterns of interaction— how these patterns develop, how they are maintained, and how they change.

As one of the social sciences, sociology has much in common with political science, economics, psychology, and anthropology. All these fields share an interest in human social behavior and, to some extent, an interest in society. In addition, they all share an emphasis on the scientific method as the best approach to knowledge. This means that they rely on **empirical research**—research based on systematic examination of the evidence—before reaching any conclusions and expect researchers to evaluate that evidence in an unbiased, objective fashion. This empirical approach is what distinguishes the social sciences from journalism and other fields that comment on the human condition. Sociology differs from the other social sciences in its particular focus. Anthropologists are primarily interested in human (and nonhuman) *culture*. For example, anthropologists have studied why rape is more common in some cultures than in others and what purposes are served by cultural celebrations like bar mitzvahs, high school graduation parties, Mardi Gras, and *quinceañeras*. Psychologists focus on individual behavior and thought patterns, such as why some individuals experience more anxiety or gamble more than others. Political scientists study political systems and behaviors, such as how dictatorships rise and fall, and economists study how goods and services are produced, distributed, and consumed, such as why cell phones with cameras are so popular. Although sociologists, too, study culture, individual behavior, politics, and the economy, their focus is always on how these and other issues affect and are affected by social groups and social interactions.

The Emergence of Sociology

Sociology emerged as a field of inquiry during the political, economic, and intellectual upheavals of the eighteenth and nineteenth centuries. Rationalism and science replaced tradition and belief as methods of understanding the world, leading to changes in government, education, economic production, and even religion and family life. The clearest symbol of this turmoil is the French Revolution (1789), with its bloody uprising and rejection of the past.

Although less dramatic, the Industrial Revolution had an even greater impact. Within a few generations, traditional rural societies were replaced by industrialized urban societies. The rapidity and scope of the change resulted in substantial social disorganization. It was as if society had changed the play without bothering to tell the actors, who were still trying to read from old scripts. Although a few people prospered mightily, millions struggled desperately to make the adjustment from rural peasantry to urban working class.

This turmoil provided the inspiration for much of the intellectual effort of the nineteenth century, such as Charles Dickens's novels and Karl Marx's revolutionary

sociology and you

Given current economic conditions, it's likely that you know one or more persons who have lost their homes to foreclosure. It's possible that they used poor judgment and took on more mortgage debt than they could reasonably expect to pay. But if you use the sociological imagination, you might also question whether other forces were at play: Did they lose their homes because they worked in construction or in another field that has crashed? Did mortgage lenders pressure them to take on unreasonably high levels of debt? Did recent changes in lending laws allow lenders to charge them very high rates of interest? The sociological imagination suggests that to truly understand how the world works, we need to analyze the broader social structure as well as individual behaviors and characteristics.

Empirical research is research based on systematic, unbiased examination of evidence.

theories. It also inspired the empirical study of society. These were the years in which scientific research was a new enterprise and nothing seemed too much to hope for. After electricity, the telegraph, and the X-ray, who was to say that researchers could not discover how to eliminate crime, poverty, or war? Many hoped that the tools of empirical research could help in understanding and controlling a rapidly changing society.

The Founders: Comte, Spencer, Marx, Durkheim, and Weber

The upheavals in nineteenth-century Europe stimulated the development of sociology as a discipline. We will look at five theorists—Auguste Comte, Herbert Spencer, Karl Marx, Emile Durkheim, and Max Weber—who are often considered the founders of sociology.

August Comte (1798–1857)

Auguste Comte, 1798–1857

The first major figure in the history of sociology was the French philosopher Auguste Comte. He coined the term *sociology* in 1839, and many regard him as the founder of this field.

Comte was among the first to suggest that the scientific method could be applied to social events (Konig 1968). The philosophy of positivism, which he developed, asserts that the social world can be studied with the same scientific accuracy and assurance as the natural world. Once scientists figured out the laws of social behavior, he and other positivists believed, they would be able to predict and control it. Although thoughtful people wonder whether we will ever be able to predict human behavior as accurately as we can predict the behavior of molecules, the scientific method remains central to sociology.

Another of Comte's lasting contributions was his recognition that an understanding of society requires a concern for both the sources of order and continuity and the sources of change. These concerns remain central to sociological research, under the labels of social structure (order) and social process (change).

Herbert Spencer (1820–1903)

Another pioneer in sociology was the British philosopher-scientist Herbert Spencer. Spencer argued that evolution led to the development of social, as well as natural, life. He viewed society as similar to a giant organism: Just as the heart and lungs work together to sustain the life of the organism, so the parts of society work together to maintain society.

These ideas led Spencer to two basic principles that still guide the study of sociology. First, he concluded that each society must be understood as an adaptation to its environment. This principle of adaptation implies that to understand society, we must focus on processes of growth and change. It also implies that there is no "right" way for a society to be organized. Instead, societies will change as circumstances change.

Spencer's second major contribution was his concern with the scientific method. More than many scholars of his day, Spencer was aware of the importance of objectivity and moral neutrality in investigation. In essays on the bias of class, the bias of patriotism, and the bias of theology, he warned sociologists that they must suspend their own opinions and wishes when studying society (Turner & Beeghley 1981).

Herbert Spencer, 1820–1903

Karl Marx (1818–1883)

Karl Marx was born in Germany in 1818. A philosopher, economist, and social activist, he received his doctorate in philosophy at the age of 23. Because of his radical views, however, he never became a professor and spent most of his adult life in exile and poverty (McLellan 2006).

Marx was repulsed by the poverty and inequality that characterized the nineteenth century. Unlike other scholars of his day, he refused to see poverty as either a natural or a God-given condition of the human species. Instead, he viewed poverty and inequality as human-made conditions fostered by private property and capitalism. As a result, he devoted his intellectual efforts to understanding—and eliminating—capitalism. Many of Marx's ideas are of more interest to political scientists and economists than to sociologists, but he left two enduring legacies to sociology: the theories of economic determinism and the dialectic.

ECONOMIC DETERMINISM Marx began his analysis of society by assuming that the most basic task of any human society is to provide food and shelter to sustain itself. Marx argued that the ways in which society does this—its modes of production—provide the foundations on which all other social and political arrangements are built. Thus, he believed that economic relationships *determine* (that is, cause) the particular form that family, law, religion, and other social structures take in a given society. Scholars call this idea **economic determinism**.

A good illustration of economic determinism is the influence of economic conditions on marriage choices. In traditional agricultural societies where the older generation owns the only economic resource—land—young people often remain economically dependent upon their parents until well into adulthood. To survive, they must remain in their parents' good graces; this means, among other things, that they cannot marry without their parents' approval. In societies where young people can earn a living without their parents' help, however, they can marry whenever and whomever they please. Marx would argue that this shift in mate selection practices is the result of changing economic relationships.

Because Marx saw all human relations as stemming ultimately from the economic systems, he suggested that the major goal of a social scientist is to understand economic relationships: Who owns what, and how does this pattern of ownership affect human relationships?

THE DIALECTIC Marx's other major contribution to sociology was a theory of social change. Many nineteenth-century scholars applied Darwin's theories of biological evolution to society; they believed that social change was the result of a natural and more or less peaceful process of adaptation. Marx, however, argued that the basis of change was conflict between opposing economic interests, not adaptation.

Marx's thinking on conflict was influenced by the German philosopher Georg Hegel. Hegel argued that for every idea (thesis), a counter idea (antithesis) develops to challenge it. The conflict between thesis and antithesis then produces a new idea (synthesis). The process through which thesis and antithesis lead to synthesis is called the **dialectic** (Figure 1.1).

Marx's contribution was to apply this model of change to economic and social systems. Within capitalism, Marx suggested, the capitalist class was the thesis and the working class was the antithesis. He predicted that conflicts between them would lead to a new synthesis. That synthesis would be a communistic economic system. Indeed, in his role as social activist, Marx hoped to encourage conflict and ignite the

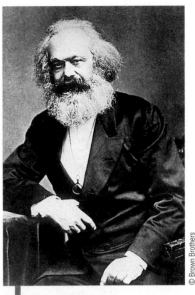

Karl Marx, 1818–1883

© Brown Brothers

FIGURE 1.1 The Dialectic
The dialectic model of change suggests that change occurs through conflict and resolution rather than through evolution.

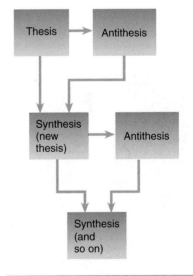

Economic determinism means that economic relationships provide the foundation on which all other social and political arrangements are built.

Dialectic philosophy views change as a product of contradictions and conflict between the parts of society.

Emile Durkheim, 1858–1917

revolution that would bring about the desired change. The workers, he declared, "have nothing to lose but their chains" (Marx & Engels 1967, 258).

Although few sociologists are revolutionaries, many accept Marx's ideas on the importance of economic relationships and economic conflicts. Much more controversial is Marx's argument that the social scientist should also be a social activist, a person who not only tries to understand social relationships but also works in the courts and the streets to change those relationships.

Emile Durkheim (1858–1917)

Emile Durkheim's life overlapped with that of Marx. While Marx was starving as an exile in England, however, Durkheim spent most of his career as a professor at the Sorbonne, the most elite university in France. Far from rejecting society, Durkheim embraced it. His research focused on understanding how societies remain stable and how stable societies foster individual happiness. Whereas Marx's legacy is a theory that highlights social conflict and social change, Durkheim's legacy is a theory that highlights social stability. Together they allow us to understand both order and change.

Durkheim's major works are still considered essential reading in sociology. These include his studies of suicide, education, divorce, crime, and social change. Two enduring contributions are his ideas about the balance between individual goals and social rules and about social science methods.

One of Durkheim's major concerns was the balance between social regulation and personal freedom. He argued that community standards of morality, which he called the *collective conscience*, not only confine our behavior but also give us a sense of belonging and integration. For example, many people complain about having to dress up; they complain about having to shave their faces or their legs or having to wear a tie or pantyhose. "What's wrong with jeans?" they want to know. At the same time, most of us feel a sense of satisfaction when we appear in public in our best clothes. We know that we will be considered attractive and successful. Although we may complain about having to meet what appear to be arbitrary standards, we often feel a sense of satisfaction in being able to meet those standards successfully. In Durkheim's words, "institutions may impose themselves upon us, but we cling to them; they compel us, and we love them" ([1895] 1938, 3). This beneficial regulation, however, must not rob the individual of all freedom of choice.

In his classic study *Suicide*, Durkheim identified two types of suicide that stem from an imbalance between social regulation and personal freedom. *Fatalistic* suicide occurs when society provides too little freedom and too much regulation: when we find our behavior so confined by social institutions that we feel trapped ([1897] 1951, 276). One example would be the young mother with several children and a job who feels overburdened by the demands of work, household, and family. *Anomic* suicide, on the other hand, occurs when there is too *much* freedom and too *little* regulation: when society's influence does not check individual passions ([1897] 1951, 258). Durkheim believed that this kind of suicide was most likely to occur in times of rapid social change. When established ways of doing things have lost their meaning, but no clear alternatives have developed, individuals feel lost. For example, many scholars attribute high rates of alcohol abuse among contemporary Native Americans to the weakening of traditional social regulation.

Durkheim was among the first to stress the importance of using reliable statistics to logically rule out incorrect theories of social life and to identify more promising theories. He strove to be an objective observer who only sought the facts. As sociology

Some Christians baptize infants by sprinkling a few drops of holy water on their foreheads. Others baptize adults by fully immersing them in flowing water. To sociologists following in Weber's footsteps, the *fact* that different Christians use different forms of baptism is less important than the *meaning* these practices have for them.

became an established discipline, this ideal of objective observation replaced Marx's social activism as the standard model for social science.

Max Weber (1864–1920)

Max Weber (vay-ber), a German economist, historian, and philosopher, provided the theoretical base for half a dozen areas of sociological inquiry. He wrote on religion, bureaucracy, method, and politics. In all these areas, his work is still valuable and insightful. Three of Weber's more general contributions were an emphasis on the subjective meanings of social actions, on social as opposed to economic causes, and on the need for objectivity in studying social issues. Weber believed that knowing patterns of behavior was less important than understanding the meanings people attach to behavior. For example, Weber would argue that it is relatively meaningless to compile statistics such as how many marriages end in divorce now compared with 100 years ago. More critical, he would argue, is understanding how the *meaning* of divorce has changed over that time period. Weber's emphasis on the subjective meanings of human actions has been the foundation of scholarly work on topics as varied as religion and immigration.

Weber trained as an economist, and much of his work concerned the interplay of things economic and things social. He rejected Marx's idea that economic factors determine all social relationships. In a classic study, *The Protestant Ethic and the Spirit of Capitalism* ([1904–05] 1958), Weber tried to show how social and religious values can affect economic systems. This argument is explained more fully in Chapter 12, but its major thesis is that the religious values of early Protestantism (self-discipline, thrift, and individualism) were the foundation for capitalism.

One of Weber's more influential ideas was that sociology must be **value-free.** Weber argued that sociology should be concerned with establishing what is and not what ought to be. Weber's dictum is at the heart of the standard scientific approach that is generally advocated by modern sociologists. Thus, although one may study poverty or racial inequality because of a sense of moral outrage, such feelings must be set aside to achieve an objective grasp of the facts. This position of neutrality is directly contradictory to the Marxist emphasis on social activism, and sociologists who

Max Weber, 1864–1920

Value-free sociology concerns itself with establishing what is, not what ought to be.

W. E. B. DuBois, 1868–1963

Jane Addams, 1860–1935

adhere to Marxist principles generally reject the notion of value-free sociology. Most modern sociologists, however, try to be value-free in their scholarly work.

Sociology in the United States

Although U.S. sociology has the same intellectual roots as European sociology, it has some distinctive characteristics. Most importantly, European sociologists are more likely to focus on constructing broad, philosophical theories of how society works, whereas U.S. sociologists more often focus on collecting systematic, empirical data. As this suggests, U.S. sociologists more often stress identifying, understanding, and solving social problems.

One reason that U.S. sociology developed differently from European sociology is that our social problems differed. Between the 1860s and the 1920s, slavery, the Civil War, and high immigration rates made racism and ethnic discrimination much more salient issues in the United States than in Europe. One of the first sociologists to study these issues was W. E. B. DuBois, who received his doctorate in 1895 from Harvard University, devoted his career to developing empirical data about African Americans, and used those data to combat racism.

The work of Jane Addams, another early sociologist and recipient of the 1931 Nobel Peace Prize, also illustrates the emphasis on social problems and social reform within early U.S. sociology. Addams was the founder of Hull House, a famous center for social services and community activism located in a Chicago slum. She and her colleagues used quantitative social science data to lobby successfully for legislation mandating safer working conditions, a better juvenile justice system, improved public sanitation, and services for the poor (Linn & Scott 2000).

Today, many U.S. sociologists continue to focus on how race, class, and gender—both individually and jointly—affect all aspects of social life. More broadly, an interest in helping to solve crucial social problems is central to the work of most U.S. sociologists. They hope to change the world for the better by systematically studying social life and making their research findings available to others. In addition, some sociologists work in social movements or for social change organizations to try more directly to alleviate social problems. Finally, a small but growing number of U.S. sociologists take their research directly to the public and policy makers: appearing on *Oprah* and *The Today Show*, publishing in the *New York Times* and on *Slate.com*, and testifying in court and before Congress regarding the nature of social issues and how best to address them.

As sociological research came of age, sociology also became a part of mainstream higher education. Almost all colleges and universities now offer an undergraduate degree in sociology. Most universities offer a master's degree in the subject, and approximately 125 offer doctoral degree programs. Graduate sociology programs are more popular in the United States than in any other country in the world.

Current Perspectives in Sociology

As this brief review of the history of sociology has demonstrated, there are many ways of approaching the study of human social interaction. The ideas of Marx, Weber, Durkheim, and others have given rise to dozens of theories about human behavior. In this section, we summarize the three dominant theoretical perspectives in sociology today: structural-functional theory, conflict theory, and symbolic interaction theory. The Concept Summary on Major Theoretical Perspectives describes these three perspectives.

concept summary

Major Theoretical Perspectives

	Structural Functionalism	Conflict Theory	Symbolic Interactionism
Nature of society	Interrelated social structures that fit together to form an integrated whole	Competing interests, each seeking to secure its own ends	Interacting individuals and groups
Basis of interaction	Consensus and shared values	Constraint, power, and competition	Shared symbolic meanings
Major questions	What are social structures? Do they contribute to social stability?	Who benefits? How are these benefits maintained?	How do social structures relate to individual subjective experiences?
Level of analysis	Social structure	Social structure	Interpersonal interaction

Structural-Functional Theory

Structural-functional theory (or *structural functionalism*) addresses the question of how social organization is maintained. This theoretical perspective has its roots in natural science and in the analogy between society and an organism. In the same way that a biologist may try to identify the parts (structures) of a cell and determine how they work (function), a sociologist who uses structural-functional theory will try to identify the structures of society and how they function.

The Assumptions behind Structural-Functional Theory

All sociologists are interested in researching how societies work. Those who use the structural-functionalist perspective, however, bring three major assumptions to their research:

1. *Stability*. The chief evaluative criterion for any social pattern is whether it contributes to the maintenance of society.
2. *Harmony*. Like the parts of an organism, the parts of society typically work together harmoniously for the good of the whole.
3. *Evolution*. Change occurs through evolution—the mostly peaceful adaptation of social structures to new needs and demands and the elimination of unnecessary or outmoded structures.

Using Structural-Functional Theory

Sociologists who use structural-functional theory focus on studying the *nature* and *consequences* of social structures. Structural-functional sociologists refer to the positive (beneficial) consequences of social structures as **functions** and to the negative (harmful) consequences of social structures as **dysfunctions**. They also draw a distinction between **manifest** (recognized and intended) consequences and **latent**

Structural-functional theory addresses the question of social organization (structure) and how it is maintained (function).

Functions are consequences of social structures that have positive effects on the stability of society.

Dysfunctions are consequences of social structures that have negative effects on the stability of society.

Manifest functions or dysfunctions are consequences of social structures that are intended or recognized.

Latent functions or dysfunctions are consequences of social structures that are neither intended nor recognized.

Team sports offer a graphic metaphor of social structure. Each person on the team occupies a different status, and each plays a relatively unique role. Structural functionalists focus on the benefits that these statuses and roles and the institution of sports itself provide to society.

© Bettmann Archive/Corbis

(unrecognized and unintended) consequences. Because these concepts are very useful, they are also used by other sociologists who do not share the underlying assumptions behind structural-functional theory.

Consider, for example, the concept of the *"battered-woman syndrome."* This is a medical diagnosis that suggests a woman who is repeatedly battered will become mentally ill. This diagnosis has been used in courts as a legal defense by battered women who assault or kill their abusers, allowing them to plead not guilty by reason of temporary insanity.

What are the consequences of this new social structure (that is, this new diagnosis)? Its *manifest function* (intended positive outcome) is, of course, to give legal recognition to the devastating psychological consequences of domestic violence. The *manifest dysfunction* is that some women might use the diagnosis as an excuse for a malicious, premeditated assault. A *latent dysfunction* is that women who are acquitted of legal charges on the basis of a temporary insanity plea could lose custody of their children, given the stigma attached to mental illness.

Another latent outcome may be the perpetuation of the view that women are irrational—that they stay with men who beat them because they are incapable of logically thinking through their options, and that they only leave when they "snap" mentally. But is this a function or a dysfunction? Remember that structural-functional analysis typically starts from the assumption that any social action or structure that contributes to the maintenance of society and preserves the status quo is functional and that any action or structure that challenges the status quo is dysfunctional. Because perpetuating the view that women are irrational would reinforce existing gender roles, this would be judged a latent *function*, not a dysfunction (Table 1.1).

As this example suggests, a social pattern that contributes to the maintenance of society may benefit some groups more than others. A pattern may be functional—that is, it may help maintain the status quo—without being either desirable or equitable. In general, however, structural-functionalists emphasize how social structures work together to create a society that runs smoothly.

TABLE 1.1 **A Structural-Functional Analysis of the Battered-Woman Syndrome**
Structural-functional analysis examines the intended and unintended consequences of social structures. It also assesses whether the consequences are positive (functional) or negative (dysfunctional). There is no moral dimension to the assessment that an outcome is positive; it merely means that the outcome contributes to the stability of society.

	Manifest	Latent
Function	Gives legal recognition to the psychological consequences of domestic violence.	Encourages the view that women are irrational.
Dysfunction	May serve as an excuse for violence against abusers.	Makes it more difficult for victims of domestic violence to retain custody of children.

Conflict Theory

Whereas structural-functional theory sees the world in terms of consensus and stability, conflict theory sees the world in terms of conflict and change. Conflict theorists contend that a full understanding of society requires a critical examination of competition and conflict in society, especially of the processes by which some people become winners and others become losers. As a result, **conflict theory** addresses the points of stress and conflict in society and the ways in which they contribute to social change.

Assumptions behind Conflict Theory

Conflict theory is derived from Marx's ideas. The following are three primary assumptions of modern conflict theory:

1. *Competition.* Competition over scarce resources (money, leisure, sexual partners, and so on) is at the heart of all social relationships. Competition rather than consensus is characteristic of human relationships.
2. *Structural inequality.* Inequalities in power and reward are built into all social structures. Individuals and groups that benefit from any particular structure strive to see it maintained.
3. *Social change.* Change occurs as a result of conflict between competing interests rather than through adaptation. It is often abrupt and revolutionary rather than evolutionary and is often helpful rather than harmful.

Using Conflict Theory

Like structural functionalists, conflict theorists are interested in social structures. However, conflict theorists focus on studying which groups benefit most from existing social structures and how these groups maintain their privileged positions.

A conflict analysis of domestic violence, for example, would begin by noting that women are battered far more often and far more severely than are men, and that the popular term *domestic violence* hides this reality. Conflict theorists' answer to the question "Who benefits?" is that battering helps men to retain their dominance over women. These theorists go on to ask how this situation developed and how it is maintained. Their answers would focus on issues such as how some religions traditionally have taught women to submit to their husbands' wishes and to accept violence within marriage, how until recently the law did not regard woman battering as a crime, and how some police officers still consider battering merely an unimportant family matter.

---

sociology and you

Whether or not you attended a senior prom in high school, you probably recognize some of the functions they serve. If you attended, you may have felt that your prom memories would help preserve your bonds with your high school friends. You also may have felt that the prom was a rite of passage, signaling that you were becoming an adult. Similarly, your parents' decisions regarding whether or not to let you attend unsupervised after-prom events functioned as a signal of their faith—or lack of faith—in your ability to behave responsibly. If you did not attend, on the other hand, you might have concluded that proms serve primarily to highlight who is most popular and who can afford the most expensive clothes and cars.

Conflict theory addresses the points of stress and conflict in society and the ways in which they contribute to social change.

Conflict theorists point out that unions exist because labor and management have different, competing interests. Workers want better pay and secure jobs; management wants to keep costs down.

Symbolic Interaction Theory

Both structural-functional and conflict theories focus on social structures and the relationships among them. But what does this tell us about the relationship between *individuals* and social structures? Sociologists who focus on the ways that individuals relate to and are affected by social structures often use symbolic interaction theory. **Symbolic interaction theory** (or *symbolic interactionism*) addresses the subjective meanings of human acts and the processes through which we come to develop and share these subjective meanings. The theory is so named because it studies the symbolic (or subjective) meaning of human interaction. Symbolic interaction theory is the newest of the three theoretical traditions described in this chapter.

Assumptions behind Symbolic Interaction Theory

When symbolic interactionists study human behavior, they begin with three major premises (Charon 2006):

1. *Meanings are important.* Any behavior, gesture, or word can have multiple interpretations (can symbolize many things). To understand human behavior, we must learn what it means *to the participants.*
2. *Meanings grow out of relationships.* When relationships change, so do meanings.
3. *Meanings are negotiated between people.* We do not accept others' meanings uncritically. Each of us plays an active role in negotiating the meanings that things have for us and others.

Using Symbolic Interaction Theory

These three premises direct symbolic interactionists to study how relationships and social structures shape individuals. For example, symbolic interactionists interested in violence against women have researched how boys learn to consider aggression a natural part of being male when they are cheered for hitting others during hockey games, when dads tell them to fight anyone who makes fun of them, when older brothers physically push them around, and the like. Symbolic interactionists also have explored how teachers unintentionally reinforce the idea that girls are inferior by allowing boys to take over schoolyards and to make fun of girls in the classroom. All these experiences, some researchers believe, set the stage for later violence against women.

Symbolic interactionists are also interested in how individuals actively modify and negotiate relationships. Why do two children raised in the same family turn out differently? In part, because each child experiences subtly different relationships and situations even within the same family, and each may derive different meanings from those experiences.

Most generally, symbolic interactionists often focus on how relationships shape individuals, from childhood through old age. The strength of symbolic interactionism is that it focuses attention on how larger social structures affect our everyday lives, sense of self, and interpersonal relationships and encounters.

Interchangeable Lenses

Neither symbolic interaction theory, conflict theory, nor structural-functional theory is complete in itself. Together, however, they provide a valuable set of tools for understanding the relationship between the individual and society. These three theories can be regarded as interchangeable lenses through which society may be viewed. Just as a telephoto lens is not always superior to a wide-angle lens, one sociological theory will not always be superior to another.

Symbolic interaction theory addresses the subjective meanings of human acts and the processes through which people come to develop and communicate shared meanings.

© Jeff Greenberg/ PhotoEdit.

© John Van Hasselt/Corbis

Conflict theorists typically view prostitution as an outgrowth of poverty and sexism; structural functionalists consider it functional for society. Symbolic interactionists ask questions such as how do prostitutes (such as these young women at a legal brothel in Nevada) maintain a positive identity in a stigmatized occupation?

Occasionally, the same subject can be viewed through any of these perspectives. We will generally get better pictures, however, by selecting the theoretical perspective that is best suited to the particular subject. In general, structural functionalism and conflict theory are well suited to the study of social structures, or **macrosociology**. Symbolic interactionism is well suited to the study of the relationship between individual meanings and social structures, or **microsociology**. The following sections provide three "snapshots" of female prostitution taken through the theoretical lens of structural-functional, conflict, and symbolic interaction theory.

Structural-Functional Theory: The Functions of Prostitution

Structural-functionalists who study female prostitution often begin by examining its social structure and identifying patterns of relationships among pimps, prostitutes, and customers. Then they focus on identifying the consequences of this social structure. In a still-famous article published in 1961, Kingsley Davis listed the following functions of prostitution:

- It provides a sexual outlet for poor and disabled men who cannot compete in the marriage market.
- It provides a sexual outlet for businessmen, sailors, and others when away from home.
- It provides a sexual outlet for those with unusual sexual tastes.

Provision of these services is the manifest or intended function of prostitution. Davis goes on to note that, by providing these services, prostitution has the latent function of protecting the institution of marriage from malcontents who, for one reason or another, do not receive adequate sexual service through marriage. Prostitution is the safety valve that makes it possible to restrict respectable sexual relationships (and hence childbearing and child rearing) to marital relationships, while still allowing for the variability of human sexual appetites.

Conflict Theory: Unequal Resources and Becoming a Prostitute

Conflict theorists analyze prostitution as part of the larger problem of unequal access to resources. Women, they argue, have not had equal access to economic opportunity.

Macrosociology focuses on social structures and organizations and the relationships between them.

Microsociology focuses on interactions among individuals.

In some societies, they cannot legally own property; in others, they suffer substantial discrimination in opportunities to work and earn. Because of this inability to support themselves, women have had to rely on economic support from men. They get this support by exchanging the one scarce resource they have to offer: sexual availability. To a conflict theorist, it makes little difference whether a woman barters her sexual availability through prostitution or through marriage. The underlying cause is the same.

Conflict theory is particularly useful for explaining why so many runaway boys and girls work as prostitutes. These young people have few realistic opportunities to support themselves by regular jobs: Many are not old enough to work legally and, in any case, would be unable to support themselves adequately on the minimum wage. Their young bodies are their most marketable resource.

Symbolic Interaction Theory: How Prostitutes Maintain Their Self-Concepts

Symbolic interactionists who examine prostitution take an entirely different perspective. They want to know, for example, how prostitutes learn the trade and how they manage their self-concept so that they continue to think positively of themselves despite their work. For one such study, sociologist Wendy Chapkis (1997) interviewed more than fifty women "sex workers"—prostitutes, call girls, actresses in "adult" films, and others. Many of the women she interviewed felt proud of their work. They felt that the services they offered were not substantially different from those offered by day-care workers or psychotherapists, who are also expected to provide services while acting as if they like and care for their clients. Chapkis found that as long as prostitutes are able to keep a healthy distance between their emotions and their work, they can maintain their self-esteem and mental health. As one woman described it: "Sex work hasn't all been a bed of roses and I've learned some painful things. But I also feel strong in what I do. I'm good at it and I know how to maintain my emotional distance. Just like if you are a fire fighter or a brain surgeon or a psychiatrist, you have to deal with some heavy stuff and that means divorcing yourself from your feelings on a certain level. You just have to be able to do that to do your job" (Chapkis, 79).

As these examples illustrate, many topics can be studied fruitfully with any of the three theoretical perspectives. Each sociologist must decide which perspective will work best for a given research project.

Researching Society

The things that sociologists study—for example, drug use, marital happiness, and poverty—have probably interested you for a long time. You may have developed your own opinions about why some people have good marriages and some have bad marriages or why some people break the law and others do not. Sociology is an academic discipline that critically examines commonsense explanations of human social behavior. It aims to improve our understanding of the social world by observing and measuring what actually happens. Obviously sociological research is not the only means of acquiring knowledge. Some people learn what they need to know from the Bible or the Koran or the Book of Mormon. Others get their answers from their parents, television, or the Internet. When you ask such people, "But how do you know that that is true?" their answer is simple: "My mother told me," "I heard it on *The Daily Show*," or "I read it on Wikipedia."

Sociology differs from these other ways of knowing in that it requires empirical evidence that can be confirmed by the normal human senses. We must be able to see, hear, smell, or feel it. Before social scientists would agree that they "knew" religious intermarriage increased the likelihood of divorce, for example, they would want to see evidence.

All research has two major goals: accurate description and accurate explanation. In sociology, we first seek accurate descriptions of human interactions (How many people marry and whom do they marry? Which people are mostly likely to abuse their children or to flunk out of school?). Then we try to explain those patterns (Why do people marry, abuse their children, or flunk out?).

The Research Process

At each stage of the research process, scholars use certain conventional procedures to ensure that their findings will be accepted as scientific knowledge. The procedures used in sociological research are covered in depth in classes on research methods, statistics, and theory construction. At this point, we merely want to introduce a few ideas that you must understand if you are to be an educated consumer of research results. We look at the five steps of the general research process, and in doing so review three concepts central to research: *variables*, *operational definitions*, and *sampling*.

Step One: Stating the Problem

The first step in the research process is carefully stating the issue to be investigated. We may select a topic because of a personal experience or out of commonsense observation. For example, we may have observed that African Americans appear more likely to experience unemployment and poverty than do white Americans. Alternatively, we might begin with a theory that predicts, for instance, that African Americans will have higher unemployment and poverty rates than white Americans because they experienced discrimination in schools and in workplaces. In either case, we begin by reviewing the research of other scholars to help us specify exactly what it is that we want to know. If a good deal of research has already been conducted on the issue and good theoretical explanations have been advanced for some of the patterns, then a problem may be stated in the form of a **hypothesis**—a statement about relationships that we expect to observe if our theory is correct. A hypothesis must be testable; that is, there must be some way in which data can help weed out a wrong conclusion and identify a correct one. For example, the *belief* that whites *deserve* better jobs than African Americans cannot be tested, but the *hypothesis* that whites receive better job offers than African Americans can be tested.

Step Two: Setting the Stage

Before we can begin to gather data, we first have to set the stage by selecting variables, defining our terms, and deciding exactly which people (or objects) we will study.

Understanding Variables

To narrow the scope of a problem to manageable size, researchers focus on variables rather than on people. **Variables** are measured characteristics that vary from one individual, situation, or group to the next (Babbie 2010). If we wish to analyze differences in rates of African American/white unemployment, we need information on two variables: race and unemployment. The individuals included in our study would

A **hypothesis** is a statement about relationships that we expect to find if our theory is correct.

Variables are measured characteristics that vary from one individual or group to the next.

be complex and interesting human beings, but for our purposes, we would be interested only in these two aspects of each person's life.

When we hypothesize a cause-and-effect relationship between two variables, the cause is called the **independent variable**, and the effect is called the **dependent variable**. In our example, race is the independent variable, and unemployment is the dependent variable; that is, we hypothesize that unemployment *depends on* one's race.

Defining Variables

In order to describe a pattern or test a hypothesis, each variable must be precisely defined. Before we can describe racial differences in unemployment rates, for instance, we need to be able to decide whether an individual is unemployed. The process of deciding exactly how to measure a given variable is called **operationalizing**, and the exact definition we use to operationalize a variable is its **operational definition**. Reaching general agreement about these definitions may pose a problem. For instance, the U.S. government labels people as unemployed if they are actively seeking work but cannot find it. This definition ignores all the people who became so discouraged in their search for work that they simply gave up. Obviously, including discouraged workers in our definition of the unemployed might lead to a different description of patterns of unemployment.

Sampling

It would be time consuming, expensive, and probably impossible to get information on race and employment status for all adults. It is also unnecessary. The process of **sampling**—taking a systematic selection of representative cases from a larger population—allows us to get accurate empirical data at a fraction of the cost that examining all possible cases would involve.

Sampling involves two processes: (1) obtaining a list of the population you want to study and (2) selecting a representative subset or sample from the list. The best samples are **random samples**. In a random sample, cases are chosen through a random procedure, such as tossing a coin, ensuring that every individual within a given population has an equal chance of being selected for the sample.

Once we have a list of the population, randomly selecting a sample is fairly easy. But getting such a list can be difficult or even impossible. A central principle of sampling is that a sample is only representative of the list from which it is drawn. If we draw a list of people from the telephone directory, then our sample can only be said to describe households listed in the directory; it will omit those with unlisted numbers, those with no telephones, those who use only cell phones, and those who have moved since the directory was issued. The best surveys begin with a list of all the households, individuals, or telephone numbers in a target region or group.

Step Three: Gathering Data

There are many ways of gathering sociological data, including running experiments, conducting surveys, and observing groups in action. Because this is a complex subject, we explore it in more detail later in this chapter.

Step Four: Finding Patterns

The fourth step in the research process is to look for patterns in the data. If we study unemployment, for example, we will find that African Americans are twice as likely as white Americans to be unemployed (U.S. Bureau of the Census 2009a). This finding is a **correlation**: an empirical relationship between two variables—in this case, race and employment.

The **independent variable** is the cause in cause-and-effect relationships.

The **dependent variable** is the effect in cause-and-effect relationships. It is dependent on the actions of the independent variable.

Operationalizing refers to the process of deciding exactly how to measure a given variable.

An **operational definition** describes the exact procedure by which a variable is measured.

Sampling is the process of systematically selecting representative cases from the larger population.

Random samples are samples chosen through a random procedure, so that each individual in a given population has an equal chance of being selected.

Correlation exists when there is an empirical relationship between two variables (for example, income increases when education increases).

Step Five: Generating Theories

After a pattern is found, the next step in the research process is to explain it. As we will discuss in the next section, finding a correlation between two variables does not necessarily mean that one variable causes the other. For example, even though there is a correlation between race and unemployment, many whites are unemployed and many African Americans are not. Nevertheless, if we have good empirical evidence that being black increases the *probability* of unemployment, the next task is to explain why that should be so. Explanations are usually embodied in a **theory**, an interrelated set of assumptions that explains observed patterns. Theory always goes beyond the facts at hand; it includes untested assumptions that explain the empirical evidence.

In our unemployment example, we might theorize that the reason African Americans face more unemployment than whites is because many of today's African American adults grew up in a time when the racial difference in educational opportunity was much greater than it is now. This simple explanation goes beyond the facts at hand to include some assumptions about how education is related to race and unemployment. Although theory rests on an empirical generalization, the theory itself is not empirical; it is, well, theoretical.

It should be noted that many different theories can be compatible with a given empirical generalization. We have proposed that educational differences explain the correlation between race and unemployment. Others might argue that the correlation arises because of discrimination. Because there are often many plausible explanations for any correlation, theory development is not the end of the research process. We must go on to test the theory by gathering new data.

The scientific process can be viewed as a wheel that continuously moves us from theory to data and back again (Figure 1.2). Two examples illustrate how theory leads to the need for new data and how data can lead to the development of new theory.

As we have noted, data show that unemployment rates are higher among African Americans than among white Americans. One theoretical explanation for this pattern links higher African American unemployment to educational deficits. From this theory, we can deduce the hypothesis that African Americans and whites of equal education will experience equal unemployment. To test this hypothesis, we need more data, this time about education and its relationship to race and unemployment.

A study by Lori Reid (2002) tests this hypothesis for black women. Reid asked whether educational deficits explained why African American women are more likely to lose their

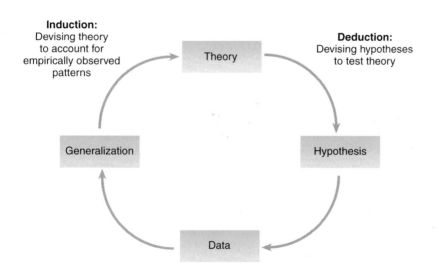

Induction:
Devising theory to account for empirically observed patterns

Deduction:
Devising hypotheses to test theory

Theory

Hypothesis

Data

Generalization

FIGURE 1.2 **The Wheel of Science**
The process of science can be viewed as a continuously turning wheel that moves us from data to theory and back again.

A **theory** is an interrelated set of assumptions that explains observed patterns.

jobs than are whites. She found that education does play a small role. However, other factors—including black women's segregation in vulnerable occupations and residence in areas where unemployment was rising—were far better predictors of unemployment.

Reid's findings could be the basis for revised theories. These new theories would again be subject to empirical testing, and the process would begin anew. In the language of science, the process of moving from data to theory is called **induction**, and the process of moving from theory to data is called **deduction**. Figure 1.2 illustrates these two processes.

Research Methods

The theories and findings reported in this book stem from a variety of research methods. This section reviews the most common methods (summarized in the Concept Summary on Comparing Research Methods) and illustrates their advantages and disadvantages, using research on alcohol use as an example.

concept summary

Comparing Research Methods

Method	Advantages	Disadvantages
Experiments	Excellent for studying cause-and-effect relationships.	Based on small, nonrepresentative samples examined under highly artificial circumstances. Many subjects cannot be ethically studied through experiments.
Surveys	Very versatile—can study anything that we can ask about; can be done with large, random samples so that results represent many people; good for studying incidence, trends, and differentials.	Subject to social-desirability bias. Better for studying individuals than for studying social contexts, processes, or meaning.
Participant Observation	Places behaviors and attitudes in context. Shows what people do rather than what they say they do.	Limited to small, nonrepresentative samples. Relies on interpretation by single researcher.
Content Analysis	Inexpensive. Useful for historical research. Researcher does not affect data.	Only useful with recorded communications. Relies on researchers' interpretations, but multiple researchers can compare their conclusions.
Use of Existing Statistics	Inexpensive. Useful for historical research. Researcher does not affect data.	Limited to available data: cannot collect data to fit research questions.

Induction is the process of moving from data to theory by devising theories that account for empirically observed patterns.

Deduction is the process of moving from theory to data by testing hypotheses drawn from theory.

Experiments

The **experiment** is a research method in which the researcher manipulates the independent variable to test theories of cause and effect. In the classic experiment, a researcher compares an **experimental group** to a **control group**. The only difference between the two groups is that only the former is exposed to the independent variable under study. If the groups are otherwise the same, comparing them should show whether the independent variable has an effect.

If we wanted to assess whether alcohol use affects grades, for example, we would need to compare an experimental group that drank alcohol with a control group that did not. We would begin by dividing a group of students randomly into two groups. If the initial pool is large enough, we could assume that the two groups are probably similar on nearly everything. For example, both groups probably contain an equal mix of good and poor students and of lazy and ambitious students. We could then ask the control group to agree not to drink alcohol for 5 weeks and ask the experimental group to drink daily during the same period. At the end of the 5 weeks, we would compare the grades of the two groups. Since the groups were similar at the start, if grades went up among the nondrinkers, we could conclude that abstaining from alcohol caused their grades to rise.

As this example suggests, experiments are a great way to test hypotheses about cause and effect. They have three drawbacks, however. First, experiments are unethical if they expose subjects to harm. For example, requiring students to drink daily might lower their course grades or turn them into heavy drinkers. Second, subjects often behave differently when they are in an experiment. For example, although alcohol consumption might normally lower student grades, the participants in our experiment might work extra hard to keep their grades up because they know we are collecting data on them. Finally, experiments occur in very unnatural environments, and so it is difficult to generalize from experiments to the real world.

Surveys

In **survey research**, the researcher asks a relatively large number of people the same set of standardized questions. These questions may be asked in a personal interview, over the telephone, online, or in a paper-and-pencil format. Because survey researchers ask many people the same questions, they can ascertain how common a behavior or pattern is (**incidence**), how the behavior or pattern has changed over time (**trend**), and how it varies from group to group (**differential**). Thus, survey data on alcohol use may allow us to say such things as the following: 80 percent of the undergraduates at Midwestern State currently use alcohol (incidence); the proportion using alcohol has remained about the same over the last 10 years (trend); and the proportion using alcohol is higher for males than for females (differential). Survey research is extremely versatile; it can be used to study attitudes, behavior, ideals, and values. If you can think of a way to ask a question about a topic, then you can study the topic with survey research.

Most researchers employing surveys in their work use a **cross-sectional design** for their research: They take a sample (or cross section) of the population at a single point in time and look at how groups differ on the independent and dependent variables. Thus, to study the potential impact of alcohol use on grades, we might begin with a sample of students and then divide them into groups according to how often they drank alcohol. We could then compare these groups to see which earn the higher grades.

The **experiment** is a method in which the researcher manipulates independent variables to test theories of cause and effect.

An **experimental** group is the group in an experiment that experiences the independent variable. Results for this group are compared with those for the control group.

A **control group** is the group in an experiment that does not receive the independent variable.

Survey research is a method that involves asking a relatively large number of people the same set of standardized questions.

Incidence is the frequency with which an attitude or behavior occurs.

A **trend** is a change in a variable over time.

A **differential** is a difference in the incidence of a phenomenon across social groups.

A **cross-sectional design** uses a sample (or cross section) of the population at a single point in time.

Understanding Spurious Relationships

If we divide students into those who do and those who don't own Macintosh laptops, we find that, on average, those who own Macs have higher grades. This does not necessarily mean that owning a Mac *causes* higher grades. In the example below, the relationship between Macs and grades is *spurious*.

Spurious relationship: Owning a Mac seems (falsely) to lead to higher grades.

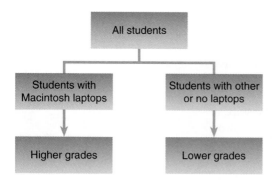

Nonspurious (true) relationship: Students who come from wealthier families are more likely to own Macintosh laptops and more likely to get higher grades. Wealth, not Mac ownership, causes higher grades.

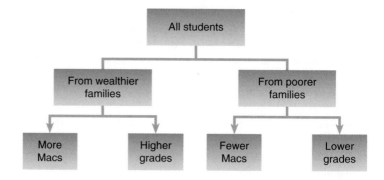

FIGURE 1.3 College Grades and Frequency of Alcohol Use

These data show that the more often a student drinks alcohol, the lower grades he or she is likely to earn. The data cannot tell us, however, whether drinking *caused* lower grades.

SOURCE: Boynton Health Service (2007).

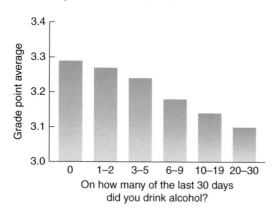

Longitudinal research is any research in which data are collected over a long period of time.

A **spurious relationship** exists when one variable *seems* to cause changes in a second variable, but a third variable is the *real* cause of the change.

In 2007, researchers in Minnesota did just that (Boynton Health Service 2007). They surveyed more than 24,000 undergraduates and then divided them according to how often they drank. Figure 1.3 shows the results: As alcohol use goes up, grades steadily (if slightly) go down.

Does this mean that drinking caused these students to get lower grades? Not necessarily. First, all we know is that the more frequently students drank, the lower their grades were. We cannot tell which is the cause and which is the effect: Did drinking cause students to get lower grades, or did getting lower grades lead students to drink? To sort this out, we would need to use **longitudinal research**, that is, to collect data over a period of time. We could either interview the same group of individuals multiple times (perhaps every month, perhaps every 5 years) or interview different groups, each randomly selected from the same population but weeks, months, or years apart. That way we could see whether students' grades began falling before or after their drinking increased.

A second problem is that we cannot be sure there is *any* cause-and-effect relationship between drinking and grades. Most likely nondrinkers and frequent drinkers differed in many ways from the start. The frequent drinkers may have been under more stress or may have grown up in neighborhoods where education was less valued. One of these variables might have caused them both to drink *and* to get lower grades. In this case, the apparent (but false) cause-and-effect relationship between drinking and lower grades would be considered a **spurious relationship**. A relationship between two variables (like drinking and grade point average) is considered *spurious* when it appears that one variable is affecting another, but in reality a third variable is affecting the first two variables. The Concept Summary on Understanding Spurious Relationships illustrates this idea.

Survey research is an excellent way of finding the relationship between two variables, such as whether drinking affects grades among college students.

To avoid being misled by a spurious relationship, we would need to use a sample large enough to allow us to test for the effects of other possible variables. For example, instead of only comparing the grade point average of drinkers versus nondrinkers, we would compare the grades of four groups: (1) drinkers under stress, (2) drinkers not under stress, (3) nondrinkers under stress, and (4) nondrinkers who were not under stress.

As our example suggests, if we really want to understand what is going on in survey research, we need to use large, longitudinal surveys. But collecting such data is very expensive, and few sociologists can afford the costs on their own. Instead, many turn to government agencies such as the U.S. Census Bureau or to nonprofit organizations such as the National Opinion Research Center, which each year collects vast amounts of data from a random sample of the U.S. population for its General Social Survey (GSS);

This is the strategy sociologist Robert Crosnoe (2006) used to understand alcohol use among adolescents. He based his research on longitudinal data collected by the federal government from almost 12,000 middle and high school students. Because the data covered multiple years, Crosnoe could tell that students tended to begin drinking after their grades went down rather than the drinking preceding the low grades. And because the study was so large, he could divide the students according to many different variables and be sure that failing grades really had affected students' alcohol use, rather than some other factor leading students to have both lower grades and higher drinking levels.

But regardless of the size or time frame of a survey, an important drawback of this technique is that respondents may misrepresent the truth. Both frequent drinkers *and* nondrinkers may lie about their habits because they fear others will look down on them. Sociologists refer to such misrepresentation as **social-desirability bias**—the tendency for people to color the truth so that they appear to be nicer, richer, and generally more desirable than they really are. Decoding the Data: Alcohol Use among Full-Time Students on the next page provides data on this topic from a large national survey. The data suggest that underage drinking—including heavy drinking—is quite common (although we need to consider whether social desirability bias might have affected the data).

Social-desirability bias is the tendency of people to color the truth so that they sound more desirable and socially acceptable than they really are.

decoding the data

Examining the Data: Can you think of a sociological explanation for why young men are more likely than young women to drink alcohol and to drink heavily? Are girls and boys taught different messages about drinking? about drinkers? How? By whom? Do the dangers of drinking and heavy drinking differ for men and women? How might this affect their levels of drinking?

Critiquing the Data: Might these data overstate the differences between men and women's drinking habits? Might men overestimate their drinking or women underestimate their drinking? Why?

Alcohol Use During Last 30 days, among Full-Time College Students Aged 18 to 20
SOURCE: The NSDUH Report (2006).

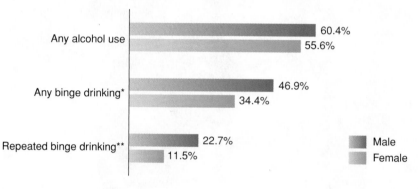

*Five or more drinks on the same occasion
**Five or more drinks on five or more days

As this example suggests, survey research is not the best strategy for studying hidden or socially unacceptable behaviors. Nor is it a good strategy for examining ideas and feelings that cannot easily be reduced to questionnaire form. Finally, survey research studies individuals outside of their normal contexts. If we want to understand the situations and social contexts in which individuals drink, we must turn instead to participant observation.

Participant Observation

Participant observation refers to research conducted "in the field" by researchers who participate in their subjects' daily life, observe daily life, or interview people in-depth about their lives. This method is particularly useful for discovering patterns of interaction and learning the meaning those patterns hold for individuals. Unlike survey researchers, who ask people about what they do or believe, participant-observers aim to *see* what people are actually doing. Participant observation is used most often by symbolic interactionists—that is, by researchers who want to understand subjective meanings, personal relationships, and the process of social life.

The three major techniques involved in participant observation are interviewing, participating, and observing. A researcher goes to the scene of the action, where she may interview people informally in the normal course of conversation, participate in whatever they are doing, observe the activities of other participants, or do all three. Researchers decide which of these techniques to use based on both intellectual and practical criteria. A participant observer studying alcohol use on campus, for example, would not need to get "smashed" every night. She would, however, probably do long, informal interviews with both users and nonusers, attend student parties and activities, and attempt to get a feel for how alcohol use fits in with certain student subcultures.

In some cases, participant observation is the only reasonable way to approach a subject. This is especially likely when we are examining behaviors that break normal social rules or groups that fall outside the mainstream of society. For example, if

Participant observation refers to conducting research by participating, interviewing, and observing "in the field."

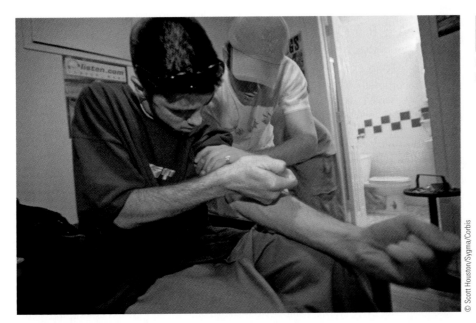

Participant observation is the best—and perhaps only—way to study highly stigmatized behaviors such as injecting illegal drugs.

© Scott Houston/Sygma/Corbis

fraternity members are asked to indicate on a survey how often they black out after drinking, they may not give an honest answer—or may not even remember the correct answer. If, on the other hand, we observe fraternity parties on campus, we may get a more accurate view of how often students black out. And if we spend weeks or months observing a fraternity and building trust, we will likely get more honest answers when we do choose to interview members.

Similarly, participant observation is often the only way to obtain information about groups that are truly outside the mainstream. If we wanted to study college students' drinking, we could mail out surveys and expect that at least some would reply. But how could we mail surveys to homeless alcoholics? And why would they reply, even if we could find them? For this reason, participant observation is often our best source of information about groups such as topless dancers, illegal drug users, and neo-Nazi skinheads.

On the other hand, a major disadvantage of participant observation is that it is usually based on small numbers of individuals who have not been selected randomly. The data tend to be unsystematic and the samples not very representative. However, we do learn a great deal about the few individuals involved. This information can help us to generate ideas that we can examine more systematically with other techniques. For this reason, researchers often use participant observation as the initial step in exploring a research topic.

Another disadvantage of participant observation is that the observations and generalizations rely on the interpretation of one researcher. Because researchers are not robots, it seems likely that their findings reflect some of their own worldview. This is a greater problem with participant observation than with survey or experimental work, but all science suffers to some extent from this phenomenon. The answer to this dilemma is **replication**, redoing the same study with another researcher or with different samples to see if the same results occur.

Focus on American Diversity: Studying Life in "The Projects" on the next page illustrates the advantages and disadvantages of using participant observation to study life in poor, African American communities.

Replication is the repetition of empirical studies by another researcher or with different samples to see if the same results occur.

focus on  AMERICAN DIVERSITY

Studying Life in "The Projects"

What is life like for extremely poor African Americans who live in segregated housing projects? Initially, sociologist Sudhir Venkatesh—then a graduate student—thought he could answer this question using standardized survey questions, including "How does it feel to be black and poor? Very bad, somewhat bad, neither bad nor good, somewhat good, or very good." He soon learned that such questions were useless at best: Some simply laughed at the questions, some responded with brief or misleading answers, and some concluded that he must be working for the police. As a result, Venkatesh realized that the only way to learn about life in the projects was to listen and watch. He did so for almost a decade, spending much of his time with members of the Black Kings street gang.

Venkatesh's research (2000, 2008, 2009) allowed him to document the extreme isolation and hardship experienced by project residents. Although the projects were owned by the city of Chicago, the residents received almost no services. Many apartments lacked running water or electricity, and many buildings had only black holes where elevators once ran. The police rarely ventured into the buildings, and emer-

gency services rarely responded when anyone dialed 911. As a result, street gangs served as quasi-governments. Building residents relied on the gangs to discipline (that is, to beat) anyone who battered a woman, robbed a resident, or (under the influence of drugs) behaved so crazily in building lobbies that they scared residents or visitors. In exchange, the gang leaders received free rein to sell drugs and to demand "protection money" from area businesspeople, whether prostitutes or grocery store owners.

At the same time, Venkatesh found, the residents showed great ingenuity in finding ways to survive in the midst of incredible hardship. For example, one group of five families survived by pooling the resources of their five apartments: one with a working stove, one with working heat, one with running water, and so on. Others augmented their small incomes with a wide variety of off-the-record home businesses, from baking pies to fixing cars to selling lottery tickets.

Venkatesh's participant observation allowed him a view into life in America's ghettoes that could not have been obtained through any other methods. At the same time, his experiences illustrate the pitfalls of participant research. Early on, he realized that so long as he "hung out" with gang members, nonmem-

In the housing project studied by Sudhir Venkatesh, the external hallways that link the apartments look more like prison cells than like balconies.

bers would not fully trust him. But if he spent time with *non*members, the gang members wouldn't trust him—and might also make it dangerous for him to visit the projects. Moreover, because he spent so much time with the gang, he naturally found that he sometimes saw the world at least partially through the gang members' eyes. Finally, because it was unsafe for him to wander around the projects on his own, he was initially only able to see what others wanted him to see. Because Venkatesh spent so many years conducting his research, however, he eventually was able to view the situation from all sides and to paint a thorough—and fascinating—picture of life in the projects.

Content Analysis

So far, all the methods we've discussed rely on observing or interviewing people. In other cases, however, sociologists focus their research not on people but on the documents that people produce. **Content analysis** refers to the systematic examination of documents of any sort.

Sociologists who use content analysis follow essentially the same procedures as those who conduct surveys. But instead of taking a sample of individuals and then asking them a list of questions, sociologists who use content analysis take a sample of *documents* and then systematically ask questions about those documents. For example, to explore how rap music portrays alcohol use, researcher Denise Herd (2005) first identified the most popular rap songs over an 18-year period. She then chose a random sample of 341 songs, read the lyrics for each song, and systematically noted

Content analysis refers to the systematic examination of documents of any sort.

whether the song mentioned alcohol and whether it linked alcohol to positive effects (like glamour or wealth) or to negative effects (like losing a girlfriend or going to jail). Herd found that rap music mentioned alcohol use more often over time and that the songs typically mentioned only alcohol's positive consequences.

Researchers can use content analysis with any type of written document: court transcripts, diaries, student papers, and so on. They can also use it with electronic "documents" such as blogs, web pages, and public comments emailed to politicians and archived online for anyone to view. In addition, sociologists may analyze not only a document's text but also its images—exploring, for example, how alcohol use is portrayed on billboards, in magazines, online, or on television.

A main advantage of content analysis is that it can be quite inexpensive: no one need spend months in the field collecting observations or spend days going door-to-door asking people to answer surveys. In addition, content analysis can be used with historical as well as contemporary documents. We could, for example, analyze the last 30 years of alcohol ads to see how the portrayal of alcohol has changed over time. Finally, because we are looking at existing documents, we cannot affect the data itself: A participant observer might affect how much the students he observes drink, but a sociologist conducting content analysis can't affect what appears in a magazine ad.

The obvious disadvantage of content analysis is that it can only be used with existing documents, and so will not work for some research topics. In addition, as with participant observation, it relies on researchers' interpretations of the data. However, with content analysis a team of researchers can look at the data and compare their conclusions, making it less likely that any one researcher's bias affects the results.

Using Existing Statistics

Regardless of which methods sociologists use, they often augment their data with existing statistics from other sources. Federal, local, and state governments provide a wealth of information to researchers, such as how house prices have changed over time, how life expectancy has risen or fallen, how cities have grown or shrunk in population, and so on. If we were studying alcohol use in a particular college, for example, we could obtain data from the U.S. Census on per capita alcohol consumption in the college's neighborhood. We could obtain data on alcohol-related car accidents from our state's Health or Motor Vehicles Department. Or we could obtain data on sexual assaults that might be linked to alcohol use from the college or local police department. We could use these data to provide a broader picture of the problem, or we could combine these data with the data we collected ourselves—exploring, for example, whether more sexual assaults occurred during years when students who answered our survey reported higher levels of drinking.

The advantages and disadvantages of using existing statistics are similar to those for content analysis. Since we are using existing data, the costs are low to nonexistent, and we can study the past as well as the present. The disadvantage is that we cannot collect data to fit our research questions but must instead rely on whatever data are available.

Sociologists: What Do They Do?

A degree in sociology can be the starting point to a successful career. Your particular career options, however, will vary depending on whether you also pursue graduate training in sociology.

Using a Bachelor's Degree in Sociology

Like other liberal arts majors, sociology provides students with the basic education needed for entry-level positions in many fields. In addition, sociology teaches students how to think critically, analyze data, and understand both social problems and human relationships. As a result, undergraduate sociology majors graduate with skills and knowledge that can serve them well in journalism, business, teaching, health care, and many other fields. In addition, undergraduate sociology training provides excellent grounding for graduate education in a variety of fields; Michelle Obama obtained an undergraduate degree in sociology before pursuing a law degree.

If you want to work as a sociologist, however, you will also need to obtain a graduate degree in sociology. For some jobs, a master's degree may be enough; for others, a Ph.D. is required.

Sociologists in Colleges and Universities

About three-quarters of U.S. sociologists with graduate degrees work as professors or lecturers in colleges and universities. At some schools, sociologists are solely expected to engage in teaching and to help with their school's administrative work. At others, they are expected to engage in both teaching and research.

Some sociology professors use their research to understand basic principles of human social behavior. Others focus more directly on addressing social problems such as violence, illness, and unemployment. For example, sociology professors who study disasters played crucial roles in helping the government, nonprofit organizations, and communities respond to the environmental damage caused by Hurricane Katrina. A particularly good example is University of New Orleans sociologist Shirley Laska, who had predicted New Orleans's vulnerability to hurricanes in a widely cited report published a year before the hurricane struck.

Sociologists in Government

Sociologists also find employment at all levels of government, from local to national. For example, sociologists at the U.S. Census Bureau measure changes in the population and help communities decide whether to build day-care centers or nursing homes. At the Department of Education, sociologists help policy makers decide whether schools should increase or decrease their use of standardized tests. And at local, state, and national health departments, sociologists have researched such topics as why students engage in unsafe sex during spring break and how schools can best encourage their students to adopt safer practices.

Sociologists in Business

Sociologists are employed in various positions in the business world. Some use their knowledge of human interaction to work in human relations departments or firms, especially with regard to issues of gender or ethnic diversity. Others work in market research. Sociologists can help businesses predict whether signing a movie star to blog about their product might increase sales or which features would help woo consumers from iPhones to a new smartphone. Sociologists' understanding of human behavior and of how to *research* human behavior are invaluable assets for those seeking positions of this type.

Sociologists in Nonprofit Organizations

Nonprofit organizations range from hospitals and clinics to social-activist organizations and private think tanks; sociologists are employed in all these types of organizations. Sociologists at the American Foundation for AIDS Research, for example, have studied the causes of unsafe sexual activity and have evaluated the effectiveness of different strategies used to encourage condom use. They also have conducted the background research needed to convince communities to adopt more controversial approaches, such as distributing clean needles to addicts to prevent the transmission of HIV.

Although most sociologists work in research, a small but growing number work for nonprofits or on their own as marriage, family, or rehabilitation counselors. The training that sociologists receive is very different from that received by psychologists, social workers, and other counselors, but it can be very useful in helping individuals understand how their personal problems connect to broader social issues and social forces.

Sociologists Working to Serve the Public

Most sociologists are committed to a value-free approach to their work as scholars. Many, however, also dedicate themselves to changing society for the better, whether they work in government, business, nonprofit organizations, or academia. As a result, sociologists have served on a wide variety of public commissions and in public offices to encourage positive social change. They work for change independently, too, both as individuals and in organizations such as Sociologists without Borders (www.sociologistswithoutborders.org), which is committed to "advancing transnational solidarities and justice." Value-free scholarship does not have to mean value-free citizenship.

Where This Leaves Us

Sociology is a diverse and exciting field. From its beginnings in the nineteenth century, it has grown into a core social science that plays a central role in university education. Its three major perspectives—structural functionalism, conflict theory, and symbolic interactionism—provide a complementary set of lenses for viewing the world, while its varied methodological approaches supply the tools needed to study social life in all its complexity. These lenses and tools position sociologists not only to understand the world, but to help change it for the better.

Summary

1. Sociology is the systematic study of social behavior. Sociologists use the concepts of role and social structure to analyze common human dramas. When we use the *sociological imagination,* we focus on understanding how social structures affect individual behavior and personal troubles.

2. The rapid social change that followed the industrial revolution was an important inspiration for the development of sociology. Problems caused by rapid social change stimulated the demand for accurate information about social processes. This social-problems orientation remains an important aspect of sociology.

3. There are three major theoretical perspectives in sociology: structural-functional theory, conflict theory, and symbolic interaction theory. The three can be seen as alternative lenses through which to view society, with each having value as a tool for understanding how social structures shape human behavior.

4. Structural functionalism has its roots in evolutionary theory. It identifies social structures and analyzes their consequences for social harmony and stability. Identification of manifest and latent functions and dysfunctions is part of its analytic framework.

5. Conflict theory developed from Karl Marx's ideas about the importance of conflict and competition in structuring human behavior and social life. It analyzes social structures by asking who benefits from them and how these benefits are maintained. This theory assumes that competition is more important than consensus and that change is a positive result of conflict.

6. Symbolic interaction theory examines the subjective meanings of human interaction and the processes through which people come to develop and communicate shared symbolic meanings. Whereas structural functionalism and conflict theory emphasize macrosociology, symbolic interactionism focuses on microsociology.

7. Sociology is a social science. This means it relies on critical and systematic examination of the evidence before reaching any conclusions and that it approaches each research question from a position of neutrality. This is called value-free sociology.

8. The five steps in the research process are stating the problem, setting the stage, gathering the data, finding patterns, and generating theory. These steps form a continuous loop called the *wheel of science.* The movement from data to theory is called induction, and the movement from theory to hypothesis to data is called deduction.

9. Any research design must identify the variables under study, specify the precise operational definitions of these variables, and describe how a representative sample of cases for studying the variables will be obtained.

10. Experiments are excellent ways of testing cause-and-effect hypotheses. However, experiments measure behavior in highly artificial conditions, and individuals may behave differently when they are in experiments. In addition, experiments can sometimes expose subjects to harm.

11. In survey research, a researcher asks a large number of people a set of standard questions. This method is useful for describing incidence, trends, and differentials for random samples, but not as good for describing the contexts of human behavior or for establishing causal relationships.

12. Participant observation is a method in which the researcher observes or interviews in depth a small number of individuals. The method is an excellent source of fine detail about human interaction and its subjective meanings. However, it typically relies on nonrepresentative samples and on one researcher's interpretations of the data, unverified by other observers.

13. Content analysis refers to the systematic study of written documents, whether contemporary or historical. Its advantage is that it is inexpensive and that the researcher cannot bias the data itself. However, it can only be used with existing documents, and it relies on researchers' interpretations of the data.

14. Sociologists often base their research on existing statistics obtained from government agencies, nonprofit organizations, and other sources. This inexpensive method permits the study of the past as well as the present, but can only be used when appropriate data is available.

15. Most sociologists teach and do research in academic settings. A growing minority is employed in government, nonprofit organizations, and business, where they do applied research. Regardless of the setting, sociological theory and research have implications for social policy.

Thinking Critically

1. Which of your own personal troubles might reasonably be reframed as public issues? Does such a reframing change the nature of the solutions you can see?

2. Consider how a structural-functional analysis of gender roles might differ from a conflict analysis. Would men be more or less likely than women to favor a structural-functionalist approach?

3. Can you think of situations in which a change of friends, living arrangements, or jobs has caused you to change your interpretations of a social issue (such as gay marriage, single motherhood, or unemployment benefits)?

4. Consider what study design you could ethically use to determine whether drinking alcohol, living in a sorority, or growing up with a single parent reduces academic performance.

Book Companion Website

www.cengage.com/sociology/brinkerhoff
Prepare for quizzes and exams with online resources—including tutorial quizzes, a glossary, interactive flash cards, crossword puzzles, essay questions, virtual explorations, and more.

CHAPTER 2

Culture

© Jon Arnold Images Ltd/Alamy

Introduction to Culture

In Chapter 1 we said that sociology is concerned with analyzing the contexts of human behavior and how these contexts affect our behavior. Our neighborhood, our family, and our social class provide part of that context, but the broadest context of all is our culture. **Culture** is the total way of life shared by members of a community.

In some places, a culture cuts across national boundaries. French Canadian people and culture, for example, can be found in both Canada and New England. In other places, two distinct cultures may coexist within a single national boundary, as French and English culture do within Canada. For this reason, we distinguish between cultures and societies. A **society** is the population that shares the same territory and is bound together by economic and political ties. Often the members share a common culture, but not always.

Culture resides essentially in nontangible forms such as language, values, and symbolic meanings, but it also includes technology and material objects. A common image is that culture is a "tool kit" that provides us with the equipment necessary to deal with the common problems of everyday life (Swidler 1986). Consider how culture provides patterned activities of eating and drinking. People living in the United States share a common set of tools and technologies in the form of refrigerators, ovens, cell phones, computers, and coffeepots. As the advertisers suggest, we share similar feelings of psychological release and satisfaction when, after a hard day of working or playing, we take a break with a cup of coffee or a cold beer. The beverages we choose and the meanings attached to them are part of our culture. Despite many shared meanings and values, however, this example also illustrates some of the difficulties inherent in any discussion of a single common culture: Although Mormon Americans and Muslim Americans share our American culture, the former do not drink coffee and neither group drinks alcohol.

Culture can be roughly divided into two categories: material and nonmaterial. *Nonmaterial culture* consists of language, values, rules, knowledge, and meanings shared by the members of a society. *Material culture* includes the physical objects that a society produces—tools, streets, sculptures, and toys, to name but a few. These material objects depend on the nonmaterial culture for meaning. For example, Barbie dolls and figurines of fertility goddesses share some common physical features, but their meaning differs greatly and depends on nonmaterial culture.

Theoretical Perspectives on Culture

As is true in other areas of sociology, structural functionalists, conflict theorists, and symbolic interactionists each have their own approach to the study of culture.

The Structural-Functionalist Approach

The structural-functionalist approach treats culture as the underlying basis of interaction. It accepts culture as a given and emphasizes how culture shapes us rather than how culture itself is shaped. Scholars taking this approach have concentrated on illustrating how norms, values, and language guide our behavior. We will return to this topic later when we discuss the carriers of culture.

Culture is the total way of life shared by members of a community. It includes not only language, values, and symbolic meanings but also technology and material objects.

A **society** is the population that shares the same territory and is bound together by economic and political ties.

The Conflict Theory Approach

In contrast, conflict theorists focus on culture as a social product. They ask why culture develops in certain ways and not others, and whose interests these patterns serve. These scholars would take an interest, for example, in how the content of television shows is affected by government versus corporate ownership.

Conflict theorists also investigate how culture can reinforce power divisions within society. They argue that **cultural capital**—upper-class attitudes and knowledge—brings power and status to individuals in the same way that *financial* capital (that is, money) does (Bourdieu 1984; Lamont & Fournier 1992). If you never learned to play golf, select a red wine, appreciate an opera, or eat a five-course meal with five different forks, your cultural deficiencies will be painfully apparent to others at upper-class events. You lack some of the cultural capital needed to marry into or work in these social circles and may be ridiculed by others if you try to do so. In this way, culture serves as a *symbolic boundary* that keeps the social classes apart.

Finally, conflict theorists analyze what happens when cultures come into conflict with each other. We will explore this topic further when we discuss subcultures, countercultures, and the battles over assimilation versus multiculturalism.

The Symbolic Interactionist Approach

Whereas conflict theorists often focus on *what* the media portray (How many blacks are in TV shows? Is violence portrayed as fun?), symbolic interactionists focus on how people *interpret* and *use* what they see in the media. They explore the meanings people derive from culture and cultural products, and how those meanings result from social interaction. For example, research in this tradition has documented how women find empowering messages in romance novels and horror films, how the rise of Viagra has changed the meaning of male sexuality, why people identify with pop music stars, and what "ethnic" foods (Chinese noodles, Italian pastas, southern biscuits) mean both to those who belong to ethnic groups and to outsiders (Loe 2004; Vares & Braun 2006; Vannini 2004; Bai 2003).

Confidently and properly ordering and eating a meal at a fine restaurant requires "cultural capital" that you may not have unless you were raised in an upper-class or at least upper middle-class home.

Bases of Human Behavior: Culture and Biology

Why do people behave as they do? What determines human behavior? To answer these questions, we must be able to explain both the varieties and the similarities in human behavior. Generally, we will argue that biological factors help explain what is common to humankind across societies, whereas culture explains why people and societies differ from one another.

Cultural capital refers to having the attitudes and knowledge that characterize the upper social classes.

Cultural Perspective

Regardless of whether they are structural functionalists, conflict theorists, or symbolic interactionists, sociologists share some common orientations toward culture: Nearly all hold that culture is *problem solving*, culture is *relative,* and culture is a *social product.*

Culture Is Problem Solving

Regardless of whether people live in tropical forests or in the crowded cities of New York, London, or Tokyo, they confront some common problems. They all must eat, they all need shelter from the elements (and often from each other), and they all need to raise children to take their place and continue their way of life. Although these problems are universal, the solutions people adopt vary considerably. For example, traditionally, the mother's brother was responsible for child rearing in the Trobriand Islands, and communal nurseries were responsible in some Israeli kibbutzim.

Whenever people face a recurrent problem, cultural patterns will evolve to provide a ready-made answer. This does not mean it is the best answer or the only answer or the fairest answer, but merely that culture provides a standard pattern for dealing with this common dilemma. One of the issues that divides conflict and functional theorists is how these answers develop. Functionalists argue that the solutions we use today have evolved over generations of trial and error, and that they have survived because they work, because they help us meet basic needs. A conflict theorist would add that these solutions work better for some people than for others. Conflict theorists argue that elites manipulate culture to rationalize and maintain solutions that work to their advantage. Scholars from both perspectives agree that culture provides ready-made answers for most of the recurrent situations we face in daily life; they disagree on who benefits from a particular solution.

Culture Is Relative

The solutions that each culture devises may be startlingly different. Among the Wodaabe of Niger, for example, mothers may not speak directly to their first- or second-born children and, except for nursing, they may not touch them. The babies' grandmothers and aunts, however, lavish affection and attention on them (Beckwith 1983). The effect of this pattern of child rearing is to emphasize loyalties and affections throughout the entire kinship group rather than just with one's own children or parent. This practice helps ensure that each new entrant will be loyal to the group as a whole.

Is it a good or a bad practice? That is a question we can answer only by seeing how it fits in with the rest of the Wodaabe culture and by taking the viewpoint of one or another social group. Does it help the people meet recurrent problems and maintain a stable society? If so, structural functionalists would say it works; it is functional. Conflict theorists, on the other hand, would want to know who is helped and who is hurt by the practice. Both sets of theorists, however, believe that each cultural trait should be evaluated in the context of its own culture. This belief is called **cultural relativity.** A corollary of cultural relativity is that no practice is universally good or universally bad; goodness and badness are relative, not absolute.

This type of evaluation is sometimes a difficult intellectual feat. For example, no matter how objective we try to be, most of us believe that infanticide, human sacrifice, and cannibalism are absolutely and universally wrong. Such an attitude reflects **ethnocentrism**—the tendency to use the norms and values of our own culture as standards against which to judge the practices of others. Ethnocentrism usually means

---

Cultural relativity requires that each cultural trait be evaluated in the context of its own culture.

Ethnocentrism is the tendency to judge other cultures according to the norms and values of one's own culture.

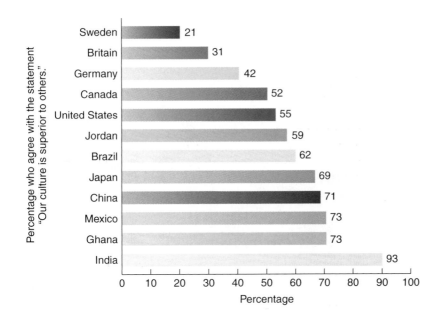

FIGURE 2.1 **Ethnocentrism around the World**
Ethnocentrism—the belief that one's culture is superior to other cultures—is more common in the United States than in some European countries, but much less common than in various other countries.
SOURCE: Pew Research Center (2007).

that we see our way as the right way and everybody else's way as the wrong way. When American missionaries first came to the South Sea Islands, for example, they found that Polynesians did many things differently from Americans. Rather than viewing Polynesian practices as merely different, however, the missionaries viewed those practices as wrong and probably wicked. As a result, the missionaries taught the islanders that the only acceptable way (the American way) to have sexual intercourse was in a face-to-face position with the man on top, the now-famous "missionary position." They taught the Polynesians that women and men should wear Western clothes, even if the clothes don't suit the Polynesian climate, that they should have clocks and come on time to appointments, and a variety of other Americanisms that the missionaries maintained to be morally right behavior. Figure 2.1 shows levels of ethnocentrism around the world.

Ethnocentrism is often a barrier to interaction among people from different cultures, leading to much confusion and misinterpretation. It is not, however, altogether bad. In the sense that it represents pride in our own culture and confidence in our own way of life, ethnocentrism is essential for social integration. In other words, we learn to follow the ways of our culture because we believe that they are the right ways; if we did not share that belief, there would be little conformity in society. Ethnocentrism, then, is a natural and even desirable product of growing up in a culture. An undesirable consequence, however, is that we simultaneously discredit or diminish the value of other ways of thinking and feeling. As a result, ethnocentrism can make it difficult for us to change our ways even if change would be in our best interests (Diamond 2005). For example, Norwegian explorers in Antarctica fared far better than did British explorers because the Norwegians adopted Inuit ("Eskimo") clothing, skis, and dogsleds, whereas the British considered such tactics beneath them—and sometimes died as a result (Huntford 2000).

Culture Is a Social Product

A final assumption sociologists make about culture is that culture is a social, not a biological, product. The immense cultural diversity that characterizes human societies results not from unique gene pools but from cultural evolution.

In 1911, a British team under Robert F. Scott and a Norwegian team under Roald Amundsen raced to become the first explorers ever to reach Antarctica. The British team's ethnocentricism led to its downfall: Scott's team relied on man-hauled sleds and perished, Amundsen's team adopted Inuit dog sleds and skiing techniques and succeeded.

AP Images

Some aspects of culture are produced deliberately. Shakespeare decided to write *Hamlet* and J. K. Rowling to write the *Harry Potter* books; marketing teams created the Geico gecko and the MacIntosh Apple icon. Governments, bankers, and homeowners commission designs for homes, offices, and public buildings from architectural firms, and people buy publishing empires so that they can spread their own version of the truth. Other aspects of culture—such as language, fashion, and ideas about right and wrong—develop gradually through social interaction. But all these aspects of culture are human products; none of them is instinctive. People *learn* culture, and, as they use it, they change it.

Culture depends on language. A culture without language cannot effectively transmit either practical knowledge (such as "fire is good" and "don't use electricity in the bathtub") or ideas (such as "God exists") from one generation to the next. With language, cultures can pass on inventions, discoveries, and forms of social organization for the next generation to use and improve.

Because of language, human beings don't need to rely on the slow process of genetic evolution to adapt to their circumstances. Whereas biological evolution may require literally hundreds of generations to adapt the organism fully to new circumstances, cultural evolution allows changes to occur much more rapidly.

Biological Perspective

As television programs on the Discovery Channel regularly demonstrate, clothing, eating habits, living arrangements, and other aspects of culture vary dramatically around the globe. It is tempting to focus on the exotic variety of human behavior and to conclude that there are no limits to what humankind can devise. A closer look, however, suggests that there are some basic similarities in cultures, such as the universal existence of the family, religion, cooperation, and warfare. When we focus on these universals, cultural explanations need to be supplemented with biological explanations.

Sociobiology is the study of the biological basis of all forms of human (and nonhuman) behavior (Alcock 2001; Wilson 1978). Sociobiologists believe that humans and all other life forms developed through evolution and natural selection. According to this perspective, species change primarily through one mechanism: Some genes reproduce more often than do others. As these genes increase in number, the species takes on the traits linked to these genes.

Which genes reproduce most often? Genes will reproduce most often if the people who carry them have more children and raise more of them until they are old enough to reproduce themselves (Alcock 2001; Daly & Wilson 1983). For example, sociobiologists suggest that parents who are willing to make sacrifices for their children, occasionally even giving their lives for them, are more successful reproducers; by ensuring their children's survival, these parents increase the likelihood that their own genes will contribute to succeeding generations. Thus, sociobiologists argue that we have evolved biological predispositions toward cultural patterns that enable our genes to continue after us.

Sociobiology provides an interesting theory about how humans evolved over tens of thousands of years. Most scholars who study the effect of biology on human behavior, however, investigate more contemporary questions, such as "How do hormones, genes, and chromosomes affect human behavior today?" Joint work by biologists and social scientists helps us to understand how biological and social factors work together to determine human behavior. For example, Booth and Osgood (1993) found that men were statistically more likely to engage in deviant behavior if they had *both* high levels of testosterone *and* low levels of social integration. Research such as this suggests that only by recognizing and taking into account the joint effects of culture and biology can we fully understand human behavior.

The Carriers of Culture

In this section, we review three vital aspects of nonmaterial culture—language, values, and norms—and show how they shape both societies and individuals. We then explore how social control pressures individuals to live within the rules of their culture.

Language

The essence of culture is the sharing of meanings among members of a society. The chief mechanism for this sharing is a common language. **Language** is the ability to communicate in symbols—orally, by manual sign, or in writing.

What does *communicate with symbols* mean? It means, for example, that when you hear the word *dog* or see the curved and straight lines that represent that word in a book, you understand that it means a four-legged domestic canine. Almost all communication occurs through the use of symbols. Even the meanings of physical gestures such as touching or pointing are learned as part of culture.

Scholars of sociolinguistics (the relationship between language and society) agree that language has three distinct relationships to culture: Language embodies culture, it is a symbol of culture, and it creates a framework for culture (Romaine 2000; Trudgill 2000).

Sociobiology is the study of the biological basis of all forms of human (and nonhuman) behavior.

Language is the ability to communicate in symbols—orally, by manual sign, or in writing.

Because language is such an important carrier and symbol of culture, protests have emerged around the world whenever people feel their language is under attack.

The **Sapir-Whorf hypothesis** argues that the grammar, structure, and categories embodied in each language affect how its speakers see reality. Also known as the *linguistic relativity hypothesis*.

Language as Embodiment of Culture

Language is the carrier of culture; it embodies the values and meanings of a society as well as its rituals, ceremonies, stories, and prayers. Until you share the language of a culture, you cannot fully participate in it (Romaine 2000; Trudgill 2000).

A corollary is that loss of language may mean loss of a culture. Of the approximately 300 to 400 Native American languages once spoken in the United States, only about 20 may survive much longer (Dalby 2003, 147–148). When these languages die, important aspects of these Native American cultures will vanish. This vital link between language and culture is why many Jewish and Chinese parents in the United States send their children to special classes after school or on weekends to learn Hebrew or Chinese. This is also why U.S. law requires that people must be able to speak English before they can be naturalized as U.S. citizens. To participate fully in Jewish or Chinese culture requires some knowledge of these languages; to participate in U.S. culture requires some knowledge of English.

Language as Symbol

A common language is often the most obvious outward sign that people share a common culture. This is true of national cultures such as French and Italian and subcultures such as youth. A distinctive language symbolizes a group's separation from others while it simultaneously symbolizes unity within the group of speakers (Joseph et al. 2003; Romaine 2000; Trudgill 2000). For this reason, groups seeking to mobilize their members often insist on their own distinct language. For example, Jewish pioneers who moved in the early 1900s from the ghettos of Europe to what was then Palestine declared that everyone within their communities must speak Hebrew. Yet no one had spoken Hebrew except in prayers for hundreds of years. Nevertheless, within a few decades, Hebrew became the national language of Israel.

Similarly, in the last two decades some Americans have opposed bilingual education and pushed to declare English the official language of the United States, while French Canadians have fought to make French the official language of Quebec (Dalby 2003). Meanwhile, government bureaucracies in Mexico and France fight to keep English words from creeping into Spanish and French. All these efforts are largely symbolic; in any country, both immigrants and native-born citizens will continue to use or will quickly adopt whichever language has the most social status and social utility (Ricento & Burnaby 1998).

Map 2.1 shows the percentage of people in different states who speak a language other than English at home. The percentages are high and rising. However, many of these individuals already speak English outside the home, and most who are now children will switch to speaking primarily English as they grow up. Moreover, history suggests that the children of these non-English speakers will speak only English (Dalby 2003).

Language as Framework

According to some linguists, languages not only symbolize our culture but also help to create a framework in which culture develops. The **Sapir-Whorf hypothesis** (also known as the **linguistic relativity hypothesis**) argues that the grammar, structure, and categories embodied in each language influence how its speakers see reality (Whorf 1956). According to this hypothesis, for example, because Hopi grammar does not have past, present, and future grammatical tenses (for example, "I had," "I have," "I will have"), Hopi speakers think differently about time than do English speakers.

This theory has come under attack in recent years. Most linguists now believe that although differences among languages influence thought in small ways, the

MAP 2.1: Percent of U.S. Residents 5 Years and Over Who Speak a Language Other Than English at Home
Almost 20 percent of U.S. residents now speak a language other than English at home, leading some Americans to worry that American culture and the English language are at risk. But many of these foreign-language speakers also speak English, and many of their children speak only English.

SOURCE: factfinder.census.gov. Calculated from 2007 American Community Survey data set. Accessed April 2009.

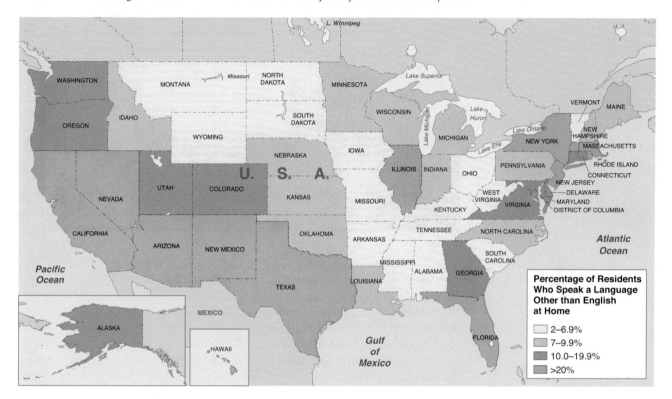

universal qualities of language and human thought far overshadow those differences. The difficulties of translating from one language to another illustrate the conceptual differences among languages; that translation is nonetheless possible shows that, despite those differences, people in all cultures have essentially the same linguistic capabilities (Trudgill 2000).

Values

After language, the most central and distinguishing aspect of culture is **values**, shared ideas about desirable goals (Hitlin & Piliavin 2004). Values are typically couched in terms of whether a thing is good or bad—desirable or undesirable. For example, many people in the United States believe that a happy marriage is desirable. In this case and many others, values may be very general. They do not, for example, specify what constitutes a happy marriage.

Some cultures value tenderness and cooperation; others value toughness and competition. Nevertheless, because all human populations face common dilemmas, certain values tend to be universal. For example, nearly every culture values stability and security, a strong family, and good health. But cultures can achieve these goals in dramatically different ways. In many traditional societies, individuals try to gain security by having many children whom they can call on for aid. In our society,

Values are shared ideas about desirable goals.

Norms that govern daily life are usually not as explicit as in this classroom. Nevertheless, most of us figure out social norms without much trouble just from observing those around us.

sociology and you

As you sit in your college classroom, you are following a long list of norms. Your very presence in the classroom reflects your acknowledgment that higher education is useful. No matter how bored you might be, you sit reasonably still and try not to fidget. If you are falling asleep, you pull your cap brim down to hide your droopy eyelids. You raise your hand rather than call out to demonstrate your respect for the teacher. And you write down whatever the teacher says, or at least write something down so it looks like you are taking notes.

Norms are shared rules of conduct that specify how people ought to think and act.

Folkways are norms that are the customary, normal, habitual ways a group does things.

individuals try to ensure security by putting money in the bank or investing in an education. Conversely, among the Kwakiutl tribe of the Pacific Northwest, individuals traditionally ensured economic security not by saving wealth but by giving it away in a custom called a *potlatch*. When one person gave a gift to another, the receiver was obligated to help out the giver in the future. In this way, poorer persons received needed gifts and wealthier persons could count on help if they should ever lose their wealth. As this suggests, although many cultures place a value on establishing security against uncertainty and old age, the specific guidelines for reaching this goal vary. These guidelines are called norms.

Norms

Shared rules of conduct are **norms**. They specify what people *ought* or *ought not* to do. The list of things we ought to do sometimes seems endless. We begin the day with "I'm awfully tired, but I ought to get up," and may end the day with "I'd like to keep partying but I'd better go to bed." In between, we ought to brush our teeth, eat our vegetables, work hard, love our neighbors, and on and on. The list is so extensive that we may occasionally feel that we have too many obligations and too few choices. Of course, some pursuits are optional and allow us to make choices, but the whole idea of culture is that it provides a blueprint for living, a pattern to follow.

Norms vary enormously in their importance both to individuals and to society. Some, such as fashions, are short-lived. Others, such as those supporting monogamy and democracy, are powerful and long-lasting because they are central to our culture. Generally, we distinguish between two kinds of norms: folkways and mores.

Folkways

The word **folkways** describes norms that are simply the customary, normal, habitual ways a group does things. Folkways is a broad concept that covers relatively permanent traditions (such as fireworks on the Fourth of July) as well as passing fads and fashions (such as wearing baggy versus tight shorts).

Folkways carry no moral value. If you choose to violate folkways by having hamburgers for breakfast and oatmeal for dinner, or by sleeping on the floor and dyeing your hair purple, others may consider you eccentric, weird, or crazy, but they will not brand you immoral or criminal.

Mores

In contrast to folkways, other norms do carry moral value. These norms are called **mores** (more-ays). Whereas eating oatmeal for dinner may lead others to consider you odd, eating your dog or spending your last dollar on liquor when your child needs shoes may lead others to consider you immoral. They may turn you in to the police or to a child protection association; they may cut off all interaction with you or even chase you out of the neighborhood. Because people who break these norms are considered immoral, we know that these norms are mores, and not simply folkways. Not all violations of mores result in legal punishment, but all result in such informal reprisals as ostracism, shunning, or reprimand. These punishments, formal and informal, reduce the likelihood that people will violate mores.

Laws

When mores are enforced and sanctioned by the government, they are known as **laws.** If laws cease to be supported by norms and values, they may be overturned or the police may simply stop enforcing them. However, laws don't always emerge from popular values. New laws forbidding driving while texting, for example, were adopted to *change* existing norms, not to reflect them.

The Concept Summary on Values, Norms, and Laws compares these three important concepts.

Social Control

From our earliest childhood, we learn to observe norms, first within our families and later within peer groups, at school, and in the larger society. After a period of time, following the norms becomes so habitual that we can hardly imagine living any other way—they are so much a part of our lives that we may not even be aware of them as constraints. We do not think, "I ought to brush my teeth or else my friends and family will shun me"; instead we think, "It would be disgusting not to brush my teeth, and I'll hate myself if I don't brush them." For thousands of generations, no human considered it disgusting to go around with unbrushed teeth. For most people in the United States, however, brushing their teeth is so much a part of their feeling about the kind of person they are that they would disgust themselves if they did not do so.

Through indoctrination, learning, and experience, many of society's norms come to seem so natural that we cannot imagine acting differently. No society relies completely on this voluntary compliance, however, and all encourage conformity by the use of **sanctions**—rewards for conformity and punishments for nonconformity. Some sanctions are formal, in the sense that the legal codes identify specific penalties, fines, and punishments meted out to individuals who violate formal laws. Formal sanctions are also built into most large organizations to control absenteeism and productivity. Some of the most effective sanctions, however, are informal. Positive sanctions such as affection, approval, and inclusion encourage normative behavior, whereas negative sanctions such as a cold shoulder, disapproval, and exclusion discourage norm violations.

Despite these sanctions, norms are not always a good guide to what people actually do, and it is important to distinguish between normative behavior (what we

Mores are norms associated with fairly strong ideas of right or wrong; they carry a moral connotation.

Laws are rules that are enforced and sanctioned by the authority of government. They may or may not be norms.

Sanctions are rewards for conformity and punishments for nonconformity.

concept summary

Values, Norms, and Laws

Concept	Definition	Example from Marriage	Relationship to Values
Values	Shared ideas about desirable goals	It is desirable that marriage include physical love between wife and husband	
Norms	Shared rules of conduct	Have sexual intercourse regularly with each other, but not with anyone else	Generally accepted means to achieve value
Folkways	Norms that are customary or usual	Share a bedroom and a bed; kids sleep in a different room	Optional but usual means to achieve value
Mores	Norms with strong feelings of right and wrong	Thou shalt not commit adultery	Morally required means to achieve value
Laws	Formal standards of conduct, enforced by public agencies	Illegal for husband to rape wife; sexual relations must be voluntary	Legally required means; may or may not be supported by norms

are supposed to do) and actual behavior. For example, our own society has powerful mores supporting marital fidelity. Yet research has shown that nearly half of all married men and women in our society have committed adultery (Laumann et al. 1994). In this instance, culture expresses expectations that differ significantly from actual behavior. This does not mean the norm is unimportant. Even norms that a large minority, or even a majority, fail to live up to are still important guides to behavior. The discrepancy between actual behavior and normative behavior—termed *deviance*—is a major area of sociological research and inquiry (see Chapter 6).

Cultural Diversity and Change

By definition, members of a community share a culture. But that culture is never completely homogeneous. In the following sections, we will look at two expressions of diversity within cultures—subcultures and countercultures—and at the processes by which cultures change.

Subcultures and Countercultures

No society is completely homogeneous. Instead, each society has within it a dominant culture, as well as subcultures and countercultures.

Subcultures share in the overall culture of society but also maintain a distinctive set of values, norms, lifestyles, and traditions and even a distinctive language. The "Greek life" of traditional (residential) fraternities and sororities offers an excellent

Subcultures are groups that share in the overall culture of society but also maintain a distinctive set of values, norms, and lifestyles and even a distinctive language.

example of a subculture. To enter a fraternity or sorority, prospective members must first demonstrate that their fashion style; partying or studying habits; and attitudes toward sex, drinking, community service, and scholarship fit the culture of a particular house as well as of the Greek system as a whole. Those who are "tapped" must then go through the ritual of hazing, an experience that can range from humorous to dangerous and that cements ties to the fraternity or sorority and its culture. After initiation, members learn the special traditions of the house, which can include songs, passwords, and other rituals.

Greek subculture does not have its own language, but it does have its own slang terms for members of other houses, among other things. It also has its own values, beginning with loyalty to fellow members. Some fraternities, for example, expect their members to tutor fraternity "brothers" when needed; other fraternities expect members to help their brothers cheat on exams. Fraternities also expect members to adopt a distinctive lifestyle: living together in sex-segregated houses and cooking, eating, and socializing primarily with other house members. Those who actively participate in this subculture gain strong, supportive bonds during college and strong social networks afterward.

Subcultures differ from the dominant culture, but they are not at odds with it. In contrast, **countercultures** are groups whose values, interests, beliefs, and lifestyles conflict with those of the larger culture. This theme of conflict is clear among one current U.S. countercultural group—punkers. Some punkers are part-timers who shave their heads and listen to death rock but nevertheless manage to go to school or hold a job. Hardcore punkers, however, emphatically reject "straight" society. They refuse to work or to accept charity; they live angry and sometimes hungry lives on the streets. They cover their arms with tattoos or stick safety pins into their clothes or eyebrows because they *want* people to know they have rejected mainstream values.

Assimilation or Multiculturalism?

Until very recently, most Americans believed it would be best if the various ethnic and religious subcultures within American society would adopt the dominant majority culture. **Assimilation** refers to the process through which individuals learn and adopt the values and social practices of the dominant group, more or less giving up their own values in the process. As discussed in more detail in Chapter 12, assimilation was, and to some extent still is, one of the major goals of our educational institutions (Spring 2004). In schools, immigrant children learn not only to read and write English but also to consider American foods, ideas, and social practices preferable to those of their own native culture or subculture. Teachers encourage children named Juan or Mei Li to go by the name John or Mary. School curricula focus on the history, art, literature, and scientific contributions of Europeans and European Americans while downplaying the contributions of U.S. minority groups and non-Western cultures.

In the last quarter century, however, more and more Americans have concluded that America has always been more of a "salad bowl" of cultures than a melting pot. Many have come to believe that this "salad bowl" is one of Americans' greatest strengths and that it should be cherished rather than eliminated. These beliefs are often referred to as **multiculturalism**. Reflecting this idea, many schools and universities now incorporate materials that more accurately reflect American cultural diversity.

Countercultures are groups whose values, interests, beliefs, and lifestyles conflict with those of the larger culture.

Assimilation is the process through which individuals learn and adopt the values and social practices of the dominant group, more or less giving up their own values in the process.

Multiculturalism is the belief that the different cultural strands within a culture should be valued and nourished.

These deaf students believe that they share a common culture and should have rights like those given to any minority culture.

Case Study: Deafness as Subculture

Most people who can hear consider deafness undesirable, even catastrophic (Dolnick 1993). At best, they see being deaf as a medical condition to be remedied. However, some deaf people maintain that deafness is not a disability but a culture (Dolnick 1993; Padden & Humphries 2006). To these individuals, the essence of deafness is not the inability to hear but a valued culture based on their shared language, American Sign Language (ASL). ASL is not just a way to "speak" English with one's hands but is a language of its own, complete with its own rules of grammar, puns, and poetry. Furthermore, ASL is learned and shared. Whereas babies who can hear begin to jabber nonsense syllables, deaf babies of parents who sign begin to "babble" nonsense signs with their fingers (Dolnick 1993). This shared language encourages, in turn, shared values and a positive group identity. Studies show, for instance, that many deaf people would not choose to join the "hearing" culture even if they could.

Thinking of deafness as a culture illustrates many of the points made earlier. For instance, culture is problem solving, and deaf culture embodies a way to solve the human problem of communication. Using ASL shapes deaf people's experiences, reminding them of their common values, norms, and cultural identity. For this reason, many deaf individuals have reacted with outrage to the increasing use of cochlear implants (Arana-Ward 1997). These devices, when surgically implanted in the ear, help some otherwise deaf persons to hear sounds. Hearing sounds, however, is not the same as understanding what they mean: Many implant recipients—especially older children who were born deaf—are frustrated by a cacophony of sounds that they cannot interpret, even after months or years of training. Some deaf activists argue that most children who receive implants waste their formative years in an often futile struggle to fit into the hearing world, when they could instead have become native speakers of ASL and valued members of the deaf community. These activists, therefore, view cochlear implants not as a neutral medical technology but as an example of the ethnocentrism of hearing persons.

At the same time, because deaf Americans function *within* American culture (reading newspapers, purchasing clothes at the mall, working alongside people who can hear), it is most accurate to consider deafness a *sub*culture rather than a culture. Those who believe that even deaf children who receive cochlear implants should learn ASL are arguing in favor of a multicultural model in which children can feel comfortable in both the deaf and the hearing worlds. Those who argue that deaf children will only learn how to function in the modern world if they receive implants, receive constant training in speech and hearing, and never learn to sign are arguing that these children are best served by full assimilation into the hearing world.

Sources of Cultural Diversity and Change

Culture provides solutions to common and not-so-common problems. The solutions devised are immensely variable. Among the reasons for this variability are environment, isolation, cultural diffusion, technology, exposure to mass media, and dominant cultural themes.

Environment

Why are the French different from Australian aborigines, the Finns different from the Navajo? One obvious reason is the very different environmental conditions in which they live. These conditions determine which kinds of economies can flourish, which kinds of clothes and foods are practical, and, to a significant extent, the degree of scarcity or abundance.

Isolation

When a culture is cut off from interaction with other cultures, it is likely to develop unique norms and values. Where isolation precludes contact with others (such as in the New Guinea highlands until recently), a culture can continue on its own course, unaltered and uncontaminated by others. Since the nineteenth century, however, almost no cultures have been able to maintain their isolation from other cultures.

Cultural Diffusion

If isolation is a major reason why cultures remain both stable and different from each other, then cultural diffusion is a major reason why cultures change and become more similar over time. Cultural diffusion is the process by which aspects of one culture or subculture become part of another culture. For example, not only have many residents of Mexico City become regular consumers of McDonald's hamburgers, but belief in the value of fast food is gradually replacing Mexicans' traditional belief in the value of long, family-centered meals. Meanwhile, salsa now outsells ketchup in the United States, and Heinz now offers a green ketchup specifically to compete with salsa.

At its broadest level, cultural diffusion becomes the globalization of culture, in which cultural elements (including fashion trends, musical styles, and cultural values) spread around the world. Nowadays, taxi drivers in Bombay, Senegal, and Peru blare U.S. popular music from their radios, while Americans relish the chance to eat in

Diffusion of modern technology is particularly rapid when new tools enhance a society's ability to meet basic human needs at the same time that they are consistent with existing cultural patterns. Leaders, regardless of time, place, or the cultural bases of their authority, share a common need to communicate effectively with large numbers of followers.

Cultural diffusion is the process by which aspects of one culture or subculture are incorporated into another.

French and Chinese restaurants. The globalization of culture is likely to proceed even more rapidly in the future due to the Internet.

The globalization of culture is part of the broader topic of globalization, which we discuss further at the end of this chapter.

Technology

The tools available to a culture will affect its norms and values and its economic and social relationships. Facebook, for example, has dramatically changed attitudes toward privacy, especially among young people. Many young people now consider it perfectly normal to post intimate thoughts, updates on daily activities, and candid photos online, even though their parents may be horrified. At the same time, this new technology has affected not only *attitudes* toward privacy but also *access* to privacy. For example, a nude photo or description of a wild party posted for friends may later be discovered by a parent, professor, or potential employer. Finally, Facebook has increased access to social relationships ("friending") while raising delicate new questions about culturally appropriate ways to manage unwanted relationships ("unfriending").

Mass Media

The mass media are an example of **popular culture**: aspects of culture that are widely accessible and broadly shared, especially among ordinary folks. (In contrast, **high culture** refers to aspects of culture primarily limited to the middle and upper classes, such as opera, modern art, or modernist architecture.) The mass media includes movies; television; genre fiction such as romances, mysteries, or science fiction; and popular music styles like country or hip-hop.

An important question for researchers is whether the mass media simply reflect existing cultural values or whether the media can change values. The answer is that media probably do both. For example, for much of the twentieth century, movies and television usually portrayed African Americans as lazy or foolish and unmarried women as evil, disturbed, or unhappy (Entman & Rojecki 2000; Levy 1990). These depictions reflected American cultural beliefs of the time. Yet these days, Denzel Washington can play a romantic lead, an action hero, or a smart lawyer in movies. Social change in American culture allowed the actor to get these roles, but seeing him in them also creates more cultural change, by suggesting to white Americans that African Americans can be attractive, ethical, smart, and professional. White Americans' acceptance of Denzel Washington as a movie hero may thus have helped Barack Obama win election as president. As this suggests, exposure to mass media can be a source of cultural change. Focus on Media and Culture: The Media and Self-Esteem addresses how the media affect the self-concepts of young men and women.

Dominant Cultural Themes

Popular culture refers to aspects of culture that are widely accessible and commonly shared by most members of a society, especially those in the middle, working, and lower classes.

High culture refers to the cultural preferences associated with the upper class.

Cultures generally contain dominant themes that give them a distinct character and direction. Those themes also create, in part, a closed system. New ideas, values, and inventions can gain acceptance only when they can fit into the existing culture without too greatly distorting existing patterns. Sioux culture, for example, readily adopted rifles and horses because those tools meshed well with its hunting-based culture. But Sioux culture rejected Anglo-American cultural preferences for wood houses and private land ownership because those preferences clashed with the nomadic and communal Sioux way of life.

The Media and Self-Esteem

Over the last several decades, the average American has grown considerably heavier. Yet magazines, movies, television, and even video games increasingly celebrate an extremely rare female body type, far slimmer than that of the typical American girl or woman (Wykes & Gunter 2005; Grogan 2008). Meanwhile, media images of boys' and men's bodies also now idealize a body that is both muscular and slender-waisted, with no extra fat (Pope et al. 2000). The net result is that the gap between media images and actual male and female bodies has increased substantially. How has this affected American culture and the self-concept of young men and women? And has this had a different effect on nonwhite and Hispanic Americans, who are more rarely—and more narrowly—portrayed in the media?

Many scholars believe that unrealistic images in the media have altered cultural notions about what constitutes attractiveness and have damaged self-esteem among young men and women. As a result, they argue, young people often try to lose fat or build muscle through dangerously unhealthy eating patterns, steroid use, or exercise (Wykes & Gunter 2005; Grogan 2008). In fact, numerous surveys have shown that the more exposure individuals have to media, the more likely they are to be dissatisfied with their bodies. Males as well as females are affected, although less strongly, apparently because males realize that their appearance is less important to others than is female appearance (Wykes & Gunter 2005; Grogan 2008). Finally, surveys suggest that body dissatisfaction has also become more common among nonwhites and Hispanics. This trend seems linked to two factors: (1) Media portrayals of these groups have become more common, and (2) social interaction

2009/Jupiter Images

Unrealistic media images have altered our cultural ideas about attractiveness and now threaten the self-esteem of both men and women.

between these groups and white Americans has become more common (Grogan 2008).

Other scholars argue that both culture and young people are more resilient than this. Some argue that the link between media watching and body dissatisfaction may be a spurious correlation, and that something else may cause individuals both to watch media and to be dissatisfied with their bodies. Others suggest that individuals may critically evaluate what they see and read in the media rather than adopting media values automatically.

To explore these issues, sociologists have used interviews to examine how individuals *use* media. Melissa Milkie (1999), for example, found that both African American and white girls believe the images of female beauty shown in

girls' magazines are unrealistically thin. The white girls, however, tried to live up to those images because they assumed that their friends and boyfriends would judge them based on those images. In contrast, the African American girls believed that the media images reflected only white culture and assumed that *their* friends and boyfriends felt the same. As a result, they were less concerned about meeting media standards.

Taken together, Milkie's results suggest that (1) individuals are active consumers of media messages, (2) different audiences interpret the same media messages differently, and (3) media do shape both culture and individual beliefs and actions, at least in part because we judge ourselves through the "media-filled" eyes of others who matter to us.

Case Study: American Consumer Culture

U.S. culture is a unique blend of complex elements. It is a product of the United States' environment, its immigrants, its technology, and its place in history. These days, one of the ways in which U.S. culture diverges most strongly from other cultures is in its exceptionally strong emphasis on consumerism.

Consumerism is a philosophy that says "buying is good." In turn, this philosophy reflects the belief that "we are what we buy," and that through buying certain goods we can assert or improve our social status. In American consumer culture, children attempt to improve their social status by buying the "hottest" toys, teenagers by buying T-shirts from their favorite bands, and adults by buying BlackBerries. Ironically, consumers also believe they are asserting their individuality through their purchases, rarely noticing that millions of others are buying the same goods for the same reasons.

How did this consumer culture develop? The simple answer is that more consumer goods became available and affordable than ever before. But this is only a partial answer. Research suggests that the most important cause was a change in the comparisons people used in deciding whether to make a purchase (Schor 1998). Advertising now permeates our lives more than ever before—billboards adorn public buses and sports stadiums, movie theaters show advertisements before the films, ads pop up at popular Internet sites, schools broadcast television programs laced with commercials in the classroom, and so on. All of this has instilled in children and adults the belief that they need certain products to be a certain sort of person (Quart 2003).

Similarly, as the number of hours Americans watch television per week soared, so did their desire for the goods they saw on television. Instead of deciding what kind of shoes to wear or what kind of kitchen appliances to buy by looking at what their classmates or neighbors owned, Americans sought out consumer goods like those used by their favorite television characters. In fact, for every hour of television watched each week, individuals' annual spending on consumer goods increased by more than $200 (Schor 1998).

Finally, in the past, women (who do most family shopping) typically compared their belongings with those of their neighbors, whose family incomes were usually similar to their own. Now that a majority of women work outside the home, most compare their belongings with those of their fellow workers, including supervisors with much higher incomes. As a result, families now spend higher percentages of their income on consumer goods, both big and small. For example, the median house size has increased from 1500 square feet in 1973 to 2300 square feet in 2007—with prices to match (U.S. Bureau of the Census 2009a).

Consumer culture affects our lives in many ways. Even though the recent downturn in the U.S. economy has reduced consumer spending, shopping (or at least window-shopping) remains a major form of recreation, and shopping malls have replaced parks, athletic fields, church basements, and backyards as popular gathering spots. College students put their grades at risk by working extra hours, in many cases to buy the latest gadgets or fashions. Moreover, despite these extra hours working, the average debt of graduating students rose (in constant dollars) by more than 50 percent between 1993 and 2007, to an average of almost $22,000 among those who had debts (Project on Student Debt 2008). Students who have $40,000 in debt must think twice before taking a job at a nonprofit organization, taking a year off to travel, or pursuing a graduate degree. Meanwhile, adults carry heavy debts and risk bankruptcy to buy expensive cars and houses as a way to "prove" their success and improve their social status. Figure 2.2 illustrates the rising gap between household debt and savings in American households between 1989 and 2007. Since then, consumer debt has held

Consumerism is the philosophy that says "buying is good" because "we are what we buy."

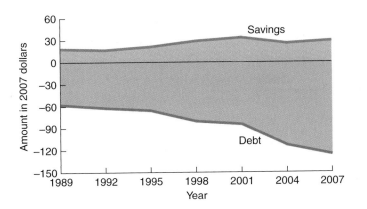

FIGURE 2.2 **Debt versus Savings in U.S. Households with Savings***
Because of both economic hard times and growing consumer desires, Americans' debt has increased more rapidly than have their savings. Moreover, this chart does not include households with no savings at all, which doubled from 5 percent of households in 1989 to 10 percent in 2007. Nor does it show the many Americans who have lost their jobs and savings since 2007.

*In thousands of 2007 dollars. Mean debt and median savings.

SOURCE: Federal Reserve (2009b).

steady (Federal Reserve 2009a), but savings have fallen even further due to the many newly unemployed Americans who must use their savings to pay their bills.

Consequences of Cultural Diversity and Change

No culture remains isolated forever, and none remains forever unchanged. Although cultural diversity and change often help societies cope with existing problems, they can also create new problems. Two such problems are cultural lag and culture shock.

Cultural Lag

Whenever one part of a culture changes more rapidly than another part, social problems can arise. This situation is known as **cultural lag** (Brinkman & Brinkman 1997). Most often, cultural lags occur when social practices and values do not keep up with technological changes.

The rise in "sexting"—sending sexually suggestive photos via cell phones—illustrates the problems that occur when law, values, and social practices lag behind technological change. Sexting has become an increasingly accepted part of life for young people. According to one large (but nonrandom) survey, 20 percent of teenagers and 33 percent of young adults between the ages of 20 and 26 have engaged in sexting (Hamill 2009). But neither cultural values among older adults nor laws (as interpreted by older adults) have kept pace with this change. As a result, some young "sexters" have found themselves arrested on charges of child pornography—even if they were only sending photos of themselves to friends.

Even in the absence of legal sanctions, sexting carries risks: In the same survey, more than a third of teens and almost half of people ages 20 to 26 stated that they commonly share with others suggestive photos that are sent to them. As a result, individuals who send suggestive self-portraits often lose control over who sees the photos—and potentially over their reputations.

As this example illustrates, serious social problems can arise when technological changes leave members of a society without agreed-upon social values, clear legal decisions, or standard social practices defining how they should act.

Cultural lag occurs when one part of a culture changes more rapidly than another.

Culture Shock

In the long run, cultural diversity and cultural change often result in improvements in quality of life. In the short run, however, people often find both diversity and change unsettling. **Culture shock** refers to the disconcerting and unpleasant experiences that can occur when individuals encounter a different culture. For example, U.S. citizens who work in Greece often are surprised by the Greek customs of hugging acquaintances and standing very close (by American standards) to anyone they are speaking with. Greeks who work in the United States are similarly confused by American customs that limit greetings to simple handshakes and dictate maintaining considerable physical distance during conversations. As a result, Americans sometimes conclude that Greeks are pushy or even sexually aggressive, while Greeks sometimes conclude that Americans are elitist or emotionally cold.

Globalization

As this discussion of cultural shock suggests, cultural change can occur not only within one society but also across societies. At its broadest, this change is referred to as the **globalization of culture**. More generally, **globalization** refers to the process through which ideas, resources, practices, and people increasingly operate in a worldwide rather than local framework. Because globalization is having such an impact on the world and its cultures, we devote this section to exploring its sources and effects—economic and political, as well as cultural.

The Sources of Globalization

Globalization stems from a combination of technological and political forces. The rise of the Internet, e-mail, cell and satellite phones, fax machines, and the like all made it easier, cheaper, and faster for corporations and individuals to invest, work, and sell their goods internationally. So, too, did the decline over time in shipping and airfare costs.

Political changes also contributed to globalization. The collapse of the Soviet Union in 1991 made it possible for the nations that emerged in its wake (like Belarus and Estonia) as well as the nations that had been restrained by its political power (like Poland and Armenia) to move toward more capitalistic economic systems. To do so, they needed to seek out economic, political, and cultural ties to other nations that could either serve as sources of raw goods and labor or markets for their products.

The collapse of the Soviet Union also reduced political tensions that had pressed nations to adopt international trade barriers. Now that the nations of Europe are no longer fearful of Soviet might, they have combined into what is in some ways a continental government, in the form of the European Union (EU). Within this Union, goods, individuals, and services can flow more freely than ever before. Polish doctors can now seek higher-paying jobs in Finland, Finnish doctors seek work in Sweden, and Swedish doctors seek work in England, with little concern about visas or immigration laws. German factories can transport and sell their products in Spain, and Spanish factories can send their products to Greece with minimal paperwork or tariffs to pay. Similarly, the North American Free Trade Agreement (NAFTA) was adopted in 1994 to reduce trade barriers between Canada, the United States, and Mexico.

Although the recent economic downturn has led some nations to *increase* trade barriers as a means of protecting farmers and manufacturers in their own countries, globalization remains a powerful force.

The Impact of Globalization

Globalization is a powerful force. Around the world, it is affecting culture, economics, and politics, as well as other aspects of social life.

Cultural Impact

In an African urban nightclub, young people listen to American hip-hop music and drink Pepsi. In New York City, young people go to Jamaican reggae concerts and read *Harry Potter* books. In India, Hollywood films compete with "Bollywood" (Bombay-produced) films. Also, people everywhere loved the film *Slumdog Millionaire*: directed by a British citizen, filmed in India with an Indian cast, and the winner of an American Academy Award. All of these are examples of the global spread of culture, as movies, television shows, music, literature, and other arts increasingly are distributed and enjoyed around the world.

These elements of popular culture carry with them not only entertainment but also cultural values. As Indian adolescents watch American films, they not only learn about the latest U.S. fashions and music but also learn to question traditional Indian practices and beliefs like arranged marriages, the subservience of women, obedience to parents, and the idea that the family is more important than the individual. As a result, many people around the world question the impact of globalization on their—and, especially, their children's—cultural values.

As globalization spreads American products and American cultural values around the world, it can challenge the cultures of other societies. As a result, globalization can sharply increase tensions both within nations and between the United States and other nations.

Economic Impact

Globalization has also had a striking economic impact on both the selling and producing of goods. Increasingly, economic activity takes place between people who live in different nations as goods and services are sold internationally. These days, Russians and Chinese buy Coca-Cola, and Americans buy Volvos and Toyotas. Globalization also exists when goods are *produced* internationally. A transnational corporation such as Toyota, for example, may buy raw goods in one country, process them into car parts in a second country, assemble its cars in a third country, arrange for data processing to occur in a fourth country, and then sell its cars worldwide.

Observers differ greatly in their assessments of the possible effects of such international economic enterprises (Wade 2001; Bordo et.al. 2003). Some hope that ties of international finance will create a more interdependent (and peaceful) world, while stimulating economic growth and improving everyone's standard of living (Stiglitz 2003). Others argue that transnational corporations are harming poorer nations by extracting their raw materials, paying substandard wages to local people, and sending all the profits to the wealthier nations (Petras & Veltmeyer 2001; Wallerstein 2004). In addition, these critics allege that moving labor-intensive work to less developed nations exposes workers in those countries to dangers banned by law in Western nations (Moody 1997).

Critics have also raised questions about the impact of economic globalization even within the developed nations. In the United States, hundreds of thousands of workers lost their jobs when corporations found it cheaper to move those jobs overseas

decoding the data

International Disapproval of Aspects of Globalization

SOURCE: Pew Research Center (2007).

	Percentage who agree		
	Growing trade and business ties between other countries and our country is bad for our country	Large companies from other countries are having a bad influence on things in our country	Our way of life needs to be protected against foreign influence
Americas			
United States	36%	45%	62%
Canada	15	44	62
Argentina	19	47	70
Brazil	25	25	77
Mexico	19	32	75
Peru	15	28	50
Europe			
Britain	15	41	54
France	21	55	52
Germany	13	48	53
Italy	20	49	80
Sweden	9	39	29
Poland	15	31	62
Slovakia	15	24	69
Middle East & Asia			
Lebanon	15	24	75
Pakistan	4	26	81
Malaysia	5	11	85
China	5	22	70
India	8	24	92
Japan	17	32	64
Africa			
Ghana	4	8	80
Senegal	4	9	85
South Africa	9	18	85

Explaining the Data: Based on these data, which citizens are more likely to disapprove of trade ties with other countries: those in wealthy countries or those in poor countries? Which citizens are more likely to fear the impact of large companies from other countries? to fear foreign influence on their way of life? What might explain these patterns?

Critiquing the Data: Researchers collected these data through telephone and face-to-face interviews. Can you think of any reasons why, within each country, poor people would have been less likely to participate in the interviews? How might this affect the findings?

("NAFTA" 2003). Other workers have been forced to accept cuts in benefits or pay to keep their jobs (Bonacich et al. 1994). The question is whether this global movement of jobs raises incomes overall by shifting work from wealthier to poorer countries or merely depresses incomes overall to the level of the cheapest bidders. Decoding the Data: International Disapproval of Aspects of Globalization presents attitudes towards globalization around the world.

Political Impact

How has globalization affected the balance of political power within and across nations? Some observers have noted that transnational corporations now dwarf many national governments in size and wealth. Their ability to move capital, jobs, and prosperity from one nation to another gives them power that transcends the laws of any particular country (Sassen 2001). When a nation's economy depends on a transnational corporation, that nation can't afford to alienate the corporation. For example, Guatemala has limited ability to constrain the labor practices of the United Fruit Company because the corporation could cripple the country's economy if it wanted to (Amaro et al. 2001).

Another aspect of globalization is the sharp increase in the number of international organizations (such as the World Bank, the International Monetary Fund, and the United Nation's International Criminal Court). The underlying premise of these organizations is that they will diminish the independent power of national governments and press nations to conform to international goals (such as promoting free markets, ending torture of political prisoners, prosecuting war criminals, or reducing trade barriers). Many individuals in both wealthier and poorer nations have questioned the impact of these organizations. The data suggest that the poorer nations have indeed lost some of their political and economic autonomy and occasionally have suffered as a result (Stiglitz 2003; Khor 2001; Wade 2003; Rajagopal 2003).

Where This Leaves Us

Most of the time, we think of culture simply as something that we have, in the same way that those of us who have a home or two arms take them for granted. As this chapter has shown, though, culture is dynamic: constantly changing as the world—and the balance of power within that world—changes around us. Languages, eating habits, fashions, and the rest evolve, spread, or die: Ask your parents about the clothing they wore as children, the slang they spoke as teenagers, or the first time they ate a bagel or a tortilla.

Culture is also active, a force that changes us as it changes the world in which we live. The rise of American consumer culture is only one example of how culture changes and of how cultural changes affect all aspects of our lives, from how many hours we work each day to how we define ourselves as individuals.

Summary

1. Culture is a design for living that provides ready-made solutions to the basic problems of a society. Some describe it as a tool kit of material and nonmaterial components that help people adapt to their circumstances. Because of this, as the concept of cultural relativity emphasizes, cultural traits must be evaluated in the context of their own culture.

2. Most sociologists emphasize that culture is socially created. However, sociobiologists emphasize that human culture and behavior also have biological roots.

3. Language, or symbolic communication, is a central component of culture. Language embodies culture, serves as a framework for perceiving the world, and symbolizes common bonds among a social group.

4. Values spell out the goals that a culture finds worth pursuing, and norms specify the appropriate means to reach them.

5. The cultures of large and complex societies are not homogeneous. Subcultures and countercultures with distinct lifestyles and folkways develop to meet unique regional, class, and ethnic needs.

6. The most important factors accounting for cultural diversity and change are the physical and natural environment, isolation from other cultures, cultural diffusion, level of technological development, mass media, and dominant cultural themes.

7. Cultural diversity and change can lead to culture shock and cultural lag. Culture shock refers to the disconcerting experiences that accompany rapid cultural change or exposure to a different culture. Cultural lag occurs when changes in one part of the culture do not keep up with changes in another part.

8. Consumer culture—the philosophy that buying is good, and we are what we buy—now plays a major role in American culture.

9. Globalization refers to the process through which ideas, resources, practices, and people increasingly operate in a worldwide rather than local framework. Globalization has had enormous political, cultural, and economic effects.

Thinking Critically

1. What features of U.S. society might explain why children are raised in small nuclear families rather than in extended kin groups?

2. Can you think of an example from U.S. culture for which values, norms, and laws are not consistent with each other? What are the consequences of these inconsistencies?

3. How do environment, isolation, technology, and dominant cultural themes contribute to the maintenance and diffusion of youth subcultures?

4. Identify three white Anglo-Saxon Protestant (WASP) American ethnic foods. (If you have trouble conceptualizing this, think about why this is difficult.) If you are not a WASP, also identify a favorite ethnic food from your own culture. What do these foods mean to you? What do they mean to others? When and where do you feel comfortable eating and talking about these foods? Why?

Book Companion Website

www.cengage.com/sociology/brinkerhoff
Prepare for quizzes and exams with online resources—including tutorial quizzes, a glossary, interactive flash cards, crossword puzzles, essay questions, virtual explorations, and more.

Socialization

John Howard

This infant monkey and others studied by psychologist Harry Harlow grew up locked in individual cages without social contact. As a result, they had difficulty learning how to have sexual intercourse or raise their own babies. These experiments suggest that even apparently innate behaviors must be developed through interaction.

© Martin Rogers/Stock Boston Inc.

What Is Socialization?

At the heart of sociology is a concern with *people*. Sociology is interesting and useful to the extent that it helps us explain why people do what they do. It should let us see ourselves, our family, and our acquaintances in a new light.

In this chapter we deal directly with individuals, focusing on **socialization:** the process through which people learn the rules and practices needed to participate successfully in their culture and society. Socialization is a lifelong process. It begins with learning and coming to accept the rules and practices of our family and our subculture. As we grow older, join new groups, and take on new identities (as parent, worker, video gamer, or anything else), we learn new norms and redefine our identities.

Learning to Be Human

But how much of what we do and believe is learned and how much is built into our genes? Are we born with a tendency to cooperate or to fight? With a love for hip-hop music, country music, or no music at all? The question of the basic nature of humankind has been a staple of philosophical debate for thousands of years. It continues to be a topic of debate because it is so difficult (some would say impossible) to separate the part of human behavior that arises from our genetic heritage from the part that is developed after birth. The one thing we are sure of is that nature is never enough.

Each of us begins life with a set of human potentials: the potential to walk, to communicate, to love, and to learn. By themselves, however, these natural capacities are not enough to enable us to join the human family. Without nurture—without love and attention and hugging—the human infant is unlikely to survive, much less prosper. The effects of neglect are sometimes fatal and, depending on severity and length, almost always result in retarded intellectual and social development.

Monkeying with Isolation and Deprivation

How can we determine the importance of nurture? In a classic series of experiments, psychologist Harry Harlow and his associates studied what happened when they raised monkeys in total isolation (Blum 2002). The infants lived in individual cages with a mechanical mother figure that provided milk. Although the infant monkeys' nutritional needs were met, their social needs were not. As a result, both their physical and social growth suffered. They exhibited bizarre behavior that resembled that of some autistic children, such as staring blankly, biting themselves, and hiding in corners.

Socialization is the process of learning the roles, statuses, and values necessary for participation in social institutions.

As adults, these monkeys refused to mate; if artificially impregnated, the females would not nurse or care for their babies (Harlow & Harlow 1966). These experiments provided dramatic evidence of the importance of social contact; even apparently innate behaviors such as sexuality and maternal behavior did not occur unless developed through social interaction.

Harlow's attempts to socialize monkeys reared in isolation for 6 months produced mixed results, with many showing no improvement at all. The best results occurred when he subsequently placed these 6-month-old monkeys with 3-month-old monkeys and their mothers, giving those raised in isolation a second chance to be socialized along with the younger monkeys (Harlow & Suomi 1971).

The Necessity of Nurture for Humans

Learning to be a monkey, however, is quite different from learning to be a human, and so the Harlow experiments can only suggest what would happen to children raised without nurturing. Of course, we cannot ethically isolate children away from caring adults to see what would happen. Unfortunately, sometimes that happens nonetheless.

Some of the clearest evidence regarding the consequences of raising children with little or no nurture comes from studies of children raised in low-quality orphanages. In these orphanages, the children's physical needs were met but they received little true nurturing. Many of these children were devastated by the experience. Some withdrew from the social world, neither crying nor showing interest in anything around them. Others became violent toward themselves or others. Even if later adopted into good homes, they were significantly more likely to experience difficulties in thinking or learning. They were also more likely to experience problems in social relationships—either engaging in indiscriminate friendliness or withdrawing into autistic or near-autistic behaviors. These effects are illustrated by a study that compared children adopted by British parents either from high-quality British orphanages or from low-quality Romanian orphanages. The researchers found that 12 percent of the

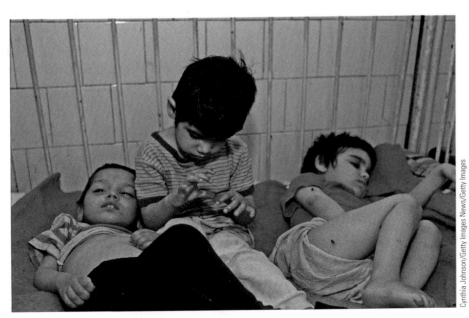

When children (such as these South African orphans) receive little true nurturing, their ability to learn, think, and develop normal human relationships may permanently decline.

Cynthia Johnson/Getty Images News/Getty Images

Romanian-born children exhibited autistic or near-autistic patterns, whereas none of the children from the British orphanages did (Rutter et al. 1999).

Other evidence on the importance of nurture comes from rare cases of children whose parents raised them in situations of extreme physical and emotional deprivation. The true story of Genie (a pseudonym) illustrates the consequences of severe deprivation (Newton 2004). Until the age of 13, Genie's abusive father kept her tied to a chair and locked in a small room. Her mother—blind, disabled, and cowed—could do nothing to help her. Genie was never spoken to or socialized in any way. When her mother finally took Genie and ran away, Genie could not talk, walk, or even use a toilet. After years of therapy, her abilities improved, but they remained far below the level needed for her to live on her own.

Genie's case is extreme. But milder forms of deprivation occur in homes in which parents fail to provide adequate social and emotional stimulation. Children who have their physical needs met but are otherwise ignored by their parents often exhibit problems similar to those of Genie and of orphans and monkeys raised without nurturing. The bottom line is that for children, as for monkeys, physical and social development depend on interaction with others of their species. Walking, talking, loving, and laughing all depend on socialization through sustained and intimate interaction.

Theoretical Perspectives on Socialization

To become a functioning member of society, each of us must be socialized. But how does socialization occur? What are the processes through which, as children and adults, we learn the rules, values, and behaviors of our society? In the following pages we look at some psychological and sociological theories of socialization.

Freudian Theory

The first modern theory of socialization was developed by psychoanalyst Sigmund Freud at the beginning of the twentieth century. Freud's theory of socialization links social development to biological cues. According to Freud, to become mentally healthy adults, children must develop a proper balance between their **id**—natural biological drives, such as hunger and sexual urges—and their **superego**—internalized social ideas about right and wrong. To find that balance, children must respond successfully to a series of developmental issues, each occurring at a particular age and linked to biological changes in the body.

For Freud, the years from 3 to 6 are especially important because this is when he believed the superego developed. According to Freud, during this stage children first start noticing genitalia. When boys learn that girls lack penises, they conclude that girls must have been castrated by their fathers as punishment for some wrongdoing. To avoid this fate, boys quickly adopt their father's rules and values, thus developing a strong superego. In contrast, Freud argued that because girls need not fear castration they can never develop a strong superego (Freud [1925] 1971, 241–260).

Freud based his theory on his personal interpretations of his patients' lives and dreams, rather than on scientific research. Nevertheless, his conception of human nature and socialization continues to permeate American culture and social science: The theory is still used (in a much revamped form) by some psychologists and even some sociologists (e.g., Chodorow 1999).

The **id** is the natural, unsocialized, biological portion of self, including hunger and sexual urges.

The **superego** is composed of internalized social ideas about right and wrong.

Piaget and Cognitive Development

Another influential psychological theory of socialization is cognitive development theory. This theory has its roots in the work of Swiss psychologist Jean Piaget (1954). Piaget developed his theory through intensive observations of normal young children. His goal was to identify the stages that children go through in the process of learning to think about the world.

Piaget's observations led him to conclude that there are four stages of cognitive development. In the first stage, children learn to understand things they see, touch, feel, smell, or hear, but they do not understand cause and effect. So, for example, very young children love playing peekaboo because it is a delightful surprise each time the person playing with them removes his or her hands to reveal his or her presence. In later stages, children may learn to use language, symbols, and numbers; to understand cause and effect; and to understand abstract concepts such as truth or justice. Piaget recognized, however, that some children lack the capacity to reach the highest stages of development.

Critics of Piaget's work suggest that Piaget's model is too simplistic. They argue that in addition to individual differences among children, cultural and gender differences may also affect the nature and trajectory of cognitive development (Gilligan 1993). They also question whether Piaget's ideas reflect only one culture's definitions of what it means to have high cognitive development.

Structural-Functional Theory

The starting premise of all structural-functionalist analyses is that in a properly functioning society, all elements of society work together harmoniously for the good of all. The same is true for structural-functionalist analyses of socialization.

As we'll see in later chapters, structural functionalists believe that schools, religious institutions, families, and the other social arenas in which children are socialized are designed to integrate the young smoothly into the broader culture, avoiding

Structural functionalists point out that schools teach children not only to read and write but also to obey authority and conform to society's rules.

altrendo images/Altrendo/Getty Images

conflict or chaotic social change. In families, children learn to mind their manners, and in schools they learn to be on time and obey the rules. In places of worship, children learn to practice accepted rituals (lighting candles at Mass or on the Sabbath, praying to the east or on Sundays) and to respect traditional ideas about good and bad, right and wrong.

From the perspective of structural functionalism, this socialization is all for the good: Through socialization, young people learn how to become happy and productive members of society. And in the best situations, socialization does work as structural functionalists claim: Children learn both to fit into the world of their elders and to think for themselves, so that they can adapt to any changes the future brings. Critics of this theory, on the other hand, point out that in socializing children to the world as it is, we also teach them to accept existing inequalities and make it difficult for them to see how to change the world for the better. This is the perspective taken by conflict theorists.

Conflict Theory

Conflict theory's approach to socialization is the opposite of structural functionalism's. Whereas structural functionalism assumes that socialization benefits everyone, conflict theory assumes it benefits only those in power.

Conflict theorists focus on how socialization reinforces unequal power arrangements. Some look at how parents socialize children to consider girls less valuable than boys by requiring girls to wash dishes after dinner but allowing boys to go outside to play. Others investigate how teachers socialize working-class children to fill working-class jobs by punishing signs of creativity and rewarding strict obedience. Still others explore how priests, ministers, rabbis, and other religious leaders may socialize congregants to believe that the privileges of the wealthy and of dominant ethnic groups have been granted by God.

Conflict theory is useful for understanding how socialization can quash dissent and social change and reproduce inequalities. It is less useful for explaining the sources and benefits of a stable social system.

Symbolic Interaction Theory

Sociologists who use symbolic interaction theory begin with three basic premises:

1. To understand human behaviors, we must first understand what those behaviors mean to individuals.
2. Those meanings develop within social relationships.
3. Individuals actively construct their self-concepts, within limits imposed by social structures and social relationships.

In addition, symbolic interaction theorists use two central concepts—the *looking-glass self* and *role taking*—to understand how individuals construct their self-concepts. The next sections explore these two concepts.

The Looking-Glass Self

The **self-concept** is our sense of who we are as individuals.

The **looking-glass self** is the process of learning to view ourselves as we think others view us.

Charles Horton Cooley (1902) provided a classic description of how we develop our self-concept. The **self-concept** is our sense of who we are as individuals, in terms of both our personalities and our position in society. Cooley proposed that we develop our self-concept by learning to view ourselves as we think others view us. He called this the **looking-glass self** (see the Concept Summary on the

Looking-Glass Self). According to Cooley, there are three steps in the formation of the looking-glass self:

1. We imagine how we appear to others.
2. We imagine how others judge us based on those appearances.
3. We ponder, internalize, or reject these judgments.

For example, an instructor whose students doze during class may well conclude that the students consider him a bad teacher. He may internalize their view of his teaching abilities and conclude that he needs to seek another line of work. Alternatively, however, he may recall colleagues who have complimented him on his teaching and other classes that seemed to appreciate his style. As a result, he may instead conclude that this semester's students are simply not smart enough to appreciate his teaching.

As this suggests, our self-concept is not merely a mechanical reflection of the views of those around us; rather it rests on our interpretations of and reactions to their judgments. We engage actively in defining our self-concept, choosing whose looking-glass we want to pay attention to and using past experiences to aid us in interpreting others' responses.

concept summary

The Looking-Glass Self

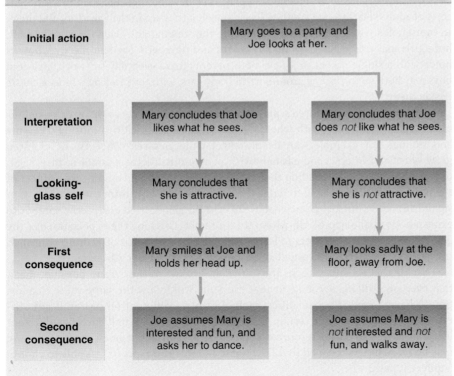

Symbolic interaction considers subjective interpretations to be extremely important determinants of the self-concept. This premise of symbolic interactionism is apparent in W. I. Thomas's classic statement: "If men define situations as real, they are real in their consequences" (Thomas & Thomas 1928, 572). People interact with others based on their subjective interpretations of how others think about them and about the world. Thus subjective interpretations have real consequences—whether or not they accurately reflect what others are thinking.

Role Taking

The most influential contributor to symbolic interaction theory during the last century was George Herbert Mead (1934). To Mead, the self had two components, which he referred to as the *I* and the *me*. In English grammar, we use the word *I* when *we* do something and use the word *me* when someone else does something to us ("*I* disobeyed my mom. Then she punished *me*."). Similarly, Mead used the word *I* to refer to the spontaneous, creative part of the self and the word *me* to describe the part of the self that responds to others' expectations.

Mead argued that we learn to function in society and to balance the desires of the *I* with the social awareness of the *me* through the process of **role taking**. This means learning how others important to us see the world and gradually adopting their perspectives.

According to Mead, role taking begins in childhood, when we learn the rights and obligations associated with being a child in our particular family. To understand what is expected of us as children, we must learn our mother's and father's views. We must learn to see ourselves from our parents' perspective and to evaluate our behavior from their point of view. Only when we have learned their views as well as our own will we really understand what our own obligations are.

Mead maintained that children develop their knowledge of how to function in society by playing games. When children play, they develop their ideas of how different sorts of adults relate to one another, based on what they see in the world around them. In households where moms are responsible for cooking and dads for home repair, little girls may enjoy playing with toy stoves and boys with toy hammers. In households with a single parent or where both parents leave each day for work, girls and boys may both enjoy driving around in their toy cars, with toy briefcases or tool chests beside them.

As this suggests, children's play often focuses on the behavior of their **significant others**—individuals with whom they have close personal relationships. Parents and siblings, for example, can deeply affect children's self-concepts. As children grow older and interact increasingly with people outside their families, they begin to learn what others—including their teachers, neighbors, and employers—expect of them. Eventually, they come to judge their behavior not only from the perspective of significant others but also from what Mead calls the **generalized other**—the composite expectations of all people with whom they interact. Learning the expectations of the generalized other is equivalent to learning the norms and values of a culture. Through this process, we learn how to act like an American or a Pole or a Nigerian.

Saying that everyone learns the norms and values of the culture does not mean that everyone will behave alike or that everyone will follow the same rules. In addition to having unique personalities, each of us has a different set of significant others, each grows up within certain cultures and subcultures, and each has different levels of access to social resources such as education and money. As a result, we may be more or less inclined to follow society's rules and more or less able to choose a different path.

Role taking involves imagining ourselves in the role of others in order to determine the criteria they will use to judge our behavior.

Significant others are the role players with whom we have close personal relationships.

The **generalized other** combines the expectations of all with whom we interact.

Agents of Socialization

Socialization is a continual process of learning. Each time we encounter new experiences, we must reassess who we are and where we fit into society. This challenge is most evident when we undergo important life transitions—when we leave home for the first time, join the military, change careers, or get divorced, for example. Each of these shifts requires us to expand our skills, adjust our attitudes, and accommodate ourselves to new realities.

Socialization takes place in many contexts. We learn what others expect of us from our parents, teachers, bosses, religious leaders, and friends, as well as from television, movies, and even comic books. These **agents of socialization**—the individuals, groups, and media that teach us social norms—profoundly affect our personalities, self-concepts, values, and behaviors, especially if the messages learned in one setting are reinforced elsewhere. (See Focus on Media and Culture: Girls' Hair, Girls' Identities.) Each of these agents of socialization is discussed more fully in later chapters. They are introduced here to illustrate the importance of social structures for learning.

Family

Perhaps the most important agent of socialization is the family. As the tragic cases of child neglect and the monkey experiments so clearly demonstrate, the initial warmth and nurturance we receive at home are essential to normal cognitive, emotional, and physical development. In addition, our family members—usually our parents but sometimes our grandparents, stepparents, or others—are our first teachers. From them we learn not only how to tie our shoes and hold a crayon but also beliefs and goals that may stay with us for the rest of our lives.

The activities required to meet the physical needs of a newborn provide the initial basis for social interaction. Feeding and diaper changing give opportunities for cuddling, smiling, and talking. These nurturant activities are all vital; without them, the child's social, emotional, and physical growth will be stunted (Handel, Cahill, & Elkin 2007; Blum 2002; Rutter et al. 1999).

In addition to these basic developmental tasks, the child has a staggering amount of learning to do before becoming a full member of society. Much of this early learning occurs in the family as a result of daily interactions: The child learns to talk and communicate, to play house, and to get along with others (Handel, Cahill, & Elkin 2007). As the child becomes older, teaching is more direct, and parents attempt to produce conformity and obedience, impart basic skills, and prepare the child for life outside the family. Families differ, however, in the means they use to impart these values and skills: Some will try to rely only on hugs and praise, others will consider a "good spanking" a useful tool, and a small percentage will beat a child who disobeys. Decoding the Data: Attitudes toward Spanking explores these differences.

One reason the family is the most important agent of socialization is that the self-concept formed during childhood has lasting consequences. In later stages of development, we pursue experiences and activities that integrate and build on the foundations established in the primary years. Although our personalities and self-concepts do not take final, fixed form in childhood, childhood experiences set the stage for our later development.

The family is also an important agent of socialization in that the parents' religion, social class, and ethnicity influence the child's behaviors, beliefs, self-concept,

The **agents of socialization** are all the individuals, groups, and media that teach social norms.

Girls' Hair, Girls' Identities

Why does a "bad hair day" matter so much to girls and women that some will just stay home; some will go through their day cranky, unconfident, or depressed; and most will sacrifice time and money to avoid this fate? This is the question that led Rose Weitz, one of the authors of this textbook, to write the book *Rapunzel's Daughters: What Women's Hair Tells Us about Women's Lives* (2004). A good part of the answer to this question, she found, lies in girls' socialization.

As Weitz discovered, parents, teachers, friends, neighbors, and even strangers passing on the street all teach girls to consider their hair central to their identity and to their position in the world. Parents praise their daughters when their hair is neatly styled, refuse to take them to church or the mall when it isn't, and drag them to beauty parlors even when their daughters could care less. Teachers will pull out a comb and fix girls' hair when they consider it too unruly, and strangers will comment on how a girl's "beautiful blonde curls" (if she is white) or naturally long, straight hair (if she is African American) will surely garner her a rich husband.

Girls are also socialized to consider their hair central to their identity through material culture and the mass media. Through toys and other gifts, girls learn to consider hair work both fun and meaningful. Barbie dolls are an especially clear example. In addition to garden-variety Barbies, girls can get (among many others) Fashion Queen Barbie, which comes with blonde, brunette, and "titian-haired" wigs; Growin' Pretty Hair Barbie, whose hair can be pulled to make it longer; and Totally Hair Barbie, the most popular Barbie ever, which comes with hair to her toes, styling gel, a hair pick, and a styling book. The Barbie "Styling Head," which consists of

A vast array of toys teach young girls to consider their hair central to their identies and to consider it a source of fun, pleasure, and personal meaning.

nothing but a head with long hair, is also popular. Similarly, although few would think it appropriate to give boys curling irons or blow dryers as gifts, many parents, aunts, and uncles give such gifts to girls, further reinforcing the importance of their hair.

The importance of girls' hair is also reinforced by the mass media. Time after time, in movies like *Pretty Woman* or *America's Sweethearts*, apparently plain women get the guy once they get a new hairstyle (and ditch the glasses). If you see a girl or woman in a movie with bad hair, she is either the villain, the comic sidekick, or about to get both a makeover and the guy. Even children's cartoons follow this pattern: Smurfette, the only female on the show *The Smurfs*, was created by the wicked wizard Gargamel to be an evil, conniving seductress who would cause the Smurfs' downfall. When Papa Smurf changed her into a good Smurfette, her messy, medium-length, brown hair became long, smooth, and blonde.

Magazines aimed at teenagers more directly socialize girls to focus on their hair and appearance. Weitz found that about half of advertisements and articles in recent issues of *Seventeen* (the most popular teen magazine) focused on how girls could change their hair or bodies. Similarly, makeover stories—in which ugly ducklings become swans by changing their hair, makeup, clothes, and even bone structure—are a regular feature on television talk shows and provide the entire focus of various television programs.

Although some girls are more immune than others to media messages, few escape their effects fully. As discussed in the previous chapter, even when individual girls reject the idea that they should define themselves through their appearance, they still feel obligated to *act* as if they accept that idea because they believe that others accept it and judge them on that basis (Milkie 1999).

decoding the data

Attitudes Toward Spanking

Caption: Education significantly affects Americans' attitudes toward spanking children: Those who have graduated college are less likely than others to approve of spanking.

Question: Do you strongly agree, agree, disagree, or strongly disagree that it is sometimes necessary to discipline a child with a good, hard spanking?

SOURCE: General Social Survey (2009). **http://sda.berkeley.edu**. Accessed June 2009.

	Less than 12 Years' Education	High School Graduate	Some College	College Graduate
Strongly Agree	27%	31%	24%	17%
Agree	49	45	48	45
Disagree	17	19	24	30
Strongly Disagree	7	5	4	8

Explaining the Data: Although you probably haven't discussed the merits of spanking in any of your classes, something about attending college may make you less likely to approve of spanking by the time you graduate. How is college changing your ideas about families and personal relationships? How is college changing your ideas about how people learn? Your ideas about "proper" ways to behave? Might any of these changes affect your attitudes toward spanking?

Critiquing the Data: Are there any reasons why college graduates might be less likely than others to *admit* that they approve of spanking?

College graduates differ from others in many ways. Most importantly, they disproportionately come from middle- or upper-class families and themselves have middle- or upper-class income. Can you think of any reasons why higher-income persons would be less likely to rely on spanking? Might income, rather than college, explain the difference in attitudes?

and position in society. They influence the expectations that others have for the child, and they determine the groups with which the child will interact outside the family. Thus, the family's race, class, and religion shape the child's initial experiences in the neighborhood, at school, and at work.

Peers

In past centuries, and in some parts of the world today, children often lived on isolated farms where their families remained almost the only important agent of socialization throughout their childhood. For the last several decades, however, compulsory education together with the late age at which most youths become full-time workers have led to the emergence of a youth subculture in modern societies. In recent years, this development has been accelerated by the tendency for both parents to work outside the home, creating a vacuum that may be filled by interaction within a *peer group* (Osgood et al. 1996). The **peer group** refers to all individuals who share a similar age and social status; each member of the peer group is referred to as a **peer.** Most children

The **peer group** refers to all individuals who share a similar age and social status.

A **peer** is a member of a peer group.

Alyson Aliano/Riser/Getty Images

All peer groups—from cheerleaders to gang members—pressure their members toward dressing, thinking, and behaving similarly.

place a high value on peer acceptance and quickly adopt peer culture (Harris 1998; Handel, Cahill, & Elkin 2007).

What are the consequences of peer interaction for socialization and the development of the self-concept? Because kids who hang out together tend to dress and act similarly, peer pressure creates conformity to the peer group—whether the group is cheerleaders, honor students, or gang members. As a result, conformity to peer values and lifestyles can be a source of family conflict when, for example, your friends urge you to pierce your tongue and your parents express horror at the idea. The more time you have to hang out with friends unsupervised by adults, the more likely it is that your friends will affect you (Haynie & Osgood 2005).

The impact of peers is so great that some scholars now believe it is stronger than that of family (Corsaro 2003, 2004). Because the judgments of one's peers are unclouded by love or duty, they are particularly important in helping us get an accurate picture of how we appear to others. In addition, the peer group is often a mechanism for learning behaviors and values different from those of adults. For example, peer groups teach their own cultural norms about everything from whether one should share with another child to whether one should smoke or drink.

However, the effects of peer pressure are often overestimated (Haynie & Osgood 2005). First, it may appear that kids share attitudes and behaviors because they hang out together, when in fact they chose to become friends because they *already* shared attitudes and behaviors. Mormons seek other Mormons, ravers seek other ravers, and heavy drinkers seek other heavy drinkers. Second, adolescents remain concerned about their parents' opinions as well as their friends'. Even if they engage in behavior with their friends that their parents disapprove of, they usually do so only if they think their parents won't find out.

Schools

Around the world, schools serve as important agents of socialization—for those who can attend. In poorer countries, many children attend school for only a few years at most, as Map 3.1 shows. Even in these circumstances, schools provide the opportunity to learn basic reading, writing, and arithmetic skills that can enormously improve individuals' economic prospects.

In wealthier nations such as the United States, schooling has become accepted as a natural part of childhood. The central function of schools in these nations is to impart the skills and abilities necessary to function in a highly technological society.

In both poor and wealthy nations, schools teach much more than just basic skills and technical knowledge. They also transmit society's central values. For minorities and immigrants, this typically means learning the values of the dominant culture (Rothstein 2004; Spring 2004; Handel, Cahill, & Elkin 2007). In addition, whereas families typically treat their children as special persons with unique needs and problems, teachers must deal with children en masse and so cannot afford to offer individualized attention. Partly as a result, schools place a high value on teaching children

MAP 3.1: **Percentage of First-Graders Who Continue through Fifth Grade**
In Western countries, almost all students who enter first grade continue their schooling at least through fifth grade. In other countries, especially in Africa, far fewer children do so. Note that this map does not include those—usually girls—who never enter school at all.

SOURCE: Human Development Reports. Literacy and Enrollment Statistics. **http://hdrstats.undp.org/indicators/117.html**. Accessed June 2009.

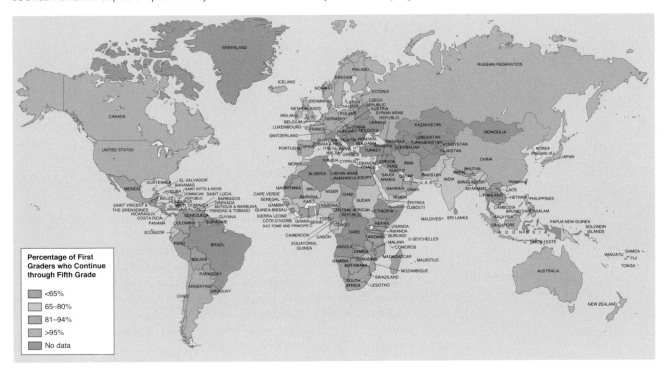

Percentage of First Graders who Continue through Fifth Grade

- <65%
- 65–80%
- 81–94%
- >95%
- No data

(especially if they are poor or working-class) to sit still, follow orders, and otherwise fit in (Gatto 2002). In addition, schools may teach children to compete with others and to evaluate themselves and others according to their level of achievement. In all these ways, then, schools serve as training grounds for the workplace, the military, and other bureaucracies.

Mass Media

Throughout our lives we are bombarded with messages from television, websites, podcasts, magazines, films, billboards, and other mass media. The **mass media** are communication forms designed to reach broad audiences. (The term *mass medium* refers to any *one* mode of mass media.) The most important mass medium for socialization is undoubtedly television. Nearly every home has one, and the average person in the United States spends many hours a week watching it (although many young people now spend even more time each day using other forms of media, as Table 3.1 shows).

Scholars continue to vigorously debate the effects of television viewing. Many suggest that the media promote violence, sexism, racism, and other problematic ideas and behaviors, but the evidence is contradictory (Felson 1996). The most universally accepted conclusion is that the mass media can be an important means of supporting and validating what we already know. Through a process of selective perception, we

sociology and you

If you have chosen your major, you likely have begun professional socialization. Whether your major is English or engineering, your professors have begun teaching you technical concepts and skills. They also have stressed certain ways of thinking about the world: to place more value on working with numbers versus working with words, working with your hands versus with your mind, and working collaboratively versus competitively.

The mass media are all forms of communication designed to reach broad audiences.

TABLE 3.1 Daily Hours and Minutes of Media Usage among 8- to 18-Year-Olds
On average, both boys and girls are exposed to media during about half of their waking hours. Boys spend more time playing video games, while girls spend more time listening to music.

Medium	Boys	Girls	Total
Television	3:17	3:20	3:18
Movies	1:01	:53	:57
Video games	1:12	:25	:45
Print media	:40	:45	:43
Music media	1:29	2:00	1:45
Computers	1:00	1:04	1:02
Total	8:38	8:27	8:33

SOURCE: Kaiser Family Foundation (2005).

tend to give special notice to material that supports our beliefs and self-concept and to ignore material that challenges us.

Television, however, may play a more active part than this. Studies suggest that characters seen regularly on television can become role models whose imagined opinions become important as we develop our own beliefs and behaviors (Felson 1996). For example, adolescents might watch *Grey's Anatomy* for ideas about how to deal with the opposite sex. They then can use the show to supplement knowledge about U.S. norms gained through their own experiences. These findings imply that the content of television can have an important influence.

Religion

In every society, religion is an important source of individual direction. The values and moral principles in religious doctrine give guidance about appropriate values and behaviors. Often the values we learn through religion are compatible with the ideals we learn through other agents of socialization. For example, the golden rule ("Do unto others as you would have them do unto you") taught in religious education fits easily with similar messages heard at home and at school.

The role of religion, however, cannot be reduced to a mere reinforcer of society's norms and values. As we point out in Chapter 12, participation in religions can change individuals' beliefs, self-concepts, and social position, and political movements based on religious differences can change whole societies. Moreover, even within modern U.S. society, there are important differences in the messages delivered by, say, the Mormon, Jewish, and Baptist religions, as well as differences between the conservative and liberal wings of each of these religions. These differences account for some significant variability in socialization experiences.

Case Study: Learning Social Class at the Toy Store

Socialization can occur in many different places and forms. During the 12 weeks sociologist Christine L. Williams (2006) spent working at two toy stores, she observed how children learned to understand their own and others' social class. Most basically, children learned how many toys their parents would buy for them, learned to compare what they received to what was available and to what others were purchasing, and learned to view those purchases as one measure of their worth. White parents also, if unconsciously, taught their children that they (both parents and children) were more important than store clerks by expecting the clerks to put aside their other work, follow the parents around the store, and wait patiently until the parents needed them. Meanwhile, poor children who came to the stores on their own were quickly shooed outside by store employees, thus teaching these children that others considered them unwanted or even dangerous.

Socialization through the Life Course

As our discussion of agents of socialization suggested, socialization occurs throughout life, beginning in childhood and continuing throughout our adult lives, even into old age.

Childhood

Early childhood socialization is called **primary socialization**. It is primary in two senses: It occurs first, and it is most critical for later development. During this period, children develop personality and self-concept; acquire motor abilities, reasoning, and language skills; and begin learning the values and behaviors considered appropriate in their society. The Concept Summary on Types of Socialization illustrates primary and other types of socialization.

During the period of primary socialization, children also are expected to learn and embrace the norms and values of society. Most learn that conforming to social rules is an important key to gaining acceptance and love, first from their family and then from others. Because young children are so dependent on the love and acceptance of their

concept summary

Types of Socialization

Type	Definition	Example
Primary socialization	Socialization in earliest childhood	Tiffany's two moms hug, hold, and talk to her often, teaching her the basics of language and social interaction.
Anticipatory socialization	Socialization that prepares us for anticipated future social positions	Manuel's parents buy him a toy medical kit, and he plays at being a doctor.
Professional socialization	Socialization to the values, behaviors, and skills of a profession	Jody's law professors teach her to analyze legal cases and to compete, rather than cooperate, with her fellow students.
Resocialization	Socialization—often involuntary—to replace previously learned values and behaviors with new ones	Mark is ordered by a judge to take a course in anger management.

Primary socialization is personality development and role learning that occurs during early childhood.

focus on   A GLOBAL PERSPECTIVE

Preschool Socialization in Japan and the United States

B ecause each culture holds differ- ent values and traditions, each culture socializes its children differently. Compare, for example, Japanese and American kindergartens (Small 2001, 129–132).

A central value of Japanese culture is the sense of belonging to a unified, homogeneous nation with common goals. Reflecting this, kindergartens across Japan are state-regulated, have similar facilities, and use similar cur- ricula, which have changed little over the years. In this way, the country en- sures that all children—across genera- tions and regions—are socialized into the same values, and that all have more or less equal access to the resources they need to be successfully socialized and to succeed in their later studies. In contrast, American culture values in- dividual rights vis-á-vis the state, and states' rights vis-á-vis the national gov- ernment. As a result, we expect and ac- cept great differences in how and what young children are taught—even sub- sidizing parents who choose to school their children on their own.

Another central value of Japanese culture is an emphasis on cooperation and group accomplishment over indi- vidual achievement. Japanese adults are expected to take pride in the suc- cesses of their work groups and to humbly downplay their own successes. Similarly, national preschool curricula in Japan stress cooperation over individual

Whereas American kindergartens emphasize individual achievement and pride, Japanese kindergartens socialize children to value cooperation and group accomplishment.

© Sean Sprague/The Image Works. Reproduced by permission

achievement. Teachers speak to their students in ways that emphasize the students' "groupness"; use games, songs, and other activities designed to teach students to work together and to think of themselves as a group; and continually urge students to consider how their actions affect others. When children misbehave, teachers integrate them back into the group and into ac- ceptable behavior, rather than high- lighting the misbehaviors. In contrast, in the United States, kindergarteners are taught from the start to interpret their successes as resulting from their individual achievements rather than from group support or activity, and

they learn to take pride in those suc- cesses. Teachers goad children—or at least middle-class children—to perform better by praising those who are succeeding and by correcting or chastising those who are not. Finally, teachers quickly conclude that certain students are troublemakers, best dealt with by isolating them from the group rather than by trying to integrate them into it.

In sum, socialization in both Japanese and U.S. kindergartens both reflects and reinforces the different cultural values of these two countries.

family, they are under especially strong pressure to conform to their family's expecta- tions. This is a critical step in turning them into conforming members of society. If this learning does not take place in childhood, conformity is unlikely to develop in later life. Focus on a Global Perspective: Preschool Socialization in Japan and the United States compares the values taught in these two countries.

What we learn—or don't learn—in childhood can affect us for the rest of our lives. For example, the number of words we learn by age 3 highly predicts our reading ability and our likelihood of graduating from high school (Farkas & Beron 2003). Unfortunately, compared to more affluent and white parents, on average, poor, working-class, and African American parents speak far less to their children and use a smaller vocabulary when they do speak. As a result, by age 3 poor children know 33 percent fewer words than do working-class children and 50 percent fewer words than do middle-class children (Farkas & Beron 2003). The good news is that most of these differences evaporate when poor or African American parents expose their children to more words; many organizations now work either to change parents' behavior or to expose their children to more words in special preschool programs.

Adolescence

Adolescence serves as a bridge between childhood and adulthood. As such, the central task of adolescence is to begin to establish independence from one's parents.

During adolescence, we often engage in **anticipatory socialization**—learning the beliefs and behaviors needed to prepare us for the social positions we are likely to assume in the future. Until about 1980, for instance, all American girls were required to take "home economics" courses to learn how to sew and cook. Boys were required to take "shop" courses to learn how to fix cars and use woodworking tools. These days, boys can take cooking and girls can take woodworking. Nevertheless, teenagers' household chores, part-time jobs, and volunteer work still tend to divide along traditional lines. While boys sometimes help around the house, girls more often are expected to cook, clean house, and care for their younger siblings (Lee, Schneider, & Waite 2003). If a boy does help at home, he's likely to take on such "masculine" tasks as mowing the lawn or washing the car. Similarly, girls' part-time jobs (such as babysitting) often teach caregiving, whereas boys' work more often teaches mechanical skills. And when boys and girls take part-time jobs at restaurants, boys more often are assigned to heavy, riskier work such as running the deep fryer while girls more often are assigned to run the cash registers and required to wear skimpy clothes. In all these ways, girls and boys prepare for the family and work positions they anticipate holding as adults.

Adulthood

Because of anticipatory socialization, most of us are more or less prepared for the responsibilities we will face as spouses, parents, and workers. Goals have been established, skills acquired, and attitudes developed that prepare us to accept and even embrace the positions we are likely to hold as adults, in the family and in the world. Because anticipatory socialization is never complete, however, anyone who wants to enter a professional field must first undergo **professional socialization**. The purpose of professional socialization is to learn not only the knowledge and skills but also the *culture* of a profession. Medical training provides an example of this process.

© O'Brien Productions/CORBIS

Anticipatory socialization prepares us for the roles we will take in the future. Children everywhere play out their visions of how mommies and daddies ought to behave.

sociology and you

On television, the Internet, billboards, and elsewhere, advertising also plays an important part in socialization. Soda ads, for example, not only suggest that one brand tastes better than another, but also aim to convince us that users of that brand are funnier, wealthier, more attractive, and more popular than others. How is your favorite beverage advertised?

Anticipatory socialization is the process that prepares us for roles we are likely to assume in the future.

Professional socialization is the process of learning the knowledge, skills, and cultural values of a profession.

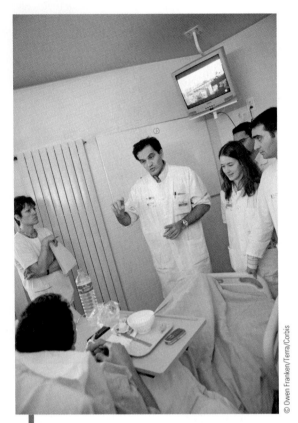

© Owen Franken/Terra/Corbis

Through their professional socialization, medical students learn both technical skills and the cultural values of the medical world.

Most commonly, people choose to become doctors out of a desire to help others. Yet one of the primary tenets of medical culture is that doctors should be emotionally detached—distancing themselves from their patients and avoiding any show of emotion (Weitz 2010). According to sociologist and medical school professor Frederic Hafferty (1991), this cultural norm is taught from the beginning of medical education. Through his observations, Hafferty discovered that when new students first enter medical school, second-year students almost invariably take them to the school's anatomy laboratory. There the second-year students proudly display the most grotesque partially dissected human cadaver available. Although officially they do so to show off the school's laboratory facilities, their true purpose seems to evoke horror and disgust in the new students so that the second-year students can make fun of them. Through this process, the second-year students both demonstrate how "tough" they have become and teach the new students that medical culture stigmatizes "weakness." This is a particularly vivid example of professional socialization, but every job change we make as adults requires some socialization to new responsibilities and demands.

Age 65 and Beyond

More and more Americans now live far beyond age 65. Some will continue holding down full-time jobs into their seventies, eighties, and even beyond, and others will go back to school and embark on socialization for a new career. But for most, growing older means developing a new identity as a retiree. To do so successfully, some individuals engage in anticipatory socialization: trying out volunteer work before retirement, developing new hobbies or educational interests to pursue after retirement, or thinking through a retirement "game plan" for where they will live and what they will do. Others must learn how to find meaning and fill their time once their days are no longer structured around work and their job is no longer central to their identity. In this process, they may ponder the choices made by friends and relatives who retired earlier, and may use those choices as models of what they should—or should not—do themselves. They may also seek services from the many nonprofit organizations that have emerged specifically to help "active retirees" learn to contribute to their communities in new ways.

For those who worked solely as homemakers before age 65, growing older presents a different set of challenges. For some, the most difficult challenge is figuring out how to be a homemaker when one's spouse is also home full-time. In these cases, both individuals may need to adapt their beliefs and behaviors. For example, despite their earlier socialization, a couple might conclude that the only way to avoid fighting is for both of them to agree on new ideas regarding how to divide household chores.

Unfortunately, many older people also must figure out who they are and what they will do with their lives after their spouse or long-term partner dies. Some will eventually find others to love, while others must learn to live alone. As in other stages of life, peers can help socialize the newly widowed into their new status and into new ways of thinking about and acting in the world, from teaching new skills (how to cook, change a light bulb, or have safe sex) to teaching how to value independence and solitude.

Older age also typically means coming to terms with declining physical abilities. As a result, individuals who learned to prize independence must come to depend on their spouses, partners, children, or paid caregivers. This transition is easiest for those who gradually adopt new ways of looking at the world, replacing older beliefs and practices with a new set that better matches their circumstances.

Resocialization

As our discussion of socialization across the life course suggests, socialization is usually a gradual process. Sometimes, though, our position in society changes abruptly and extremely, forcing us to abandon our self-concept and way of life for a radically different one. The process of learning the beliefs and values associated with a new way of life is called **resocialization**. Typically, this term refers to circumstances in which people are forced to change their way of life rapidly and against their will.

A drastic example of resocialization occurs when people become permanently disabled. Those who become paralyzed experience intense resocialization to adjust to their handicap. Their social position and capacities suddenly change, and their old self-concepts no longer cover the situation. They may lose the ability to control their bladders and bowels, to walk or dress themselves, or to function sexually as they had previously. If they are younger, they may wonder whether they will ever marry or have children; if they are older, they may have to reevaluate their adequacy as lovers, spouses, or parents. These changes require a radical redefinition of self. If self-esteem is to remain high, priorities will have to be rearranged and new, less physically active behaviors given prominence.

Resocialization may also be deliberately imposed by society. When individuals' behavior leads to social problems—as with criminals, alcoholics, and mentally disturbed individuals—society may decree that they must abandon their old identities and accept more conventional ones.

Total Institutions

Generally speaking, a radical change in self-concept requires a radical change in environment. Drug counseling one night a week is not likely to alter drastically the beliefs and behaviors of a teenager who spends the rest of the week with peers who are constantly "wasted." Thus, the first step in the resocialization process often involves isolating the individual from his or her past environment in **total institutions**—facilities in which all aspects of life are strictly controlled for the purpose of radical resocialization (Goffman 1961a). Monasteries, prisons, boot camps, and mental hospitals are good examples. Within these total institutions, inmates lose the statuses, social positions, and relationships that had formed the bases of their self-concepts. Even their clothes are taken from them, replaced by uniforms. Inmates also lose control over the structure of their days and instead are forced to follow rigid schedules set by others. Finally, inmates are often expected to engage in self-criticism to reveal the inferiority of their past perspectives, peer groups, and behaviors.

Resocialization occurs when we abandon our self-concept and way of life for a radically different one (often against our will).

Total institutions are facilities in which all aspects of life are strictly controlled for the purpose of radical resocialization.

Case Study: Resocializing Young Offenders

How should society deal with young people who commit crimes? Most Americans believe that young offenders should be treated differently from adult criminals, in part because we have more faith that young offenders can be resocialized. But how should that resocialization work?

Beginning in the 1980s, one popular model was to use prison boot camps. In these total institutions, youths were locked away from any competing influences and kept on a strict schedule of strenuous calisthenics, military drilling, hard physical labor, drug counseling, and study (Anderson 1998). To teach them to respect authority and to leave their old self-concepts behind, their heads were shaved, they were called derogatory names, and they were forbidden from even looking prison officials in the eye. As research on boot camps accumulated, however, it became clear that most of these strategies had little if any effect (MacKenzie, Wilson, & Kider 2001). Essentially, the boot camps taught young people how to follow the rules in the camps, but did not give them tools needed to succeed once they returned to ordinary life.

Because of these problems, many communities have instead begun to emphasize rehabilitation over punishment and therapy over discipline (Anne E. Casey Foundation 2009; Moore 2009). Most importantly, youths in these programs are taught nonviolent ways of handling interpersonal conflict, often while living at home or in supervised group homes, rather than in detention centers, jails, or prisons. These programs have resulted in significant declines in costs, in the numbers of youths convicted of second crimes, and in the number of youths in prison—by more than 50 percent, in some places (Anne E. Casey Foundation 2009; Moore 2009).

Where This Leaves Us

Each of us is unique, a product of our individual biology, abilities, personality, experiences, and choices. But each of us is also a social creation. Through socialization we come to learn the behaviors and values expected of us and, more often than not, to take on those behaviors and values as our own. Sometimes that process is obvious: a parent slapping a child's hand for grabbing a cookie without permission, a minister preaching a sermon on the wages of sin, one girl giving another girl pointers on how to flirt. Other times, we are no more aware of the socialization process than a fish is aware of water; it is simply a part of the life around us. The typical American, for example, now spends several hours each day watching television. During those hours we not only learn who murdered this week's victim on *CSI*, who is this season's *American Idol*, and who's sleeping with our favorite desperate housewife, but we also learn ways of looking at the world: to fear random violence and trust the police; to value success, talent, and fame; to honor wealth; and so on.

Summary

1. Socialization is the process of learning the rules and values of a given culture.
2. Although we are all a product of socialization, this does not mean that we have no choices in our lives. But unless we understand the ways in which we have been socialized, we will be unable to see our choices clearly and to turn those choices into realities.
3. Although biological capacities enter into human development, our identities are socially bestowed and socially sustained. Without human relationships, even our natural capacities would not develop.
4. Freudian theory links social development to biological cues. Freud believed that to become a healthy adult, children must develop a reasonable balance between id and superego.
5. Piaget theorized that cognition develops through a series of stages. Only in the last stage do children develop the capacity to understand and think abstractly, and some children may never reach that stage.
6. Structural functionalists theorize that socialization—in schools, religious institutions, families, and elsewhere—smoothly integrates the young into the broader culture, avoiding conflict or chaotic social change. It is most useful for explaining the benefits of a stable social system.
7. Conflict theory focuses on how socialization reinforces unequal power arrangements. It is most useful for understanding how socialization can quash dissent and social change and reproduce inequalities.
8. Symbolic interaction theory emphasizes that self-concept develops through actively interpreting our interactions with others and the images of ourselves that we glean from others. Two important concepts connected with this theory are the looking-glass self and role taking.
9. Socialization occurs across the life course. Four important types of socialization are primary socialization, anticipatory socialization, professional socialization, and resocialization.
10. The two most important agents of socialization are the family and peers. Other important agents of socialization include teachers, the mass media, and religion.

Thinking Critically

1. Think about the behaviors that teachers expect of college students. How does the socialization you received in your family make it easier or harder for you to meet teachers' expectations? How about socialization from your peers?
2. List some ways that a family's social class might influence what a child learns through socialization. Can you think of any ways that living in the city versus living in the country might matter?
3. Thinking back to your childhood, what values *might* you have learned from your two favorite television shows? *Did* you learn from them? How do you explain why you did or did not?

Book Companion Website

www.cengage.com/sociology/brinkerhoff
Prepare for quizzes and exams with online resources—including tutorial quizzes, a glossary, interactive flash cards, crossword puzzles, essay questions, virtual explorations, and more.

CHAPTER 4

Social Structure and Social Interaction

Zoran Milich/Masterfile

Intertwining Forces: Social Structure and Social Interaction

Most people who become sociologists do so because they are interested in studying particular social problems, such as homelessness, mental illness, or racial inequality. Each of these problems has roots in and consequences for both broad social structures and everyday social interactions. For example, racial inequality in the United States in part stems from the nature of our national economy and political institutions: There simply aren't enough well-paying jobs near nonwhite communities, and these communities rarely have enough political power to entice corporations to bring in good jobs. But racial inequality is also reinforced on a day-to-day basis whenever teachers spend less time with nonwhite than with white students or police officers assume that nonwhites are more likely than whites to be criminals. As this example suggests, to fully understand society and social problems, sociologists must look at both social structure and social interaction. This chapter describes these two basic features of society. As we will see, research on social structures often draws on structural-functionalist or conflict theories, whereas research on social interaction typically draws on symbolic interaction theory.

Social Structures

Many of our daily encounters occur in patterns. Every day we interact with the same people (our family or best friends) or with the same kinds of people (salesclerks or teachers). These patterned relationships are called social structures. Each of these dramas has a set of actors (mother/child or buyer/seller) and a set of norms that define appropriate behavior for each actor.

As described in Chapter 1, a social structure is a recurrent pattern of relationships. Social structures can be found at all levels in society. Baseball games, friendship networks, families, and large corporations all have patterns of relationships that repeat day after day. Some of these patterns are reinforced by formal rules or laws, but many more are maintained by force of custom.

The patterns in our lives are both constraining and enabling (Giddens 1984). If you would like to be free to set your own schedule, you will find the 9-to-5, Monday-to-Friday work pattern a constraint. On the other hand, preset patterns provide convenient and comfortable ways of handling many aspects of life. They help us to navigate heavy traffic, find dates and spouses, and raise our children.

Whether we are talking about a Saturday afternoon ball game, families, or the workplace, social structures can be analyzed in terms of three concepts: *status*, *role*, and *institution*.

Status

The basic building block of society is status—a person's position in a group, relative to other group members. Sociologists who want to study the status structure of a society examine two types of statuses: achieved and ascribed. An achieved status is a position (good or bad) that a person can attain in a lifetime. Being a father is an achieved status; so is being a convict. An ascribed status is a position generally assumed to be fixed by birth or inheritance and unalterable in a person's lifetime. For example, although

A **status** is a specialized position within a group.

An **achieved status** is optional, one that a person can obtain in a lifetime.

An **ascribed status** is fixed by birth and inheritance and is unalterable in a person's lifetime.

MAP 4.1:   **Mixed-Race People in the United States**
About 2 percent of U.S. residents—and 4 percent of U.S. children—belong to two or more races. The number of mixed-race people per 1,000 people varies enormously from state to state. This map does *not* reflect the rising number of individuals who are part Hispanic because Hispanics are not considered a race.
SOURCE: U.S. Bureau of the Census (2009b).

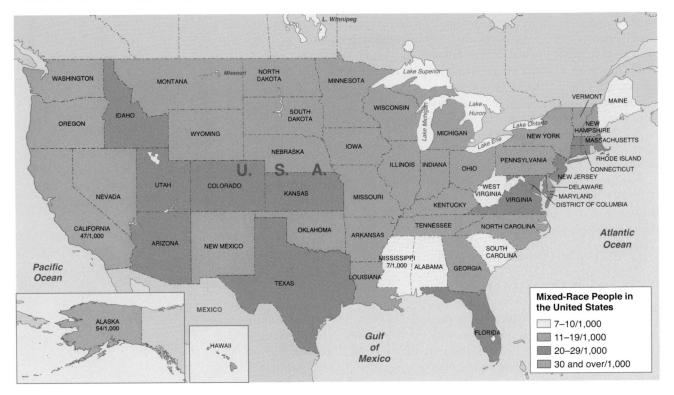

some people have gender reassignment surgery and some people "pass" as members of a different race, we assume that sex and race are unchangeable. Hence sociologists generally consider sex and race to be ascribed statuses.

Each individual holds multiple statuses simultaneously. You may be a daughter or son as well as an athlete, a Christian, a waiter, and so on. This combination of statuses is referred to as your **status set.**

Sociologists who analyze the status structure of a society typically focus on four related issues (Blau 1987): (1) identifying the number and types of statuses that are available in a society; (2) assessing the distribution of people among these statuses; (3) determining how the consequences—the rewards, resources, and opportunities—differ for people who occupy one status rather than another; and (4) ascertaining what combinations of statuses are likely or even possible.

Case Study: Race as a Status

To illustrate how our lives are structured by status membership, we apply this approach to one particular ascribed status and ask how being African American affects relationships and experiences in the United States.

To begin: How many racial statuses are there in the United States? The 1990 census asked Americans to identify themselves as belonging to one of five racial

---

Status set refers to the combination of all statuses held by an individual.

categories: white, African American, Native American, Asian, or other. For years, this same question with more or less the same list of possible answers appeared on almost every social survey. The 2000 census, however, allowed individuals to choose more than one race, thus creating the category *mixed race*; Map 4.1 shows the distribution of mixed-race people across the United States. The nearly universal concern about racial statuses alerts us to the importance of racial status in our daily lives, while the addition of the mixed-race category suggests that racial statuses—and ideas about racial statuses—do change. This concept is explored more fully in Decoding the Data: American Diversity.

It is not just the number of statuses that has consequences. The numerical distribution of the population among racial statuses also encourages or discourages certain patterns of behavior. For example, according to the latest U.S. Census, 2.1 million African Americans live in New York City, but only 3 live in Worland, Wyoming. Consequently, white New Yorkers have a far greater chance, statistically, of marrying an African American than do white residents of Worland.

Of course, numbers alone do not tell the whole story. By nearly every measure that one might choose, there is substantial inequality in the rewards, resources, and

decoding the data

American Diversity

Some surveys ask people to select the one racial group that best describes them. Some allow people to select more than one racial group, and some also ask individuals whether or not they are Hispanic (which is not considered a racial group). These U.S. Census data illustrate the different answers we get from these different questions.

The Short Answer	Percentage	A Longer Answer	Percentage
White	75.1	White Non-Hispanic	69.1
		Hispanic	12.5
African American	12.3	African American Non-Hispanic	12.1
Native American	0.9	Native American	0.7
Asian American	3.6	Asian American	3.6
Other	8.0	Other	0.3
		Mixed Race	1.6

Explaining the Data: What sociological factors—peer pressure, family ties, socialization, cultural norms—do you think would lead someone like Barack Obama, whose mother was a white American and whose father was an African, to identify as African American?

Critiquing the Data: Compare the data in these two graphs. How does allowing individuals to choose more than one race affect our image of race in America? How does combining data on race with data on Hispanic identity affect our image of American diversity?

Although racial inequality continues to plague the United States, the election of President Barack Obama demonstrates that it is possible for African Americans to succeed in this country.

Chip Somodevilla/Getty Images News/Getty Images

opportunities available to African American and white people in the United States. Of course, African Americans can succeed, as the election of President Barack Obama amply demonstrated. Nevertheless, compared to whites, African Americans are twice as likely to die in infancy, twice as likely to be unemployed, and *six* times more likely to be murdered (U.S. Bureau of the Census 2009a). Similarly, when Hurricane Katrina hit New Orleans in 2005, African Americans were far more likely than white residents to stay in the city. The cause was poverty: African Americans were far less likely than others to have transportation out of the city, money to rent hotel rooms elsewhere, and well-off relatives with large homes who could take them in for an extended stay. Obviously, racial status has enormous consequences on the structures of daily experiences.

Although racial inequality persists, racial status does not correspond as directly with occupational and educational statuses as it once did, and different combinations of statuses are possible. Forty years ago, being African American meant probably having much less education and a much lower status occupation than whites. Today, knowing a person's ascribed status (race) is not such an accurate guide to his or her achieved statuses (education or occupation). Nevertheless, 34 percent of all nurse's aides in the United States are African American, compared to only 6 percent of all physicians (U.S. Bureau of the Census 2009a). The processes through which these overlapping racial, political, and economic statuses are maintained are discussed further in Chapter 8.

Roles

Social interaction refers to the ways individuals interact with others in everyday, face-to-face situations.

The status structure of a society provides the broad outlines for **social interaction**: the ways individuals interact with others in everyday, face-to-face situations.

These broad outlines are filled in by roles. As described in Chapter 1, roles are sets of norms that specify the rights and obligations of each status. To use a theatrical metaphor, the status structure is equivalent to the cast of characters ("a young girl, her father, and their maid," for example), whereas roles are equivalent to the scripts that define how the characters ought to act, feel, and relate to one another. This language of the theater helps to make a vital point about the relationship between status and role: People occupy statuses, but they play roles. This distinction is helpful when we analyze how structures work in practice—and why they sometimes don't work. A man may occupy the status of father, but he may play the role associated with it poorly.

Sometimes people fail to fulfill role requirements despite their best intentions. It is hard to be a good provider, for instance, when there are no jobs available. Failure is also particularly likely when people face incompatible demands because of multiple or complex roles. Sociologists distinguish between two types of incompatible role demands: When incompatible role demands develop within a single status, we refer to **role strain**; when they develop because of multiple statuses, we refer to **role conflict**. For example, role strain occurs when parents don't have enough time to wash their children's clothes, cook their dinner, help them with homework, and play a game together all in the same evening. Role conflict occurs when a parent's need to take time off to care for a sick child conflicts with an employer's expectation that the parent put work obligations ahead of family obligations.

As this suggests, social roles are always changing and flexible. We do not simply play the parts we are assigned with machinelike conformity. Instead, each of us plays a given role differently, depending on our other social statuses and roles, our resources, and the social rewards or punishments that our role performances evoke from others.

Institutions

Social structures vary in scope and importance. Some, such as those that pattern a Friday night poker game, have limited application. The players could change the game to Saturday night or up the ante, and it would not have a major effect on the lives of anyone other than members of the group. If a major corporation changed seniority or family leave policies, it would have somewhat broader consequences, not only affecting employees of that firm but also setting a precedent for other firms. Still, the impact of change in this one corporation (or social structure) would likely be limited to certain sorts of businesses. In contrast, changes in other social structures have the power to shape the basic fabric of all our lives. We call these structures social institutions.

An **institution** is an enduring and complex social structure that meets basic human needs. Its primary features are that it endures for generations; includes a complex set of values, norms, statuses, and roles; and addresses basic human needs. Embedded in the statuses and roles of the family institution, for example, are enduring patterns for dating and courtship, child rearing, and care of the elderly. Because the institution of family consists of millions of separate families, however, the exact rules and behaviors surrounding dating or elder care will vary.

Despite these variations, institutions provide routine patterns for dealing with predictable problems of social life. Because these problems tend to be similar across societies, we find that every society tends to have the same types of institutions.

Role strain occurs when incompatible role demands develop within a single status.

Role conflict occurs when incompatible role demands develop because of multiple statuses.

An **institution** is an enduring social structure that meets basic human needs.

Religion is one of the basic social institutions. Although doctrines and rituals vary enormously, all cultures and societies include a structured pattern of behavior and belief that provides individuals with explanations for events and experiences that are beyond their own personal control.

Basic Institutions

Five basic social institutions are:

- The family, to care for dependents and rear children.
- The economy, to produce and distribute goods.
- Government, to provide community coordination and defense.
- Education, to train new generations.
- Religion, to supply answers about the unknown or unknowable.

These institutions are basic in the sense that every society provides some set of enduring social arrangements designed to meet these important social needs. These arrangements may vary from one society to the next, sometimes dramatically. Government institutions may be monarchies, democracies, dictatorships, or tribal councils. However, a stable social structure that is responsible for meeting these needs is common to all healthy societies.

In simple societies, all of these important social needs—political, economic, educational, and religious—are met through one major social institution, the family or kinship group. Social relationships based on kinship obligations serve as a basis for organizing production, reproduction, education, and defense.

As societies grow larger and more complex, the kinship structure is less able to furnish solutions to all the recurrent problems. As a result, some activities gradually shift to more specialized social structures outside the family. The economy, education, religion, and government become fully developed institutionalized structures that exist separately from the family. (The institutions of the contemporary United States are the subjects of Chapters 10 to 13.)

As the social and physical environments of a society change and the technology for dealing with those environments expands or contracts, the problems that individuals face also change. Thus, institutional structures are not static; new structures emerge to cope with new problems—or a society will collapse into chaos (Diamond 2005). For example, the African country of Uganda responded actively to the AIDS epidemic, providing public education on safer sex, access to condoms, and access to treatment for those already infected. As a result, its economy has held steady despite the effect of the disease. In contrast, the South African government rejected modern understandings of the disease and its prevention. Rates of AIDS infection have soared, and families, schools, and the economy are collapsing.

Institutional Interdependence

Each institution of society can be analyzed as an independent social structure, but none really stands alone. Instead, institutions are interdependent; each affects the others and is affected by them.

In a stable society, the norms and values embodied in the roles of one institution will usually be compatible with those in other institutions. For example, a society that stresses male dominance and rule by seniority in the family will also stress the same norms in its religious, economic, and political systems. In this case, interdependence reinforces norms and values and adds to social stability.

Sometimes, however, interdependence is an important mechanism for social change. Because each institution affects and is affected by the others, a change in one tends to lead to change in the others. Changes in the economy lead to changes in the family; changes in religion lead to changes in government. For example, when years of schooling become more important than hereditary position in determining occupation, hereditary position will also be endangered in government, the family, and religion.

Institutions as Agents of Stability or Inequality

Sociologists use two major theoretical frameworks to approach the study of social structures: structural functionalism and conflict theory. The first focuses on the part that institutions play in creating social and personal stability; the second focuses on the role of institutions in legitimizing inequality. Because each framework places a different value judgment on stability and order, each prompts us to ask different questions about social structures.

Structural-functional theorists begin with the question "How do institutions help to stabilize a society?" To answer this question, they focus on the ready-made, shared patterns for responding to everyday problems that institutions offer. By keeping us from having to reinvent the social equivalent of the wheel with each new encounter and each new generation, structural functionalists argue, these patterns and the institutions that underlie them allow social life to run smoothly in stable and predictable ways. Moreover, because these patterns have been sanctified by tradition, we tend to experience them as morally right. As a result, we find satisfaction and security in social institutions.

In contrast, although conflict theorists acknowledge that institutions meet basic human needs, they raise the question "Why this social pattern rather than another?" Their answers typically emphasize who benefits from existing institutions and illustrate how institutions support the interests of those already in power. Because institutions have existed for a long time, we tend to think of our familial, religious, and political systems not merely as one way of fulfilling a particular need but as the only acceptable way. Just as an eleventh-century Christian might have thought, "Of course witches should be burned at the stake," so we tend to think, "Of course women should sacrifice their careers for their children." In both cases, the cloak of tradition obscures our ability to recognize inequalities, making inequality seem normal and even desirable. As a result, conflict theorists argue that institutions stifle social change and help maintain inequality.

Types of Societies

Institutions give a society a distinctive character. In some societies, the church is the dominant institution; in others, it is the family or the economy. Whatever the circumstance, recognizing the institutional framework of a society is critical to understanding how it works.

Societies range greatly in complexity. In simple societies, we often find only one major social institution—the family or kinship group. Complex, modern societies, however, have as many as a dozen institutions. What causes this expansion of institutions? The triggering event appears to be economic change. When changes in technology, physical environment, access to resources, or social arrangements increase economic surpluses, institutions are often able to expand (Lenski 1966; Diamond 1997). In this section we sketch a broad outline of the institutional evolution that accompanied four revolutions in production.

Hunting-and-Gathering Societies

In hunting-and-gathering societies like that of the Kung Bushmen, tasks tend to be divided along gender lines. Individuals accumulate few personal possessions because there is little surplus and because possessions would be difficult to move.

Hunting-and-gathering societies are those in which people have little or no means of obtaining food other than killing wild animals or finding edible fruits, vegetables, seeds, and the like (Lee & Daly 2005). These societies are based on subsistence economies, in which people rarely can obtain or store more food than they can eat. In some years, game and fruit are plentiful, but in many years scarcity is a constant companion.

The basic units of social organization in hunting-and-gathering societies are the household and the local band, both of which are based primarily on kinship. Most hunting-and-gathering societies are organized around these units. A band rarely exceeds 50 people in size and tends to be nomadic or semi-nomadic. Because of their frequent wanderings, members of these societies accumulate few personal possessions.

The division of labor is simple, based on age and sex (Lee & Daly 1999). The common pattern is for older boys and men (other than the elderly) to participate in hunting and deep-sea fishing and for older girls and women to participate in gathering, shore fishing, and preserving. Aside from inequalities of status by age and sex, few structured inequalities exist in subsistence economies. Members possess little wealth; they have few, if any, hereditary privileges; and the societies are almost always too small to develop class distinctions. In fact, a major characteristic of subsistence societies is that individuals are homogeneous, or alike. Apart from differences occasioned by age and sex, members generally have the same everyday experiences.

All human societies originated as hunting-and-gathering societies, but few remain. Those that do are found in places like the Great Victoria Desert of Australia and the Amazon jungle. They have survived both because they have learned over the generations how to use all the resources these environments offer and because few outsiders have any interest in taking over these harsh environments.

Horticultural Societies

Around the world, the movement away from hunting-and-gathering societies began with the development of agriculture. During this "first revolution" in agriculture, people began to plant and cultivate crops, rather than simply harvesting whatever nature provided. This led to the development of **horticultural societies**—that is, societies based on small-scale, simple farming, without plows or large beasts of burden. With only digging sticks or hoes to help, horticultural societies could not grow much food. But unlike hunting-and-gathering societies, they occasionally could grow enough to have surplus food.

Once societies could grow more than they needed to survive, they changed dramatically. Although peasants still had to work full time to produce food, others—higher up on the newly emerging class hierarchy—could now live off the surplus produced by those peasants. This privileged group could now take time off from basic production and turn to other pursuits: art, religion, writing, and frequently warfare.

Because of relative abundance and a settled way of life, horticultural societies tend to develop complex and stable institutions outside the family. Some economic activity may occur outside the family, a religious structure with full-time priests may develop, and a stable system of government—complete with bureaucrats, tax collectors, and a hereditary ruler—often develops. Such societies are sometimes very large. The Inca Empire, for example, had an estimated population of more than 4 million.

Hunting-and-gathering societies are those in which most food must be obtained by killing wild animals or finding edible plants.

Horticultural societies are characterized by small-scale, simple farming, without plows or large beasts of burden.

Agricultural Societies

Approximately 5,000 to 6,000 years ago, a second agricultural revolution occurred, and the efficiency of food production was doubled and redoubled through better technology (Diamond 1997). We use the term **agricultural societies** to refer to those whose economies are based on growing food using plows and large beasts of burden.

The shift to agricultural societies was accompanied by improvements in technology such as the use of metal tools, the wheel, and better methods of irrigation and fertilization. These changes dramatically altered social institutions. Most importantly, these changes meant that fewer people were needed to produce food. As a result, some could instead move to large urban centers and find work in the growing number of new trades. Meanwhile, technology, trade, reading and writing, science, and art grew rapidly as larger and larger numbers of people could now devote full time to these pursuits.

At the same time, growing occupational diversity also brought greater inequality. In the place of the rather simple class structure of horticultural societies, a complex class system developed, with merchants, soldiers, scholars, officials, and kings—and, of course, the poor peasants who comprised the bulk of the population and on whose labor the rest all ultimately depended.

One of the common uses to which societies put their new leisure time and other new technology was warfare. With the domestication of the horse (cavalry) and the invention of the wheel (chariot warfare), military technology became more advanced and efficient. Military might was used as a means to gain greater surplus through conquering other peoples. The Romans were so successful at this that they managed to turn the peoples of the entire Mediterranean basin into a peasant class that supported a ruling elite in Italy.

Industrial Societies

The third major revolution in production was the advent of industrialization about 200 years ago in Western Europe. **Industrial societies** are those whose economies are built primarily around the mass production of nonagricultural goods using mechanical, electrical, or fossil-fuel energy. The shift from human and animal labor to mass production caused an explosive rise in cities and transformed political, social, and economic institutions. Old institutions such as education expanded dramatically, and new institutions such as science, medicine, and law emerged.

The shift to industrial societies occurred in tandem with a shift from *gemeinschaft* to *gesellschaft* (Wirth 1938). **Gemeinschaft** refers to societies in which people share close personal bonds with most of those around them. In contrast, **gesellschaft** refers to societies in which people are tied primarily by impersonal, practical bonds. This shift began with the development of agricultural societies and intensified as the move from farms to factories and cities increased.

Postindustrial Societies

During the last few decades, wealthy countries like the United States have experienced a rapid shift toward a *postindustrial society*. Whereas industrial societies are characterized by the mass production of goods such as clothes, cars, and computers, **postindustrial societies** are characterized by a focus on producing either *information* or *services*. Postindustrial jobs include researcher, doctor, and software developer as well as maid, store clerk, and Wal-Mart greeter. Meanwhile, industrial production (such as

Agricultural societies are based on growing food using plows and large beasts of burden.

Industrial societies are characterized by mass production of nonagricultural goods.

Gemeinschaft refers to societies in which most people share close personal bonds.

Gesellschaft refers to societies in which people are tied primarily by impersonal, practical bonds.

Postindustrial societies focus on producing either information or services.

sociology and you

As members of a postindustrial society, your decision to seek a college degree is a wise one. A generation ago, many people without college degrees could find well-paying, stable jobs working in factories that produced everything from clothing to cars to computers. These days, anyone not trained to work in the "information industries" is likely to end up in a low-paying service job.

manufacturing clothing and computers) has increasingly shifted to poorer countries like Bangladesh and Peru.

The shift to a postindustrial society is changing the relative strength of social institutions. Since jobs in the postindustrial world divide much more sharply between well-paying jobs requiring four or more years of higher education and poorly paying jobs for everyone else, education has become far more important. Similarly, information technology now has enormous impact on all social institutions, affecting how we communicate with our family, participate in religion, acquire an education, and so on.

Case Study: When Institutions Die

Throughout most of history, changes in production, reproduction, education, and social control occurred slowly. When these changes occurred gradually and harmoniously, institutions could continue to support one another and to provide stable patterns that met ongoing human needs. On other occasions, however, old institutions—along with old roles and statuses—disappear before new ones can evolve. When this happens, societies and the individuals within them are traumatized and may fall apart.

In 1985, Anastasia Shkilnyk chronicled just such a human tragedy in her book *A Poison Stronger Than Love*. Although the book focuses on the plight of the Ojibway Indians of Northwestern Ontario, it provides a useful framework for understanding the fate of many traditional societies faced with rapid social change.

A Broken Society

In 1976, Shkilnyk was sent by the Canadian Department of Indian Affairs to Grassy Narrows, an Ojibway community of 520 people, to advise the community on how to alleviate economic disruption caused by mercury poisoning in nearby lakes and rivers. Grassy Narrows was a destroyed community. Drunken 6-year-olds roamed winter streets when the temperatures were 40 degrees below zero. The death rate for both children and adults was very high compared with that for the rest of Canada. Nearly three-quarters of all deaths were linked directly to alcohol and drug abuse. A quote from Shkilnyk's journal evokes the tragedy of life in Grassy Narrows:

> *Friday.* My neighbor comes over to tell me that last night, just before midnight, she found 4-year-old Dolores wandering alone around the reserve, about 2 miles from her home. She called the police and they went to the house to investigate. They found Dolores's 3-year-old sister, Diane, huddled in a corner crying. The house was empty, bare of food, and all the windows were broken. The police discovered that the parents had gone to Kenora the day before and were drinking in town. Both of them were sober when they deserted their children. (Shkilnyk 1985, 41)

Like Dolores and Diane's parents, most of the adults in Grassy Narrows were binge drinkers. When wages were paid or the welfare checks came, many drank until they were unconscious and the money ran out. Often children waited until their parents had drunk themselves unconscious and then drank the liquor that was left. If they could not get liquor, they sniffed glue or gasoline.

Yet 20 years before, the Ojibway had been a thriving people. How was a society so thoroughly destroyed?

Ojibway Society before 1963

The Ojibway have been in contact with whites for two centuries. In 1873, they signed the treaty that defined their relationship with the Canadian government and established the borders of their reservation.

In the decades that followed, the Ojibway continued their traditional lives as hunters and gatherers. The family was their primary social institution. A family group could consist of a group of brothers plus their wives and children or of a couple, their unmarried children, their married sons, and the wives and children of those sons. In either case, the houses or tents of this family group would all be clustered together, perhaps as far as a half mile from the next family group.

Family groups carried out all economic activities. These activities varied with the season. In the late summer and fall, families picked blueberries and harvested wild rice; in the winter, they hunted and trapped. In all these endeavors, the entire family participated, with everybody packing up and going to where the work was. The men would hunt and trap, the women would skin and prepare the meat, and the old people would come along to care for and teach the children. The reserve served only as a summer encampment. From late summer until late spring, the family was on the move.

Besides being the chief economic and educational unit, the family was also the major agent of social control. Family elders enforced the rules and punished those who violated them. In addition, most religious ceremonies were performed by family elders. Although a loose band of families formed the Ojibway society, each family group was largely self-sufficient, interacting with other family groups only to exchange marriage partners and for other ceremonial activities.

The earliest changes brought by white culture did not disrupt this way of life particularly. Even the development of boarding schools, which removed many Indian children from their homes for the winter months, had only a limited effect on Ojibway life: The boarding schools took the children away but did not disrupt the major social institutions of the society they left behind. When the children returned home each summer, their families could still educate them into Ojibway culture and social structure.

The Change

In 1963, however, the government decided that the Ojibway should be brought into modern society and given the benefits thereof: modern plumbing, better health care, roads, and the like. To this end, they moved the entire Ojibway community from the old reserve to a government-built new community about 4 miles from their traditional encampment. The new community had houses, roads, schools, and easy access to "civilization." The differences between the new and the old were sufficient to destroy the fragile interdependence of Ojibway institutions.

First, all the houses were close together in neat rows, assigned randomly without regard for family group. As a result, the kinship group ceased to exist as a physical unit. Second, the replacement of boarding schools with a local community school meant that mothers had to stay home with the children instead of going out on the trap line. As a result, adult women overnight became consumers rather than producers, shattering their traditional relationships with their husbands and community. Because women and children could no longer leave home, men had to go out alone on the trap line. And because the men disliked leaving their families behind, they cut their trapping trips from several weeks to a few days, and trapping ceased to be a way of life for the whole family. The productivity of the Ojibway reached bottom in May 1970 when the government ordered the tribe to halt all fishing after pollution from a white-owned paper mill had caused mercury levels in the reservation's rivers to reach dangerous levels. Because of all these changes, the community became heavily dependent on government aid rather than on themselves or on each other.

The result was the total destruction of the old patterns of doing things—that is, of social roles, statuses, and institutions. The relationships between husbands and wives were no longer clear. What were their rights and obligations to each other now that their joint economic productivity had ended? What were their rights and obligations to their children when no one cared about tomorrow?

The Future of the Ojibway

In 1985, the Ojibway finally reached a $16.7 million out-of-court settlement with the government and the paper mill to compensate for damages to their way of life arising from both government policies and mercury pollution. However, environmental pollution remains a serious health and economic problem (Envirowatch 2006). In addition, mining and clear-cutting of the land by outside corporations now pose new threats to the tribe and its environment. Nevertheless, Ojibway society has begun the process of healing and recovery. It is developing school programs to teach young people the Ojibway language, using money from the settlement to develop local industries that will provide an ongoing basis for a productive and thriving society, and it is organizing politically against these new threats to its environment, health, and culture (Envirowatch 2006; Turtle Island Native Network News 2009). In the process, it is rebuilding old social institutions and creating new ones.

A Sociological Response

Unfortunately, the Ojibway are not an exceptional case. Their tragedy has been played out in tribe after tribe, band after band, all over North America. In some tribes alcoholism touches nearly every family. Compared with other Americans, Native

Both on their reservation and in front of Canada's Parliament, members of the Ojibway community continue to protest against clear-cutting and other forms of environmental devastation at Grassy Narrows.

Mike Casses/Reuters/Landov

American youths and adults are about twice as likely to report abuse of alcohol or illicit drugs (NHSDA Report 2003). As a result, they are significantly more likely to die from chronic liver disease, cirrhosis, accidents, homicide, and suicide (National Center for Health Statistics 2009). In addition, experts estimate that methamphetamine abuse is now twice as common on Indian reservations as elsewhere in the country (Wagner 2006).

High levels of alcohol and drug use are health problems, economic problems, and social problems. Among the related issues are fetal alcohol syndrome, child and spouse abuse, unemployment, teenage pregnancy, nonmarital births, and divorce. How can these interrelated problems be addressed? To paraphrase C. Wright Mills (see Chapter 1), when one or two individuals abuse alcohol or drugs, this is an individual problem, and for its relief we rightfully look to clinicians and counselors. When large segments of a population have alcohol or drug problems, this is a public issue and must be addressed at the level of social structure.

A sociological response to reducing alcohol and drug problems among Native Americans begins by asking what social structures encourage substance abuse. Conversely, why don't social structures reward those who avoid substance abuse?

The answer depends on one's theoretical framework. Structural functionalists would likely focus on the destruction of Native American institutions and the absence of harmony between their remaining institutions and those of white society. Conflict theorists would likely focus on how whites damaged or destroyed Native American societies by systematically and violently stripping them of their means of economic production.

Regardless of theoretical position, it is obvious that Native Americans are severely economically disadvantaged. Unemployment is often a way of life; on some reservations, up to 85 percent of the adults are unemployed. Lack of work is a critical factor in substance abuse in all populations. Having a steady, rewarding job is an incentive to avoid substance abuse; it also reduces the time available for drinking and drug use, which are essentially leisure-time activities. From this perspective, the solution to high levels of substance abuse among Native Americans must include changing economic institutions to provide full employment and bolstering Native American culture and pride, as well as hiring more doctors, counselors, and others to help individuals fight addiction.

In many ways, fighting substance abuse is like fighting measles. We cannot eradicate the problem by treating people after they have it; we have to *prevent* it in the first place. When substance abuse is epidemic in a community, it requires community-wide efforts for prevention. Statuses, roles, and institutions must be rebuilt so that people have a reason to avoid abusing drugs or alcohol. This is just as true when we are talking about isolated Native communities as when we are talking about college students, the subject of Focus on Media and Culture: Alcohol and Spring Break on the next page.

Social Interaction and Everyday Life

Why do people do what they do? The answer depends not only on their social roles but also on the situation and on their social status, resources, personalities, and previous experiences. Two people playing the role of physician will do so differently, and the same individual will play the role differently with different patients and in different circumstances. Social structure explains the broad outlines of why we do what we do, but it doesn't deal with specific concrete situations. This is where the sociology of

focus on  MEDIA AND CULTURE

Alcohol and Spring Break

Spring break comes in many flavors. Some students travel with their families, some work on service projects, some stay home to earn extra income or catch up on schoolwork, and some go to the beach to party with friends. Most of those partiers will return from their trips with nothing worse than bad sunburns. A few, though, will die when alcohol or drugs lead to car crashes, drownings, or falls from apartment balconies. And some will return with permanent disabilities, sexually transmitted diseases, or psychological traumas caused by sexual assault.

Students who travel together to "party beaches" for spring break typically drink more heavily, have more sexual partners, and use condoms less regularly than during the rest of the year (Grekin, Sher, & Krull 2007; Sönmez et al. 2006; Lee, Maggs, & Rankin 2006). What is it about spring break that sparks these sorts of activities?

When students go on spring break, they leave behind the social institutions—family, education, and work—that normally control their behaviors. They also leave behind the people who normally enforce institutional rules: professors, dorm counselors, bosses, parents. Once on spring break, students no longer need to meet the normal role expectations for them nor to protect their statuses as students, family members, or workers. There are no authorities around to supervise or judge their behaviors. And the students who *are*

around may come from other campuses or states, giving everyone an air of anonymity.

At the same time, the absence of normal roles, statuses, and institutions allows new norms to arise that encourage behaviors that would be unacceptable back home. For example, in one survey conducted for the American Medical Association, more than half of female college students reported that engaging in casual sex during spring break is a way to fit in (Robert Wood Johnson Foundation 2006).

Finally, these new norms are reinforced by corporations that profit from them. Video companies find easy profits in videos such as *Girls Gone Wild* that celebrate spring break as an "anything goes" party. These companies not only show the wildest side of spring break but also teach high school students to expect such activities when they go to college. Similarly, alcohol manufacturers and tour companies promote the wilder side of spring break to sell their products. For example, "Dos Equis girls" hand out free drinks while wearing string bikinis, and one tour company's website jokes, "Don't worry about the water [in Mexico] because

When students go to wild "party beaches" for spring break, they leave behind the social institutions that normally control their behaviors. As a result, many drink more heavily, have more sexual partners, and use condoms less regularly than they otherwise would.

you will be drinking beer" (Robert Wood Johnson Foundation 2006).

In sum, like New Orleans's Mardi Gras and Brazil's Carnaval, spring break offers students an opportunity to revel in freedom from everyday institutions, roles, and statuses.

everyday life comes in. Researchers who study the **sociology of everyday life** focus on the social processes that structure our experience in ordinary, face-to-face situations.

Managing Everyday Life

Much of our daily life consists of routines. For example, we all learn dozens of routines for carrying on daily conversations and can usually find an appropriate one for any occasion. Small rituals such as "Hello. How are you?" "Fine. How are you?" will carry

The **sociology of everyday life** focuses on the social processes that structure our experience in ordinary, face-to-face situations.

us through multiple encounters every day. If we supplement this ritual with half a dozen others, such as "Thanks/You're welcome" and "Excuse me/No problem," we will be equipped to meet most of the repetitive situations of everyday life.

Nevertheless, each encounter is potentially problematic. What do you do when you say "How are you?" to someone purely as a social gesture, and they then regale you with their troubles for the next 20 minutes? What do you do when your father asks where his car keys are, and you know your brother took them without permission? Although, as Chapter 2 discussed, our culture provides a tool kit of routines, each of us must constantly decide which routine to employ, how, when, and why.

At the beginning of any encounter, then, individuals must resolve two issues: (1) What is going on here—what is the nature of the action? and (2) What identities will be granted—who are the actors? All action depends on our answers to these questions. Even the decision to ignore a stranger in the hallway presupposes that we have asked and answered these questions to our satisfaction. How do we do this?

Because two people meeting in a business setting share the same *frame*, both know what to do and what it means when one extends a hand to the other.

Frames

The first step in any encounter is to develop an answer to the question, What is going on here? The answer forms a frame, or framework, for the encounter. A **frame** is roughly identical to a definition of the situation—a set of expectations about the nature of the interaction episode that is taking place.

All face-to-face encounters are preceded by a framework of expectations—how people will act, what they will mean by their actions, and so on. Even the simplest encounter—say, approaching a salesclerk to buy a pack of gum—involves dozens of expectations: In most parts of the United States we expect that the salesclerk will speak English, will wait first on the person who got to the counter first, will not try to barter with us over the price, and will not put us down if we are overweight. These expectations—the frame—give us guidance on how we should act and allow us to evaluate the encounter as normal or deviant.

Our frames will be shared with other actors in most of our routine encounters, but this is not always the case. We may simply be wrong in our assessment of what is going on, or other actors in the encounter may have an entirely different frame. The final frame that we use to define the situation will be the result of a negotiation between the actors.

Identity Negotiation

After we have put a frame on an encounter, we need to answer the second question: Which identities will be acknowledged? This question is far more complex than simply attaching names to the actors. Because each of us has a repertoire of roles and identities from which to choose, we are frequently uncertain about which identity an actor is presenting *in this specific situation.*

To some extent, identities will be determined by the frame being used. If a student's visit to a professor's office is framed as an academic tutorial, then the professor's academic identity is the relevant one. If the professor is a friend of the family, then their interaction might be framed as a social visit, and other aspects of the professor's identity (hobbies, family life, and so on) become relevant.

A **frame** is an answer to the question, what is going on here? It is roughly identical to a definition of the situation.

Typically, identities are not problematic in encounters. Although confusion about identities is a frequent device in comedy films, in real life, a few minutes chatting will usually resolve any confusion about actors' identities. In some cases, however, identity definitions are a matter of serious conflict. For example, Jennifer may want Mike to regard her as an equal, but Mike may prefer to treat her as an inferior.

Resolving the identity issue involves negotiations about both your own and the other's identity. How do we negotiate another's identity? We do so by trying to manipulate others into playing the roles we have assigned them. Mostly we handle this through talk. For example, "Let me introduce Mary, the computer whiz" sets up a different encounter than "Let me introduce Mary, the party animal." Of course, others may reject your casting decisions. Mary may prefer to present a different identity than you have suggested. In that case, she will try to renegotiate her identity.

Identity issues can become a major hidden agenda in interactions. Imagine a newly minted male lawyer talking to an established female lawyer. If the man finds this situation uncomfortable, he may try to define it as a man/woman encounter rather than a junior lawyer/senior lawyer encounter. He may start with techniques such as "How do you, as a woman, feel about this?" To reinforce this simple device, he might follow up with remarks such as "You're so small, you make me feel like a giant." He may interrupt her by remarking on her perfume. He may also use a variety of nonverbal strategies such as stretching his arm across the back of her chair to assert dominance. Through such strategies, actors try to negotiate both their own and others' identities.

Dramaturgy

The management of everyday life is the focus of a sociological perspective called *dramaturgy.* **Dramaturgy** is a version of symbolic interaction that views social situations as scenes manipulated by actors to convey their desired impression to the audience (Brissett & Edgley 2005).

The chief architect of the dramaturgical perspective is Erving Goffman (1959, 1963). To Goffman, all the world was a theater. Like actors, each of us uses our appearance to establish our character—something we do each morning as we choose which clothes to wear, how to style our hair, and whether this would be a good day to show off any tattoos or piercings that we have (e.g., Pitts-Taylor 2003). And like actors, we can use facial expressions, eye contact, posture, and other body language to enhance, reinforce, or even contradict the things we say. For example, telling a worried friend that "Your dress looks fine" doesn't mean as much if you say it without looking up from your cell phone.

Sociologists who use dramaturgy also point out that life, like the theater, has both a front region (the stage) where the performance occurs and a back region where rehearsals take place and different behavioral norms apply. For example, waiters at expensive restaurants are acutely aware of being on stage and act in a dignified and formal manner (Fine 1996). Once in the kitchen, however, they may be transformed back into rowdy college kids.

The ultimate back region for most of us, the place where we can be our real selves, is at home. Nevertheless, even here front-region behavior is called for when company comes. ("Oh yes, we always keep our house this clean.") On such occasions, a married couple functions as a team in a performance designed to manage their guests' impressions. People who were screaming at each other before the doorbell rang suddenly start calling each other "dear" and "honey." The guests are the audience, and they too play a role. By seeming to believe the team's act, they contribute to a successful visit/performance.

Dramaturgy is a version of symbolic interaction that views social situations as scenes manipulated by the actors to convey the desired impression to the audience.

Impression Management

So far, we've mostly focused on *what* people do in everyday encounters. But it's also important to ask *why* people do what they do. The answer most often supplied by scholars studying everyday behavior is that people are trying to enhance their social position and self-esteem (Owens, Stryker, & Goodman 2001; Guadagno & Cialdini 2007). These are some of the most important rewards that human interaction has to offer, and we try to manage the impression we make on others to improve our chances of getting these rewards.

The work that we do to control others' views of us is known as **impression management** (Goffman 1959). Most of the time, we use impression management to gain social approval from others. We wear fashionable clothes and hairstyles and try to behave in courteous and friendly ways. However, impression management can also be used to appear *less* socially acceptable: Punks, goths, "emos," or gang members, for example, may choose hairstyles and clothing in part because they *want* others to fear them or be repelled by them (Wilkins 2008; Pitts-Taylor 2003).

We also engage in impression management when we *explain* our behaviors and choices. Two common strategies are avoiding blame and gaining credit (Tedeschi & Riess 1981; Guadagno & Cialdini 2007).

This boy's body language radiates his dissatisfaction.

Avoiding Blame

There are many potential sources of damage to our social identity and self-esteem. We may have lost our job, flunked a class, been unintentionally rude, or said something that we immediately feared made us look stupid. When we behave in ways that make us look bad, or when we fear we are on the verge of doing so, we need to find ways to protect our social position and self-esteem.

Most of this work is done through talk. C. Wright Mills (1940, 909) noted that we learn how to justify our norm violations more or less at the same time that we learn the norms themselves. If we can successfully explain away our rule-breaking, we can present ourselves as people who normally obey norms and who deserve to be thought well of by ourselves and others. The two basic strategies we use to avoid blame are *accounts* and *disclaimers*.

Accounts

Much of the rule-breaking that occurs in everyday life is of a minor sort that can be explained away. We do this by giving **accounts**, explanations of unexpected or untoward behavior. Accounts fall into two categories: *excuses* and *justifications* (Scott & Lyman 1968). **Excuses** are accounts in which an individual admits that the act in question is bad, wrong, or inappropriate but claims he or she couldn't help it. **Justifications** are accounts that explain the good reasons the violator had for breaking the rule; often these take the form of appeals to some higher rule (Scott & Lyman, 47).

Students are often quite adept at excuses and justifications. When the website **www.rateyourstudents.com** asked professors to report their favorites, one told of a student who apologized for turning in a paper late (Troop 2007). The student's *excuse* was that he was the school mascot and had left his paper stuck in the arm of the mascot costume, which had been locked in the sports department office over the weekend.

Impression management consists of actions and statements made to control how others view us.

Accounts are explanations of unexpected or untoward behavior. They are of two sorts: excuses and justifications.

Excuses are accounts in which one admits that the act in question is wrong or inappropriate but claims one couldn't help it.

Justifications are accounts that explain the good reasons the violator had for choosing to break the rule; often they are appeals to some alternate rule.

© Jose Luis Pelaez, Inc./Surf/Corbis

focus on  AMERICAN DIVERSITY

Becoming Goth

It's never easy being a young person. No longer kids but not yet fully independent adults, young people in their teens and twenties struggle both to create their own identities and to convince others to believe in those identities. But why would someone choose an identity that seems guaranteed to lead to social rejection? To answer this question, sociologist Amy C. Wilkins (2008) spent months observing and interviewing young people who identified themselves as Goths.

Goths favor black clothes, often torn and safety-pinned; tattoos that lean more to skulls than to butterflies; dark makeup for both males and females; and black or wildly colored hair in styles that defy peer norms. Stickers, T-shirts, and other items proudly highlight Goth's enjoyment of loud and angry bands and of anything related to death, including vampires, cemeteries, or horror films.

As this description suggests, it takes work to create a Goth impression. So why would anyone wish to do so?

The Goths interviewed by Wilkins claimed that they had always been Goth in their hearts, and had simply found a community that shared their views. Wilkins, however, reached a different conclusion. The Goths, she noticed, were all white and middle-class, with no interest in athletics but considerable interest in math, computers, science, and science fiction. In other words, they were "geeks." Before they became Goths, others would pick them last for teams at recess, ridicule them in hallways, or consider them fun targets for violence. They didn't have the social status of white boys who excelled at sports or of white girls who dressed well and had fashionable hairstyles. Nor did they have the social status of African American or Hispanic kids, who are assumed to be cool by high school students who value hip-hop culture.

By adopting Goth appearances and managing others' impressions of them, Goths achieved several goals. First, they scared other people—intentionally—and thus were less likely to become targets for violence. Second, they gained respect from their peers, who recognized Goths as rebels. Third, they gained new *accounts* that justified their behaviors, interests, and appearances and allowed them to discount the views of anyone who didn't share their views. Similarly, other white, middle-class kids who don't neatly fit cultural norms—boys interested in art, poetry, or bisexuality and smart girls not inclined toward smiling—also may adopt punk or emo identities and appearances. By so doing, they can turn themselves from outcasts into "outlaws."

Adopting a Goth appearance can help marginalize young people to justify their actions and beliefs, to discount anyone who doesn't share their views, to gain respect from peers who now view them as rebels, and thus to transform themselves from outcasts into "outlaws."

Another professor submitted the following student *justification*:

I will be unable to be in class today because every year we have a Jell-O wrestling competition on campus, and it has just come to my attention that the 50 gallons of Jell-O that we previously made has spoiled. So now I have to remake the 50 gallons before 9 o'clock tonight.... I understand this is a really weird circumstance, but without the Jell-O we have no competition, and without the competition we lose all of our fund-raising. Thank you, and have a good weekend. (Troop 2007)

Accounts such as these are verbal efforts to resolve the discrepancy between what happened and what others legitimately expected to happen. When others accept our accounts, our self-identity and social status are preserved and our interactions with others can proceed normally.

Disclaimers

A person who recognizes that he or she is likely to violate expectations may preface that action with a **disclaimer**, a verbal device used in advance to defeat any doubts and negative reaction that might result from conduct (Hewitt & Stokes 1975, 3). Students often begin a query with "I know this is a stupid question, but.…" The disclaimer lets the hearer know that the speaker knows the rules, even though he or she doesn't know the answer.

Disclaimers occur before the act; accounts occur after the act. Nevertheless, both are verbal devices we use to try to maintain a good image of ourselves, both in our own eyes and in the eyes of others. They help us to avoid self-blame for rule-breaking and to reduce the chances that others might blame us for our actions. If we succeed in this impression management, we can retain a fairly good reputation and social status, despite occasional failures in meeting our social responsibilities.

The Concept Summary on Using Disclaimers and Accounts reviews the differences between these two verbal strategies.

concept summary

Using Disclaimers and Accounts

As part of the battle against terrorism, the U.S. government authorized the use of various tactics that other countries outlaw as torture, such as waterboarding: pouring water over a prisoner's face to force water inhalation, thereby causing the prisoner to experience great pain, the sensation of drowning, and sometimes brain, lung, or bone damage. The strategies used by U.S. government and military officials to avoid blame for waterboarding illustrate the ways people use disclaimers and accounts.

Strategy	Definition	Example
Disclaimers	Verbal strategies used *in advance* to ward off the possibility that others may think one is doing something wrong	We would never use torture, although of course we will need to use waterboarding and other forms of "harsh" or "enhanced" interrogation.
Accounts	Explanations offered *after the fact* to try to avoid blame for behaviors generally considered unacceptable	(See examples of excuses and justifications, both of which are types of accounts.)
Excuses	Acknowledging that a behavior is wrong, but stating that it was out of your control	We (the military) had to use waterboarding because top government officials ordered us to do so.
Justifications	Arguing that although a behavior might have seemed wrong, it was justified because of a higher moral good	We had to use waterboarding to stop the terrorists and save American lives.

A **disclaimer** is a verbal device employed in advance to ward off doubts and negative reactions that might result from one's conduct.

Like this mouse breeder showing off his awards, most of us seek ways to enhance our credit with others.

© Patrick Ward/Stock Boston Inc.

Gaining Credit

To maintain our self-esteem, we need not only to avoid blame but also to get credit for anything good we do. With this goal in mind, we employ a variety of verbal devices to associate ourselves with positive outcomes (Guadagno & Cialdini 2007). Just as there are a variety of ways to avoid blame, there are many ways we can claim credit. One way is to link ourselves to situations or individuals with high status. This ranges from dropping the names of popular students we happen to know, to wearing a baseball cap from a winning team, to making a $1,000 donation at a political fund-raiser so we can get a signed photograph of the President to hang on our wall.

Claiming credit is a strategy that requires considerable tact. Bragging is generally considered inappropriate, and if you pat yourself too hard on the back, you are likely to find that others will refuse to do so. The trick is to find the delicate balance where others are subtly reminded of your admirable qualities without your actually having to ask for or demand praise. If you do very well on an exam, for example, you might let others know how well you did while simultaneously suggesting that your high score was just a matter of luck.

Case Study: Impression Management and Homeless Kids

One of the best ways to understand impression management is to look at individuals who have what Goffman (1961b) called *spoiled identities*—identities that are extremely low in status. Examples include sex offenders, traitors, and people with disfiguring facial scars. How do people with spoiled identities sustain their self-esteem and manage the way others views them?

A study among homeless kids in transitional settings (such as shelters and motels) in San Francisco investigated just this question. Anne Roschelle spent four years volunteering at drop-in centers for homeless kids, observing their activities and conversations, and talking with them formally and informally (Roschelle & Kaufman 2004).

The kids Roschelle met were keenly aware of their spoiled identities. They knew that local newspapers often ran stories on the "homeless problem," and that local politicians gained votes by vowing to remove the homeless from the city. As one kid explained, "Everyone hates the homeless because we represent what sucks in society. If this country was really so great there wouldn't be kids like us" (Roschelle & Kaufman 2004, 30). How, then, did these kids maintain their self-esteem and try to control others' images of them?

Roschelle and her co-author, Peter Kaufman, found that the kids used two sets of strategies: fitting in and fighting back. *Fitting in* could take various forms. Kids struck up friendships with volunteers and with other homeless kids so they would feel they were valued as individuals. They also tried to fit in by dressing, talking, and acting as much like nonhomeless kids as they could: selecting the most stylish coats from the donations box rather than the warmest ones, for example. Kids also chose their words carefully to hide their homelessness. At school, they called caseworkers their "aunts," called homeless shelter staff their "friends," and referred to friends who slept three cots away as friends who lived three houses away.

Like everyone else, homeless youths try to manage others' impressions of them. This young man may well have found that owning a cute puppy encourages others to view him as less threatening and as more deserving of aid.

Janine Wiedel Photolibrary/Alamy

Homeless kids also protected their identities by *fighting back*. First, they used "gangsta" clothes, gestures, and actions to intimidate nonhomeless kids. Second, they adopted sexual behaviors and attitudes far beyond their years and took pride in their sexual "conquests." Finally, they bolstered their social position by loudly criticizing homeless street people who were more stigmatized than themselves:

> Rosita: Man, look at those smelly street people, they are so disgusting, why don't they take a shower?
>
> Jalesa: Yeah, I'm glad they don't let them into Hamilton [shelter] with us.
>
> Rosita: Really, they would steal our stuff and stink up the place!
>
> Jalesa: Probably be drunk all the time too. (Roschelle & Kaufman 2004, 37)

By contrasting themselves with more stigmatized others, Rosita, Jalesa, and other kids could feel better about themselves.

The homeless kids that Roschelle and Kaufman studied possessed many traits that typically lead to poor self-esteem and social disapproval: They were hungry, poor, ragged, and homeless in a society that values wealth and blames poverty on the poor. Yet many nevertheless managed to feel good about themselves and to control, at least in part, how others viewed them. Their experiences confirm the assumption made by the interaction school: Even in the face of a spoiled identity, we can use impression management to negotiate a positive self-concept and a more satisfying social position. But their experiences also illustrate that tactics used to do so can be harmful: Thirteen-year-olds who take pride in "seducing" 33-year-olds or in threatening others with knives and guns are likely to suffer in the long run.

Where This Leaves Us

In the 1950s, structural-functional theory dominated sociology, and a great deal of emphasis was placed on the power of institutionalized norms to determine behavior. Beginning in the 1960s, however, sociologists grew increasingly concerned that this view of human behavior reflected an "oversocialized view of man" (Wrong 1961). In 1967, Garfinkel signaled rebellion against this perspective when he argued that the deterministic model presented people as "judgmental dopes" who couldn't do their own thinking.

Since then, scholars have increasingly tended to view social behavior as more negotiable and less rule bound and have increasingly focused on how people resist rather than accommodate to social pressures (Weitz 2001). This change is obvious not only in the sociology of everyday life, but also in most other areas of sociology, including studies of hospitals, businesses, schools, and other large organizations (e.g., Jurik, Cavender, & Cowgill 2009; Bettie 2003). This does not mean that rules don't make a difference. Indeed, they make a great deal of difference, and there are obvious limits to the extent to which we can negotiate given situations. Each actor's ability to negotiate depends on his or her access to resources and power, both of which are strongly determined by social structure.

The perspective of life as problematic and negotiable is a useful balance to the role of social structure in determining behavior. Our behavior is neither entirely negotiable nor entirely determined.

Summary

1. The analysis of social structure—recurrent patterns of relationships—revolves around three concepts: status, role, and institution. Statuses are specialized positions within a group and may be of two types: achieved or ascribed. Roles define how status occupants ought to act and feel.

2. Because societies share common human needs, they also share common institutions: enduring and complex social structures that meet basic human needs. Some of those common institutions are family, economy, government, education, and religion.

3. Institutions are interdependent; none stands alone, and so a change in one results in changes in others. Structural functionalists point out that institutions regulate behavior and maintain the stability of social life across generations. Conflict theorists note that these patterns often benefit one group more than others.

4. An important determinant of institutional development is the ability of a society to produce an economic surplus.

Each major improvement in production has led to an expansion in social institutions.

5. The sociology of everyday life analyzes the patterns of human social behavior in concrete encounters in daily life.

6. Deciding how to act in a given encounter requires answering two questions: What is going on here? and Which identities will be acknowledged? These issues of framing and identity negotiation may involve competition and negotiation between actors or teams of actors.

7. Dramaturgy is a symbolic interactionist perspective pioneered by Erving Goffman. It views the self as a strategist who is choosing roles and setting scenes to maximize self-interest.

8. The desire for approval is an important factor guiding human behavior. To maximize this approval, people engage in active impression management to sustain and support their self-esteem. This work takes two forms: avoiding blame and gaining credit.

Thinking Critically

1. Is social class an achieved or ascribed status? What would a structural functionalist say? A conflict theorist? A symbolic interactionist?

2. Consider religion as an institution. How would a conflict theorist view it? What might a structural functionalist say? Which position is closest to your own view and why?

3. Pick a social problem that affects you personally; for example, alcoholism, unemployment, racism, sexism, illegal immigration. Describe a social structural solution—one that focuses on changing the underlying social structural causes of the problem rather than on improving individuals' situations one by one.

Book Companion Website

www.cengage.com/sociology/brinkerhoff
Prepare for quizzes and exams with online resources—including tutorial quizzes, a glossary, interactive flash cards, crossword puzzles, essay questions, virtual explorations, and more.

4. Describe a time when you disagreed with someone about his or her identity. What kind of situation was it, and why was the identity problematic? In the end, whose definition of identity was accepted? Why?

Groups, Networks, and Organizations

Human Relationships

At one level, sociology is the study of relationships: how they begin, function, change, and affect both individuals and the community. In this chapter we review the basic types of human relationships, from small and intimate groups to large and formal organizations, and discuss some of the consequences of these relationships.

Social Processes

Some relationships operate smoothly; others are plagued by conflict and competition. We use the term **social processes** to describe the types of interaction that go on in relationships. This section looks closely at four social processes that regularly occur in human relationships: *exchange, cooperation, competition,* and *conflict.*

Exchange

Exchange is voluntary interaction in which the parties trade tangible or intangible benefits with the expectation that all parties will benefit (Stolte, Fine, & Cook 2001). A wide variety of social relationships include elements of exchange. In friendships and marriages, exchanges usually include intangibles such as companionship, moral support, and a willingness to listen to the other's problems. In business or politics, an exchange may be more direct; politicians, for example, openly acknowledge exchanging votes on legislative bills—I'll vote for yours if you'll vote for mine.

Exchange relationships work well when people return the favors they receive, maintaining a balance between giving and taking (Molm & Cook 1995). The expectation that people maintain this balance is called the **norm of reciprocity** (Gouldner 1960; Uehara 1995). If you help your sister-in-law move, she is then obligated to you. Somehow she must pay you back. If she fails to do so, your relationship will likely suffer. By extension, it's wiser to refuse favors when you *don't* want a relationship with someone. For example, if someone you don't know well volunteers to type your term paper, you will probably be suspicious. Your first thought is likely to be, "What does this guy want from me?" If you don't want to owe this person a favor, you're better off typing your own paper. Nonsociologists might sum up the norm of reciprocity by concluding that there's no such thing as a free lunch.

Exchange is one of the most basic processes of social interaction. Almost all voluntary relationships involve the expectation of exchange. In marriage, for example, each partner is expected to provide affection and sexual access to the other.

An exchange relationship survives only if each party to the interaction gets something out of it. This doesn't mean that the rewards must be equal: They often aren't. Nor does this mean that each party to the exchange relationship has equal power; rather, the actor with greater control over a more valuable resource always has more power. In children's play groups, for example, one child may be treated badly by the other children and be allowed to play with them only if he agrees to give them his lunch or allows them to use his bicycle. If this boy has no one else to play with, he may find this relationship more rewarding than playing alone. Very unequal exchange relationships usually continue only when few good alternatives exist (Molm 2003; Stolte, Fine, & Cook 2001).

sociology and you

The norm of reciprocity also applies in dating relationships. If you are a man and buy your date dinner or a movie, you may feel that she now owes you something in return—gratitude, a good night kiss, or more. If you are the woman, you *also* may believe that you now owe your date something, and so you may do things you really don't want to do in exchange. When couples disagree on who owes what to whom, situations like these can escalate to anger, breakups, or even sexual assault.

Social processes are the forms of interaction through which people relate to one another; they are the dynamic aspects of society.

Exchange is a voluntary interaction from which all parties expect some reward.

The **norm of reciprocity** is the expectation that people will return favors and strive to maintain a balance of obligation in social relationships.

Cooperation

Cooperation occurs when people work together to achieve shared goals. Exchange is a trade: I give you something and you give me something else in return. Cooperation is teamwork: people working together to achieve shared goals. Consider, for example, an intersection with a four-way stop sign. Although we may be tempted to speed through the stop sign, we rarely (if ever) do so because we know we'll get through more safely and more quickly if we take turns. Most continuing relationships have some element of cooperation. Spouses cooperate in raising their children; children cooperate in tricking their substitute teachers.

Cooperation also operates at a much broader social level. Neighbors may work together to fight against a proposed high-rise apartment building, and a nation's citizens may support higher taxes to provide health care for the needy. Individuals are most likely to cooperate when faced with a common threat, when cooperation seems in their economic self-interest, when they share a sense of community identity, and when they value belonging to a community (Van Vugt & Snyder 2002).

Competition

But sometimes people can't reach their goals through exchange or cooperation. If our goals are mutually exclusive (for example, I want to sleep and you want to play loud music, or we both want the same job), we cannot both achieve our goals. Situations like these foster *competition* or *conflict*.

Competition is any struggle over scarce resources that is regulated by shared rules. The rules usually specify the conditions under which winning will be considered fair and losing will be considered tolerable. When the norms are violated and rule-breaking is uncovered, competition may erupt into conflict.

One positive consequence of competition is that it stimulates achievement and heightens people's aspirations. It also, however, often results in personal stress, reduced cooperation, and social inequalities (elaborated on in Chapters 7 through 9).

Because competition often results in change, groups that seek to maximize stability often devise elaborate rules to avoid the appearance of competition. Competition is particularly problematic in informal groups such as friendships and marriages. Friends who want to stay friends will not compete for anything of high value; they might compete over computer game scores, but they won't compete for each other's spouses. Similarly, most married couples avoid competing for their children's affection because they realize that such competition could destroy their marriage.

Conflict

Cooperation is interaction that occurs when people work together to achieve shared goals.

Competition is a struggle over scarce resources that is regulated by shared rules.

Conflict is a struggle over scarce resources that is not regulated by shared rules; it may include attempts to destroy, injure, or neutralize one's rivals.

When a struggle over scarce resources is not regulated by shared rules, **conflict** occurs (Coser 1956). Because no tactics are forbidden and anything goes, conflict may include attempts to neutralize, injure, or destroy one's rivals. Conflict creates divisiveness rather than solidarity.

Conflict with outsiders, however, may enhance the solidarity of the group. Whether the conflict is between warring superpowers or warring street gangs, the us-against-them feeling that emerges from conflict with outsiders causes group members to put aside their jealousies and differences to work together. From nations to schools, groups have found that starting conflicts with outsiders helps to squash conflict within their own group. For example, some critics argue that U.S. politicians voted to invade Iraq in 2003 to divert the public's attention away from economic problems at home.

When the struggle for scarce resources (including children's toys) is not regulated by norms that specify the rules of fair play, conflict often results.

Social Processes in Everyday Life

Exchange, cooperation, competition, and even conflict are important aspects of our relationships with others. Few of our relationships involve just one type of group process. Even friendships usually involve some competition as well as cooperation and exchange. Similarly, relationships among competitors often involve cooperation.

We interact with people in a wide range of relationships, both temporary and permanent, formal and informal. In the rest of this chapter, we discuss three general types of relationships: *groups*, *social networks*, and *organizations*.

Groups

A **group** is a collection of two or more people that has two special characteristics: (1) Its members interact within a shared social structure of statuses, roles, and norms, and (2) its members recognize that they depend on each other. Groups may be large or small, formal or informal; they range from a pair of lovers to the residents of a local fraternity house to Toyota employees.

The distinctive nature of groups stands out when we compare them to other collections of people. *Categories* of people who share a characteristic, such as all dorm residents, bald-headed men, or Hungarians, are not groups because most members of this category never meet, let alone interact. Similarly, *crowds* who temporarily cluster together on a city bus or in a movie theater are not groups because they are not mutually dependent. Although they share certain norms, many of those norms (such as not staring) are designed to *reduce* their interactions with each other.

The distinguishing characteristics of groups hint at the rewards of group life. Groups are the people we take into account and the people who take us into account. They are the people with whom we share many norms and values. Thus, groups can foster solidarity and cohesion, reinforcing and strengthening our integration into society. When groups function well, they offer benefits ranging from sharing basic survival and problem-solving techniques to satisfying personal and emotional needs.

A **group** is two or more people who interact on the basis of shared social structure and recognize mutual dependency.

Conversely, when groups function poorly, they create anxiety, conflict, and social stress.

Types of Groups

Almost all students belong to a family group as well as to the student body of their college or university. And as students, they also interact with many different types of groups, such as sororities, athletic teams, sociology majors, dorm residents, and honor students. Obviously, some of these groups affect their members more than others do. This section discusses three types of groups: *reference groups*, *primary groups*, and *secondary groups*.

Reference Groups

If Jim belongs to a fraternity, it's likely that he often checks that his appearance, grades, athletic skills, and so on compare favorably with those of his fraternity brothers. If Nancy's church community is central to her life, she probably compares herself to other church members her age. The fraternity is Jim's *reference group* and the church community is Nancy's. **Reference groups** are groups that individuals compare themselves to regularly. Typically, individuals choose reference groups whose members are similar to themselves. Sometimes, however, they choose reference groups because they *aspire* to belong to that group. For example, before Mike joined the fraternity, he probably first looked for a fraternity whose members dressed more or less like he did and then bought a few new items to fit in even better.

The reference groups we choose have powerful effects on our lives. For example, decades of research suggest that happiness drops when we compare ourselves to others who are better off than we are—a situation known as **relative deprivation**. Conversely, happiness increases when we compare ourselves to those who are worse off.

The impact of relative deprivation was recently demonstrated in a study on military life conducted by Jennifer Hickes Lundquist (2008). Military life is not easy: Members of the armed forces must follow strict rules for all aspects of their lives, give up control over their schedules, and leave home and family—sometimes for life-threatening assignments—whenever ordered to do so.

Lundquist found that relative satisfaction with military life was essentially the reverse of satisfaction with civilian life: African American women were most satisfied with military life, followed by African American men, Latina women, Latino men, and then white women. White men were the least satisfied with military life, even though they were the most satisfied with civilian life.

What explained these findings? Lundquist found that satisfaction with military life depended primarily on whether individuals believed their lives in the military were better than the lives of people like them—their reference group—in civilian life. In fact, African Americans, Latinos, and women face less discrimination in the military than in civilian life, with women minorities gaining a double benefit (Lundquist 2008). Members of these groups were satisfied with military life because they realized that their pay, quality of life, and opportunities for promotion were better than they would be in civilian life. In contrast, white men were most likely to believe that people like them could do well in civilian life and so were *least* happy in the military.

Primary Groups

Primary groups are characterized by face-to-face interaction, and so they are typically informal, small, and personal (Cooley [1909] 1967). The family is a primary group, as are friendship networks, co-workers, and gangs. The relationships formed in these

---

Reference groups are groups that individuals compare themselves to regularly.

Relative deprivation exists when we compare ourselves to others who are better off than we are.

Primary groups are groups characterized by intimate, face-to-face interaction.

Whether our primary group is made of punks, athletes, or committed sunbathers, we tend to dress, behave, and believe in ways similar to that of other group members, thus reinforcing our connection to each other.

Ghislain & Marie David de Lossy/Taxi/Getty Images.

groups are relatively permanent, generate a strong sense of loyalty and belongingness, constitute a basic source of identity, and strengthen our sense of social integration into society.

The major purpose of primary groups is to serve *expressive needs*: to provide individuals with emotional support and a sense of belonging to a social group. Your family and close friends, for example, probably feel obligated to help you when needed. You can call on them to listen to your troubles, to bring you soup when you have the flu, and to pick you up in the dead of night if your car breaks down.

Because we need primary groups so much, they have tremendous power to bring us into line. From society's point of view, this is the major function of primary groups: They are the major agents of social control. For example, most of us don't shoplift because we would be mortified if our parents, friends, or co-workers found out. The reason most soldiers go into combat is because their buddies are going. We tend to dress, act, vote, and believe in ways that will keep the support of our primary groups. In short, we conform. The law would be relatively helpless at keeping us in line if we weren't already restrained by the desire to stay in the good graces of our primary groups. One corollary of this, however, which Chapter 6 addresses, is that if our primary groups consider shoplifting or tax evasion acceptable, then our primary-group associations may lead us into law-breaking rather than conformity.

Secondary Groups

By contrast, **secondary groups** are formal, large, and impersonal. Whereas the major purpose of primary groups is to serve expressive needs, secondary groups usually form to serve *instrumental needs*—that is, to accomplish some specific task. The quintessential secondary group is entirely rational and contractual in nature; the participants interact solely to accomplish some purpose (earn credit hours, buy a pair of shoes, get a paycheck). Their interest in each other does not extend past this contract. The differences between these two types of groups are explored more fully in the Concept Summary: Differences between Primary and Secondary Groups on the next page.

Secondary groups are groups that are formal, large, and impersonal.

concept summary

Differences between Primary and Secondary Groups

	Primary Groups	Secondary Groups
Size	Small	Large
Relationships	Personal, intimate	Impersonal, aloof
Communication	Face-to-face	Indirect—memos, telephone, etc.
Duration	Permanent	Temporary
Cohesion	Strong sense of loyalty, we-feeling	Weak, based on self-interest
Decisions	Based on tradition and personal feelings	Based on rationality and rules
Social structure	Informal	Formal—titles, officers, charters, regular meeting times, etc.
Purpose	Meet expressive needs—provide emotional support and social integration	Meet instrumental goals—accomplish specific tasks

The major purpose of secondary groups is accomplishing specific tasks. If you want to build an airplane, raise money for a community project, or teach introductory sociology to 2,000 students a year, then secondary groups are your best bet. They are responsible for building our houses, growing and shipping our vegetables, educating our children, and curing our ills. In short, we could not do without them.

The Shift to Secondary Groups

In preindustrial society, there were few secondary groups. Vegetables and houses were produced by families, not by Del Monte or Del Webb. Parents taught their own children, and neighbors nursed one another's ills. Under these conditions, primary groups served both expressive and instrumental functions. As society has become more industrialized, more and more of our instrumental needs are met by secondary rather than primary groups.

In addition to losing their instrumental functions to secondary groups, primary groups have suffered other threats in industrialized societies. Each year, about 13 percent of U.S. households move to a new residence (U.S. Bureau of the Census 2009). This fact alone means that our ties to friends, neighborhoods, and co-workers are seldom really permanent. People change jobs, spouses, and neighborhoods. One consequence of this breakdown of traditional primary groups is that many people rely on secondary groups even for expressive needs; if they have marriage problems, for example, they may join a support group rather than talk to a parent.

Many scholars have suggested that these inroads on the primary group represent a weakening of social control; that is, the weaker ties to neighbors and kin mean that people feel less pressure to conform. They don't have to worry about what the

Many of the groups we participate in combine characteristics of primary and secondary groups. The elementary school classroom is a secondary group, yet many of the friendships developed there will last for 6, 12, or even 40 years.

© Bob Daemmrich/Stock, Boston Inc.

neighbors will say because they haven't met them; they don't have to worry about what mother will say because she lives 2,000 miles away, and what she doesn't know won't hurt her. There is some truth in this suggestion, and it may be one of the reasons that small towns with stable populations are more conventional and have lower crime rates than do big cities with more fluid populations (an issue addressed more fully in Chapter 14).

Interaction in Groups

We spend much of our lives in groups. We have work groups, family groups, and peer groups. In class we have discussion groups, and everywhere we have committees. Regardless of the type of group, its operation depends on the quality of interaction among members. This section reviews some of the more important factors that affect interaction in small groups. As we will see, interaction is affected by group size, physical proximity, and communication patterns.

Size

The smallest possible group is two people. As the group grows to three, four, and more, its characteristics change. With each increase in size, each member has fewer opportunities to share opinions and contribute to decision making or problem solving: Think of the difference between being in a class of 15 students versus a class of 500. In many instances, the larger group can better solve problems and find answers. This benefit, however, comes at the expense of individual satisfaction. Although the larger group can generate more ideas, each person's ability to influence the group diminishes. As the group gets larger, interaction becomes more impersonal, more structured, and less personally satisfying.

Physical Proximity

Interaction occurs more often when group members are physically close to one another. This effect extends beyond the laboratory. You are more likely to become friends

FIGURE 5.1 **Patterns of Communication**

Patterns of communication can affect individual participation and influence. In each figure the circles represent individuals and the lines represent the flow of communication. The all-channel network pattern provides the greatest opportunity for participation and occurs more often when participants differ little in status. The wheel pattern, by contrast, occurs most often when one individual has more status and power than do the others, such as in a classroom.

All-channel

Circle

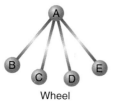

Wheel

Cohesion in a group is characterized by high levels of interaction and by strong feelings of attachment and dependency.

with the student who sits next to you in class or who rooms next to you than with the student who sits at the end of your row or who rooms at the end of your hall.

Communication Patterns

Interaction of group members can be either facilitated or hindered by patterns of communication. Figure 5.1 shows some common communication patterns for five-person groups. The communication pattern allowing the greatest equality of participation is the *all-channel network*. In this pattern, each person can interact equally with every other person. Each participant has equal access to the others and an equal ability to become the focus of attention.

The other two common communication patterns allow for less interaction. In the *circle pattern*, people can speak only to their neighbors on either side. Although this pattern reduces interaction, it doesn't give one person more power than the others. In the *wheel pattern*, on the other hand, a single, pivotal individual holds most of the power in the group. For example, in a traditional classroom students primarily interact with the teacher, who directs the flow of interaction, rather than with other students.

Communication patterns are often created, either accidentally or purposefully, by the physical distribution of group members. When committee members sit at a roundtable, all-channel network or circle communication patterns easily emerge. When members instead sit at a rectangular table, the people at the two ends and in the middle of the long sides have more chance of participating in and influencing the group's decisions.

Cohesion

Another characteristic of groups is their degree of **cohesion**, or solidarity. A cohesive group is characterized by higher levels of interaction and by strong feelings of attachment and dependency. Because its members feel that their happiness or welfare depends on the group, the group can make extensive claims on the individual members (Hechter 1987). Cohesive adolescent friendship groups can enforce unofficial dress codes on their members; cohesive youth gangs can convince new male members to commit random murders and can convince new female members to submit to gang rapes.

Marriage, church, and friendship groups differ in their cohesiveness. What makes one marriage or church more cohesive than another? Among the factors are small size, similarity, frequent interaction, long duration, a clear distinction between insiders and outsiders, and few ties to outsiders (McPherson, Popielarz, & Drobnic 1992; McPherson & Smith-Lovin 2002). Although all legal marriages in our society are the same size (two members), a marriage in which the partners are more similar, spend more time together, and so on will generally be more cohesive than one in which the partners are dissimilar and see each other for only a short time each day.

Group Conformity

When a man opens a door for a woman, do you see traditional courtesy or sexist condescension? When you listen to Lil Wayne, Kelly Clarkson, or Coldplay, do you hear good music or irritating noise? Like taste in music, many of the things we deal with and believe in are not true or correct in any absolute sense; they are simply what our groups have agreed to accept as right. Researchers who look at individual decision making in groups find that group interaction increases conformity. This was famously demonstrated in two classic experiments by Solomon Asch (1955) and Stanley Milgram (1974).

The Asch Experiment

In Asch's experiment, the group consisted of nine college students, all supposedly unknown to each other. The experimenter told the students that they would be tested on their visual judgment. For example, in one test, the experimenter showed the students two cards. Card A showed only one very tall line; Card B showed one very tall line and two much shorter lines. The experimenter then asked the students to choose the line on Card B that most closely matched the (very tall) line on Card A. This was not a difficult task: Anyone with decent vision could tell that the two very tall lines were the best match.

Each group of students viewed pairs of cards like these 15 times. The first few times, all the students agreed on the obviously correct answer. In subsequent trials, however, the first eight students—in reality, all paid stooges of the experimenter—all gave the same, obviously wrong, answer. The real test came in seeing what the last student—the real subject of the experiment—would do. Would he go along with everybody else, or would he publicly disagree? Photographs of the experiment show that the real subjects wrinkled their brows, squirmed in their seats, gaped at their neighbors, and, 37 percent of the time, agreed with the wrong answer (Asch 1955).

The Milgram Experiment

The Milgram (1974) experiment provided even more troubling evidence of the power of groups to instill conformity. For this experiment, subjects were told that they would act as teachers in an experiment on learning. An experimenter instructed the "teacher" to read a list of word pairs to a "learner," who was expected to memorize them. Then the teacher would read the list a second time, providing only the first word in each pair plus four possible correct answers. If the learner gave the wrong answer, the teacher was instructed to give the correct answer and then administer an electric shock to the learner (placed in another room, out of sight of the teacher). The voltage of the shock increased with each wrong answer, and teachers were told by the experimenter (who stayed in the room throughout the experiment) to continue reading the list until the learner got all answers correct.

In reality, both the experimenter and the learner were working with Milgram. The real question was what the "teacher" would do. The results were horrifying: Two-thirds of the teachers continued giving shocks until stopped by the experimenter. Yet by this point the teachers had turned the dial on the "shock" machine past a point marked *Danger: Severe Shock* to one marked simply with three large red Xs. Meanwhile, the learners had first demanded to be let free, then screamed in pain, and then eventually fell silent.

In later experiments, Milgram tested the effect of putting the learner and teacher in the same room, having the experimenter leave the room, and having other teachers—all confederates—perform the same tasks as the teacher who was really under study. Conformity was highest in the presence of the experimenter and of other teachers who appeared to go along with the experimenter, and was lowest when the teacher had to physically hold the learner's hand on the electric shock equipment.

Sadly, later experiments in the United States and elsewhere continue to find that between 61 and 66 percent of individuals will inflict pain on others if instructed to do so by an experimenter (Blass 1999; Burger 2009). These results make it easier to understand why U.S. soldiers—already trained in obedience and in group solidarity—would severely mistreat prisoners when ordered to do so in Abu Ghraib, Guantánamo, and elsewhere.

Subjects in the Milgram experiments were ordered to administer electric shocks that they believed were dangerous to others. Although subjects found the experience stressful—note the subject's clenched fist in this photo—most obeyed orders.

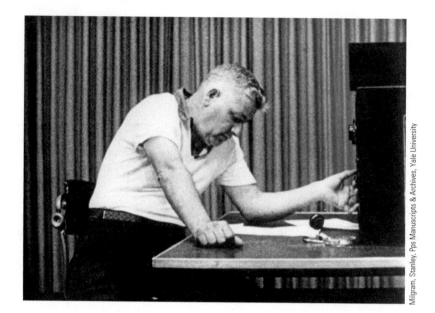

Milgram, Stanley, Pps Manuscripts & Archives, Yale University

Understanding Small Group Conformity

In both the Asch and Milgram experiments, some subjects probably became convinced that they just couldn't see the lines clearly or that giving electric jolts to experimental subjects was acceptable. Others probably went along not because they were persuaded by the group but because they decided not to make waves. When the object being judged is subjective—whether Jennifer Hudson is better than Britney Spears, or football more interesting than basketball—the group is likely to influence not only public responses but also private views. Whether we go along because we are really convinced or because we are avoiding the hassles of being different, we all have a strong tendency to conform to the norms and expectations of our groups.

Yet small groups rarely have access to legal or formal sanctions—they usually can't throw those who disagree with them in jail or the like—so why do individuals so often go along with the group's opinions? First, all of us like to believe that we understand what's going on in the world around us. But this isn't always easy. A simple thermometer can tell you whether or not the temperature outside is above 90 degrees, but there's no way to know whether Iran will bomb Israel in the next year, for example. Individuals are especially likely to adopt group views when they are not sure their own knowledge or views are correct (Levine 2007). Second, individuals adopt group views because they fear being rejected by others if they don't (Levine 2007). The major weapons that groups use to punish nonconformity are ridicule and contempt, but their ultimate sanction is exclusion from the group. From "you're fired" to "you can't sit at our lunch table anymore," exclusion is one of the most powerful threats we can make against others. This form of social control is most effective in cohesive groups, but the Asch and Milgram experiments show that fear of rejection and embarrassment can induce conformity even among strangers.

Group Decision Making

One of the primary research interests in the sociology of small groups is how group characteristics (size, cohesion, and so on) affect group decision making. This research has focused on a wide variety of actual groups: flight crews, submarine crews, protest organizers, business meetings, and juries, to name a few (e.g., Gastil, Burkhalter, & Black 2007; Ghaziani & Fine 2008).

Generally, groups strive to reach consensus; they would like all their decisions to be agreeable to every member. As the size of the group grows, consensus requires lengthy and time-consuming interaction so that everybody's objections can be clearly understood and incorporated. Thus, as groups grow in size, they often adopt the more expedient policy of majority rule. This policy results in quicker decisions, but often at the expense of individual satisfaction. It therefore reduces the cohesiveness of the group.

Choice Shifts

One of the most consistent findings of research on small groups is the tendency for group members' opinions to converge (or become more similar) over time. For example, in one experiment, the experimenter first asked several subjects to take a seat in a darkened room (Sherif 1936). The experimenter then flashed a dot of light on the front wall of the room and asked each subject to record his or her estimate of how far the dot moved during the experimental period. In reality, the dot didn't move at all. Afterwards, the experimenter asked the participants to share their answers. These answers varied considerably. Then the experimenter repeated the experiment four times. Each time the estimates grew closer. The final estimate given by each participant closely approximated the average of all participants' initial estimates.

Although groups typically move toward convergence, they do not always converge on a middle position. Instead, groups may reach consensus on an extreme position. This is called the *risky shift* when the group converges on a risky option and the *tame shift* when the choice is extremely conservative. Sometimes these choice shifts depend on persuasive arguments put forward by one or more members, but often they result from general norms in the group that favor either conservatism or risk (Davis & Stasson 1988; Jackson 2007). For example, one might expect a church steering committee to choose the safest option and a terrorist group to choose the riskiest option.

A special case of choice shift is *groupthink* (Janis 1982; Street 1997; Jackson 2007). **Groupthink** refers to situations in which the pressures to agree are so strong that they stifle critical thinking. For example, sociologist Diane Vaughan (1996) showed how groupthink contributed to the tragic 1986 explosion of the space shuttle *Challenger*. The engineers working on the *Challenger* all knew before the launch that the shuttle's O-rings probably would suffer some damage. But political pressures to launch the shuttle, coupled with a culture within NASA that rewarded risk taking, created a situation in which the engineers essentially convinced each other that the O-ring had little chance of failing. As this example illustrates, groupthink often results in bad decisions.

Social Networks

Each of us belongs to a variety of primary and secondary groups. Through these group ties we develop a **social network**. This social network is the total set of relationships we have. It includes our family, our insurance agent, our neighbors, some of our classmates and co-workers, and the people who belong to our clubs. Our social networks link us to hundreds of people in our communities and perhaps across the country and around the world.

Groupthink exists when pressures to agree are strong enough to stifle critical thinking.

A **social network** is an individual's total set of relationships.

focus on A GLOBAL PERSPECTIVE

Talking about AIDS in Mozambique

In addition to offering us friendship, job prospects, and help studying for exams, social networks have the potential to save our lives. This was a topic explored by sociologists Victor Agadjanian and Cecilia Menjívar in their research on AIDS in Mozambique, a country in southern Africa. Mozambique is among the poorest countries in the world and has one of the highest rates of infection with HIV, the virus that causes AIDS (Agadjanian and Menjívar 2002).

Because of Mozambique's poverty, most residents have very little access to information of any sort: Many live on scattered family farms where they rarely see a newspaper or interact even with neighbors, and few own even a radio, let alone a television or computer. Agadjanian and Menjívar's research examined how church membership affected individuals'

access to information about AIDS. Interestingly, they did *not* focus on the effect of religious beliefs or practices. Instead, they studied church congregations as *social networks*. Agadjanian and Menjívar found that Mozambique's churches divided into two basic types: large, *mainline* churches affiliated with international denominations such as Methodists, and smaller, *peripheral* churches that evolved in Africa and hold Pentecostal-type beliefs. Mainline churches offered a broad network of weak ties, while peripheral churches offered more strong ties.

The researchers found that both types of church memberships and social ties improved individuals' access to information about AIDS. Because mainline churches included doctors, nurses, and other educated people, the *weak* ties among members gave everyone access to relatively good information about AIDS. In addition, due to mainline churches' relatively liberal religious

views, they were willing to host occasional events for members on AIDS education.

On the other hand, the *strong* ties between members of *peripheral* churches made it easier for these individuals to talk about the need to prevent infection. For example, one peripheral church member explained:

> Those who aren't religious are at greater risks [of contracting HIV] because they have no one who can advise and tell them 'Hey, beware of AIDS.' Because here, among us, when I see that something's wrong, I say 'You go out [and have sex] at night... and get involved with women—you'll rot. Didn't you see what happened to so-and-so? He was buried because of AIDS. Hmm!'

Another told a researcher how he and his friends talk about AIDS during breaks in church services or while walking home from church: "We say, 'Hey, to protect yourself from AIDS you

Your social network does not include everybody with whom you have ever interacted. Many interactions, such as those with some classmates and neighbors, are so superficial that they cannot truly be said to be part of a relationship at all. Unless contacts develop into personal relationships that extend beyond a brief hello or a passing nod, they would not be included in your social network.

Social networks serve vital functions for individuals and for society. Strong social networks lead to lower risks of suicide and depression, better health, and longer life expectancy (Bearman & Moody 2004; Smith & Christakis 2008). They also increase the odds that individuals will care about and participate in political and civic issues (Putnam 2000; Wellman 1999). Thus the study of social networks is an important part of sociology.

Strong and Weak Ties

Although our insurance agent and our mother are both part of our social network, there is a qualitative difference between them. We can divide our social networks into two general categories of intimacy: *strong ties* and *weak ties*. **Strong ties** are relationships characterized by intimacy, emotional intensity, and sharing. **Weak ties** are relationships characterized by low intensity and emotional distance (Granovetter 1973). Co-workers, neighbors, fellow club members, distant cousins, and in-laws generally fall in this category. If you and the person you sit next to in class often chat about how you spent the weekend, and occasionally trade notes,

Strong ties are relationships characterized by intimacy, emotional intensity, and sharing.

Weak ties are relationships characterized by low intensity and lack of intimacy.

should stay with one girlfriend, use condoms. If you play a lot, have six, seven girlfriends, you won't even know how you'll get infected.'"

As these quotes suggest, even though both mainline and peripheral churches officially supported only abstinence or marital fidelity as a means of preventing

AIDS, membership in either type of church increased individuals' exposure to the idea that condoms could also reduce their risk of infection.

MAP 5.1: Number of Persons Infected with HIV per 1,000 Residents, Ages 15 to 49
The rate of people infected with HIV (the virus that causes AIDS) is far higher in southern Africa (including Mozambique) than anywhere else in the world. In contrast, in the United States only 6 people per 1,000 are infected.
SOURCE: Population Reference Bureau (2008)

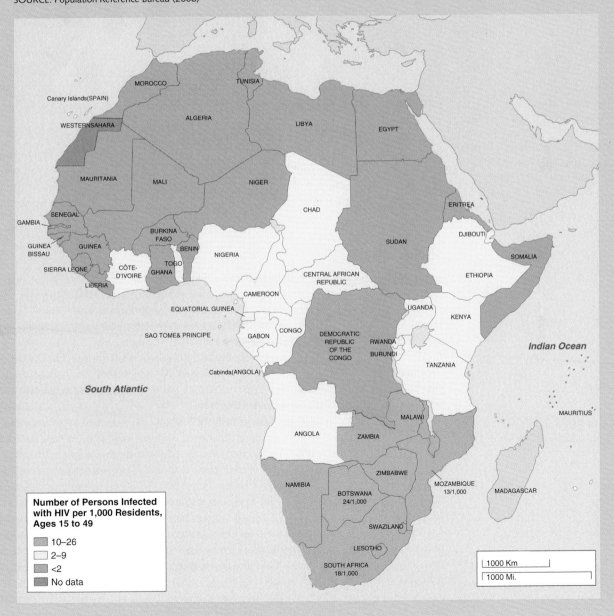

Strong ties to close friends and family are crucial for social life. All of us depend heavily on those with whom we have strong ties, in both good times and bads.

but never get together outside of class, you have a weak tie. If the two of you often hang out together, and you'd feel comfortable asking him or her for advice on your romantic relationships, you have a strong tie. In sum, strong ties *bond* us to those who are close to us, and weak ties *bridge* the gap between us and others with whom we are less closely tied.

Strong Ties

Strong ties are crucial for social life. If you are sick, or broke, or your car breaks down just when you need to get to campus for a final exam, it is your strong ties you will call on for help. These are the people who care the most about you, and whom you are most likely to care deeply about. Strong ties give us emotional support, financial help, and all sorts of practical aid when needed. However, strong ties can't always be relied on: When those you turn to are also financially or emotionally stressed to the limit, they may not be able to give you the help you need (Menjívar 2000). Not surprisingly, this problem is most severe among poor people, who need the most assistance but whose strong ties are least able to afford to help.

Across socioeconomic groups, Americans' strong ties decreased dramatically between 1985 and 2004 (McPherson, Smith-Lovin, & Brashears 2006). When, in 1985, a national random sample of Americans were asked to name the people with whom they had discussed matters important to them during the previous six months, the most common response was to give three names. When the question was repeated with a similar sample in 2004, the most common response was "No one." This is a dramatic shift in only 19 years. In both surveys, the most common confidants were friends and spouses, but reliance on friends declined while reliance on spouses increased. Most telling, respondents were far less likely in 2004 to report that they turned to parents, children, siblings, co-workers, neighbors, or co-members of groups.

Several factors affect the number and composition of strong ties (McPherson, Smith-Lovin, & Brashears 2006). The most important of these factors is education.

People with more education have more strong ties, have a greater diversity of strong ties, and rely less on kinship ties. People with more education are also more likely to have strong ties to influential people—lawyers and doctors rather than plumbers and mechanics. Nonwhites have fewer strong ties than do whites, especially with regard to kinship ties. Neither age nor gender affects the average number or type of strong ties (McPherson, Smith-Lovin, & Brashears 2006). Decoding the Data: Strong Ties explores these issues further.

decoding the data

Strong Ties

Periodically, surveys ask Americans the number of people during the last six months with whom they had discussed matters important to them. The number who answered *zero* has increased substantially over time and is more common among some groups than among others.

SOURCE: General Social Survey. **http://sda.berkeley.edu**. Accessed May 2009.

Percentage Who Discussed Matters Important to Them with No One	
Race	
Whites	19.9
African Americans	37.6
Years of education	
0–8 years education	30.6
9–12 years	26
13 or more	20.1
Sex	
Males	24.2
Females	21.4
Year of survey	
1985	8.3
2004	22.6

Explaining the Data: Can you think of any sociological (not personality) reasons why African Americans are more likely than whites to have no one with whom they discuss important matters? What might explain why people with more education have more confidants? Why males have fewer confidants than do females? How did family life, work life, and social life change between 1985 and 2004? How might those changes have led to a decrease in confidants?

Critiquing the Data: Is the ability to discuss important matters with others a good way to measure strong ties? Can you think of any other measure that might better capture the nature of strong ties among African Americans, males, or less educated persons?

sociology and you

Being a college student affects your strong and weak ties. If you moved from home to go to college, your strong ties to family and high school friends probably weakened, especially if you moved far away. If you belong to a fraternity or sorority or live in a dorm, you have certainly added more weak ties and probably more strong ties as well. Moreover, your new ties to college students may serve you well in the future, as these new friends are likely to enter professional careers and to be good resources for you in many ways.

FIGURE 5.2 Jill's Ties and Groups
Everyone belongs to both primary and secondary groups. Within these groups we each have both weak and strong ties. Jill has strong ties to four of her sorority sisters, three of her classmates, her boyfriend, her father, and her brother. She has weak ties to her mother, her sister, her other sorority sisters, and her other classmates.

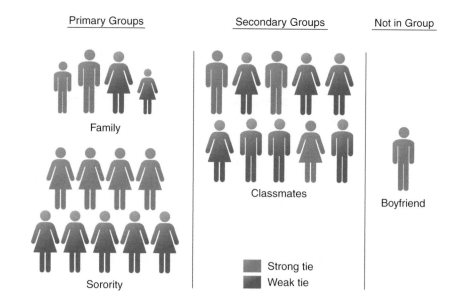

Weak Ties

Weak ties are also important to social life. For example, research indicates that many people first hear about jobs and career opportunities through weak ties (Granovetter 1974; Newman 1999b). In this and other instances, the more people you know, the better off you are.

As this suggests, weak ties are crucial whenever you need to learn or obtain something that requires a broad network. If you have a question about Microsoft Word, for example, you may well have a strong tie with someone who can answer it. If you have a question about Linux software, however, you'll probably need to turn to your large network of weak ties to find an answer.

One of the best sources of weak ties is the Internet: If you have a rare disease, enjoy an unusual hobby, or love an obscure band, you can easily create weak ties with others who share your needs or interests.

Ties versus Groups

The distinction between strong and weak ties obviously parallels the distinction between primary and secondary groups. The difference between these two sets of concepts is that strong and weak apply to one-to-one relationships, whereas primary and secondary apply to the group as a whole. We can have both strong and weak ties within the primary as well as the secondary group. (See Figure 5.2 for an illustration.)

For example, the family is obviously a primary group; it is relatively permanent, with strong feelings of loyalty and attachment. We are not equally intimate with every family member, however. We may be very close to our mother but estranged from our brother. Similarly, although the school as a whole is classified as a secondary group, we may have developed an intimate relationship, a strong tie, with one of our schoolmates. *Strong* and *weak* are terms used to describe the relationship between two individuals; *primary* and *secondary* are characteristics of the group as a whole.

Voluntary Associations

In addition to relationships formed with individuals, many of us voluntarily choose to join groups and associations. We may join a Bible study group, a soccer team, the Elks, or the Sierra Club. These groups, called **voluntary associations**, are nonprofit organizations designed to allow individuals an opportunity to pursue their shared interests collectively. They vary considerably in size and formality. Some—for example, the Elks and the Sierra Club—are very large and have national headquarters, elected officers, formal titles, charters, membership dues, regular meeting times, and national conventions. Others—for example, soccer teams and knitting groups—are small, informal groups that draw their membership from a local community or neighborhood.

Functions of Voluntary Associations

Voluntary associations are an important mechanism for enlarging our social networks. Most of the relationships we form in such associations will be weak ties. But voluntary associations also can introduce us to people with whom we will develop strong ties as close friends and intimates.

Voluntary associations perform an important function for individuals. Studies document that people who participate in them generally report greater personal happiness, longer life, more political participation, and a greater sense of community (Stalp, Radina, and Lynch 2008; Borgonovi 2008; Walker 2008; McFarland & Thomas 2006).

The correlation between high participation and greater satisfaction does not necessarily mean that joining a voluntary association is the road to happiness. At least part of the relationship between participation and happiness is undoubtedly due to the fact that happy people who feel politically effective and attached to their communities are more likely than others to join voluntary associations. It also appears to be true, however, that greater participation can be an avenue for achievement and can lead to feelings of integration and satisfaction.

Participation in Voluntary Associations

Although some social critics have argued that membership in U.S. voluntary associations has declined—a thesis popularized in the book *Bowling Alone*, by Robert Putnam (2000)—most observers believe that, if anything, participation has increased (Rich 1999). It is true that some large voluntary associations, such as the Elks and bowling leagues, have seen declines in membership. Other groups, however, are burgeoning, especially small local associations, groups focused on ethnicity or gender issues, alternative religious organizations, and Internet-based groups (Rich 1999).

Americans belong to an average of two voluntary associations, considerably above the average for industrialized nations (Curtis, Baer, & Grabb 2001). Among those who report membership, a large proportion are passive participants who belong in name only. They buy a membership in the Parent-Teacher Association (PTA) when pressured to do so, but they don't go to meetings. Similarly, anyone who subscribes to *Audubon* magazine is automatically enrolled in the local Audubon Club, but few subscribers become active members. Because so many of our memberships are superficial, they are also temporary. Nevertheless, most people in the United States maintain continuous membership in at least one association.

Membership in voluntary associations is highest among middle-aged, married, well-educated, and middle-class individuals (Curtis, Grabb, & Baer 1992). In addition, having school-age children draws both men and women into youth-related groups and

Voluntary associations are nonprofit organizations designed to allow individuals an opportunity to pursue their shared interests collectively.

Joining an amateur baseball league or other voluntary association is guaranteed to increase our network of weak ties. If we become close friends with any fellow members or teammates, we also increase our network of strong ties.

© 2009 Jupiterimages.

so increases voluntary association membership (Rotow 2000). Interestingly, marriage increases men's participation in associations but not women's, primarily by drawing men into church-related groups. Conversely, full-time employment increases women's participation but not men's, primarily by drawing women into job-related groups. Taken together, these findings suggest that individuals are more likely to participate in voluntary associations when their neighborhood, work, children, or some other aspect of their lives provides them with opportunities to do so.

Community

In everyday life, we often hear about the benefits of having "community." Yet we rarely hear a clear definition of what this means. According to sociologists, a **community** is a collection of individuals characterized by dense, cross-cutting social networks (Wellman 1999). A community is strongest when all members connect to one another through complex overlapping ties.

Yet network ties need not be strong to have important consequences for individuals and the community. For example, research shows that even when neighbors share only weak ties, they often help each other in many ways—loaning tools, picking up the mail when a family is out of town, and the like (Wellman & Wortley 1990). Similarly, neighborhoods experience substantially less crime and delinquency when neighbors enjoy weak ties and so believe they have both the right and the obligation to sanction teenagers who throw trash, shout profanities, or otherwise misbehave (Sampson & Raudenbush 1999; Sampson, Morenoff, & Earls 1999; Sampson, Morenoff, & Gannon-Rowley 2002).

Computer Networks and Communities

With the exponential rise in use of the Internet, many individuals now seek and find online networks and communities (DiMiaggio et al. 2001; Wellman 1999; Wellman et al. 1996). The Internet's potential for promoting both strong and weak ties has been

A **community** is a collection of individuals characterized by dense, cross-cutting social networks.

Participating in computer games can increase social networks by helping each individual make new friends and by cementing existing friendships.

© Mark Peterson/Corbis

most impressively demonstrated by the spectacular rise of "social networking" sites such as Twitter, Facebook, and MySpace. Similarly, online discussion groups and chat rooms allow anyone to quickly send out a comment or request to a large and diverse audience. Although the quality of information and relationships obtained via the Internet can vary widely, the Internet does provide a wide network of weak ties to many people who might otherwise be isolated. Furthermore, because individuals often forward the comments or requests they receive to others, this network of weak ties can grow both broadly and quickly.

Although less common, online networks also can provide strong ties and a true sense of community. Even when individuals initially enter online groups simply to obtain information, those who stay typically do so because they enjoy the social support, companionship, and sense of community the group offers. Relatively strong online communities can form over anything from organizing political efforts to writing and sharing personal journals, each group fulfilling a different combination of instrumental and expressive functions. Many of the most popular online groups link people who share a health problem. Within these groups, individuals share not only suggestions regarding medical treatment but also their fears, sorrow, and triumphs as they grapple with their injuries or illnesses.

Interestingly, even participating in video and computer games—seemingly a highly individual activity—can *increase* individuals' social networks. This topic is explored more fully in Focus on Media and Culture: Gaming and Social Life on the next page.

Complex Organizations

Few people in our society escape involvement in large-scale organizations. Unless we are willing to retreat from society altogether, a major part of our lives is organization-bound. Even in birth and death, large, complex organizations (such as hospitals and

focus on MEDIA AND CULTURE

Gaming and Social Life

According to various surveys, about two-thirds of college students play computer or video games at least occasionally and half of teenagers play a video game daily (Lenhart et al. 2008; Jones 2003). Many adults react to data like these with horror: Why, they ask, are young people spending hours sitting by themselves and staring at screens? And aren't these young people losing their connection to people and to society when they do so?

The short answer is probably no. Rather than isolating individuals from the social life around them, gaming is often a highly social pastime. In surveys, most young gamers report that gaming either increases or doesn't affect the time they spend with family and friends. Even frequent gamers spend no less time interacting with friends than do others. Instead, they report, gaming—whether online or offline—has helped them make new friends and cement existing friendships (Lenhart et al. 2008; Jones 2003). Both in their bedrooms and in college computer labs, young people often find that trading

tips on new games or game strategies is a great way to share time with friends or start a conversation with a potential new friend (Lenhart et al. 2008; Jones 2003). Students also interact with other friends and potential friends in online message boards or using chat options on interactive, multi-player games. Gaming also can offer a low-key way for young people to "hang out" with parents and other adult relatives—and to get a chance to shine whenever they can help an older relative understand how to use a game. Thus gaming, it seems, more often increases rather than decreases social ties.

Gaming also may increase engagement with the broader society (or *civic engagement*). One of the most important predictors of whether individuals become active in their communities and society is whether they have opportunities to engage in activities that press them to think about others and about the greater good, such as helping others and debating ethical issues. Simulation games such as *The Sims* most obviously provide these opportunities, but even a violent, sexist, racist game can give gamers the opportunity to help other gamers. Young people

who play games that give them opportunities for thinking about others and about the greater good are significantly more likely than other gamers to raise money for charity, participate in a protest, seek information online about current events, or persuade others to vote for a specific candidate—all measures of civic engagement (Lenhart et al. 2008). Unfortunately, these data can't tell us which is cause and which is effect: Does gaming lead to civic engagement, or are more-engaged students drawn to certain sorts of games? Regardless, it does seem clear that gaming does not *reduce* civic engagement.

Similarly, players of *The Sims* often adopt avatars (online alter-identities) that differ greatly from their real-life identities. Although players could use these avatars to explore all sorts of behaviors that they would never consider in real life, instead they typically have their avatars obey everyday norms for politeness and courtesy, essentially acting the same as they would if invited to dinner at someone else's home (Martey & Stromer-Galley 2007). Thus playing *The Sims* reinforces rather than challenges basic rules of social life.

---

vital statistics bureaus) make demands on us. Throughout the in-between years, we are constantly adjusting to organizational demands.

Sociologists use the term **complex organizations** to refer to large, formal organizations with elaborate status networks (Handel 2002). Examples include universities, governments, corporations, churches, and voluntary associations such as fraternities or the Kiwanis Club.

These complex organizations make a major contribution to the overall quality of life within society. Because of their size and complexity, however, they don't supply the cohesion and personal satisfaction that smaller groups do. In fact, members often feel as if they are simply cogs in the machine rather than important people in their own right. This is nowhere more true than in a bureaucracy.

Bureaucracy is a special type of complex organization characterized by explicit rules and a hierarchical authority structure, all designed to maximize efficiency. In popular usage, bureaucracy often has a negative connotation: red tape, silly rules, and unyielding rigidity. In social science, however, it is simply an

Complex organizations are large, formal organizations with elaborate status networks.

Bureaucracy is a special type of complex organization characterized by explicit rules and hierarchical authority structure, all designed to maximize efficiency.

organization in which the roles of each actor have been carefully planned to maximize efficiency.

The "Ideal Type" of Bureaucracy: Weber's Theory

Most large, complex organizations are bureaucracies: IBM, the federal government, U.S. Steel, the Catholic Church, colleges, and hospitals. The classic description of an "ideal type" of bureaucracy was outlined a century ago by Max Weber ([1910] 1970a). By "ideal type," Weber did *not* mean that this is the *best* form of bureaucracy, merely that it is what bureaucracies are *expected* to be like. According to Weber, bureaucracies are expected to be characterized by the following:

1. *Division of labor.* Bureaucratic organizations employ specialists in each position and make them responsible for specific duties. Job titles and job descriptions specify who is to do what and who is responsible for each activity.

2. *Hierarchical authority.* Positions are arranged in a hierarchy so that each one is under the control and supervision of a higher position. Frequently referred to as chains of command, these lines of authority and responsibility are easily drawn on an organization chart, often in the shape of a pyramid.

3. *Rules and regulations.* All activities and operations of a bureaucracy are governed by abstract rules or procedures. These rules are designed to cover almost every possible situation that might arise: hiring, firing, and the everyday operations of the office. The object is to standardize all activities.

4. *Impersonal relationships.* Theoretically, interactions in a bureaucracy are guided by rules rather than by personal feelings, with the goal of eliminating favoritism and bias.

5. *Careers, tenure, and technical qualifications.* Candidates for bureaucratic positions are supposed to be selected on the basis of technical qualifications such as education, experience, or high scores on civil service examinations. Once selected for a position, individuals should advance in the hierarchy by means of achievement and seniority, and should be able to keep their jobs as long as their performance holds up.

6. *Efficiency.* Bureaucratic organizations are intended to maximize efficiency by coordinating the activities of a large number of people in the pursuit of organizational goals. From the practice of hiring on the basis of credentials rather than personal contacts to the rigid specification of duties and authority, the whole system is constructed to keep individuality, whim, and favoritism out of the operation of the organization.

Weber realized that few if any bureaucracies totally meet this description: Workers often must do tasks beyond those they are assigned; lines of authority are sometimes unclear; environments often change before new rules evolve to deal with those changes; biases like sexism and racism certainly can lead to the hiring of unqualified or less qualified persons; and organizations can, at times, be wildly inefficient. In addition, over the last quarter century American corporations, a major form of bureaucracy, have downsized and now contract out many services. In this new environment, workers have less guarantee of tenure, and corporations can't be as hierarchical, since they can't exert as much control over contracted workers—especially if the workers live half a world away (Scott 2004). Still, Weber's list of bureaucratic characteristics helps us understand the *expected* role and nature of bureaucracies.

Bureaucracies like McDonalds depend on hierarchical (top-down) control, a clear division of labor between different types of workers, and strict rules for how each type of worker should do his or her job.

Matthias Schrader/dpa/Landov

Real Bureaucracies: Organizational Culture

Weber's classic theory of bureaucracy almost demands not individuals, but robots who will follow every rule to the letter. Yet when workers really do follow every rule, no matter how nonsensical or unnecessary, work quickly grinds to a halt. In fact, in cities where police cannot legally strike, police unions sometimes instead protest through "slowdowns," in which officers follow every rule for the purpose of throwing the system into chaos. Not surprisingly, therefore, few organizations try to be totally bureaucratic. Instead, they strive to create an atmosphere of goodwill and common purpose among their members so that they all will apply their ingenuity and best efforts to meeting organizational goals (Kunda 1993). This goodwill is as essential to efficiency as are the rules.

Sociologists use the term **organizational culture** to refer to the pattern of norms and values that structures how business is actually carried out in an organization (Kunda 1993). The key to a successful organizational culture is cohesion, and most organizations strive to build cohesion among their members. They do this by encouraging interaction and loyalty among employees, and by such tactics as providing lunchrooms; sponsoring after-hours sports leagues and company picnics; and promoting unifying symbols such as company mascots. For example, Google is famous for such on-site "perks" as free massages, free gourmet meals, and volleyball courts. But many other organizations use less expensive versions of the same strategies. When organizational managers succeed at motivating loyalty, workers may be willing to skip vacations, work extremely long hours, and sacrifice time with family and friends; it is not uncommon for workers at "fun" companies like Google to work 12, 15, or even 24 hours a day to meet a deadline.

Compared with a business like Chrysler Motor Company, businesses like Google also stand out for their emphasis on flexibility and informal decision making. Why would this be so? Companies that develop software require creativity from their

Organizational culture refers to the pattern of norms and values that structures how business is actually carried out in an organization.

employees and must change strategies rapidly in response to changes in the broader environment and changes made by their competitors. In contrast, changes came slowly to the factory line at Chrysler—which may partly explain why it was forced to file for bankruptcy in 2009. The degree of bureaucratization in an organization is related to the degree of uncertainty in the organization's activities. When activities tend to be routine and predictable, the organization is likely to emphasize rules, central planning, and hierarchical chains of command. This explains why, for example, classrooms tend to be less bureaucratic and factories more bureaucratic.

Critiques of Bureaucracies

Bureaucracy is the standard organizational form in the modern world. Organizations from churches to governments are run along bureaucratic lines. Yet despite the widespread adoption of this organizational form, it has several major drawbacks. Three of the most widely acknowledged are as follows:

1. *Ritualism.* Rigid adherence to rules may mean that a rule is followed regardless of whether it helps accomplish the purpose for which it was designed. The rule becomes an end in itself rather than a means to an end. For example, individuals may struggle to arrive at 8 A.M. and leave at 4 P.M. when they could work more effectively from 10 to 6. Although the existence of a bureaucracy per se doesn't always breed rote adherence to rules (Foster 1990), an overemphasis on bureaucratic rules can stifle initiative and prevent the development of more efficient procedures.

2. *Alienation.* The emphasis on rules, hierarchies, and impersonal relationships can sharply reduce the cohesion of the organization. Reduced cohesion results in several drawbacks: It reduces social control, reduces member satisfaction and commitment, and increases staff turnover. All of these may interfere with the organization's ability to reach its goals.

3. *Structured inequality.* Critics charge that the modern bureaucracy with its multiple layers of authority is a profoundly antidemocratic organization. Bureaucracies concentrate power in the hands of a few people, whose decisions then pass down as orders to subordinates.

In addition to these concerns, more recent criticism has focused on the dangers of *McDonaldization* (Ritzer 1996). **McDonaldization** refers to the process through which a broad range of bureaucracies adopt management goals derived from the fast-food restaurant industry.

Not surprisingly, the McDonald's restaurant chain exemplifies the central management goals of McDonaldization: efficiency, calculability, predictability, and control. McDonald's streamlined its procedures to serve customers extremely rapidly (efficiency) and shifted from advertising how good its burgers taste (something that can't be measured) to advertising how many ounces of meat the burgers contain (calculability). McDonald's guarantees that a Big Mac in New York tastes exactly like a Big Mac in Des Moines (predictability). And each McDonald's restaurant requires its employees to follow strict guidelines for work procedures (control) and pressures customers to order and leave quickly by offering limited menus and uncomfortable seats (control and efficiency). These principles have now been adopted by all kinds of bureaucracies around the world, from drop-off laundries to "telephone-sex" businesses.

Ironically, the attempt to rationalize bureaucratic structures through McDonaldization often produces *irrational* consequences for the society as a whole. The disadvantages of McDonaldization stem directly from each supposed advantage.

McDonaldization is the process by which the principles of the fast-food restaurant—efficiency, calculability, predictability, and control—are coming to dominate more sectors of American society.

For instance, although it is more efficient for businesses to use voice-mail systems instead of operators, it is less efficient for the customers who must listen to a series of menus, hoping that they will reach the department they seek before getting disconnected. Moreover, businesses lose customers when customers hang up in frustration after getting lost in voice-mail mazes. Similarly, businesses like McDonald's make decisions based on calculations of how they can best generate a profit, but they do not calculate the impact of their business decisions on the environment or on the quality of life of their customers or workers. The predictability that chain stores and restaurants offer makes the world a less interesting place, as large national businesses drive out unique local businesses. And the control that McDonaldized organizations offer is frequently dehumanizing—something you have probably experienced every time your name and identity have been replaced by an institutional identification number.

Where This Leaves Us

Humans are social beings. We live our lives within relationships, groups, networks, and—whether we like it or not—complex organizations. Without these human connections we cannot survive, let alone thrive. Groups, networks, and organizations help us obtain the very basics of life—food, clothing, work, shelter, companionship, love. They also enable us to make our mark on the world, as we raise children within families, create better communities through voluntary associations, strive for success within complex organizations from schools to corporations, and so on. Yet working with others also carries risks, for exchange and cooperation can turn into competition and conflict, and groups can affect our ideas and behaviors in ways we may not even recognize. A sociological understanding of groups, networks, and organizations an help us understand, prevent, and, where necessary, counteract these effects.

Summary

1. Relationships are characterized by four basic social processes: exchange, cooperation, competition, and conflict.
2. Groups differ from crowds and categories in that group members take one another into account, and their interactions are shaped by shared expectations and interdependency.
3. Reference groups are groups that individuals compare themselves to regularly. Relative deprivation—which occurs when we compare ourselves to others who are better off than we are—can reduce happiness.
4. Primary groups are characterized by intimate, face-to-face interaction. They are essential to individual satisfaction and integration, and they are also primary agents of social control in society. Secondary groups are large, formal, and impersonal. They are generally task oriented and perform instrumental functions for societies and individuals.
5. Group size, proximity, and communication patterns all affect group interaction. Group interaction can lead to conformity and consensus among group members, sometimes around obviously incorrect decisions. The amount of interaction in turn affects group cohesion.
6. Each person has a social network that consists of both strong and weak ties. The number of strong ties is generally greater for individuals who are white and who have more years of education.
7. Strong ties are the people we can count on when we really need help of some sort. Weak ties, however, are more useful when we need to reach out to a broad social network, such as when searching for work.

8. Voluntary associations are nonprofit groups that bring together people with shared interests. They combine some of the expressive functions of primary groups with the instrumental functions of secondary groups.

9. When individuals are linked by dense, cross-cutting networks, they form a community. Communities have important influences on members, even when social ties within the community are relatively weak.

10. Complex organizations are large, formal organizations with elaborate status networks. Bureaucracies are complex organizations whose goal is to maximize efficiency. Bureaucracies are expected to be characterized by a division of labor; hierarchical authority; rules and regulations; impersonal social relations; an emphasis on careers, tenure, and technical qualifications; and an emphasis on efficiency.

11. Although most contemporary organizations are built on a bureaucratic model, many are far less rational than the classic model suggests. Critics of McDonaldization suggest that the bureaucratic emphasis on rationality can have irrational consequences. In addition, all effective bureaucracies must rely on organizational culture to inspire employees to give their best efforts and to help meet organizational goals.

Thinking Critically

1. Do social networking sites like Facebook serve as primary groups or secondary groups? Do they enforce group conformity? Explain, with examples.

2. Can you think of a situation in your life in which your behavior was more affected by a secondary than by a primary group?

3. Suppose you were trying to get help for a family member's substance-abuse problem. What would be the advantage of turning to your strong ties? your weak ties?

4. From your experience, what are some of the functions of bureaucracy? What are some of the problems?

Book Companion Website

www.cengage.com/sociology/brinkerhoff
Prepare for quizzes and exams with online resources—including tutorial quizzes, a glossary, interactive flash cards, crossword puzzles, essay questions, virtual explorations, and more.

CHAPTER 6

Deviance, Crime, and Social Control

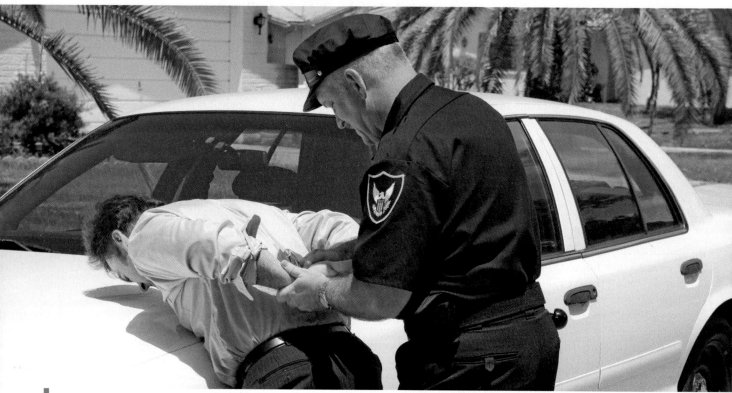

Image copyright Lisa F. Young, 2009. Used under license from Shutterstock.com.

Conformity and Deviance

Understanding Conformity
Defining Deviance

Theoretical Perspectives on Deviance

Structural-Functional Theories
Conflict Theory
Symbolic Interaction Theories
Case Study: Medicalizing Deviance

Crime

Property Crimes and Violent Crimes
Victimless Crimes
White-Collar Crimes
Correlates of Crime: Age, Sex, Class, and Race
Fear of Crime

The Criminal Justice System

Why Punish?
The Police

The Courts
Prisons
Other Options

Where This Leaves Us

Conformity and Deviance

In providing a blueprint for living, our culture supplies norms and values that structure our behavior. These norms and values tell us what we ought to believe in and what we ought to do. Because we are brought up to accept them, for the most part we do what we are expected to do and think as we are expected to think. Only "for the most part," however, because none of us follows all the rules all the time.

Previous chapters concentrated on how norms and values structure our lives and how we learn them through socialization. This chapter considers some of the ways individuals break out of these patterns—from merely eccentric behaviors to serious violations of others' rights.

Understanding Conformity

To understand why people *break* social norms, we first must understand why most people, most of the time, conform. The forces and processes that encourage conformity are known as **social control**. Social control takes place at three levels:

- Through internalized self-control, we police ourselves.
- Through informal controls, our friends and intimates reward us for conformity and punish us for nonconformity.
- Through formal controls, the state or other authorities discourage nonconformity.

Self-control occurs because individuals internalize the norms of their group, making them part of their basic belief system and their very identity. Most of us do not murder, rape, or rob, not because we fear arrest but because it would never occur to us to do these things; they would violate our sense of self-identity.

This self-control is reinforced by **informal social control**: all the small and not-so-small ways that friends, co-workers, and others around us informally keep us from behaving improperly. Thus, even if your own values do not prevent you from breaking into your professor's office to steal the answers to your midterm test, you might decide against doing so because you fear how others will respond if they find out. Your friends might consider you a cheat, your family would be disappointed in you, your professor might publicly embarrass you by denouncing you to the class.

If none of these considerations is a deterrent, you might be scared into conformity by the thought of **formal social controls**: administrative sanctions such as fines, expulsion, or imprisonment. Those who steal test answers, for example, face formal sanctions such as automatic failing grades, loss of scholarships, and dismissal from school.

Whether we are talking about cheating on examinations or murder, social control rests largely on self-control and informal social controls. Few formal agencies have the ability to force compliance to rules that are not supported by individual or group values. Sex is a good example. In many states, sex between unmarried persons is illegal, and theoretically you could be fined or imprisoned for it. Even if the police devoted substantial effort to stamping out illegal sex, however, they would probably not succeed. These days, relatively few unmarried adults feel ashamed about having sexual relations—some even brag about it, and some find that their friends cheer them on. In such conditions, formal sanctions cannot enforce conformity. Prostitution, marijuana use, underage drinking—all are examples of situations in which laws unsupported by public consensus have not produced conformity.

Social control consists of the forces and processes that encourage conformity, including self-control, informal control, and formal control.

Informal social control is self-restraint exercised because of fear of what others will think.

Formal social controls are administrative sanctions such as fines, expulsion, or imprisonment.

Defining Deviance

People may break out of cultural patterns for a variety of reasons and in a variety of ways. Whether your nonconformity leads others to consider you deviant or merely eccentric depends, among other things, on the seriousness of the rule you violate. If you wear bib overalls to church or carry a potted palm with you everywhere, you will be challenging the rules of conventional behavior. Probably nobody will care too much, however; these are minor kinds of nonconformity. Norm violations only become **deviance** when they exceed the tolerance level of the community and bring negative sanctions. Deviance is behavior of which others disapprove to such an extent that they believe something significant ought to be done about it.

Defining deviance as behavior of which others disapprove has an interesting implication: It is not the *act* that is important but the *audience*. The same act may be deviant in front of one audience but not another, deviant in one place but not another.

Few acts are intrinsically deviant. Even taking another's life may be acceptable in war, police work, or self-defense. Whether an act is regarded as deviant often depends on the time, the place, the individual, and the audience. For this reason, sociologists stress that *deviance is relative*. For example, alcohol use is deviant for adolescents but not for adults, having two wives is deviant in the United States but not in Nigeria, wearing a gun in town (if you are a civilian) is deviant now but wasn't 150 years ago, and wearing a skirt is deviant for American men but not for American women.

As these examples suggest, deviance can be divided into criminal and noncriminal activities. When deviance is against the law, it is crime (a subject we discuss in more detail later in this chapter). But many types of deviance are not against the law, such as burping in public, refusing to shower for a month, or publicly declaring oneself an atheist. This topic is discussed in more detail in Focus on Media and Culture: Extreme Body Modification.

The sociology of deviance has two overarching concerns: how rules become established and why people break the rules of their time and place. In the next section, we review several major theories that address these questions.

Theoretical Perspectives on Deviance

Biological and psychological explanations for deviant behavior typically focus on how processes within the individual lead to deviance. Such theories often look for the causes of deviance in genetics, neurochemical imbalances, or childhood failures to internalize appropriate behavior or attitudes. Most sociologists agree that biology and psychology play a role in causing deviance but consider social forces even more important. Sociological theories, therefore, search for the causes of deviance within the social structure rather than within the individual (see the Concept Summary on Theories of Deviance on page 130).

Structural-Functional Theories

In Chapter 1, we said that the basic premise of structural-functional theory is that the parts of society work together like the parts of an organism. From this point of view, deviance can be useful for a society—at least up to a point. Consider spring break: It's easier to settle down to your final papers and exams in May if you got a break from

Deviance refers to norm violations that exceed the tolerance level of the community and result in negative sanctions.

Extreme Body Modification

Recent years have seen an explosion in "extreme body modification": "full-sleeve" tattoos, large piercings, brands scarred into the flesh with hot metal, and ornamental scars carved with razors or knives. Moreover, these modifications now appear on the face, neck, and other parts of the body where they are intended to be seen.

Although "everyday" tattooing—a delicate butterfly atop a woman's breast, a dragon on a man's bicep—has become increasingly accepted, extreme body modification remains a form of deviance. Many Americans consider such modifications not only unattractive but also repugnant: Western culture regards bodily fluids as contaminated and so typically stigmatizes any (nonmedical) practices that break through the skin and allow blood or pus to seep out (Pitts-Taylor 2003). The stigma is strongest against women body modifiers, since our cultural norms identify smooth skin as key to female attractiveness.

So why do people engage in extreme body modification? The practice is most common in certain subcultures, such as "modern primitives," skaters, and skinheads (Atkinson 2003; Pitts 2003).

Within these subcultures, body modifications are regarded both as attractive and as a valued sign of group membership. As one modern primitive said:

In other cultures, getting a tattoo means that you're "one of us." It's a mark of pride, a comin\g of age that no one can take away. I love that about my tattoos, I feel as if I'm a member of a tribe, one of the pack. (Atkinson & Young 2001, 129–30)

Others seek out the pain of body modification to recover their sense of control and personal strength after illness, chemotherapy, surgery, or the like. For example, one young woman described how getting a tattoo allowed her to "reclaim" her body after being raped:

I cried the whole time I was being tattooed, all of the fear, and hate, and sorrow came to the surface, and every time the needles struck me I relived the pain of the rape. I don't think any amount of talk, with whoever, could have forced me to get back in touch with my body like that.... I consider that day my second birthday, the day I really started to move on with my life. (Atkinson & Young 2001, 131)

Finally, individuals can use extreme body modification as a political statement. Modern primitives, for example, adopt large-scale Polynesian or Maori tattoos to declare their rejection of mainstream culture and their commitment to what they view as a more authentic and natural way of life.

Janine Wiedel Photolibrary/Alamy

the work in March. In addition, according to structural-functionalists, deviance can help nudge a society toward needed, incremental social changes. But when deviance becomes extreme, they argue, it is *dysfunctional* (disruptive) to the society.

This perspective was first applied to the explanation of deviance by Emile Durkheim. Durkheim recognized the potential benefits of minor deviance. In his classic study of suicide ([1897] 1951), however, he focused on the causes of dysfunctional, extreme deviance. To explore this issue, Durkheim raised the question of why people in industrialized societies are more likely to commit suicide than are people in agricultural societies. He suggested that in traditional societies the rules tend to be well known and widely supported. As a society grows larger, becomes more diverse, and experiences rapid social change, the norms of society may become unclear or no longer apply. Durkheim called this situation **anomie** and believed it was a major cause of suicide in industrializing nations.

Anomie is a situation in which the norms of society are unclear or no longer applicable to current conditions.

concept summary

Theories of Deviance

	Major Question	Major Assumption	Cause of Deviance	Most Useful for Explaining Deviance Among
Structural-Functional Theory				
Strain theory	Why do people break rules?	Deviance is an abnormal characteristic of the social structure.	A dislocation between the goals of society and the means to achieve them.	The working and lower classes who cannot achieve desired goals by prescribed means.
Conflict Theory				
Conflict Theory	How does unequal access to scarce resources lead to deviance?	Deviance is a normal response to competition and conflict over scarce resources.	Inequality and competition.	All classes: Lower class is driven to deviance to meet basic needs and to act out frustration; upper class uses deviant means to maintain its privileges.
Symbolic Interaction Theories				
Differential association theory	Why is deviance more characteristic of some groups than others?	Deviance is learned like other social behaviors.	Subcultural values differ in complex societies; some subcultures hold values that favor deviance; these are learned through socialization.	Delinquent gangs and those integrated into deviant subcultures and neighborhoods.
Deterrence theories	When is conformity not the best choice?	Deviance is a choice based on cost/benefit assessments.	Failure of sanctioning system (benefits of deviance exceed the costs).	All groups, but especially those lacking a "stake in conformity."
Labeling theory	How do acts and people become labeled deviant?	Deviance is relative and depends on how others label acts and actors.	People whose acts are labeled deviant and who accept that label become career deviants.	The powerless who are labeled deviant by more powerful individuals.

Importantly, Durkheim and later structural-functional theorists define deviance as a social problem rather than a personal trouble; it is a property of the social structure, not of the individual (Passos & Agnew 1997). As a consequence, the solution to deviance lies not in reforming the individual deviant but in changing the dysfunctional aspects of the society.

Explaining Individual Deviance: Strain Theory

The classic structural-functionalist theory of crime is Robert Merton's (1957) **strain theory**. Strain theory begins by noting that most of us are conformists, who (as Merton defined the term) accept both our society's culturally approved *goals* and its culturally approved *means* for reaching these goals. Strain theory argues that deviance results when individuals cannot reach culturally approved goals using culturally approved means. This theory is most commonly used to explain lower-class crime.

Strain theory suggests that deviance occurs when culturally approved goals cannot be reached by culturally approved means.

concept summary

Merton's Modes of Adaptation

Merton's strain theory of deviance suggests that deviance results whenever there is a disparity between goals and the institutionalized means available to reach them. Individuals caught in this dilemma may reject the goals or the means or both. In doing so, they become deviants.

Modes of Adaptation	Cultural Goals	Institutional Means
Innovation	Accepted	Rejected
Ritualism	Rejected	Accepted
Retreatism	Rejected	Rejected
Rebellion	Rejected/replaced	Rejected/replaced

sociology and you

As a college student, you are using a culturally accepted means—attending college—to achieve a culturally accepted goal—a well-paying career. In Merton's terms, you are a conformist. If you cheat on an exam to achieve your goals, Merton would consider you an innovator and your professors will consider you deviant (because you have broken *their* cultural norms). If your peers consider cheating acceptable, however, you will not be a deviant within peer culture.

American culture places strong emphasis on economic success. Although this goal is widely shared by Americans, the means to obtain it are not. Few lower-class Americans are able to achieve success through culturally approved means, such as attending school to become a lawyer or computer programmer. According to Merton, lower-class persons turn to crime not because they *reject* American values but because they *accept* them: They believe that only through crime can they achieve our shared cultural goal of economic success.

Of course, few people who find society's norms inapplicable to their situation respond by turning to a life of crime. Merton identifies four ways in which people adapt to anomie without becoming criminals: innovation, ritualism, retreatism, and rebellion. These four strategies are illustrated in the Concept Summary on Merton's Modes of Adaptation.

In Merton's terms, *innovation* refers to people who accept society's goals but reject accepted institutional means, instead using illegitimate means to achieve their goals. Innovators include poor teenagers who steal flashy cars, students who cheat on tests, and athletes who use steroids to boost their performance. *Ritualism* refers to people who continue to use culturally approved means for achieving socially desired goals even though they have rejected—or at least given up on—those goals. A primary example of the ritualist is the worker who follows all bureaucratic procedures just to keep his or her job, not to get ahead. *Retreatism* refers to those who have given up on both society's goals and its accepted means. They are society's dropouts: the vagabonds, drifters, and street people. Like retreatism, *rebellion* also refers to those who abandon society's goals and means, but rebels additionally adopt alternative values. These are people like revolutionaries, Rastafarians, or the Rainbow Tribe who hope to create an alternative society.

Explaining Neighborhood Crime Rates: Collective Efficacy Theory

Whereas strain theory attempts to explain why some *individuals* are more likely to engage in crime than are others, collective efficacy theory attempts to explain why some *neighborhoods* have higher rates of crime than others (Sampson & Raudenbush 1999; Sampson, Morenoff, & Earls 1999; Sampson, Morenoff, & Gannon-Rowley 2002). Collective efficacy theory is also a structural-functionalist theory because it, too, assumes that crime or deviance occurs when the parts of a society no longer work together smoothly.

In Merton's terms, homeless alcoholics are *retreatists*: they have given up on both culturally accepted goals and culturally accepted means for reaching those goals.

bobhdeering/Alamy

Collective efficacy refers to the extent to which individuals in a neighborhood share the expectation that neighbors will intervene and work together to maintain social order. If your neighbors believe it is important to work together to control neighborhood crime and delinquency and are likely to call the police when teenagers race cars down the block or scrawl graffiti on a wall, then you live in a neighborhood with high collective efficacy. Collective efficacy is most common in neighborhoods that experience few structural disadvantages: They have high rates of employment and home ownership, many residents whose work and incomes give them a sense of control over their lives, and police and municipal services that they can count on for help when needed. According to collective efficacy theory, crime is most likely in neighborhoods that suffer extreme structural disadvantage and, as a result, experience low collective efficacy. This theory has strong empirical support and is growing in influence.

Conflict Theory

Structural-functional theory suggests that deviance results from a lack of integration among the parts of a social structure (norms, goals, and resources); it is viewed as an abnormal state produced by extraordinary circumstances. Conflict theorists, however, see deviance as a natural and inevitable product of competition in a society in which groups have different access to scarce resources. They suggest that the ongoing processes of competition should be the real focus of deviance studies (Lemert 1981).

Conflict theory proposes that deviance results from competition and class conflict. Class conflict affects deviance in two ways (Reiman 2005): (1) Class interests determine how the criminal justice system defines and responds to crime, and (2) economic pressures can lead to crime, particularly property crime, among the poor.

Collective efficacy refers to the extent to which individuals in a neighborhood share the expectation that neighbors will intervene to stop social disorder and deviance and will work together to maintain social order.

Defining and Responding to Crime

Conflict theorists argue that the law is a weapon used by the ruling class to maintain the political and economic status quo (Arrigo 1998; Liska, Chamlin, & Reed 1985; Reiman 2005). Supporters of this position argue that the very definitions of crime sometimes reflect the interests of the wealthy. Corporations can kill or injure thousands when they sell cars, contact lenses, or other goods that they know are harmful. They can endanger workers when they cut corners on factory safety, and they endanger whole communities when they dump dangerous chemicals into the water or soil. They also can impoverish workers and investors through shady business practices, even while their executives earn multimillion-dollar salaries. Yet these actions are often defined by the courts as ordinary and necessary business practices rather than as crimes.

Similarly, conflict theorists argue that the criminal justice system's response to behaviors labeled criminal also reflects the interests of the wealthy. Our system spends more money deterring muggers than embezzlers and more money arresting prostitutes than arresting their clients. Except in rare, high-profile cases, courts typically impose much more severe sentences for street crimes than for corporate crimes and impose much heavier sentences against those who use drugs favored by the poor (such as "crack" cocaine) than against those who use drugs favored by the more affluent (such as other forms of cocaine). Police are more likely to arrest those who assault members of the ruling class (well-off whites) than those who assault the powerless (nonwhites and the poor) (Reiman 2005). Finally, even when people from the upper and lower classes commit similar crimes, those from the lower class are more likely to be arrested, prosecuted, and sentenced (Reiman 2005).

As this suggests, most conflict theorists reject structural functionalism's assumption that poor people are unusually likely to commit crimes. Instead, and as research suggests, most poorer people adjust their goals downward sufficiently so that they can meet their goals through respectable means (Simons & Gray 1989). Meanwhile, many highly successful individuals adjust their goals so far upward that they cannot reach them by legitimate means. Recent court cases that reveal Microsoft's illegal attempts to gain a monopoly over Internet services and tobacco manufacturers' attempts to make cigarettes more addictive provide clear evidence that the means-versus-goals discrepancy is not limited to the lower class. Conflict theorists argue that it only appears that rich people commit fewer crimes because rich people control the state, schools, and courts, and so are often able to avoid criminal labels (Reiman 2005).

Lower-Class Crime

Although the preceding view of the way crime is defined would be accepted by all conflict theorists, some believe that individuals in the lower class really are more likely to commit criminal acts. One critical criminologist has declared that crime is a rational response for the lower class (Quinney 1980). These criminologists generally agree with Merton that a means/ends discrepancy is particularly acute among the poor and that it may lead to crime (Reiman 2005). They believe, however, that this is a natural condition of an unequal society.

Symbolic Interaction Theories

Symbolic interaction theories of deviance suggest that it is learned through interaction with others and involves the development of a deviant self-concept. Deviance is believed to result not from broad social structure but from specific face-to-face

© Alonv Reininger/Contact Press Images

Differential association theory argues that people who grow up in crime-ridden neighborhoods are more likely to become criminals themselves. It is easy to see how this theory applies to gang members like these, but can you think of how it might also apply to white-collar criminals?

Differential association theory argues that people learn to be deviant when more of their associates favor deviance than favor conformity.

Deterrence theory suggests that deviance results when social sanctions provide insufficient rewards for conformity.

Labeling theory is concerned with the processes by which labels such as *deviant* come to be attached to specific people and specific behaviors.

interactions. This argument takes three forms: *differential association theory, deterrence theory,* and *labeling theory.*

Differential Association Theory

Not surprisingly, researchers have found that those who have more delinquent friends are more likely to become delinquent themselves (Haynie & Osgood 2005). **Differential association theory**, first proposed by Edwin Sutherland, explains this finding by arguing that people *learn* to be deviant through their associations with others.

How does differential association encourage deviance? There are two primary mechanisms. First, if our interactions are mostly with deviants, we may develop a biased image of the generalized other. We may learn that, "of course, everybody steals" or, "of course, you should beat up anyone who insults you." The norms we internalize may differ greatly from those of conventional society. Second, if we interact mostly within a deviant subculture, that subculture will reward us not for *following* conventional norms but for *violating* them. Through these mechanisms, we can learn that deviance is acceptable and rewarding.

Deterrence Theory

Differential association theory can only explain deviance that occurs in settings and groups that encourage it. Deterrence theory provides a broader explanation of deviance. This theory suggests that individuals will engage in deviance when they believe it will offer more rewards than will conformity *and* when they believe the potential risks and costs of deviance are low. **Deterrence theory** combines elements of structural-functional and symbolic interaction theories. Although they place the primary blame for deviance on an inadequate (dysfunctional) system of rewards and punishments, they also believe that individuals actively make a cost/benefit decision about whether to engage in deviance (McCarthy 2002; Paternoster 1989; Piliavin et al. 1986). When social structures do not provide adequate rewards for conformity, more people will choose deviance.

For example, people who lack jobs or who have only dead-end jobs are more likely than others to believe they have little to lose and much to gain from crime or other forms of deviance, especially if they believe the risk of arrest is low (Crutchfield 1989; Devine, Sheley, & Smith 1988; McCarthy 2002). Conversely, those who have strong bonds with their parents, do well in school, feel a part of their school, and hold good jobs are more likely to avoid deviance because they feel they have too much to lose (Haynie & Osgood 2005).

Labeling Theory

A third theory of deviance, which combines symbolic interaction and conflict theories, is labeling theory. **Labeling theory** focuses on how and why the label *deviant* comes to be attached to specific people and behaviors. This theory takes to heart the maxim that deviance is relative. As the chief proponent of labeling theory puts it, "Deviant behavior is behavior that people so label" (Becker 1963, 90).

EXPLAINING INDIVIDUAL DEVIANCE The process through which a person becomes labeled as deviant depends on the reactions of others toward nonconforming behavior. The first time a child acts up in class, it may be owing to high spirits or a bad mood. This impulsive act is *primary deviance*. What happens in the future depends on how others interpret the act. If teachers, counselors, and other children label the child a troublemaker and if the child accepts this definition as part of her self-concept, then she may take on the role of a troublemaker. Continued rule violation because of a deviant self-concept is called *secondary deviance*.

The major limitations of labeling theory are that (1) it doesn't explain why primary deviance occurs, and (2) it cannot explain repeated deviance by those who haven't been caught—that is, labeled. For this reason, it is less popular today as an explanation of why individuals become deviants.

EXPLAINING DEVIANCE LABELING Labeling theory is more useful as an explanation of how behaviors become labeled as deviant. Many labeling theorists take a conflict perspective in exploring this topic. They argue that groups sometimes try to label the behavior of other groups as deviant as a means of increasing their own power and status. Because groups try to "sell" their moral ideas about who should be labeled deviant, just as entrepreneurs sell their ideas for new businesses, sociologists refer to those who attempt to create new definitions of deviance as **moral entrepreneurs**. Typically, the more power a group has, the more successful it will be in branding others as deviant. This, labeling theorists allege, explains why lower-class deviance is more likely to be subject to criminal sanctions than is upper-class deviance.

But groups can fight back against those who would label them deviant. For example, the Parents Television Council (PTC) is a nonprofit organization that campaigns against television shows that offend its conservative morality. One of its targets is the World Wrestling Federation, which PTC has lambasted for its violent and sexually explicit shows (Lowney 2003). The Federation responded in two ways. First, it attacked with humor by forming a wrestling team called the Right to Censor. This team pretended to preach the PTC's moral values while brazenly cheating during fights. Second, it successfully sued the PTC for libel and slander. Through both these strategies, the Federation protected its public image and fended off the PTC's efforts to label the Federation's shows as deviant in the public's eyes.

Stockbyte/White/PhotoLibrary

According to deterrence theory, individuals such as prom queens and kings are unlikely to engage in deviance because they have too much to lose if they do so.

Case Study: Medicalizing Deviance

In recent years, more and more behaviors once labeled *deviant* have become labeled *mental illnesses*. Labeling theory's emphasis on subjective meanings and conflict theory's emphasis on the power to define the situation give us a framework for understanding this shift.

Five hundred years ago, the most powerful social institution in Western society was the church. At that time, those who routinely became drunk in public were regarded as sinners and publicly castigated by ministers (the moral entrepreneurs of the time). But by the 1800s, the state and the criminal justice system had become more

sociology and you

Does your college or university forbid smoking in campus buildings? If so, you have witnessed the work of moral entrepreneurs. Anti-tobacco activists across the country have worked to outlaw smoking in public buildings, in private restaurants and bars, and even on the street. They also have fought to stigmatize smokers through advertising campaigns that portray smokers as ugly, stupid, and selfish. By so doing, they have created new definitions of morality and of deviance.

Moral entrepreneurs are people who attempt to create and enforce new definitions of morality.

powerful than the church. Although ministers still railed against those who drank alcohol, public drunks were now treated as criminals and thrown into jails.

These days, churches and judges vie for power with doctors and pharmaceutical companies. Individuals whose drinking gets publicly out of control will still be regarded as criminals by some and as sinners by others. Still others, however, argue that these individuals suffer from the disease of alcoholism. The behavior hasn't changed, and it's still considered deviant. But a different group (doctors) now define what is deviance. As with heavy drinking, other criminal behaviors like child abuse, gambling, murder, and rape are also regarded by some as signs of mental illness, better treated by doctors than by sheriffs (Conrad 2007). In addition, a wide range of human variations in behavior, appearance, and personality have also been redefined as illness. Doctors now propose cosmetic surgery to "cure" low self-esteem among women with small breasts and pharmaceutical companies declare that their drugs can cure shy people of "Social Anxiety Disorder" (Conrad 2007; Lane 2007; Sullivan 2001). Similarly, pharmaceutical companies now encourage doctors to diagnose people who become understandably sad following job loss or a death in the family as having "Major Depression" and to treat them with powerful drugs (Horwitz & Wakefield 2007). This process of redefining "badness," oddness, or ordinary human variation into illness is referred to as **medicalization** (a topic we discuss further in Chapter 10).

What happens when a behavior is medicalized? Individuals who acquire the *ill* label rather than the *bad* or *odd* label are more likely to receive treatment and sympathy rather than punishment or stigma (Conrad 2007). As you might expect—and as labeling and conflict theory would both predict—people in positions of power more often succeed in claiming the sick label. The upper-class woman who shoplifts is treated for obsessive-compulsive disorder, whereas the lower-class woman who does so is arrested for theft. The middle-class boy who acts up in school is medicated for hyperactivity, whereas the lower-class boy is jailed for juvenile delinquency.

Crime

Most deviant behavior is subject only to informal social controls. When deviance becomes labeled crime, it becomes subject to legal penalties. This is, in fact, the definition of **crime: behavior considered so unacceptable that it is subject to legal penalties.** Most, though not all, crimes violate social norms and are subject to informal as well as legal sanctions. In this section, we briefly discuss the different types of crimes, look at crime rates in the United States, and describe who is most likely to commit these crimes.

Property Crimes and Violent Crimes

Each year the federal government publishes the Uniform Crime Report (UCR), which summarizes the number of criminal incidents known to the police for five major crimes (Federal Bureau of Investigation 2009):

- *Murder and nonnegligent manslaughter.* Overall, murder is a rare crime. But it affects some segments of society much more than others. Almost 50 percent of all murder victims are African American and three-quarters are male (Federal Bureau of Investigation 2009).

Medicalization is the process through which something becomes defined as a medical problem.

Crime is behavior that is subject to legal or civil penalties.

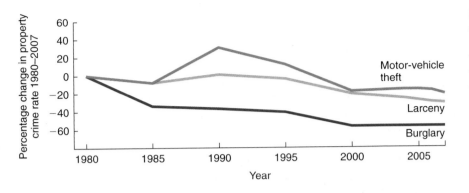

FIGURE 6.1 **Percentage Change over Time in Property Crime Rates** Burglary, larceny, and motor vehicle theft have all declined steadily since the 1990s and are now less common than they were in 1980.
SOURCE: U.S. Department of Justice (1995), Federal Bureau of Investigation (2009).

- *Rape.* Rape accounts for about 6 percent of all reported violent crimes (Federal Bureau of Investigation 2009). Even though most rapes go unreported, about 80,000 women each year report being raped. The best survey on the topic using a large, national, random sample found that 15 percent of all American women and 2 percent of all men had been raped at some point in their lives (Tjaden & Thoennes 1998).
- *Robbery.* Robbery is defined as taking or attempting to take anything of economic value from another person by force or threat of force. Unlike simple theft or larceny, robbery involves a personal confrontation between the victim and the robber. The rate of robbery has fallen almost 50 percent since 1990.
- *Assault.* Aggravated assault is an unlawful attack for the purpose of inflicting severe bodily injury. Because of this definition, most assaults involve a weapon.
- *Property crimes (burglary, larceny-theft, motor-vehicle theft, and arson).* Property crimes are much more common than are crimes of violence. They account for almost 90 percent of the crimes covered in the UCR (Federal Bureau of Investigation 2009).

Figure 6.1 shows the trend in property crimes since 1980. All major property crimes declined substantially between 1980 and 2000, and all are considerably lower than they were 30 years ago. The causes of this decline are hotly debated. However, most observers agree that a major reason is that young people commit most crimes, and there are now fewer young people than in earlier generations.

Violent crimes, too, are now less common than they were in 1980. However, and as Map 6.1 on the next page shows, violent crimes remain most common in the southern states, as well as in states with many poor, young people.

The unusually high rates of violent crime in the southern states—a long-standing trend—appear to reflect that region's "culture of honor." These states were initially settled by emigrants from poor, isolated, border regions of Scotland and northern England. Growing up in these areas, the emigrants had learned from childhood that they could not count on the law to protect them or their sheep from outlaws. As a result, a culture developed that encouraged young men to respond aggressively to any perceived threat against their property or honor. Aspects of this culture continue to this day in the southern United States, especially in rural areas (Gladwell 2008; Shackelford 2005).

Victimless Crimes

The so-called **victimless crimes**—such as drug use, prostitution, gambling, and pornography—are voluntary exchanges between persons who desire illegal goods or

Victimless crimes such as drug use, prostitution, gambling, and pornography are voluntary exchanges between persons who desire illegal goods or services from each other.

MAP 6.1: **Violent Crime Rate by State**

In general, violent crime (as well as property crime) is most common in the South, followed by the Southwest. It is also more common in other areas with many poor young people.

SOURCE: Federal Bureau of Investigation (2009)

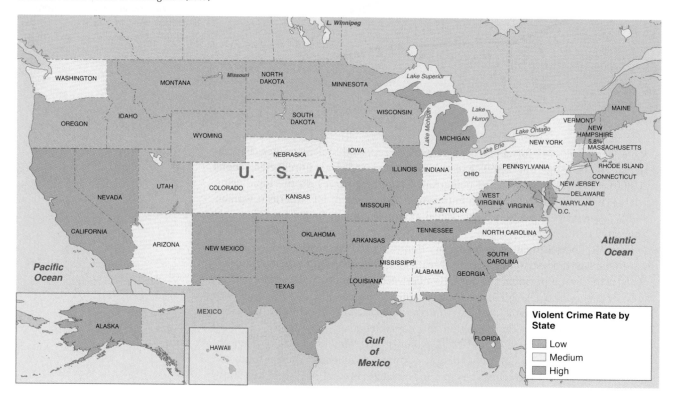

services from one another (Schur 1979). They are called victimless crimes because participants in the exchange typically do not see themselves as being victimized or as suffering from the transaction: There are no complaining victims.

There is substantial debate about whether these crimes are truly victimless. Some argue that prostitutes, drug abusers, and pornography models *are* victims (e.g., Weitzer 2007) because individuals usually enter these situations only if they feel they have no reasonable alternatives. Others believe that such activities are legitimate areas of free enterprise and free choice (Gould 2001; Gray 2000). These observers argue that although prostitutes and drug users might benefit from laws against pimping or selling contaminated drugs, they are only further victimized by laws against prostitution or drug use per se.

Because there are no complaining victims, these crimes are difficult to control. The drug user is generally not going to complain about the drug pusher, and the illegal gambler is unlikely to bring charges against a bookie. In the absence of a complaining victim, the police must find not only the criminal but also the crime. Efforts to do so are costly and divert attention from other criminal acts. As a result, laws relating to victimless crimes are irregularly and inconsistently enforced, most often in the form of periodic crackdowns and routine harassment.

The topic of victimless crimes is explored further in Decoding the Data: Legalizing Marijuana.

decoding the data

Legalizing Marijuana

According to national random surveys conducted by the Gallup Poll, about one-third of Americans now support legalizing marijuana. Support is highest among men, younger people, non-churchgoers, and college-educated people.

Explaining the Data: What about the culture, socialization, social position, or social experiences of men might make them more sympathetic than women to legalizing marijuana?

What might explain why younger people are more sympathetic? non-churchgoers? college-educated people?

Critiquing the Data: How could the question have been reworded to *increase* the percentage who supported loosening legal restrictions? (For example, we might ask whether people think marijuana should be made legal when needed for medical reasons.)

How could the question have been reworded to *reduce* the percentage who supported loosening legal restrictions?

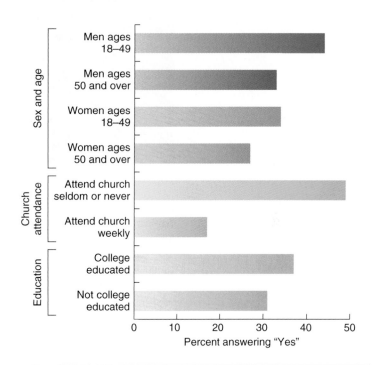

"Do you think the use of marijuana should be made legal or not?"
SOURCE: Caroll (2005).

White-Collar Crimes

Crimes committed by respectable people of high social status in the course of their work are called **white-collar crimes** (Sutherland 1961; Shover 2006). White-collar crimes can be committed by either individuals or companies. Individuals, for example, may embezzle money from their firms or defraud clients. The amounts involved can be staggering: In 2009, Bernie Madoff—former chairman of the NASDAQ stock exchange and founder of Bernard L. Madoff Investment Securities—pleaded guilty to defrauding his investment clients out of almost $65 *billion*.

When white-collar crimes are committed by companies, they are sometimes referred to as *corporate crimes*. Corporate crimes include such practices as price fixing, selling defective products, evading taxes, or polluting the environment. For example, accountants, auditors, and executives working for Enron Corporation worked together to hide the company's debts, exaggerate its profits, and pull in money from investors whom they tricked into buying their stock for much more than it was worth (Eichenwald 2005). Meanwhile, corporate executives took home multimillion-dollar

White-collar crimes refers to crimes committed by respectable people of high status in the course of their occupation.

Like drug peddlers and thieves, white-collar criminals can seriously harm individuals and society.

salaries. When its false bookkeeping became known and the company was forced into bankruptcy, Enron retirees lost their pensions, 4,000 Enron employees lost their jobs, and thousands of small investors lost their life savings.

White-collar crimes bring heavy costs to society. As the Madoff case suggests, the dollar loss due to corporate crimes can dwarf that lost through street crime (Hagan 2002). In addition to the economic cost, there are social costs as well. Exposure to repeated tales of corruption breeds distrust and cynicism and, ultimately, undermines the integrity of social institutions. If you think that all members of Congress are crooks, then you quit voting. If you think that police officers can be bought, then you cease to respect the law. Finally, white-collar crimes can cost lives when manufacturers sell cars with bad brakes, ignore safety precautions on factory lines, or dump toxic chemicals into rivers. Thus, the costs of white-collar crimes go beyond the actual dollars involved in the crimes themselves.

The reasons for white-collar crimes are similar to those for street crimes: People (and companies) want more than they can legitimately get and think the benefits of a crime outrun its potential costs (Shover 2006). Differential association also plays a role. In some corporations, organizational culture winks at or actively encourages illegal behavior. Sometimes the crimes are paltry, as when workers take home office supplies for personal use. Other times the consequences are far higher. For example, in 2009 the pharmaceutical company Eli Lilly was fined $1.4 billion by the federal government for illegally marketing the drug Zyprexa for sleep problems, depression, agitation, aggression, and hostility, even though the drug had only been approved for treating schizophrenia and was known to cause obesity and to increase the risk of diabetes (U.S. Attorney 2009). Eli Lily's sales of Zyprexa had soared after mid-level managers, following instructions of company executives, began instructing their sales personnel to disregard the law, training them in how to counter physician objections and concerns, and creating a culture in which sales-at-all-costs were valued.

The magnitude of white-collar crime in our society challenges the popular image of crime as a lower-class phenomenon. Instead, it appears that people of different statuses simply have different opportunities to commit crime. Those in lower statuses have no opportunity to engage in price fixing, stock manipulation, or tax evasion. They can, however, engage in high-risk, low-yield crimes such as robbery and larceny. In contrast, higher-status individuals have the opportunity to engage in low-risk, high-yield crimes (Reiman 2005; Shover 2006).

The lenient treatment received by most convicted white-collar criminals mocks the idea of equal justice. White-collar criminals are far less likely than are street criminals to be sentenced to prison and receive far shorter sentences when they are imprisoned (Shover 2006). However, recent high-profile cases such as those of Bernie Madoff and the Enron executives signal an increased awareness (at least among government prosecutors) of the importance of white-collar crime. Similarly, the number of corporations convicted of white-collar crime and the dollar amount of fines imposed have increased over the last decade (Shover 2006).

Correlates of Crime: Age, Sex, Class, and Race

Each year, less than half of all violent crimes reported in the UCR and less than one-quarter of property crimes are "cleared" by an arrest (that is, resulted in an arrest) (Federal Bureau of Investigation 2009). Murder is the crime most likely to be cleared, and burglary is least likely. This means that the people arrested for the criminal acts summarized in the UCR represent only a sample of those who commit these crimes; they are undoubtedly not a random sample. Nor do they represent at all those who commit white-collar crimes, which are not included in the UCR. As a result, we must be cautious in generalizing from arrestees to the larger population of criminals.

With this caution in mind, we note that the persons arrested for criminal acts are disproportionately male, young, and from minority groups. Figure 6.2 shows the pattern of arrest rates by sex and age. As you can see, crime rates for both men and women peak during ages 15 to 24, although during these peak crime years, men are about three to four times more likely to be arrested than are women. Minority data are not available by age and sex, but the overall rates show that African Americans and Hispanics are more than three times as likely as whites to be arrested.

What accounts for these differentials? Can the theories reviewed earlier help explain these patterns?

Age Differences

The age differences in arrest rates noted in Figure 6.2 on the next page are both long-standing and characteristic of nearly every nation in the world that gathers crime statistics (Cook & Laub 1998). Researchers disagree over the reasons for the high arrest rates of young adults, but deterrence theories have the most promise for explaining this age pattern.

In many ways, adolescents and young adults have less to lose than other people. They don't have a "stake in conformity"—a career, a mortgage, or a credit rating (Steffensmeier et al. 1989). When young people do have jobs and especially when they have good jobs, their chances of getting into trouble are much less (Allan & Steffensmeier 1989).

Delinquency is basically a leisure-time activity. It is strongly associated with spending large blocks of unsupervised time with peers (Haynie & Osgood 2005). When young people have "nothing better to do," a substantial portion will get their fun by causing trouble. Conversely, deviance is deterred by having a close attachment to parents or school.

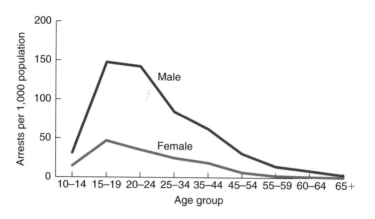

FIGURE 6.2 **Arrests per 1,000 Population by Age and Sex**
Arrests rates in the United States and most other nations show strong and consistent age and sex patterns. Arrest rates peak sharply for young people ages 15 to 24; at all ages, men are considerably more likely than women to be arrested.
SOURCE: Federal Bureau of Investigation (2009).

Sex Differences

The sex differential in arrest rates has both social and biological roots. Women's smaller size and lesser strength make them less able to engage in the types of crimes emphasized in the UCR; they have learned that, for them, these are ineffective strategies. Evidence linking male hormones to aggressiveness indicates that biology also may be a factor in women's lower inclination to engage in violent behavior.

Among social theories of deviance, deterrence theory seems to be the most effective in explaining these differences. Generally, girls are supervised more closely than boys, and they are subject to more social control, especially in less affluent families (Chesney-Lind & Shelden 2004; Hagan, Gillis, & Simpson 1985; Thompson 1989). Whereas parents may let their boys wander about at night unsupervised, they are much more likely to insist on knowing where their daughters are and with whom they are associating. The greater supervision that girls receive increases their bonds to parents and other conventional institutions; it also reduces their opportunity to join gangs or other deviant groups.

These explanations raise questions about whether changing roles for women will affect their participation in crime. Will increased equality in education and labor-force participation and increased smoking and drinking also carry over to greater equality of criminal behavior? So far, the answer appears to be no (Chesney-Lind & Shelden 2004; Steffensmeier & Allan 1996). Although the crime rate for women has increased, most of this increase is in minor property crimes and drug possession (as opposed to drug dealing) (Chesney-Lind & Shelden; Maher & Daly 1996). Meanwhile, the gender gap in rates of violent and major property crime has actually increased.

This pattern of change lends support to feminist theories of crime. Whereas deterrence theory argues that men's higher crime rates reflect their relatively weaker bonds to conventional *authority*, feminist sociologists argue that those rates reflect men's strong bonds to conventional *gender roles* (Bourgois 1995; Katz 1988; Messerschmidt 1993). According to these theories, to be considered "masculine," boys and men must challenge authority and act aggressively or even violently, at least in certain times and places. This theory is particularly useful for explaining crimes against women by groups of men, such as gang rapes (Lefkowitz 1997; Sanday 1990).

Feminist sociologists also have noted that *victimization* of females by males explains a significant proportion of crime among females (Chesney-Lind & Shelden 2004). Girls and women who have been sexually or physically abused by men (including male relatives) are more likely to run away from home, turn to drugs, enter prostitution, and respond violently to their abusers and others.

Social-Class Differences

The effect of social class on crime rates is complex. Braithwaite's (1985) review of more than 100 studies leads to the conclusion that lower-class people commit more of the direct interpersonal types of crimes normally handled by the police than do people from the middle class. These are the types of crimes reported in the UCR. Middle-class people, on the other hand, commit more of the crimes that involve the use of power, particularly in the context of their occupational roles: fraud, embezzlement, price fixing, and other forms of white-collar crime. There is also evidence that the social-class differential may be greater for adult crime than for juvenile delinquency (Thornberry & Farnworth 1982).

Nearly all the deviance theories we have examined offer some explanation of the social-class differential. Strain theorists and some conflict theorists suggest that the lower class is more likely to engage in crime because of blocked avenues to achievement, which explains why crime rises along with unemployment (Grant & Martinez-Ramiro 1997). Deterrence theorists argue that the lower classes commit more crimes because they receive fewer rewards from conventional institutions such as school and the labor market. All these theories accept and seek to explain the social-class pattern found in the UCR, where the lower class is overrepresented.

Labeling and conflict theories, on the other hand, argue that this overrepresentation is not a reflection of underlying social-class patterns of deviance but of bias in the law and within social control agencies (Williams & Drake 1980). Evidence suggests, for instance, that the disproportionately high homicide rates found among the lower social classes in most modern societies result from governmental failure to provide the least privileged with the legal means of conflict resolution available to the social elite (Cooney 1997). Overrepresentation of the lower class also reflects the particular mix of crimes included in the UCR; if embezzlement, price fixing, and stock manipulations were included, we would see a very different social-class distribution of criminals.

Race Differences

Although African Americans compose only about 12 percent of the population, they make up 34 percent of those arrested for rape, 34 percent of those arrested for assault, and 50 percent of those arrested for murder (Federal Bureau of Investigation 2009). Hispanics, who compose about 12.5 percent of the total population, represent about 28 percent of those imprisoned for violent crimes. These strong differences in arrest and imprisonment rates are explained in part by social-class differences between minority and white populations. Even after this effect is taken into account, however, African Americans and Hispanics are still much more likely to be arrested for committing crimes.

The explanation for this is complex. As we will document in Chapter 8, race and ethnicity continue to represent a fundamental cleavage in U.S. society. The continued and even growing correlation of minority status with poverty, unemployment, inner-city residence, and female-headed households reinforces the barriers between nonwhites and whites in U.S. society. An international study confirms that the larger the number of overlapping dimensions of inequality, the higher the "pent-up aggression which manifests itself in diffuse hostility and violence" (Messner 1989). The root cause of higher minority crime rates, from this perspective, is the low quality of minority employment—which leads directly to unstable families and

© Joel Gordon

Even when nonwhites and whites engage in the *same* illegal behaviors, minorities are more likely to be cited, arrested, prosecuted, and convicted.

neighborhoods (Newman 1999b; Sampson & Raudenbush 1999; Sampson, Morenoff, & Earls 1999).

Poverty and segregation combine to put African American children in the worst neighborhoods in the country, where getting into trouble is a way of life and where lack of resources makes conventional achievement almost impossible (Newman 1999b). Differential association theory thus explains a great deal of the racial difference in arrest rates. Deterrence theory is also important. Compared to non-Hispanic whites, African American children are much more likely to live in a fatherless home and Hispanic children are somewhat more likely to do so, leaving them without an important social bond that might deter deviant behavior.

But these differences in crime rates between minorities and nonminorities are to some extent more apparent than real. It is true that on average minorities commit more crimes than do whites. But when we compare minorities and whites who engage in the *same* behavior—from causing trouble in school to committing murder—minorities are more likely than whites to be cited, arrested, prosecuted, and convicted (Austin & Allen 2000; Cureton 2000). As a result, UCR rates overestimate the percentage of crime actually committed by minorities.

Fear of Crime

Since the 1990s, crime rates have dropped dramatically. Yet each year approximately two-thirds of Americans interviewed by the Gallup Pool say that they believe crime is *increasing* (Federal Bureau of Investigation 2009). Where has this "culture of fear" come from?

Many groups and individuals benefit from promoting fear. Politicians use fear-mongering to get votes, businesses "sell" fear to sell products (guns, alarm systems, anti-phishing software), and advocacy groups promote fear to gain support for their causes (criminalizing drunk driving, or offering more treatment for drug addicts). But when asked why their fears have grown, most Americans point to the media (Blendon & Young 1998). Television reporters and producers, especially, seek stories that can be told in a 3-minute spot: emotionally gripping, visually exciting, and with clear villains and victims (Altheide 2002, 2006). As a result, they often choose their headline stories according to the principle "If it bleeds, it leads."

How do the media teach people to overestimate the dangers of crime? Sociologist Barry Glassner (2004) offers three answers. First, the media *misidentify* isolated events as trends, such as describing the massacres at Virginia Tech in 2007 and at Northern Illinois University in 2008 as part of a broad pattern of school homicides, even though no such pattern exists. Second, the media *misdirect* us, making crime seem important by ignoring more serious problems. For example, television news shows spend far more time discussing the very rare school homicides than discussing, for example, the millions of American children who leave home hungry each morning to go to dilapidated, segregated schools. Third, the media *repeat* exaggerated claims of dangers so often that we believe them. True, the media do sometimes try to debunk myths of dangers, but such stories appear infrequently and are often buried far back in newspapers or later in newscasts.

Of course, crime can cause extraordinary suffering for its victims and can make any and all preventive actions seem worthwhile. But fear of crime also can cause problems. One consequence is the enormous growth in our enormously expensive prison system (discussed below). Another consequence is a deterioration in public life. Elderly people sometimes become so afraid of crime that they don't leave their homes, leading to social isolation, lack of exercise, and physical and mental deterioration.

Similarly, parents may forbid their children from playing on the street, taking public transit, or otherwise learning how to explore and enjoy the world on their own (Skenazy 2009).

The Criminal Justice System

The responsibility for dealing with crime rests with the criminal justice system, the subject of this section. Any assessment of this system must begin with the question, Why punish?

Why Punish?

Traditionally, there have been four major rationalizations for punishment:

- *Retribution*. Society punishes offenders to avenge the victim and society as a whole.
- *Prevention*. By imprisoning, executing, or otherwise controlling offenders, society keeps them from committing further crimes.
- *Deterrence*. Punishment is intended to scare both previous offenders and non-offenders away from a life of crime.
- *Reform*. By building character and improving skills, former criminals are enabled and encouraged to become law-abiding members of society.

Today, social control agencies in the United States represent a mixture of these different philosophies and practices. However, the increasing emphasis since the 1970s has been on long—even lifelong—sentences. For example, under "three strikes and you're out" laws passed around the country, individuals convicted of three felonies, regardless of the circumstances, must serve at least 25 years in prison without probation. These laws don't differentiate between a serial killer and someone who breaks into a store to steal food. The shift toward mandatory, long sentences, combined with the dearth of educational programs and psychological counseling in jails and prisons, suggests that reformation is only a minor goal of our criminal justice system.

In the United States, this system consists of a vast network of agencies—police departments, probation and parole agencies, rehabilitation agencies, criminal courts, jails, and prisons—set up to deal with persons who deviate from the law.

The Police

Police officers occupy a unique and powerful position in the criminal justice system. They can make arrests even if no one has filed a complaint against an individual, and even if no one is there to oversee their actions. Although they are supposed to enforce the law fully and uniformly, everyone realizes that this is neither practical nor possible. In 2007, there were 3.6 full-time police officers for every 1,000 persons in the nation (Federal Bureau of Investigation 2009). This means that the police ordinarily must give greater attention to more serious crimes. Minor offenses and ambiguous situations are likely to be ignored.

Police officers have a considerable amount of discretionary power in determining the extent to which the policy of full enforcement is carried out. Should a drunk and disorderly person be charged or sent home? Should a juvenile offender be charged or only reported to his or her parents? Should a strong odor of marijuana in an otherwise orderly group be overlooked or investigated? Unlike decisions meted out in courts of law, decisions made by police officers on the street are relatively invisible and thus hard to evaluate.

The Courts

Once arrested, an individual starts a complex journey through the criminal justice system. This trip can best be thought of as a series of decision stages. A significant proportion of those who are arrested are never prosecuted. Of those who are prosecuted for felonies, however, about two-thirds are eventually convicted, with almost all convictions resulting from pretrial negotiations rather than public trials (U.S. Department of Justice 2009). Thus, the pretrial phases of prosecution are far more crucial to arriving at judicial decisions of guilt or innocence than are court trials themselves. Like the police, prosecutors have considerable discretion in deciding whom to prosecute and what charges to file.

Throughout the entire process, the prosecution, the defense, and the judges participate in negotiated plea bargaining. They encourage the accused to plead guilty in the interest of getting a lighter sentence, a reduced charge, or, in the case of multiple offenses, the dropping of some charges. In return, the prosecution is saved the trouble and cost of a trial. As a result, court decisions reflect much more—and much less—than simple guilt and innocence.

Prisons

For many people, getting tough on crime means locking criminals up and throwing away the key. Presidential politics, the strength of the Republican Party, the rise of conservative religious denominations, and overall public opinion have contributed to rapid expansion in the law enforcement sector and a rapid rise in imprisonment (Curry 1996; Jacobs & Helms 1997; Kraska & Kappeler 1997). As of 2009, there are 2.3 million people in U.S. federal and state prisons—several times higher than the number 30 years ago (International Centre for Prison Studies, 2008). Rates of imprisonment are now higher in the United States than anywhere else in the world. This is primarily due to harsher sentencing policies, especially for drug-related crimes, such as "mandatory minimums" and "three strikes and you're out" laws (Figure 6.3).

Prison residents are disproportionately young men who are uneducated, poor, and African American. About 40 percent of all prisoners are African American males (U.S. Department of Justice 2009). Even more shockingly, 7.3 percent of *all* African American males ages 25 to 29 are in prison, compared with 2.6 percent of Hispanics and 1.1 percent of non-Hispanic whites. Finally, African Americans are disproportionately represented on death row, a topic discussed further in Focus on American Diversity: Capital Punishment and Racism on page 148.

The sharp increase in the use of imprisonment has resulted in a crisis in prison (and jail) conditions. Many facilities house twice as many inmates as they were designed to hold, in inhumane conditions; in Arizona, jail inmates—most of whom have not even been tried yet, let alone convicted—are housed in tents in the desert with temperatures rising up to 125 degrees. When inmates consider these conditions unjust, they become a major cause of prison riots (Useem & Goldstone 2002). As a

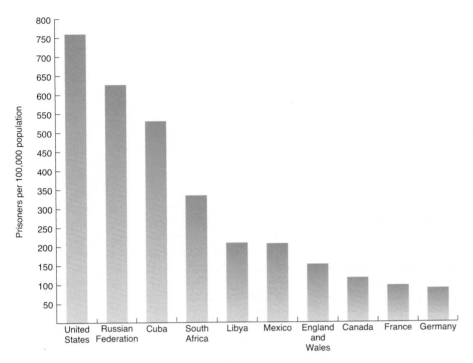

FIGURE 6.3 **Number of Prisoners per 100,000 Population**
The United States leads the world in imprisoning its own population. Not only do we imprison more people than do similar countries like Canada, we even imprison more people than do dictatorships like Libya and Cuba.

SOURCE: **www.prisonstudies.org** (2008).

result, prisons in more than 30 states are under court order to reduce crowding and improve conditions.

At the same time, severe budget crises have made it increasingly difficult for states to pay the costs of imprisoning so many people—even if prisoners are kept in poor conditions. Consequently, many states are now working to eliminate mandatory minimum sentences, to reduce sentences for those now in prison, and to find effective alternatives to imprisonment for those convicted of less dangerous crimes (Steinhauer 2009).

Other Options

Do we really need to spend billions and billions of dollars to build more prisons to warehouse a growing proportion of those accused or convicted of crime? Maybe not.

A growing number of empirical studies demonstrate that the certainty of getting caught deters crime more effectively than do long sentences (McCarthy 2002). These findings suggest that we are pursuing the wrong strategy. Rather than building more prisons to warehouse criminals for longer periods of time, we need to put more money into law enforcement.

Another approach to solving the prison crisis is to change the way we deal with convicted criminals. As the cost of imprisoning larger numbers of people balloons to crisis proportions, community-based corrections has emerged as an alternative to long prison sentences. New intensive supervision probation programs are being used across the country to safely release convicts from prison earlier. They include curfews, mandatory drug testing, supervised halfway houses, mandatory community service, frequent reporting and unannounced home visits, restitution, electronic surveillance, and split sentences (incarceration followed by supervised probation).

focus on  AMERICAN DIVERSITY

Capital Punishment and Racism

In 1972, in the case of *Furman v. Georgia*, three African American defendants appealed their death sentences to the U.S. Supreme Court on the grounds that capital punishment constituted cruel and unusual punishment (Ogletree & Sarat 2006). Their argument was that other defendants, many of whom were white, committed equally or more serious crimes but were not sentenced to death. After reviewing the data, the Supreme Court agreed with the defendants, holding that the uncontrolled discretion of judges and juries reflected racist biases and denied defendants constitutionally guaranteed rights to due process.

The *Furman* decision put a temporary stop to capital punishment and led states to give judges and juries less discretion in death penalty cases. Nevertheless, the number of people executed continued to rise sharply until 1999 (Death Penalty Information Center 2009a).

More recently, new concerns over equity and new research suggesting that the death penalty is not an effective deterrent to crime (Fagan 2006; Weisberg 2005) have resulted in a drop in the number of executions. But has this shift eliminated racial biases in capital punishment?

Unfortunately, no. Studies continue to show that race strongly predicts who is sentenced to death. African Americans and Hispanics now account for 53 percent of all Americans awaiting execution (Death Penalty Information Center 2009b). This is much higher than the percentage of African Americans and Hispanics among the general population and among convicted murderers. Moreover, black defendants who look stereotypically black are twice as likely as other black defendants to be convicted of murders (Eberhardt et al. 2006).

The race of the victim also affects the likelihood of receiving a death sentence: Those convicted of killing whites are significantly more likely to receive the death penalty than are those convicted of

killing African Americans (U.S. General Accounting Office 1996; Williams & Holcomb 2001; Death Penalty Information Center 2009a). These statistics, too, suggest that the criminal justice system regards the lives of whites as more valuable than those of nonwhites.

The importance of eliminating racism (and other biases) from death penalty cases is highlighted by the growing realization that innocent people can and do get convicted. Between 2000 and 2009, 170 people were exonerated based on DNA testing (Innocence Project 2009). The most common cause of false convictions is mistaken identifications by witnesses, which is most common when the witness is white and the accused is nonwhite. In addition, most defendants in these cases were poor, many lacked proper legal representation, and many were pressed by the police into making false confessions (Innocence Project 2009; Ogletree & Sarat 2006).

Overcrowded prisons in which inmates are depersonalized by assigned numbers, identical uniforms, and unvarying routines breed anger, violence, boredom, and further deviance.

A review conducted for the U.S. National Institute of Justice found that when these programs included treatment for drug addiction and other supportive services, they increased the chances of rehabilitation and reduced overall costs to the system (Petersilia 1999).

Where This Leaves Us

The conservative approach to confronting deviance and crime has generally been to make deviance illegal and to increase penalties for convicted criminals. This approach has dominated since the 1970s, which is why prison populations have soared. An alternative approach is, first, to develop greater tolerance for victimless crimes and other forms of deviance that are relatively inconsequential. For more serious forms of deviance and crime, we can address the social problems that give rise to these activities. A leading criminologist (Currie 1998) advocates five major strategies for doing so:

• Reduce inequality and social impoverishment.
• Replace unstable, low-wage, dead-end jobs with decent jobs.
• Prevent child abuse and neglect.
• Increase the economic and social stability of communities.
• Improve the quality of education in all communities.

These strategies would require a massive commitment of energy and money. They are not only expensive but also politically risky. Whereas law-and-order advocates want to get tough on crime by sending more criminals to jail, a policy incorporating these five strategies would channel dollars and beneficial programs into high-crime neighborhoods. Such a policy calls for more teachers and good jobs rather than for more police officers and prisons.

Observers from all sociological perspectives and all political parties recognize that social control is necessary. They recognize that rape, assault, and drug-related crimes are serious problems that must be addressed. The issue is how to do so. The sociological perspective suggests that crime can be addressed most effectively by changing the social institutions that breed crime rather than by focusing on changing individual criminals after the fact.

Summary

1. Most of us conform most of the time. We are encouraged to conform through three types of social control: (1) self-restraint through the internalization of norms and values, (2) informal social controls, and (3) formal social controls.

2. Nonconformity occurs when people violate expected norms of behavior. Acts that go beyond eccentricity, challenge important norms, and result in social sanctions are called deviance. Crimes are deviant acts that are also illegal.

3. Deviance is relative. It depends on society's definitions, on the circumstances surrounding an act, and on one's groups and subcultures.

4. Structural functionalists use strain theory to explain how individual deviance is linked to social disorganization. They use collective efficacy to explain why some neighborhoods have higher rates of crime than others. Symbolic interactionists propose differential association, deterrence, and labeling theories, which link deviance to interaction patterns that encourage deviant behaviors

and a deviant self-concept. Conflict theorists locate the cause of deviance, and of laws defining what is criminally deviant, in inequality and class conflict.

5. Rates of violent and property crimes rose from 1960 to 1990 but have fallen steadily since then.

6. Many arrests are for victimless crimes—acts for which there are no complainants. Laws relating to such crimes are the most difficult and costly to enforce.

7. The high incidence of white-collar crimes—those committed in the course of one's occupation—indicates that crime is not merely a lower-class behavior.

8. Males, minority-group members, lower-class people, and young people are disproportionately likely to be arrested for crimes. Some of this disparity is due to their greater likelihood of committing a crime, but it is also explained partly by their differential treatment within the criminal justice system.

9. The criminal justice system includes the police, the courts, and the correctional system. Considerable discretion in the execution of justice is available to authorities at each of these levels.

10. The "get-tough" approach to crime has left U.S. prisons filled beyond capacity. Evidence suggests that longer sentences may not be necessary. Alternatives to imprisonment include community-based corrections and social change to reduce the causes of crime.

Thinking Critically

1. Explain how differential association theory can or cannot explain why some children who grow up in bad neighborhoods do not become delinquent.

2. Why do you think most Americans view street crime as more serious than corporate crime? What would a conflict theorist say? A structural functionalist?

3. Describe a deviant whom you have known well—someone who got in trouble with the law or should have. Evaluate the theories of deviance in light of this one person. Which theory best explains why your acquaintance deviated rather than conformed? Which theory best explains whether or not your acquaintance was arrested and imprisoned for his or her behavior?

4. Devise a strategy for deterring white-collar or corporate crime, keeping in mind what you have read in this chapter.

5. From a sociological perspective, why would the race of the *victim* be as important as the race of the *defendant* in predicting whether a convicted killer will be sentenced to death? What does this tell us about racial and ethnic relations in our society? If racial discrimination exists in death sentencing, is that a good reason to stop capital punishment altogether? Why or why not?

Book Companion Website

www.cengage.com/sociology/brinkerhoff

Prepare for quizzes and exams with online resources—including tutorial quizzes, a glossary, interactive flash cards, crossword puzzles, essay questions, virtual explorations, and more.

CHAPTER 7

Stratification

© 2009/Jupiterimages

Structures of Inequality

Inequality exists all around us. Much of sociological research focuses on one particular kind of inequality called stratification. **Stratification** is an institutionalized pattern of inequality in which those who hold some social statuses get more access to scarce resources than do others. For example, giving a son more financial help than a daughter because the son is nicer is *not* stratification. But if a son receives more help simply because he is male, that *is* an example of stratification.

Inequality becomes stratification when two conditions exist:

- The inequality is *institutionalized*, backed up both by social structures and by long-standing social norms.
- The inequality is based on membership in a group (such as oldest sons or blue-collar workers) rather than on personal attributes.

The scarce resources that we focus on when we talk about inequality are generally of three types: *prestige*, *power*, and *money*. **Prestige**, like status, refers to the amount of social honor or value afforded one individual or group relative to another. **Power** refers to the ability to influence or force others to do what you want them to do, regardless of their own wishes. When inequality in prestige, power, or money is supported by social structures and long-standing social norms, and when it is based on group membership, then we speak of stratification.

Types of Stratification Structures

Stratification exists in every society. All societies have norms specifying that some categories of people ought to receive more money, power, or prestige than others. There is, however, wide variety in how inequality is structured.

A key difference among structures of inequality is whether the categories used to distribute unequal rewards are based on ascribed or achieved statuses. As noted in Chapter 4, *ascribed statuses* are unalterable statuses determined by birth or inheritance. *Achieved statuses* are statuses that a person can obtain in a lifetime. Being African American or male, for example, is an ascribed status; being a convict, an ex-convict, or a physician is an achieved status.

Every society uses some ascribed and some achieved statuses in distributing scarce resources, but the balance between them varies greatly. Stratification structures that rely largely on ascribed statuses as the basis for distributing scarce resources are called **caste systems**; structures that rely largely on achieved statuses are called **class systems**.

Caste Systems

In a caste system, whether you are rich or poor, powerful or powerless, depends almost entirely on who your parents are (Smaje 2000). Whether you are lazy and stupid or hardworking and clever makes little difference. Instead, your parents' position determines your own. If you are male, you are expected to enter your father's occupation or become a beggar if he was one; if you are female, you are expected to follow in your mother's footsteps as a housewife, beggar, or worker. Moreover, in a caste system you can only marry someone whose social position matches yours, and thus your children become locked in to the same status that you and your spouse hold.

India provides the best-known example of a caste system. Under its caste system, all Hindus (the majority religion) are divided into castes, roughly comparable to

Stratification is an institutionalized pattern of inequality in which social statuses are ranked on the basis of their access to scarce resources.

Prestige refers to the amount of social honor or value afforded one individual or group relative to another. Also referred to as status.

Power is the ability to direct others' behavior even against their wishes.

Caste systems rely largely on ascribed statuses as the basis for distributing scarce resources.

Class systems rely largely on achieved statuses as the basis for distributing scarce resources.

Most of India's Dalits ("untouch-ables") continue to experience poverty and discrimination.

occupational groups, which differ substantially in prestige, power, and wealth; caste systems also are common in some of India's Muslim and Christian communities. Caste membership is unalterable: It marks one's children and one's children's children.

The caste system was officially outlawed in 1950, when the new nation of India adopted its first Constitution. Since 1990, some of the nation's 200 million Dalits, or "untouchables," have moved out of abject poverty, boosted both by India's improving economy and by government policies designed to benefit them (Sengupta 2008). But most still suffer from discrimination, extreme poverty, and, sometimes, ethnic violence directed against them.

Class Systems

In a class system, achieved statuses are the major basis of unequal resource distribution. Occupation remains the major determinant of rewards, but it is not fixed at birth. Instead, you can achieve an occupation far better or far worse than those of your parents. The rewards you receive depend on your own talent, ambition, and work—or lack thereof.

The primary difference between caste and class systems is not the level of inequality but the opportunity for achievement. The distinctive characteristic of a class system is that it permits **social mobility**—a change in social class, either upward or downward. Mobility can occur between one generation and another; if you graduate college, and your parents didn't, you will likely experience upward social mobility. A change in social class can also occur within one's lifetime. For example, a middle-aged engineer whose job is "downsized" and who ends up working as a Wal-Mart greeter has obviously experienced downward social mobility.

Even in a class system, ascribed characteristics matter. Your religion, age, sex, and ethnicity, among other things, will likely influence which doors open for you and which barriers you have to surmount. Nevertheless, these factors have much less impact in a class society than in a caste society. Because class systems predominate in the modern world, the rest of this chapter is devoted to them.

Social mobility is the process of changing one's social class.

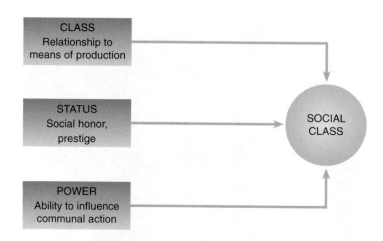

FIGURE 7.1 Weber's Model of Social Class
Weber identified three important and independent dimensions that together determine where people rank in a stratification system. The combination of these three measures is sometimes referred to as *social class*.

Classes—How Many?

A class system is an ordered set of statuses. Which statuses are included? And how are they divided? Two theoretical answers and two practical answers to these questions are presented in this section.

Marx: The Bourgeoisie and the Proletariat

Karl Marx (1818–1883) believed that there were only two classes. We could call them the haves and the have-nots; Marx called them the *bourgeoisie* (boor-zhwah-zee) and the *proletariat*. The **bourgeoisie** are those who own the tools and materials necessary for their work—the means of production. The **proletariat** are those who do not. The latter must therefore support themselves by selling their labor to the former. In Marx's view, one's **class** depends entirely on one's relationship to the means of production.

Relationship to the means of production obviously has something to do with occupation, but it is not the same thing. According to Marx, your college instructor, the manager of the Sears store, and the janitor are all proletarians because they work for someone else. If your garbage collector works for the city, he is also a proletarian; if he owns his own truck, however, he is a member of the bourgeoisie. The key factor is not how much money a person has or what type of job he does but rather whether he controls his own tools and his own work.

Marx, of course, was not blind to the fact that in the eyes of the world, store managers are regarded as more successful than truck-owning garbage collectors. Probably managers think of themselves as being superior to garbage collectors. In Marx's eyes, this is **false consciousness**—a lack of awareness of one's real position in the class structure. Marx, a social activist as well as a social theorist, hoped that managers and janitors would develop **class consciousness**—an awareness of their true class identity. If they did, he believed, a revolutionary movement to eliminate class differences would be likely to occur.

Weber: Class, Status, and Power

Several decades after Marx wrote, Max Weber developed a more complex system for analyzing classes. Instead of Marx's ranking system, which identified only two classes, Weber proposed three independent dimensions that determine where people rank in a stratification system (Figure 7.1). One of them, as Marx suggested, is class. The second is power, and the third is status, which, like prestige, means social honor or social value. Individuals who share a similar status typically form a community

The **bourgeoisie** is the class that owns the tools and materials for their work—the means of production.

The **proletariat** is the class that does not own the means of production. They must support themselves by selling their labor to those who own the means of production.

Class, in Marxist theory, refers to a person's relationship to the means of production.

False consciousness is a lack of awareness of one's real position in the class structure.

Class consciousness occurs when people understand their relationship to the means of production and recognize their true class identity.

of sorts. They invite one another to dinner, marry one another, engage in the same kinds of recreation, and generally do the same things in the same places.

Weber argued that although status and power often follow economic position, they may also stand on their own and have an independent effect on social inequality. In particular, Weber noted that status often stands in opposition to economic power, depressing the pretensions of those who "just" have money. Thus, for example, a member of the Mafia may have a lot of money and may own the means of production (a brothel, a heroin manufacturing plant, or a casino), but he will not have honor in the broader community.

Measuring Social Class

Marx and Weber provide us with theoretical concepts we can use in understanding class systems. Modern researchers, however, need practical ways of measuring class, not just theoretical definitions. These days, most researchers focus not on class (as Marx defined it) but on *social* class. A **social class** is a category of people who (as Weber suggested) share roughly the same class, status, and power and who have a sense of identification with one another. When we speak of the upper class or of the working class, we are speaking of social class in this sense.

The most direct way of measuring social class is simply to ask people what social class they belong to. The results of the 2008 General Social Survey are presented in Figure 7.2. As you can see, only tiny minorities see themselves as belonging to the upper and lower classes. The rest are split nearly evenly between those who identify as working- or middle-class. Studies show that the difference between working- and middle-class identification has important consequences, affecting what church you go to, how you vote, and how you raise your children.

Another common way to measure social class is by **socioeconomic status**. Socioeconomic status refers to education, occupation, income, or some combination of these. Socioeconomic status does *not* measure how people identify their own class position. Instead, these measures rank the population from high to low on criteria such as years of school completed, family income, or the prestige of one's occupation (as ranked by surveys of the population).

Inequality in the United States

Stratification exists in all societies. In Britain, India, and China, social structures ensure that some social classes routinely receive more rewards than do others. This section considers how stratification works in the United States.

Economic Inequality

One very important type of inequality is income inequality. **Income** refers to all money received in a given time period by a person or family. Income can include salaries, wages, pensions, dividends and interest, as well as money received from the government (through Social Security, for example). **Income inequality** refers to the extent to which incomes vary within a given population.

Income inequality is very high in all class systems but is especially high in the United States. Of the 29 industrialized nations that participate in the long-term Luxembourg Income Study (2000), only two, Mexico and Russia, have more income inequality than the United States.

FIGURE 7.2 Social Class Identification in the United States
Social class is a very real concept to most Americans. They are aware of their own social-class membership. They feel that, in a variety of important respects, they are similar to others in their own social class and different from those in other social classes.
SOURCE: General Social Survey. **http://sda. berkeley.edu.** Accessed May 2009.

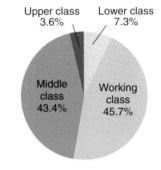

Upper class 3.6% Lower class 7.3%
Middle class 43.4% Working class 45.7%

Social class is a category of people who share roughly the same class, status, and power and who have a sense of identification with each other.

Socioeconomic status (SES) is a measure of social class that ranks individuals on income, education, occupation, or some combination of these.

Income refers to money received in a given time period.

Income inequality refers to the extent to which incomes vary within a given population.

FIGURE 7.3 Income Inequality in the United States

Imagine dividing all U.S. citizens into five equal-size groups (quintiles). If all income in the country was also divided equally, each quintile (20 percent) of Americans would receive 20 percent of all income. In reality, the richest 20 percent (quintile) of Americans receives half of all the income, and the poorest 20 percent of the population receives less than 4 percent.

SOURCE: U.S. Bureau of the Census (2006).

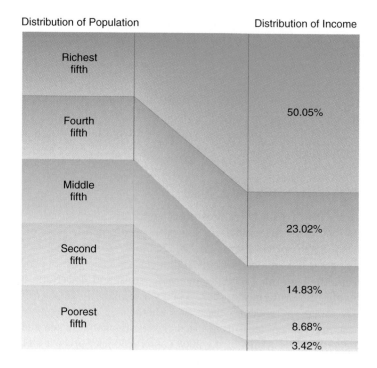

Distribution of Population

Distribution of Income

Richest fifth

Fourth fifth

Middle fifth

Second fifth

Poorest fifth

50.05%

23.02%

14.83%

8.68%

3.42%

Income inequality in the United States has increased steadily since 1970 (DeNavas-Walt & Cleveland 2002; Isaacs, Sawhill, & Haskins 2008). It has increased most, however, at the two ends of the income spectrum: The poorest 10 percent of the population has become significantly poorer, while the richest 10 percent has become significantly richer. When we divide the U.S. population into five equal-sized groups (quintiles), we find that the poorest 20 percent of American households now receive only 3.4 percent of all personal income, whereas the richest 20 percent receive 50 percent of income—more than 14 times as much (Figure 7.3). In contrast, in Sweden, for example, doctors and lawyers earn on average only about twice what waitresses and gas station mechanics earn.

The rise in income inequality stems from changes in the U.S. economic structure coupled with changes in government policy (Massey 2007; Morris & Western 1999). As we will discuss in more detail in Chapter 13, 80 percent of all Americans now work in service or retail jobs. These jobs typically pay far less than the manufacturing jobs that once dominated the U.S. economy. Meanwhile, across all economic sectors, employers are laying off permanent employees and replacing them with lower-paid temporary or part-time workers. Other employers are replacing well-paid American workers with cheaper workers either in Southern states or, increasingly, in foreign countries. At the same time, government policies have 1) made it more difficult for unions to gain members and influence, 2) cut taxes for the wealthy, decreased benefits for the poor, and 3) allowed the value of the minimum wage to decline, thus keeping down the incomes of poor and working-class Americans (Massey 2007).

As bad as income inequality is, looking only at that measure actually understates the levels of economic inequality in the United States. For a more accurate measure of inequality, we need instead to look at wealth. **Wealth** refers to the sum value of money and goods owned by an individual or household at a given point in time (including savings, investments, homes, land, cars, and other possessions). The richest

Wealth refers to the sum value of money and goods owned by an individual or household.

When children grow up in very unequal backgrounds, they are likely to end up leading very different and unequal lives.

20 percent of households by income now own *69 percent* of all wealth (McClain 2005). Historical research suggests that this unequal distribution of wealth is a long-standing pattern in the United States, dating back to at least 1810. However, wealth inequality has increased over the last two decades and is now higher in the United States than in any other industrialized nation (Mahler & Jesuit 2006).

The Consequences of Social Class

Almost every behavior and attitude we have reflects our social class at least somewhat. Do you prefer bowling or tennis? foreign films or American? beer or sherry? These choices and nearly all the others you make are influenced by your social class. Knowing a person's social class will often tell us more about an individual than any other single piece of information. This is why "Glad to meet you" is often followed by "What do you do for a living?"

But social-class differences go beyond mere preferences to real consequences. Consider the following examples:

- People with incomes of less than $7,500 a year are *four* times as likely to have been the victim of a violent crime as those with incomes over $75,000 (U.S. Bureau of the Census 2009a).
- Infants whose mothers fail to graduate from high school are 50 percent more likely to die before their first birthday than infants whose mothers attend college (National Center for Health Statistics 2009).
- Compared to those from more affluent homes, students from poor and working-class homes are much more likely to attend community colleges rather than four-year colleges and to drop out regardless of which type of college they attend (Correspondents of the New York Times 2005).

As these examples suggest, individuals who have more money enjoy a higher quality of life overall.

Theoretical Perspectives on Inequality

According to *Forbes* (2008), Steven Spielberg is now worth $3.0 *billion* and earns many millions each year. Meanwhile, the average police officer earns about $47,000, and 20 percent of American families have annual incomes below $20,291 (U.S. Bureau of Labor Statistics 2009a; U.S. Bureau of the Census 2008a). How can we account for such vast differences in income? Why isn't anybody doing anything about it? We begin our answers to these questions by examining the social structure of stratification—that is, instead of asking about Steven Spielberg or Officer Malloy, we ask why some *groups* routinely get more scarce resources than others. After we review these general theories of stratification, we will turn to explanations about how individuals are sorted into these various groups.

Structural-Functional Theory

The structural-functional theory of stratification begins (as do all structural-functional theories) with the question, Does this social structure contribute to the maintenance of society? The classic statement of this position was given by Kingsley Davis and Wilbert Moore (1945), who argued that stratification is necessary and justifiable because it contributes to the maintenance of society. Their argument begins with the premise that each society has essential tasks (functional prerequisites) that must be performed if it is to survive. The tasks associated with shelter, food, and reproduction are some of the most obvious examples. Davis and Moore argue that we need to offer high rewards as an incentive to make sure that people are willing to perform these tasks. The size of the rewards must be proportional to three factors:

- *The importance of the task.* When a task is very important, very high rewards are justified to ensure that the task is completed.
- *The pleasantness of the task.* When the task is relatively enjoyable, there will be no shortage of volunteers, and high rewards need not be offered.
- *The scarcity of the talent and ability necessary to perform the task.* When relatively few have the ability to perform an important task, high rewards are necessary to motivate this small minority to perform the necessary task.

From this perspective, it makes sense to pay doctors more than childcare workers: Although both fields are necessary, far fewer people have the intelligence, skills, and talent needed to enter medicine, especially since it requires long years of training and long hours of work in sometimes unpleasant and stressful circumstances. To motivate people who have this relatively scarce talent to undertake such a demanding and important task, Davis and Moore would argue that we must hold out the incentive of very high rewards in prestige and income. Society is likely to decide, however, that there will always be plenty of people willing and able to take care of children, even if the wages are low. To structural functionalists, then, the fact that doctors are paid more than childcare workers is a rational response to a social need.

The Concept Summary on Two Models of Stratification compares the structural-functional model of social stratification with the competing conflict model of stratification, which we discuss below.

Criticisms

This theory has generated a great deal of controversy. Among the major criticisms are these:

concept summary

Two Models of Stratification

Basis of Comparison	Structural-Functional Theory	Conflict Theory
1. Society can best be understood as:	Groups cooperating to meet common needs	Groups competing for scarce resources
2. Social structures:	Solve problems and help society adapt	Maintain current patterns of inequality
3. Causes of stratification are:	Importance of vital tasks, unequal ability, pleasantness of tasks	Unequal control of means of production maintained by force, fraud, and trickery
4. Conclusion about stratification:	Necessary and desirable	Unnecessary and undesirable, but difficult to eliminate
5. Strengths:	Consideration of unequal skills and talents and necessity of motivating people to work	Consideration of conflict of interests and how those with control use the system to their advantage
6. Weaknesses:	Ignores importance of power and inheritance in allocated rewards; functional importance overstated	Ignores the functions of inequality and importance of individual differences

1. High demand (scarcity) can be artificially created by limiting access to good jobs. For example, keeping medical schools small and making admissions criteria unnecessarily stiff reduce supply and increase demand for physicians.
2. Social-class background, sex, and race or ethnicity probably have more to do with who gets highly rewarded statuses than do scarce talents and ability.
3. Many highly rewarded statuses (rock stars and professional athletes, but also plastic surgeons and speechwriters) are hardly necessary to the maintenance of society.

Sociologists continue to research these issues and to debate the usefulness of structural-functional theory for understanding inequality.

The Conflict Perspective

Conflict theorists take a very different approach to inequality. They argue that inequality results not from consensus over how to meet social needs but from class conflict.

Karl Marx provided the classic conflict theory of inequality. He argued that inequality grew naturally from the private ownership of the means of production. Those who own the means of production seek to maximize their own profit by minimizing the amount of return they must give to the proletarians, who have no choice but to sell their labor to the highest bidder. In this view, stratification is neither necessary nor justifiable. Inequality does not benefit society; it benefits only the rich.

Like classic Marxist theory, modern conflict theory recognizes that the powerful can oppress those who work for them by claiming the profits from their labor (Wright 1985). It goes beyond Marx's focus on ownership, however, by considering how control also may affect the struggle over scarce resources and how class battles

Removing garbage is both unpleasant and absolutely essential to modern life, yet most garbage collectors are paid low wages. Structural-functional theory attributes their low wages to their lack of skill, whereas conflict theory attributes it to their lack of power.

©Michael Dwyer/Stock Boston Inc.

play out in governmental politics (Grimes 1989; Massey 2007). In addition, modern conflict theory looks at noneconomic sources of power, especially gender and race. These theorists argue, for example, that in the same way that capitalists benefit from the productive labor of workers, men gain benefit from the "reproductive" labor of women. The term **reproductive labor** describes traditionally female tasks such as cooking, cleaning, and nurturing—those tasks that often make it possible for others to work and play. Modern conflict theorists point out that in most families, those with the least power do the most reproductive labor; as a result, these individuals end up having fewer opportunities to earn the good incomes that might otherwise increase their power within the family (Cancian & Oliker 2000).

Criticisms

There is little doubt that people who have control (through ownership or management) systematically use their power to extend and enhance their own advantage. Critics, however, question the conclusion that this means that inequality is necessarily undesirable and unfair. First, people *are* unequal. Some people are harder working, smarter, and more talented than others. Unless forcibly held back, these people will pull ahead of the others—even without force, fraud, and trickery. Second, coordination and authority *are* functional. Organizations work better when those trying to do the coordinating have the power or authority to do so.

Symbolic Interaction Theory

Unlike structural-functional theory and conflict theory, symbolic interaction theory does not attempt to explain why some social groups are so much better rewarded than others. Instead, it asks *how* these inequalities are perpetuated in everyday life.

One of the major contributions of symbolic interaction theory is its identification of the importance of **self-fulfilling prophecies**. Self-fulfilling prophecies occur when something is *defined* as real and therefore *becomes* real in its consequences. This social dynamic is one of the ways that social-class statuses are reinforced.

Reproductive labor refers to traditionally female tasks such as cooking, cleaning, and nurturing that make it possible for a society to continue and for others to work and play.

Self-fulfilling prophecies occur when something is *defined* as real and therefore *becomes* real in its consequences.

For example, when teachers assume that lower-class students are less intelligent and less able to do intellectual work, the teachers are less likely to spend time helping them learn. Instead, teachers may shuffle lower-class students off to vocational classes that emphasize discipline and mechanical skills rather than intellectual skills. After several years of such "schooling," lower-class students may, in fact, have fewer intellectual skills than do others.

Symbolic interaction theory also helps us understand how everyday interactions reinforce inequality by constantly reminding us of our place in the social order. For example, in most restaurants, waiters and waitresses must enter through the back door. They often must use separate bathrooms that are far less pleasant than those used by customers, take their breaks in windowless rooms that lack air conditioning, and wear clothes that make them look like maids and butlers. Customers often speak rudely (or crudely) to serving staff, who are expected to smile in response. And, at the end of the evening, the customer decides whether the waiter or waitress deserves a tip. In all these ways, normal restaurant interactions reinforce customers' sense of social superiority and servers' sense of social inferiority.

The Determinants of Social-Class Position

With each generation, the social positions in a given society must be allocated anew. Some people will get the good positions and some will get the bad ones; some will receive many scarce resources and some will not. In a class system, this allocation process depends on two things: the opportunities available to specific individuals and the overall opportunities available in a society's labor market. We refer to these, respectively, as micro- and macro-level factors that affect achievement.

Microstructure: Individual Opportunities

Unlike in a caste system, in the United States social position is not directly or completely inherited. Yet people tend to belong to a social class the same as or similar to that of their parents. How does this come about? The best way to describe the system is as an **indirect inheritance model**. Parents cannot fully determine their children's social status, but they strongly affect whether their children will have the opportunities needed to obtain or maintain a higher social status.

The best single predictor of your eventual social class is your parents' income (Corcoran 1995; Isaacs, Sawhill, & Haskins 2008). Your parents' income affects your life chances in many ways (Corcoran 1995; Harris 1996; Bettie 2003). If your parents are middle or upper class, you are more likely to be born healthy and more likely to get good nutrition and health care during childhood. As a result, you are less likely to have mental or physical disabilities that might reduce your potential income (Weitz 2010). Your parents will have the time and money to give you a stimulating environment in which your intellectual capacities can thrive and you will most likely attend good schools in which teachers assume their students are "college material." Similarly, as we discussed in Chapter 2, your parents will have endowed you with cultural capital: values, interests, knowledge, and social behavior patterns that mark you as middle or upper class.

Class differences in home environment and in parents' support for school also have important effects on children's success. Bright and ambitious lower-class children

The **indirect inheritance model** argues that children have occupations of a status similar to that of their parents because the family's status and income determine children's aspirations and opportunities.

Wealthier children who can study on their laptops in quiet, private bedrooms find it far easier to succeed academically than do poorer children who have no computers of any sort and who must study in busy, noisy rooms surrounded by younger brothers and sisters.

BDI Stock/Photo Spin

often find it hard to do well in school when they have to study at a noisy kitchen table, have no funds for SAT tutoring or extracurricular activities, have to work part-time jobs to help support their family, and know their parents need them to get full-time employment as soon as possible (Newman & Chen 2008). In contrast, middle-class children who grow up in supportive environments often find it hard to fail even if their ambitions and talents are modest.

In addition, if your parents went to college or have middle-class jobs, you probably always assumed that you would go to college and automatically signed up for algebra and chemistry in high school. Your parents may have given you money to take an SAT prep course, to visit colleges around the country, and to pay for as many applications as you chose to submit. If your parents didn't attend college, they may have encouraged you to start earning an income right away rather than seeking further education. Your high school advisor, too, is more likely to have encouraged you to register for shop or sewing rather than algebra or other courses needed for college entrance (Bettie 2003). If you later decided you wanted to go to college, you first had to overcome all these barriers.

If your parents graduated college, the benefits to you will continue even after you graduate college yourself. Your parents are likely to have both the income and the contacts that will help you get into a good school. After you graduate, they are likely to know people who can help you get good jobs. They may also help you buy clothes for your job interviews, purchase your first home, or pay for family vacations, allowing you to invest your earnings in a new business. They might even invest in the business themselves. All these factors make parents' income a powerful predictor of their children's eventual income (Corcoran 1995; Newman & Chen 2008).

Macrostructure: The Labor Market

The indirect inheritance model explains how some people come to be well prepared to step into good jobs, whereas others lack the necessary skills or credentials. By themselves, however, skills and credentials do not necessarily lead to class, status, or power. The other variable in the equation is the labor market: If there is a major economic depression, you will not be able to get a good job no matter what your education, motivation, or aspirations. Indeed, most observers believe that changes in the nation's economic structure and labor market will offer fewer opportunities for upward mobility over the next generation.

The proportion of positions at the top of the U.S. occupational structure has increased dramatically over the last century. Not everyone, however, has benefited equally from these new opportunities for upward mobility. Although women and minorities now have an easier time entering high-earning occupations, they tend to find themselves in the lower-earning positions within those occupations. They are more likely to be public defenders than corporate lawyers, more likely to be pediatricians than surgeons.

Labor market theorists suggest that the United States has a *segmented labor market*: one labor market for good jobs (usually in the big companies) and one labor market for poor jobs (usually in small companies). Women and minorities are disproportionately directed into companies with low wages, low benefits, low security, and short career ladders.

The American Dream: Ideology and Reality

In any stratification system, there are winners and losers. Why do the losers put up with it?

The answer lies in **ideology**. Ideology refers to any set of beliefs that strengthen and support a social, political, economic, or cultural system. Each stratification system has an ideology that rationalizes the existing social structure and motivates people to accept it. In India, for example, the Hindu religion teaches that if you are in a low caste, you must have behaved poorly in a previous life, but that if you live morally in this life, you can expect to be born into a higher caste in the next life. This ideology offers individuals an incentive to accept their lot in life.

In the United States, the major ideology that justifies inequality is the *American Dream*. This ideology proposes that equality of opportunity exists in the United States and that anyone who works hard enough will get ahead. Conversely, anyone who does *not* succeed must be responsible for his or her own failure. Belief in this ideology is considerably stronger in the United States than anywhere else in the world (Kohut & Stokes 2006; Isaacs, Sawhill, & Haskins 2008). Yet ironically, social mobility is *lower* in the United States than in most comparable Western nations (Figure 7.4 on the next page). For example, compared to the United States, social mobility is 1.4 times greater in Germany, 2.5 times greater in Canada, and more than 3 times greater in Denmark (Isaacs, Sawhill, & Haskins 2008). Studies consistently find that about 50 percent of individual Americans' incomes can be explained by their parents' incomes (Isaacs, Sawhill, & Haskins 2008). So, for example, two-fifths of those born into the poorest 20 percent of families and two-fifths of those born into the richest 20 percent of families remain in the same bracket as their parents when they grow up (Isaacs, Sawhill, & Haskins 2008).

Nevertheless, the American Dream is, for some, a reality. One-third of Americans are upwardly mobile (Isaacs, Sawhill, & Haskins 2008). For immigrants especially, the United States remains a land of opportunity: Well-educated immigrants on average earn more than other Americans, and poorly educated immigrants earn considerably more than they would have if they had stayed in their home countries. But another one-third of Americans are downwardly mobile, and the rest remain in the same social class as their parents.

An **ideology** is a set of norms and values that rationalizes the existing social structure.

FIGURE 7.4 Income Inequality and *Lack* of Social Mobility
Both income inequality and *lack* of social mobility are much higher in the United States than in most other comparable nations. The red bar for each country shows the extent of income inequality; the blue bar shows the extent to which men's incomes match their fathers' incomes (that is, the extent to which the country *lacks* social mobility).

SOURCE: Isaacs, Sawhill, & Haskins 2008; Luxembourg Income Study 2009.

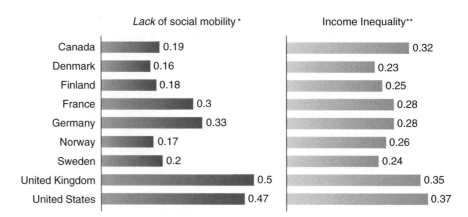

*Extent of match between incomes of fathers and their adult sons.
**Scale ranges from 0 (no income inequality) to 0.6 (highest inequality).

Explaining Upward Mobility

A major reason that the American Dream ideology can survive is because there is, indeed, some upward social mobility. Given all the social forces that hinder mobility, how can we explain why some people do indeed rise above their parents' social class?

It would be easy to assume that the reason some rise and others don't is because of intelligence and hard work, and certainly these factors matter. Most importantly, poor children who graduate college have much more upward mobility than do other poor children (Isaacs, Sawhill, & Haskins 2008). But many very intelligent poor children have no chance of going to college. And some of the hardest-working people earn the lowest incomes.

Sociologist Julie Bettie's ethnographic research is particularly useful for understanding upward social mobility. Bettie (2003) spent nine months intensively observing and interviewing at a predominantly working-class high school. Overwhelmingly, she found, teachers and schools treated students in ways that reinforced the students' class status: Middle-class "preps" were tracked into advanced classes and celebrated for their academic achievements, students from stable working-class homes were encouraged to take vocational classes, and students from poorer homes were ignored, marginalized, and expected to fail. In addition, minority students also suffered discrimination and low teacher expectations, whether they were middle class or poorer.

Nonetheless, some working-class students seemed destined for upward social mobility. All of these upwardly mobile students were smart and hard-working. But they also benefited from resources not available to other working- and lower-class students. Some had become part of middle-class peer groups and received "middle-class treatment" from teachers and advisors because they belonged to mostly middle-class athletic teams or had attended middle-class elementary schools. Some had older siblings who had gone to college and could help them both financially and culturally (by, for example, explaining the importance of earning a four-year degree). All benefited from attending a high school that included college-track, middle-class students rather than a school that was uniformly working- or lower-class. Finally, some students were the children of immigrants who had belonged to the middle class before coming to this country. Although these students lacked the financial resources available to middle-class students, they still had the cultural resources that come with college-educated parents.

Similarly, Dalton Conley (2004) found that differential access to resources explains differences in social mobility *within* families. A son who is already in college when his parents divorce or his father loses his job is more likely to graduate college than is his younger brother who was still in high school when these events occurred. Conversely, when parents' incomes rise over time, they are better able to support their last child through school than their first child. By the same token, when parents lack the money to invest in all their children's education, they may pay for their sons' education but not their daughters' education, pay for their first child but run out of money for the rest, or invest only in the child who seems most likely to succeed. Those who receive the most help from their parents are the ones most likely to experience upward social mobility. As with the students studied by Julie Bettie, social mobility depends on access to resources.

Social Class and Social Life

To a large extent, your social class determines how you live your life. This section briefly reviews the special conditions of each of the classes in the United States.

The Poor

Each year, the U.S. government sets an official *poverty level*, or poverty line: the minimum amount of money a family needs to have a decent standard of living. The poverty level adjusts for family size, and as of 2009 is $21,834 for a family of two adults and two children. In 2007, 37 million people—12.5 percent of Americans—lived in households that earned below the poverty level and were classified by the government as poor (U.S. Bureau of the Census 2009c). Undoubtedly the rate has increased since then, given current economic conditions, but these are the latest data available as of 2009.

Who Are the Poor?

Poverty cuts across several dimensions of society. It exists among white Americans as well as among nonwhites, in small towns and big cities, among those with and without full-time jobs, and in traditional nuclear families as well as in female-headed households. But poverty does not affect all groups equally. As Table 7.1 on the next page indicates, African Americans and Hispanics are far more likely to be poor than are whites or Asians; children are more likely to be poor than are middle-aged or elderly persons; noncitizens (whether native-born or not) are more likely to be poor than are citizens; and households run by single mothers are more likely to be poor than are households run by single fathers or by two parents (U.S. Bureau of the Census 2009c).

Those who live in poverty face crises every day: Parents go hungry so their children can eat, finding clothing for growing children is a nightmare, and a simple cold can easily turn into pneumonia because everyone is under stress and undernourished and no one can afford a doctor's visit. The worst off of the poor have nowhere to call home: About 3.5 million Americans—almost 40 percent of them children—are homeless, and this number is likely to increase, given current economic conditions (National Coalition for the Homeless 2008).

Poor Americans suffer not only because of their *individual* poverty but also because most live in areas of **concentrated poverty**. In these areas—whether rural or urban—schools are typically lower quality, community services are low-quality or non-existent, and jobs are few and far between. As a result, young people not only can't find work, but have few models to suggest that doing so is a reasonable goal.

Concentrated poverty refers to areas in which very high proportions of the population live in poverty.

TABLE 7.1 **Americans Living below the Poverty Level**

	Millions of People	Percentage of Group in Poverty
Total	37.3	12.5
Ethnicity		
White non-Hispanic	16	8.2
African American	9.2	24.5
Hispanic	9.9	21.5
Asian/Pacific	1.3	10.2
Age		
Under 18	13.3	18.0
18–64	20.4	10.9
65 and older	3.6	9.7
Citizenship/nativity		
Native-born	31.1	11.9
Naturalized citizen	1.4	9.5
Noncitizen	4.7	21.3
Household composition		
Married couple	2.8	4.9
Female-headed, no husband	4.1	28.3
Male-headed, no wife	0.7	13.6

SOURCE: U.S. Bureau of the Census 2009c.

In addition, concentrated poverty breeds violence, drug abuse, and alcohol abuse. These problems lead parents to keep their children inside and to stick to themselves, putting the whole social network of a community at risk.

Causes of Poverty

Earlier in this chapter, we said that both micro- and macro-level processes determine social-class position. The causes of poverty are simply a special case of these larger processes. At the micro level, some believe poverty can be explained by various "cultures of poverty"; at the macro level, some believe poverty is better explained by the lack of adequate opportunities.

THE CULTURE OF POVERTY The idea that poverty is caused (or perpetuated) by a **culture of poverty** was first promoted by anthropologist Oscar Lewis (1969). Lewis argued that poor people hold a set of values—the culture of poverty—that emphasizes living for the moment rather than thrift, investment in the future, or hard work. Recognizing that success is not within their reach and that no matter how hard they work or how thrifty they are, they will not make it, the poor come to value living for the moment.

Other scholars have argued that families remain in poverty over generations because a lack of "family values" promotes teen pregnancy and single motherhood or

The **culture of poverty** is a set of values that emphasizes living for the moment rather than thrift, investment in the future, or hard work.

because children raised on welfare conclude that it's smarter to have babies and stay on welfare than to seek employment (Mead 1986, 1992; Murray 1984). Still others argue that poor youths (especially nonwhites) grow up to be poor adults because they actively reject work, education, and marriage as symbols of a middle-class culture that they despise.

Comprehensive reviews of 30 years of research on poverty provide little support for any of these culture of poverty theories (Corcoran 1995; Small & Newman 2001; Newman & Massengill 2006). Researchers have found that poor people overwhelmingly share the same attitudes toward welfare, work, education, and marriage as do middle-class people. This research suggests that teen pregnancy and a "live for the moment" culture is a *result* of poverty, not a cause (Edin & Kefalas 2006; Newman & Massengill 2006).

THE CHANGING LABOR MARKET The culture of poverty theories implicitly blame the poor for perpetuating their condition. Critics of these theories suggest that we cannot explain poverty by looking at micro-level processes. To understand poverty, they argue, we need to look at the changing labor market. If there are no well-paying jobs available, then we don't need to psychoanalyze people in order to figure out why they are poor.

The changing labor market is particularly critical for understanding contemporary poverty. During the first decades of the twentieth century, the shift from an agricultural to an industrial society allowed many people to move upward in social class. In recent decades, however, the *de*industrialization of the United States has eliminated many of the jobs that once paid good wages to people who had little education (Newman 1999a, 1999b; Newman & Chen 2008). Instead of the good union jobs that their parents and grandparents held, today's high school dropouts and graduates often find themselves working at dead-end jobs, with no benefits, at minimum wages that pay too little

Although we usually associate poverty with minorities, most Americans who live in poverty are white.

The loss of many good jobs has forced increasing numbers of Americans into poverty.

Joe Raedle/Getty Images

to pull someone out of poverty. In sum, a major cause of poverty is the absence of good jobs.

The Near Poor

The **near poor** are those who live in households that earn from just above the poverty level to twice the federal poverty level, that is, from about $22,000 to $44,000 currently. Most observers believe that they should also be considered poor, since they still find it very hard to maintain a decent standard of living. However, the lives of the 57 million near-poor Americans differ in important ways from those with incomes below the poverty level (Newman & Chen 2008).

Compared to the poor, the near poor live in safer neighborhoods with better schools (although near and thus exposed to the dangers of poor neighborhoods). Unlike the poor, near-poor adults typically work full-time (or even two jobs), have a roof over their heads, and usually have enough food to eat. On the other hand, their jobs do not pay well, offer few or no benefits, and offer little security (Newman & Chen 2008). They are unable to save much, and so a lost job, work furlough, or week off due to illness may leave them unable to pay their bills. Because they have little or no health insurance, they have to think long and hard before going to a doctor, and often will lose some teeth because they can't afford dental care. Although the government does not consider them poor, they lack much of what others regard as a decent standard of living.

The near poor have been at the heart of the current financial crisis (Newman & Chen 2008). Because near-poor persons live and work around people with higher incomes, they experience relative deprivation (discussed in Chapter 5) whenever they compare themselves to those others. At the same time, they live in neighborhoods with few reputable banks or lending services. As a result, near-poor Americans have been especially targeted by "payday" check-cashing services and by "predatory lenders,"

The **near poor** live in households earning from just above the federal poverty level to twice the poverty level.

such as credit card companies that charge exorbitant fees and mortgage brokers that charge exorbitant interest rates to people they know will end up losing their houses (Newman & Chen 2008).

The Working Class

Who are the members of the working class? The answer is determined partly by income but mostly by occupation, education, self-definition, and lifestyle. Generally, the working class includes those who work in blue-collar industries and their families. They are the men and women who work in factories, on loading docks, and in beauty parlors; they drive trucks, work as secretaries, build houses, and work for maid services. Although they sometimes receive excellent wages and benefits, it is the working class that has 10 to 30 percent unemployment during economic recessions and slumps. And although a majority are high school graduates, an eleventh-grade education is more common than a year of college.

Quite a few members of the working class have incomes as good as or better than those at the lower end of the middle-class spectrum. Truck drivers, for example, often make more than do nurses and public school teachers. As a result, working-class families may live in the same neighborhoods as middle-class families. Their economic prospects differ, however, in three ways.

1. Working-class people have little or no chance of promotion, and their incomes rarely rise much over their lifetimes.
2. Working-class jobs are rarely secure, especially now that the American economy is shifting away from manufacturing to service industries.
3. Working-class people are much less likely than members of the middle class to receive pensions, health insurance, and other benefits.

For these reasons, working-class people are much less likely to have savings or other assets. As a result, layoffs, illnesses, or injuries can quickly drive working class families into poverty (Newman 1999a).

As a result of low prospects and economic uncertainty, members of the working class tend to place a higher value on security than do others. Whereas middle- and upper-class people typically associate having choices with having freedom and control, working-class people associate having choices with insecurity, doubts, and fear (Schwartz, Markus, & Snibbe 2006). So, for example, middle-class Americans more often enjoy rock music and its celebration of individual freedom, whereas working-class Americans more often enjoy country music, which frequently warns about the dangers of choices (such as when George Jones sang "Now I'm living and dying with the choices I've made."). Cultural differences between working- and middle-class Americans are explored further in Focus on Media and Culture: Karaoke Class Wars on the next page.

The Middle Class

The middle class is a large and diverse group. Ranging from professionals with graduate degrees to some salespersons and administrative assistants, middle-class workers have widely varying incomes, with some earning less than the typical working-class individual. Compared with those in the working class, however, middle-class workers tend to have more job security and more opportunities for promotions and raises. Until recently, middle-class workers also could expect to have important benefits such as health insurance and sick leave. The middle class is also united by having at least a high school education and, in most cases, at least some college.

focus on  MEDIA AND CULTURE

Karaoke Class Wars

Prior to the 1990s, sociologists consistently found that the cultural taste of the middle and working classes not only differed significantly, but that middle-class Americans used their cultural taste to distinguish themselves from the working class. Having an original oil painting on the wall, for example, or listening to classical music, not only *showed* that someone was middle-class but was *intended* to have that effect (if only subconsciously).

By the 1990s, however, analysts noticed that middle-class Americans seemed increasingly to be adopting aspects of working-class culture (Peterson & Simkus 1992; Brooks 2000). Listening to hip-hop or country music and wearing "bohemian" clothes signaled that a person might be middle-class, but was still "hip."

Research by Rob Drew (2005), however, questions the extent to which these two cultures are actually blurring. Based on participant-observation at 30 karaoke bars around the country, as well as questionnaires and many informal conversations with participants, Drew found that when middle-class individuals adopt working-class culture (in this case, performing karaoke), they do so in ways that identify them as really middle class.

Karaoke first took root in the United States in working-class neighborhoods. Middle-class commentators reacted with scorn, lambasting the "no-talent" singers and the "death" of true music (Drew 2005). More recently, however, karaoke has become increasingly popular among middle-class Americans. As Drew notes, though, whereas working-class karaoke singers and audiences regard karaoke as a skill deserving of

respect, middle-class participants regard karaoke as acceptable only if treated as an object of humor. They typically sing in comic voices, sing parodies of the lyrics rather than the real lyrics, sing in intentionally inappropriate styles (for instance, singing a ballad in a hard rock voice), or simply burst into laughter throughout their performances. In all these ways, Drew concluded, middle-class performers not only make fun of the songs (most of which come from "working-class" genres like country music and heavy metal), but also make fun of the very idea of a karaoke singer.

In sum, far from suggesting the blurring of class boundaries in cultural taste, middle-class adoption of karaoke has reinforced those boundaries.

Middle-class culture differs from both elite upper-class culture and working-class culture. Compared with working-class individuals, middle-class individuals are less likely to decorate their homes with religious icons or to belong to bowling leagues and are more likely to value education and equality between the sexes; compared with upper-class individuals, members of the middle class are less likely to decorate their homes with modern art or to belong to golf leagues (Halle 1993). Middle-class parents spend time explaining to their children why they need to follow rules and consider themselves responsible for shepherding their children to activities and providing them with entertainment (Lareau 2003). In contrast, working-class parents more often expect children to entertain themselves and to obey orders.

The Upper Class

In 2007, a family living in the United States required an income of $177,000 to be in the richest 5 percent of Americans (U.S. Bureau of the Census 2009a). Thus, a variety of more or less ordinary salespersons, doctors, lawyers, and managers in towns and cities across the nation qualify as very rich compared with the majority. Although their incomes are nothing to sneeze at, most of this upper 5 percent is still middle class. Like members of the working class, they would have a hard time making their mortgage payments if they—or their spouses—lost their jobs and were out of work for a few months. This is because although their current income is quite high, their wealth—their investments, savings, and assets they could easily sell—may not add up to much more than their debts.

Only a small elite ever have the opportunity to drink champagne and eat hors d'ouevres in a skybox at a football game.

© Gerd Ludwig/Woodfin Camp & Associates

The true upper class, on the other hand, consists of two overlapping groups: those whose families have had high incomes and statuses for more than a generation and those who themselves earn incomes in the millions of dollars. The central institution that cements the first group, whose upper-class status is inherited from their parents, is the private preparatory school, especially New England boarding schools such as Andover, Exeter, and Choate (Higley 1995). Many graduates of these schools attend Ivy League colleges, such as Harvard, Yale, and Princeton. After graduation, they are likely to join selective country clubs and high-status Episcopalian or perhaps Presbyterian churches, and to serve on the boards of high-culture organizations such as art museums, symphonies, opera companies, and the like.

Unlike those who inherit their millions, other members of the true upper class earned at least part of their wealth. There are about a half million millionaires in the United States. Few went from rags to riches, however. Most had middle- or upper-class parents who sent them to excellent schools and helped them financially in many ways (Table 7.2 on the next page).

Social Class and Public Policy

If the competition is fair, inequality is acceptable to most people in the United States. The question is how to ensure that no one has an unfair advantage. Politicians, activists, and social scientists have promoted various approaches to fostering equality. Two of these are fair wage movements and increasing educational opportunities.

Fair Wage Movements

One obvious way to foster income equality is to add income to those on the lower end of the social scale. Since the nineteenth century, labor unions have worked to increase wages for American workers, especially in working-class occupations

TABLE 7.2 **The Ten Richest People in the United States**
Four of these fabulously wealthy individuals inherited their fortunes. Five played a large role in generating their vast wealth but also began their careers with many advantages. Only one—Lawrence Ellison—is truly a self-made man.

Rank	Name	Net Worth ($ million)	Source of Wealth
1	William Henry Gates III	57,000	Microsoft, affluent parents
2	Warren Buffett	50,000	Berkshire Hathaway, affluent parents
3	Lawrence J. Ellison	27,000	Oracle Corporation
4	Jim C. Walton	23,400	Wal-Mart inheritance
5	S. Robson Walton	23,300	Wal-Mart inheritance
6	Alice L. Walton	23,200	Wal-Mart inheritance
7	Christy Walton	23,200	Wal-Mart inheritance
8	Michael Bloomberg	20,000	Bloomberg media companies, affluent parents
9	Charles Koch	19,000	Inherited and greatly expanded Koch Industries
10	David Koch	19,000	Inherited and greatly expanded Koch Industries

SOURCE: The Forbes Four Hundred (2008).

(Lichtenstein 2003). Unions have used such tactics as boycotts, strikes, and collective bargaining to pressure employers to meet what the unions consider fair demands for fair wages. Unions played a major role in improving the working and living conditions of workers during the first half of the twentieth century. Since the 1970s, however, manufacturing industries have declined, taking many union jobs with them.

Currently, many who are interested in income equality are focusing on raising the federal minimum wage (Waltman 2000). After adjusting for the effects of inflation, the value of the minimum wage ($7.25 per hour as of 2009) is now worth about 10 percent *less* than in 1979 (Economic Policy Institute 2009).

Individuals who work full-time, year-round at minimum wage jobs earn far less than is needed to move themselves out of poverty, let alone to support even a small family. Raising the minimum wage would at least lighten their burdens.

Increasing Educational Opportunities

Research suggests that education is key to reducing income inequality (Isaacs, Sawhill, & Haskins 2008). Pre-kindergarten classes designed to provide intellectual stimulation for children from deprived backgrounds, special education courses for those who don't speak standard English, and loan and grant programs to enable the poor to go as far in school as their ability permits—all these are designed to increase the chances of students from lower-class backgrounds getting an education.

These programs have had some success: Colleges and universities have many more students from disadvantaged backgrounds than they used to. Because children spend only 35 hours a week at school, however, and another 130 hours a week with their families and neighbors, the school cannot overcome the entire deficit that hinders disadvantaged children. For example, researchers have found that during the school year, poor children and better-off children perform at almost the same level in first- and second-grade mathematics. For poor children, however, every summer means a loss in learning, whereas every summer means a gain for wealthier children (Entwisle & Alexander 1992). The home environments of poorer children rarely include trips to the library or other activities that might encourage them to use and remember what they learn in school. As a result, many scholars and activists now support year-round schools or summer enrichment programs for students from poorer families.

Inequality Internationally

In the same way that inequality can exist *within* a nation, inequality also exists *between* nations. Indeed, a central fact in our world today is the vast international inequality. For example, gross domestic product per capita is $41,890 in the United States but only $806 in Sierra Leone (United Nations Development Programme 2007). Average life expectancy in the United States is 77 years; the average in Sierra Leone is 41. The massive disparities not only in wealth and health but also in security and justice are the driving mechanism of current international relationships.

Because massive inequality leads to political instability and to unjustifiable disparities in health and happiness, nearly every nation—whether more or less developed—supports reducing international inequality. The most accepted way to do this is through development—that is, by raising the standard of living of the less-developed nations.

What is development? First, development is *not* the same as Westernization. It does not necessarily entail monogamy, three-piece suits, or any other cultural practices associated with the Western world. **Development** refers to the process of increasing the productivity and the standard of living of a society, leading to longer life expectancies, better diets, more education, better housing, and more consumer goods.

Importantly, development is not a predictable, unidirectional process. Some countries, such as South Korea, have developed faster than others. Other countries, such as Russia and Argentina, have become *less* developed over time or have fluctuated over the years.

Three Worlds: Most- to Least-Developed Countries

Almost all societies in the world have development as a major goal: They want more education, higher standards of living, better health, and more productivity. Just as social scientists often think of three social classes in the U.S. stratification system—upper, middle, and lower—nations of the world can also be stratified into roughly three levels.

The **most-developed countries** are those rich nations that have relatively high degrees of economic and political autonomy. Examples include the United States, the Western European nations, Japan, Canada, Australia, and New Zealand. Taken together,

Development refers to the process of increasing the productivity and standard of living of a society— longer life expectancies, more adequate diets, better education, better housing, and more consumer goods.

Most-developed countries are rich nations with considerable economic and political autonomy.

MAP 7.1: Most- to Least-Developed Nations

The most-developed nations lie in Europe, North America, and some parts of Asia. The least-developed nations primarily lie in Africa and other parts of Asia.

SOURCE: United Nations Statistics Division (2009).

these nations make up roughly 20 percent of the world's population, produce about 80 percent of the world's gross product, and own about 90 percent of the world's cars (Population Reference Bureau 2008).

Less-developed countries include the countries of Central and South America, plus various countries in Asia and elsewhere. These nations hold an intermediate position in the world political economy. They have far lower living standards than the most-developed nations but are substantially better off than the poorest tier of nations.

The remaining 75 percent or so of the world's population live in the **least-developed countries**. These countries are characterized by poverty and political weakness. Although they vary in population, political ideologies, and resources, they are considerably behind on every measure of development.

The Human Development Index

The differences among the world's nations are obvious: In the most-developed countries, people are healthier, more educated, and richer. But how important are these differences to the average person's life?

One approach to answering this question is to develop an index that measures the average achievements of a country along the basic dimensions of human experience: life expectancy, educational attainment, and a decent standard of living. Another approach not only focuses on these three aspects of development but also takes into account the unequal opportunities of men and women. Map 7.1 shows the location of the most- to least-developed countries worldwide. Table 7.3 compares several basic quality-of-life

Less-developed countries are those nations whose living standards are worse than those in the most-developed countries but better than in the least-developed nations.

Least-developed countries are characterized by poverty and political weakness and rank low on most or all measures of development.

TABLE 7.3 **The Extent of International Inequality**

Type of Country	GDP Per Capita (U.S. dollars)	Life Expectancy at Birth	Infant mortality Rate/1,000 live births	Human Development Ranking*	Gender-Related Development Ranking*
Most-developed countries					
Norway	$53,690	79.3	3	2	3
United States	$41,890	77.4	6	12	16
Canada	$33,375	79.8	5	4	4
Japan	$31,267	81.9	3	8	13
Rep. of Korea	$22,029	77.0	5	26	26
Russian Federation	$10,845	64.8	14	67	59
Less-developed countries					
Saudi Arabia	$15,711	71.6	21	61	70
Brazil	$8,402	71.0	31	70	60
China	$6,757	72.0	23	81	73
El Salvador	$5,255	70.7	23	103	92
India	$3,452	62.9	56	128	113
Kenya	$1,240	51.0	79	148	127
Least-developed countries					
Haiti	$1,663	58.1	84	146	NA
Rwanda	$1,206	43.4	118	161	140
Congo, Dem. Rep.	$714	45.0	129	168	148
Ethiopia	$1,055	50.7	109	169	149
Sierra Leone	$806	41.0	165	177	157
Nigeria	$1,128	46.6	100	158	139

SOURCE: United Nations Development Programme (2007).
*Out of 177 countries

indicators for various most-developed, less-developed, and least-developed countries. In addition to information about longevity and economic productivity, Table 7.3 also includes each country's overall ranking on the composite Human Development Index and the Gender-Related Development Index. These indexes are based on information about adult literacy rates and educational attainment, life expectancy, and per capita gross domestic product; the greater the disparity between men's and women's quality of life, the lower a country's Gender-Related Development ranking will be compared with its overall Human Development ranking.

Women and children are particularly at risk in poor nations. A half-million women in the developing nations die each year during pregnancy or childbirth, at rates up to 100 times those found in the most-developed nations. Worldwide, one out of every three preschool children suffer from malnutrition (United Nations Development Programme 2007). International inequality is indeed dramatic.

In general, the more productive a nation is, the better its quality of life. Norway, with a per capita gross domestic product of $53,690, has one of the lowest infant mortality rates in the world: Each year, for every 1,000 live births in Norway, 3 children die before their first birthday. In contrast, in the world's poorest nation—Niger—per capita gross domestic product is only $630, and 81 out of every 1,000 children die before their first birthday (United Nations Development Programme 2007).

But economic productivity and quality of life do not align perfectly. For example, GDP per capita is 25% higher in the United States than Canada, but according to the United Nations Development Programme (2007), Canadians enjoy a higher quality of life than do Americans. In large part, Canada's rankings on human development reflect the fact that access to health care, education, and adequate nutrition is more universally available there than in the more affluent United States.

No nation wants to be poor and underdeveloped. Why are some nations poor, and what can be done about this? We examine two general theories of development—modernization and world-systems theory—and their implications for reducing global inequality.

Structural-Functional Analysis: Modernization Theory

Modernization theory sees development as the natural unfolding of an evolutionary process in which societies go from simple to complex economies and institutional structures. This is a structural-functional theory based on the premise that adaptation is the chief determinant of social structures. According to this perspective, developed nations are merely ahead of the developing nations in a natural evolutionary process. Given time, the developing nations will catch up.

Modernization theory emerged in the 1950s and 1960s, when many believed that developing nations would follow pretty much the same path as the developed nations. Greater productivity through industrialization would lead to greater surpluses, which could be used to improve health and education and technology. Initial expansion of industrialization would lead to a spiral of ever-increasing productivity and a higher standard of living. These theorists believed this process would occur more rapidly in the least-developed nations than it had in Europe because of the direct introduction of Western-style education, health care, and technology (Chodak 1973).

Events have shown, however, that development is far from a certain process. Some "developed" nations, such as those in the former Soviet Union, have regressed over time. Thailand and the Republic of Korea have modernized quickly, Haiti has modernized hardly at all, and Mexico has gone through wild economic upswings and downturns. The least-developed countries have not caught up with the developed world, and, in many cases, the poor have simply become poorer, while the rich have become richer.

Why haven't the less-developed nations followed in the footsteps of developed nations? The primary reason is that they encounter many obstacles not faced by nations that developed earlier: population pressures of much greater magnitude, environments ravaged by the developed nations since they were colonial powers, and the disadvantage of being latecomers to a world market that is already carved up. These formidable obstacles have given rise to an alternative view of world modernization—world-systems theory.

Conflict Analysis: World-Systems Theory

Conflict theorists' interpretations of modernization begin by arguing that the entire world is a single economic system, dominated by capitalism for the past 200 years.

Modernization theory sees development as the natural unfolding of an evolutionary process in which societies go from simple to complex economies and institutional structures.

| Because of their economic and political power, transnational corporations based in the most-developed nations are able to capture markets in less-developed nations, as FedEx is trying to do in Vietnam.

Nation-states and large **transnational corporations** (that is, corporations that produce and distribute goods in more than one country) are the chief actors in a free-market system in which goods, services, and labor are organized to maximize profits (Chirot 1986; Turner & Musick 1985). This system includes an international division of labor in which some nations extract raw materials and others fabricate raw materials into finished products.

Nation-states can pursue a variety of strategies to maximize their profits on the world market. They can capture markets forcibly through invasion, they can manipulate markets through treaties or other special arrangements (such as NAFTA), or they can simply do the international equivalent of building a better mousetrap. The Japanese auto industry (indeed, all of Japanese industry) is a successful example of the last strategy.

World-systems theory is a conflict analysis of the economic relationships between developed and developing countries. It looks at this economic system with a distinctly Marxist eye. Developed countries are the bourgeoisie of the world capitalist system, and underdeveloped and developing countries are the proletariat. The division of labor between them is supported by a prevailing ideology (capitalism) and kept in place by an exploitive ruling class (rich countries and transnational corporations) that seeks to maximize its benefits at the expense of the working class (underdeveloped and developing countries).

World-systems theory distinguishes two classes of nations: *core societies* and *peripheral societies*. **Core societies** are rich, powerful nations that are economically diversified and relatively free from outside control. They arrive at their position of dominance, in part, through exploiting other (peripheral) societies.

Peripheral societies, by contrast, are poor and weak, with highly specialized economies over which they have relatively little control (Chirot 1977). Some of the poorest countries rely heavily on a single cash crop for their export revenue. For example, 80 percent of export earnings for the island nation of Sao Tome and Principe come from cocoa (Central Intelligence Agency 2008). The economies of these and

Transnational corporations are large corporations that produce and distribute goods internationally.

World-systems theory is a conflict perspective of the economic relationships between developed and developing countries, the core and peripheral societies.

Core societies are rich, powerful nations that are economically diversified and relatively free from outside control.

Peripheral societies are poor and weak, with highly specialized economies over which they have relatively little control.

focus on A GLOBAL PERSPECTIVE

Water and Global Inequality

Without water, human life is unsustainable. And without *clean* water (unpolluted by human waste), diseases such as cholera, typhoid fever, and dysentery soon follow. Similarly, unclean water leads to fatal attacks of diarrhea in 1.8 million children each year (Prüss-Üstün, Bos, Gore, & Bartram 2008).

About one-third of the world's population, almost all of them living in less-developed nations, have limited or no access to clean water. Yet even in these nations, the wealthy have easy access to safe bottled water. In many cases, they also have access to safe, municipal water systems. Due largely to government indifference, however, poor people—whether in rural or urban areas—lack such access (Watkins 2006). Unfortunately, water inequality

in less-developed nations is increasing, as wealthy corporations and neighborhoods pressure governments to divert water from poor, rural areas to their factories and homes (Watkins 2006).

Inequality between males and females also plays a role in water inequality. Men and boys typically receive more of their family's water supplies, even though women and girls often spend hours each day walking to rivers or water pumps and carrying water back to their homes (Watkins 2006). This situation reinforces as well as reflects inequality: When girls spend hours daily in search of water, it is impossible for them to go to school, let alone play or relax.

Ironically, although people in the United States have ready access to safe, virtually free water from their taps, millions instead purchase bottled water. Our ability to do so reflects global inequality: Only members of a very rich

country can afford to buy something that they could get for free and more safely from the government (National Resource Defense Council 2008). Producing, shipping, and disposing of all those bottles requires great amounts of oil; releases great quantities of dangerous chemicals into the air, land, and water; and adds significantly to greenhouse gas emissions and global warming. Many of those processes are occurring in less-developed nations where the water is pumped, the bottles are produced, and, increasingly, the empty bottles along with other waste is sent for disposal (National Resource Defense Council 2008).

In sum, water inequality both reflects and reinforces inequality within the less- and least-developed nations and between these nations and the most-developed nations.

sociology and you

If you have traveled to a less-developed country, you have seen the consequences of economic dependence. You probably were warned not to drink the water, because these nations lack the economic resources to provide safe drinking water. Because wages are so low in these nations, you could buy meals, clothes, and souvenirs very cheaply and could purchase services like taxis that are too expensive to use at home. And because even at these low wages, many can find no jobs at all, you might have seen beggars or prostitutes and been warned about thieves.

A **war** is an armed conflict between a national army and some other group.

many other developing nations are vulnerable to conditions beyond their control: world demand, crop damage from infestation, flooding, drought, and so on.

A key element of world-systems theory is the connectedness between core society prosperity and peripheral society poverty. According to this theory, our prosperity is their poverty. In other words, our inexpensive shoes, transistors, bananas, and the rest depend on someone in a least-developed nation receiving low wages, often while working for a company based in one of the developed nations. Were their wages to rise, our prices would rise, and our standard of living would drop.

The interconnection between poverty and wealth around the world is explored further in Focus on a Global Perspective: Water and Global Inequality.

Global Inequality and Armed Conflict

Inequality can lead to armed conflict both when those who hold power use their resources to seek even more resources (as when most-developed nations invade less-developed nations to obtain oil, gas, or other valued commodities) and when those who lack power rise in revolt.

Global Inequality and War

Fights between two street gangs in Chicago, or between Hindu and Muslim citizens in India, may result in many deaths, but they are not wars. A **war** is an armed conflict in which at least one side is organized into an army working directly for a government (Kestnbaum 2009). In the past, scholars tended to define a conflict as a war only if two national armies were involved. However, much current warfare is being fought between an army on one side and groups of armed civilians on the other.

War always reflects changing relations among three groups: the government, the armed forces, and the public (Kestnbaum 2009; Paret 1992; Geyer 2002). Governments can only engage in warfare when the armed forces support them or when the public is willing to take up arms to defend a government under siege. The armed forces can only engage in warfare when they have either the support of the government and public or sufficient power to ignore or kill their opponents. The public will support the army and government if it believes the army and government care about the nation's people; the public may rise up in resistance if it regards the army and government as its enemies and if it believes resistance is worth the cost. Public awareness of vast inequalities within nations have led to violent class, ethnic, or political struggles around the world (Kerbo 2005). During the last decade, such conflicts have occurred in Israel, Macedonia, Mexico, Afghanistan, and elsewhere.

In addition to reflecting changing relations *within* a country, contemporary warfare—including civil warfare—often reflects changing relationships *between* countries (Kestnbaum 2009; Hironaka 2005). Sometimes one country seeks to grab resources directly from another country or from groups within its own country. These days, however, warfare often results when governments and business interests in the most-developed nations use their resources to bolster the power of corrupt, weak governments in the less-developed nations. Such situations have fostered armed revolts in Vietnam, Iran, and Iraq, among other places.

At the same time, governments in the less-developed countries are increasingly angry at the ways the most-developed nations have affected their economies, culture, and politics. Many observers believe that the greatest threats to the United States are posed by less-developed nations that either have or are attempting to develop nuclear weapons, such as Iran, North Korea, and Pakistan.

Global Inequality and Terrorism

On September 11, 2001, terrorists rammed two jets into the World Trade Center in New York City and a third into the Pentagon, outside Washington, D.C., killing almost 3,000 people. In November 2008, terrorists killed almost 200 civilians in a coordinated series of attacks in Mumbai, India. In July 2005, terrorists left bombs in the London public transit system that killed 52 people and injured about 700.

What do we mean when we call actions *terrorism*? To scholars, **terrorism** refers to the deliberate and unlawful use of violence against civilians for political purposes. Terrorism, however, is a social construction: One group's "terrorists" are another group's "freedom-fighters" or "martyrs" (Turk 2004). Which label sticks depends in part on who wins and gets to write the history books: The American Revolution looked very different to the British (who called it "The Rebellion"). The U.S. government typically has labeled only foreigners as terrorists, preferring to treat violence by Americans in the United States (such as murders of abortion providers and the 1995 Oklahoma City bombings) as the actions of individual, deranged, criminals rather than as terrorism (Turk 2004).

Why do individuals engage in terrorism? Surprisingly, *personal* experience of inequality plays little role: The poor are *less* likely than others to engage in terrorism. Instead, terrorism is largely rooted in perceived threats to *national* or *cultural* pride (Turk 2004). People from less-developed nations most often embrace international

Terrorism is the deliberate and unlawful use of violence against civilians for political purposes.

Anti-American sentiment in other countries reflects fear and resentment of American cultural, economic, political, and military power.

terrorism when they believe their culture is being corrupted by Western culture, and their nation's economy and government are being unethically pressured or controlled by Western nations. They become willing to engage in terrorism when socialized to consider it the only means to right these injustices (Turk 2004).

Case Study: Islamic Terrorism

Around the world, Christian, Jewish, and Hindu terrorists have engaged in politically motivated violence. Recently, though, Islamic terrorism has dominated the headlines. But although religious ideology can play a role in terrorism, global inequality is also highly important (Amanat 2001; Barber 2001; Jacquard 2002; Stern 2003).

One underlying cause of Islamic terrorism is the deepening belief among many Muslims that their nations and religion are under political attack. This belief has roots in the Russian invasion of Afghanistan and the civil war in Bosnia that pitted Muslims against Christians. Actions taken by the United States have also played a large role in creating this sense of victimization among many Muslims. The United States consistently has supported Israel against the Palestinians. It has also invaded Iraq, used economic blockades against Iran, and based military troops in Saudi Arabia near Islam's holiest sites. These actions have wounded the pride of many Muslims and left them with a sense that both Muslim governments and Islam itself are under attack.

Problems within the Muslim countries of the Middle East and Asia also have contributed to terrorism. Many of these countries have been wracked by war on and off for the last century. Poverty is very high, inequality is extreme, and governments by and large are corrupt. As a result, poor children often can only afford to attend free Islamic schools (*madrasas*) that teach little beyond extreme, fundamentalist versions of Islam. These forces have made it easier for the leaders of Al Qaeda and other similar groups to find foot soldiers for their battles. Meanwhile, although wealthier residents of Muslim nations are protected from the worst impacts of these forces, they still live in a culture of alienation, despair, and wounded pride. These are the individuals, like

Osama bin Laden and the 19 terrorists who attacked the United States on 9/11, who become the leaders and lieutenants in global terrorism.

Finally, because of the mass media and information technology, Muslims throughout the world are now inundated with American culture. In countries and cultures where women are expected to cover themselves from head to toe, American television shows display nearly naked women. Hip-hop music boasts of sexual conquests, and Hollywood romances feature independent women and men whose lifestyles are the antithesis of traditional Muslim values. Around the world, America has become the symbol of the good life, but also a symbol of materialism, violence, promiscuity, and the attack on traditionalism. Islamic terrorism has emerged in part as a way to counter all these facets of American culture.

Where This Leaves Us

Research on stratification leads to one basic conclusion: As long as some people are born in tenements to poorly educated parents who lack the time, money, and cultural resources needed to provide their children with intellectual stimulation, while others are born to wealthy, educated parents with excellent connections and "cultural capital," there can never be true equality of opportunity. Similarly, as long as some nations have a greater share of resources—money, oil, media outlets, good schools— other nations can never flourish. Those nations that lack power will have lower life expectancies, many homeless people, and many who experience malnutrition or even starvation. Thus the only way to create equal opportunity, either within or across nations, is to attack these underlying problems. The question for Americans is, should we care?

To answer this question from a moral perspective, we might point out that quality of life among poor and affluent Americans and between poor and wealthy nations are directly related to each other. Wealthy Americans enjoy a good life because they can cheaply hire maids and taxi drivers and can buy houses and other goods produced by poorly paid American workers. Similarly, citizens of the United States and other developed nations enjoy raw goods and products obtained cheaply from countries where people work for pennies an hour, and enjoy the security of knowing that other nations cannot challenge our military and economic power. By the same token, American culture is spreading around the world in part because of our economic and political power and because other cultures lack the power to oppose it.

But this is a sociology textbook, not an ethics textbook. From a sociological perspective, perhaps the most important issue is how both individuals and societies can benefit from reducing inequality. Of course, few wealthy individuals or nations want to give resources away. On the other hand, inequality costs everyone. In nations where inequality is high, everyone—including the wealthy—experiences more stress, more crime, worse health, and lower life expectancies (Marmot 2004; Wilkinson 1996, 2005). It's just not as much fun being wealthy or even middle class when you have to lock your doors all the time, worry about crime, fear that you might lose your job to a cheaper worker, and fear that your standard of living might plummet if you were ill or injured. Similarly, wealthy nations can never relax their guard when other nations envy their economic and cultural position. The events of 9/11 demonstrate what can happen when resentment of wealthy nations rises.

Summary

1. Stratification differs from simple inequality in that (a) it is based on membership in social categories rather than on personal characteristics, and (b) it is supported by norms and values that justify unequal rewards.

2. There are two types of stratification systems. In a caste system, your social position depends entirely on your parents' position. In a class system, your social position is based on educational and occupational attainment and so you may wind up in a higher or lower position than that of your parents.

3. Marx believed that there was only one important dimension of stratification: class. Weber added two further dimensions, and most sociologists now rely on his three-dimensional view of stratification: class, status, and power.

4. Inequality in income and wealth is substantial in the United States and has increased steadily since 1970. Income inequality is higher in the United States than in any other industrialized nation. Wealth inequality is even greater than income inequality in the United States.

5. Structural-functional theorists argue that inequality is a necessary and justifiable way of sorting people into positions. Conflict theorists believe that inequality arises from conflict over scarce resources, in which those with the most power manipulate the system to enhance and maintain their advantage. Symbolic interaction theory focuses on how social status is reinforced through self-fulfilling prophecies.

6. Allocation of people into statuses includes macro and micro processes. At the macro level, the labor market sets the stage by creating demands for certain statuses. At the micro level, the status attainment process is largely governed by indirect inheritance.

7. Despite high levels of inequality, most people in any society accept the structure of inequality as natural or just. In the United States, the ideology that teaches people to accept inequality is the *American Dream*, which suggests that success or failure is the individual's choice.

8. Upward mobility is most common among those who have better access to economic, educational, and cultural resources, whether compared with their siblings or with children from other families.

9. Currently 12.5 percent of the U.S. population falls below the poverty level. Although some have argued that "cultures of poverty" explain why people stay poor, research suggests that the shrinking options provided by the current labor market is a more likely explanation.

10. Among the approaches proposed for reducing poverty are fair wage movements and increasing educational opportunities.

11. International inequality is a key factor in today's world. Reducing this disparity through the development of less-developed and least-developed countries is a common international goal. Development is not the same as Westernization; it means increasing productivity and raising the standard of living.

12. The world's nations can be divided into the rich, diversified, independent, most-developed core nations; the least-developed nations of the periphery; and the less-developed nations, which fall in between these two extremes.

13. The Human Development Index and the Gender-Related Development Index use literacy and educational attainment, life expectancy, and economic productivity to assess quality of life overall, as well as quality of life adjusted for the effects of gender inequality.

14. Modernization theory, a functionalist perspective of social change, argues that less-developed countries will evolve toward industrialization by adopting the technologies and social institutions used by the developed countries.

15. World-systems theory, a conflict perspective, views the world as a single economic system in which the industrialized countries, known as core societies, control world resources at the expense of the less-developed, peripheral societies.

16. Inequality within nations can lead to nationalist revolutions or violent class or ethnic struggles.

17. War always reflects changing relations among three groups within a country—the government, the armed forces, and the public—and can reflect changing relationships between countries.

18. Terrorism is the deliberate and unlawful use of violence against civilians for political purposes. It is also a social construction: The winners usually decide who is labeled a *terrorist*. Terrorism typically results from perceived threats to national or cultural pride.

Thinking Critically

1. Can you think of any ways in which the U.S. system of stratification resembles a caste system?
2. To what social class do you belong? How do you know? How are you affected by your social class?
3. What wealth does your family own? What cultural capital does your family have? How have your family's wealth and cultural capital, or lack of wealth and cultural capital, affected you?
4. You are a hundred times better off than the average person in Haiti. Is this a necessary and just reflection of your greater contribution to society? How do you benefit from this inequality? How are you harmed by it?
5. Critically evaluate the components of the Human Development Index. Which seems most important to you? Could this index be used to understand group differences in quality of life within the United States?

Book Companion Website

www.cengage.com/sociology/brinkerhoff
Prepare for quizzes and exams with online resources—including tutorial quizzes, a glossary, interactive flash cards, crossword puzzles, essay questions, virtual explorations, and more.

CHAPTER 8

Racial and Ethnic Inequality

Mario Tama/Getty Images

Race and Ethnicity

Race and ethnicity are ascribed characteristics that define categories of people. Each has been used in various times and places as bases of stratification; that is, cultures have thought it right and proper that some people receive more scarce resources than others simply because they belong to one category rather than another. In the following section, we provide a basic framework for looking at racial and ethnic inequality before focusing on the situation in the United States.

Understanding Racial and Ethnic Inequality

How is it possible for groups to interact on a daily basis within the same society and yet remain separate and unequal? In this section, we begin by introducing some basic concepts needed to understand racial and ethnic inequality: the social construction of race and ethnicity, how disadvantages multiply, the concepts of majority and minority groups, and the basic patterns of interaction among majority and minority groups.

The Social Construction of Race and Ethnicity

A **race** is a category of people treated as distinct because of *physical* characteristics to which *social* importance has been assigned. An **ethnic group** is a category whose members are thought to share a common origin and important elements of a common culture—for example, a common language or religion (Marger 2003). Both race and ethnicity are inherited from one's parents.

Although many assume that race and ethnicity are genetic traits, all humans, regardless of their race or ethnicity, share virtually the same pool of genes. Both race and ethnicity are based loosely if at all on physical characteristics, such as skin color. For this reason, sociologists talk of the **social construction of race and ethnicity**: the process through which a culture defines what constitutes a race or an ethnic group. As this suggests, this process is based more on social ideas than biological facts; indeed, biologists are almost unanimous in believing that race has no biological reality.

How are racial and ethnic identities socially constructed? Consider the changes in racial definitions that emerged during the 1930s. Before this, the modern concept of a "white race" really didn't exist. Instead, people talked of multiple races, including an Anglo-Saxon race, a Mediterranean race (Italians and Greeks), a Hebrew race (Jews), and Slavic races (Jacobson 1998). Around 1930, doctors, politicians, lawyers, anti-immigrant activists, journalists, and others dropped these distinctions and instead began describing whites as a single racial group. Sociologists would say that these professionals and activists, whether or not they realized it, were engaging in the social construction of whiteness as a racial category. At the same time, the U.S. Bureau of the Census declared that Mexican Americans would be classified in the census as nonwhite. The Mexican government complained, and the Bureau reversed itself. Currently the Census Bureau defines Hispanic Americans as an ethnic group whose members can belong to any race. Similarly, the shift from using the term *black* to using the term *African American* reflected changing social ideas about race

A **race** is a category of people treated as distinct on account of *physical* characteristics to which *social* importance has been assigned.

An **ethnic group** is a category whose members are thought to share a common origin and important elements of a common culture.

The **social construction of race and ethnicity** is the process through which a culture (based more on social ideas than on biological facts) defines what constitutes a race or an ethnic group.

Since the breakup of the Soviet Union, people in Lithuania and other former Soviet republics can now celebrate and highlight their ethnicity.

Austrophoto/F1online digitale Bildagentur GmbH/Alamy

and ethnicity, not any new information about the biological origins of that group. Each of these examples illustrates the social construction of race and ethnicity: a political process in which groups compete over how racial and ethnic categories should be defined.

Elsewhere in the world, new ethnic identities arise as national borders shift. Only during the twentieth century did Sicilians, Napolitanos, Milanese, and others begin developing a common Italian language, culture, and ethnic identity. Conversely, since the break-up of the former Soviet Union, Lithuanians, Latvians, Kazakhs, Abkhazians, and others have worked to rebuild ethnic, linguistic, and cultural traditions and identities that had been suppressed or even abandoned during the Soviet years. As these examples illustrate, racial and ethnic statuses can fluctuate. Over time, individuals may change their racial and ethnic identification, and society, too, may change the statuses it recognizes and uses.

Majority and Minority Groups

In addition to talking specifically about whites and African Americans or Jews and Arabs, sociologists interested in race and ethnicity also talk more broadly about majority and minority groups. A **majority group** is one that is culturally, economically, and politically dominant. A **minority group** is a group that, because of physical differences, is regarded as inferior and is kept culturally, economically, and politically subordinate.

Although minority groups are often smaller than majority groups, that is not always the case. For example, by the late twentieth century, whites comprised only 15 percent of the population in South Africa. However, whites controlled all major political and social institutions until apartheid (legal segregation) was abolished in 1994. Sociologically, then, whites were the majority group under apartheid. Similarly, some scholars regard women as a minority group because, based on physical sex

A **majority group** is a group that is culturally, economically, and politically dominant.

A **minority group** is a group that is culturally, economically, and politically subordinate.

TABLE 8.1 **Income and Wealth of Families by Ethnicity**
The United States is stratified by both race and class. Within each racial or ethnic group, the richest 20 percent receive about half of all income for that group, indicating real social class differences. At the same time, whites as a group have considerably more income and wealth than do African Americans or Hispanics, indicating real racial and ethnic differences.

Income Quintile	Percent of Total Income Received, within Ethnic Group		
	African American	Hispanic	White Non-Hispanic
Poorest fifth	3	4	4
Second fifth	9	9	9
Third fifth	15	15	15
Fourth fifth	24	23	23
Richest fifth	49	49	50
Medians			
Median Income	$34,192	$35,054	$53,256
Median Wealth	$6,166	$6,766	$67,000

SOURCE: U.S. Bureau of the Census 2005, 2008b, 2008c.

differences, they have been economically, politically, and culturally subordinate to men.

Multiplying Disadvantages

Most contemporary scholars use some form of conflict theory to explain how racial and ethnic inequalities—or more generally, inequalities between majority and minority groups—are developed and maintained. This theory suggests that in the conflict over scarce resources, historical circumstances such as access to technology and the existence of slavery gave some groups advantages while holding other groups back. To maintain their power, those who have advantages work to keep others from getting access to them (Tilly 1998). These inherited advantages have left us with two stratification systems, one based on class and one based on race and ethnicity.

These two stratification systems work together to multiply disadvantages and inequality. We can see how this works in Table 8.1. As the table shows, the different racial and ethnic groups display very similar patterns of *internal* inequality: *Within* each of these three groups, the wealthiest 20 percent of families receive half of all income. On the other hand, comparing *across* the three groups, we see that the median income of white non-Hispanic families is about 1.5 times that of Hispanic and African American families.

The differences become even more extreme when we look not at income but at *wealth*. As Table 8.1 shows, the median net worth of white non-Hispanic families was $67,000. This is about *ten times* higher than the median for African American and Hispanic families. These racial differences in wealth, not racial differences in income, form the roots of the continuing U.S. racial divide (Shapiro 2004).

Case Study: Environmental Racism

One example of how poverty and racism combine to multiply inequality is *environmental racism*. The term **environmental racism** refers to the disproportionately

Environmental racism refers to the disproportionately large number of health and environmental risks that minorities face daily in their neighborhoods and workplaces.

large number of health and environmental risks that minorities, especially if they are poor, face daily in their neighborhoods and workplaces (Bullard, Warren, & Johnson 2001; Camacho 1998). For example, landfills for hazardous waste are disproportionately located in African American and Hispanic communities. Farmworkers and their children, most of whom are Hispanic and very poor, are exposed to poisons whenever the crops they pick are sprayed with pesticides. On poor Native American reservations where uranium mining is often the only well-paid job, mining has poisoned thousands of workers, as well as their spouses and children, when mine waste seeps into the water or is blown into the air. This unequal environmental burden exists because manufacturers, mining companies, and the like find it easiest to locate polluting industries in poor minority communities that lack the political power to enforce environmental restrictions and that are desperate for jobs, no matter the environmental cost.

The best predictor of exposure to environmental pollution is race; the second best predictor of exposure is poverty (Brulle & Pellow, 2006). These environmental hazards reinforce as well as reflect ethnic and class inequality: Children exposed to toxic chemicals or air pollution, for example, risk mental retardation, developmental delays, and physical illnesses such as asthma that can lead them to miss school days. As a result, these children are less likely to succeed in school and more likely to continue to live in poverty as adults.

Patterns of Interaction

Relations between racial and ethnic groups can take one of three general forms: *pluralism*, *assimilation*, or *conflict*.

Pluralism

When two or more groups coexist as separate and equal cultures in the same society, we speak of **pluralism**. In a truly pluralist society, each of the different cultures is valued, each has its own equally valued institutions, membership in one or another culture does not affect individuals' social position, and all value their shared membership in the same society.

In reality, separate rarely means equal, whether we are talking about white and African Americans, English- and French-speaking Canadians, or Shiite and Sunni Muslims in Iraq. Nevertheless, nations that consider themselves pluralistic give at least outward support to the idea of equality.

Although the United States has not achieved true pluralism, it has done much better than most other societies (Alba & Nee 2003). White and nonwhite Americans increasingly go to the same schools, live in the same neighborhoods, belong to the same social groups, and are willing to marry one another.

Assimilation

As we saw in Chapter 2, assimilation is the process through which members of a minority culture lose their defining cultural features and adopt those of the majority culture. For example, most immigrants to the United States quickly stop wearing the distinctive clothing of their native lands, and most children of immigrants speak only English. Many Jewish Americans now celebrate Christmas (or at least have Christmas trees and lights), and most Irish Americans eat corned beef and cabbage only on St. Patrick's Day, if at all.

When assimilation is complete, the traces of a minority group may all but disappear. For example, many white Americans suspect they have a Native American ancestor but have no knowledge of that ancestor's culture.

Pluralism is the peaceful coexistence of separate and equal cultures in the same society.

Although the United States is not fully pluralistic, in many settings children from different races and ethnic groups interact easily.

Image copyright Monkey Business Images, 2009. Used under license from Shutterstock.com

Conflict

Relations between minority and majority groups often take the form of conflict. For much of the twentieth century, racial and ethnic conflict in the United States was reflected in laws and customs that forbade social, political, or economic participation by minorities. In other times and places, racial and ethnic conflict has taken the form of slavery, driving minority groups into concentration camps, or expelling minorities from a country altogether. At the extreme, conflict can result in **genocide**: mass killing aimed at destroying a population (Jones 2006). Genocides have occurred throughout history; recent genocidal attacks have occurred in Afghanistan, Pakistan, Darfur, and elsewhere. The killings in Darfur are discussed further in Focus on a Global Perspective: Genocide in Darfur on the next page.

THE STAGES OF GENOCIDE No society goes straight from tolerance to genocide. Instead, societies typically follow a predictable set of stages (Stanton 2009). First, they classify individuals into different groups (Christians and Muslims, Hutus and Tutsis). Then they use symbols such as clothing or tattoos to mark the different classifications. The next step is dehumanization: convincing the general public to believe that the minority group is less than human. For example, during the nineteenth century American politicians and military officers often described Native Americans as less-than-human savages.

Once a minority group has been dehumanized, the risk of genocide is high (Hagan & Rymond-Richmond 2008; Stanton 2009). Either the government, the military, or groups of civilians may begin spreading hate propaganda, forcing segregation, and organizing plans for mass killings. After this happens, it is relatively easy to deport the intended victims to death camps or famine-starved regions and to find people willing to do the killing.

The good news is that at each of these stages, appropriate interventions can keep genocide from happening (Genocide Watch 2009). For example, during World War II,

Genocide refers to mass killings aimed at destroying a population.

focus on A GLOBAL PERSPECTIVE

Genocide in Darfur

Racial inequality is not solely an American problem. Discrimination and prejudice in other countries also deny minority groups their rights and opportunities. The genocide in Darfur offers a recent example (Hagan & Rymond-Richmond 2008). Although the political process that underlies Darfur's ethnic strife may be unique to that society, the economic processes appear to be typical of those accompanying ethnic conflict in societies throughout history: Racial and ethnic hostilities are most pronounced when economic resources are scarce and the majority group's economic advantage is threatened.

Darfur is a region in western Sudan, the largest country in Africa. Like most African countries, Sudan was cobbled together by a colonial power—in this case, Britain—during the nineteenth century. The country is overwhelmingly composed of Sunni Muslims who use Arabic as their lingua franca, but northern Sudan is primarily Arab, while the

rest of the country is now considered black African. Ironically, the physical differences between these two groups are slight enough that westerners typically cannot distinguish Sudanese Arabs from Sudanese Africans. Indeed, prior to this conflict, ethnic identity was fluid and relatively unimportant, intermarriage was common, and the distinction between "Arab" and "African" was rarely used. African farmers coexisted easily with Arab herders, since each benefited from trading with the other. Moreover, Arab herders sometimes became farmers, and African farmers sometimes became herders, depending on their shifting economic circumstances.

Since Sudan achieved independence in 1956, Arabs from northern Sudan have dominated the country's economy and government. Yet their home territories hold few of Sudan's agricultural lands, oil deposits, or other natural resources. Moreover, global warming, growing human and livestock populations, and damaging agricultural practices are all contributing to the "desertification" of northern Sudan.

To maintain their dominance over the country and its natural wealth, the Sudanese Arabs who run the country's government have used military repression, political repression, and economic strangulation against their perceived enemies. In response, since the 1980s armed resistance by Sudanese Africans, in both southern and western Sudan, has increased, as has repression by the central government.

Beginning in 2003, however, the Sudanese government moved from repression to what most observers describe as genocide against the people of Darfur (and, increasingly, against Africans in neighboring Chad). To facilitate this policy, the government embarked on a campaign to dehumanize Sudanese Africans in the minds of Sudanese Arabs (Hagan & Rymond-Richmond 2008). They then formed and armed local Arab militias, known as "Janjaweed." Janjaweed members were recruited from nomadic and semi-nomadic Arab tribes who hoped to gain not only war loot, but also access to increasingly scarce water sources,

Nazi German forces occupied Denmark. However, unlike in the rest of occupied Europe, the Danish government refused to order Denmark's Jews to wear yellow stars on their clothing, the Danish police helped Jews to hide, and a flotilla of Danish fishermen helped to ferry Jews to Sweden, which was outside Nazi control. As a result, only 17 percent of Danish Jews were killed (United States Holocaust Memorial Museum 2009; Bergen 2003). In contrast, in Poland, where much of the government, police, and the public supported killing—or at least removing—Jews, more than 90 percent of Jews died.

Map 8.1 on page 192 shows the countries in which genocidal killings are now being planned or are occurring.

Maintaining Racial and Ethnic Inequality

This section looks at how segregation, prejudice, and discrimination work together to maintain social distance and thus racial and ethnic inequality.

pasture for livestock, and arable lands (Human Rights Watch 2006). Although officially the war is now over, the deaths and destruction continue.

By 2009, the Sudanese army and the Janjaweed had killed more than 100,000 civilians and driven almost 6 million refugees away from their homes (U.S. Department of State 2009). In addition, the Janjaweed have engaged in aerial carpet bombing, systematic torture, mass amputations with machetes, and mass rape—all aimed overwhelmingly at civilians rather than at resistance fighters. To justify these actions, the government and Janjaweed have encouraged racial stereotyping of African Sudanese as inferior. As a result, Sudanese civilians increasingly identify themselves as Arab or as African, rather than as Sudanese or as members of a specific tribe. Meanwhile, both intraethnic and interethnic violence is exploding.

In March 2009, the International Criminal Court charged Sudanese President Omar al-Bashir with war crimes and crimes against humanity.

Since 2003, and in a bid to control valuable lands and water supplies, the Sudanese government has encouraged racial stereotyping of African Sudanese as inferior and has promoted the slaughter of Sudanese Africans by Sudanese Arabs like these "Janjaweed" militia members. Yet, before this conflict, ethnic identity in Sudan was fluid and relatively unimportant, intermarriage was common, and people rarely distinguished between "Arab" and "African" Sudanese.

Segregation

One easy way to maintain social distance and inequality is through **segregation**—the physical separation of minority- and majority-group members. Thus, most societies with strong divisions between racial or ethnic groups have ghettos, barrios, and Chinatowns where, by law or custom, members of the minority group live apart.

Historical studies suggest that high levels of residential segregation of Hispanic, Asian, and African Americans in the United States have existed since at least 1940. Such segregation is no longer established in law, but it is no historical accident. Segregation continues for two reasons: economic differences across racial/ethnic groups and continuing prejudice and discrimination (Iceland & Wilkes 2006).

Economic differences certainly matter. Lower-income Hispanics, African Americans, and Asians are all more likely to live in ethnically segregated neighborhoods than are wealthier members of the same groups. This suggests that if minorities' social class increases, segregation will decline.

But economic differences alone can't explain segregation. Whereas Hispanics and Asians are significantly less likely to live in segregated neighborhoods if they are at least middle class, this is not true for African Americans. Similarly, even when African

Segregation refers to the physical separation of minority- and majority-group members.

MAP 8.1: Genocide and Genocide Risk Internationally
Many countries around the world are now engaging in genocide or are at high or very high risk of doing so. Those at *high risk* have begun organizing mass killings and spreading hate propaganda. Those at *very high risk* have begun drawing up death lists or sending minority-group members to death camps located in famine-starved areas.
SOURCE: Genocide Watch (2009)

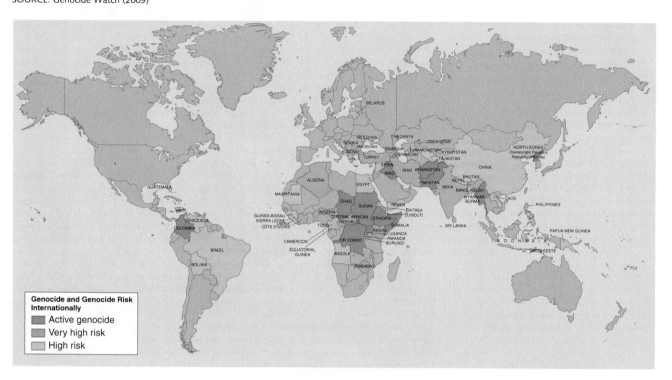

Americans are educated, affluent, and move to the suburbs, they remain substantially less likely than whites to escape "distressed" neighborhoods (Crowder, South, & Chavez 2006; Iceland & Wilkes 2006; Alba, Logan, & Stults 2000). This suggests that prejudice and discrimination continue to foster segregation of African Americans. Studies find that, compared with others with similar incomes, African Americans are less likely to be shown homes in "nicer" areas by real estate agents and are more likely to be turned down for mortgages or to face hostility from potential neighbors (Iceland & Wilkes 2006; Ross & Turner 2005).

Current data suggest that we should be guardedly optimistic. Real estate agents and mortgage brokers are less likely to discriminate against African Americans than in the recent past, and African American segregation has declined somewhat since the 1970s. Segregation of other groups shows little decline, but this is mostly explained by recent immigration from Asia and Latin America (Iceland & Wilkes 2006).

Prejudice

Segregation is typically justified based on prejudices and stereotypes. A **prejudice** is a negative view of a group of people not based on evidence. Prejudice exists despite the facts rather than because of them. A person who believes that all Italian Americans have ties to the Mafia will ignore any instances of the law-abiding behavior of Italian

Prejudice is an irrational, negative attitude toward a category of people.

Racial segregation remains a fact of life in the United States. Even among the middle class, African Americans are more likely than European Americans to live in a poor neighborhood.

Americans. If confronted with an exceptionally honest man of Italian descent, the bigot will rationalize him as the exception that proves the rule. **Racism** is a form of prejudice. It is the belief that inherited physical characteristics associated with racial groups determine individuals' abilities and are a legitimate basis for unequal treatment.

These days, explicit racism has become less common than in the past. Instead, we more often see color-blind racism. **Color-blind racism** refers to the belief that all races are created equal, that racial equality has been achieved, and that therefore any minorities who do not succeed have only themselves to blame. This belief leads many white Americans to oppose policies designed to combat racism or to improve opportunities for minorities and to oppose politicians who support such policies (Bobo & Kluegel 1993; Quillian 1996; Herring 2003; Bonilla-Silva 2006). An important criticism of the concept of color-blind racism is that it can be hard to tell whether people oppose these policies because they are *racist* or because they are *conservatives* who would oppose any government interventions (Quillian 2006).

The basic building blocks of prejudice are stereotypes. A **stereotype** is a preconceived, simplistic idea about the members of a group. For example, you may know someone who believes that all athletes and cheerleaders are dumb or that all Latinos are good dancers. Stereotyping does have its uses. It's probably a good idea to assume you should stay away from someone who is waving a gun in the air and mumbling to himself, and it's probably a safe bet that a very fashionably dressed woman can give you directions to a high-end shopping mall. Life would be very difficult if we had to start absolutely from scratch in every social interaction, with no idea of how this individual might be similar to or different from others we've met (or heard about) in the past.

On the other hand, stereotypes also *hinder* social interactions when they lead us to make false assumptions about others. The man waving the gun around might be an actor, and the fashionably dressed woman might be wearing clothes her sister chose for her. Some Asians are good at math, and some aren't. Some men are good at sports, and some are utterly uninterested. Some computer jocks are also punk rockers, and some punk rockers also enjoy knitting.

Racism is the belief that inherited physical characteristics associated with racial groups determine individuals' abilities and characteristics and provide a legitimate basis for unequal treatment.

Color-blind racism refers to the belief that all races are created equal and that racial equality has already been achieved.

A **stereotype** is a preconceived, simplistic idea about the members of a group.

sociology and you

Prejudice and stereotypes are not limited to ethnic group relationships. If you have ever assumed that older people are more interested in playing cards than in having sex, you have engaged in *stereotyping*. If stereotypes like this one lead you to conclude that older people are less capable and worthy than are younger people, you would be exhibiting *prejudice*. If those prejudices led you to decide against hiring an older person, you would be engaging in *discrimination*.

An **authoritarian personality** is submissive to those in authority and antagonistic toward those lower in status.

Scapegoating occurs when people or groups who are blocked in their own goal attainment blame others for their failures.

Explaining Prejudice

What causes prejudice? Scholars most often answer this question by pointing to the effects of one personality factor—the *authoritarian personality*—and three social factors: *socialization*, *scapegoating*, and *competition for scarce resources*.

THE AUTHORITARIAN PERSONALITY A long research tradition has documented that people who have an authoritarian personality are more likely to be prejudiced. Someone with an **authoritarian personality** tends to be submissive to those in authority and antagonistic to those lower in status (Stenner 2005). Americans with authoritarian personalities tend to be strongly prejudiced against African Americans, Jews, gay people, and women.

SOCIALIZATION We learn to hate and fear in the same way we learn to love and admire. Prejudice is a shared meaning that we develop through our interactions with others. Most prejudiced people learn prejudice when they are very young, along with other social norms. This prejudice may then grow or diminish, depending on whether groups and institutions encountered during adulthood reinforce these early teachings (Wilson 1986).

Prejudice is also learned when we look at the society around us. If we live in a very unequal society and observe that no one pays highly for a group's labor or no one "like us" wants to be around people "like them," we are likely to conclude that the members of that group are not worth much. Through this learning process, members of the minority as well as the majority group learn to devalue the minority group (Wilson 1992).

SCAPEGOATING Although everyone is socialized into some prejudicial views, certain conditions can reinforce those views. One is the experience of frustration. When individuals find it difficult to achieve their own goals, they are more likely to look for others to blame for their problems. This practice, called **scapegoating**, has appeared time and again. For example, anti-Semitism exploded in Nazi Germany during the Great Depression of the 1930s, when the German economy collapsed and many Germans were left jobless and impoverished.

COMPETITION FOR SCARCE RESOURCES Competition over scarce resources (such as good jobs, nice homes, and admission to prestigious universities) also increases prejudice. For all racial and ethnic groups, prejudicial attitudes are closely associated with the belief that gains for other racial and ethnic groups will spell losses for one's own group (Bobo & Hutchings 1996).

Maintaining Prejudice: The Self-Fulfilling Prophecy

In Chapter 7, we introduced the concept of the self-fulfilling prophecy—where acting on the belief that a situation exists causes the situation to become real. The self-fulfilling prophecy is one very important mechanism for maintaining prejudice. A classic example is the situation of American women until the last few decades. Because women were considered inferior and capable of only a narrow range of social roles, they were given limited education and barred from participation in the institutions of the larger society. That they subsequently knew little of science, government, or economics was then taken as proof that they were indeed inferior and suited only for a role at home. In fact, many women were unsuited for any other role: Being treated as inferiors had made them ignorant and unworldly. The same process reinforces boundaries between racial and ethnic groups. For example, if we believe that Jews think they are better than others, then we don't invite them to our homes.

decoding the data

Race and Job Interviews

SOURCE: Bertrand & Sendhil (2004).

	Percent Receiving Call-Backs for Job Interviews	
	"White" names	"African American" names
Among group as a whole	10.1%	6.7%
Among more-qualified applicants	11.3%	7.0%

When identical, fictitious resumes are sent to employers, with some "applicants" assigned names like Emily and Brad and others assigned names like Lakisha and Kareem, those with white-sounding names receive 50 percent more call-backs for interviews. When the resumes are tweaked to give the applicants better qualifications (such as more education), whites get more call-backs but African Americans do not.

Explaining the Data: Based on what you have read in this chapter, how would you explain why those with white names are called back for interviews more often than those with African American names? How would you explain why adding qualifications improves the chances of white applicants but not of African American applicants?

Critiquing the Data: Might employers have incorrectly identified the African American names as coming from a different ethnic or racial group? If so, how might this have affected the results?

Most African Americans do not have distinctively African American names. Would employers be more or less likely to discriminate against someone with a distinctively African American name? Why? Given this, would these data likely *under*estimate or *over*estimate discrimination against African Americans?

Some employers may not look at names on resumes or may not realize that a name suggests an individual's race. Given this fact, would these data likely underestimate or overestimate discrimination against African Americans?

When we subsequently observe that they associate only with one another, we take this as confirmation of our belief.

Discrimination

Treating people unequally because of the categories they belong to is **discrimination**. Prejudice is an attitude; discrimination is behavior. Most of the time the two go together: If your boss thinks that African Americans are less intelligent than whites (prejudice), he will likely pay his African American workers less (discrimination). Some people, however, are inconsistent, usually because their own values differ from others around them. They may be prejudiced, but they nonetheless avoid discriminating because they don't want to be sued for unfair treatment. Or they might *not* be prejudiced but nonetheless discriminate because it is expected of them—perhaps by a boss who opposes hiring minorities, or by a parent who opposes interracial romance. Decoding the Data: Race and Job Interviews looks at the effect of race on job applicants.

Most anti-racist public policies seek to reduce discrimination and segregation rather than to reduce prejudice. As Martin Luther King, Jr., remarked, "The law may not make a man love me, but it can restrain him from lynching me, and I think that's pretty important" (as quoted in Rose 1981, 90).

Discrimination is the unequal treatment of individuals on the basis of their membership in categories.

Although being Irish has little impact on most Irish Americans' lives these days, many still enjoy celebrating their cultural heritage, like these boys at a St. Patrick's Day parade.

© Sandy Felsenthal/Corbis

Institutionalized Racism

Finally, racial and ethnic inequality is also maintained by institutionalized racism. **Institutionalized racism** refers to situations in which everyday practices and social arrangements are assumed to be fair, but in fact systematically reproduce racial or ethnic inequality. For example, almost all Gypsy children in the Czech Republic are placed in special schools for the mentally handicapped, and almost all children in these schools are Gypsy (New York Times 2006). Czech school authorities argue that Gypsy children are placed in these schools based on standardized evaluations, but this policy effectively makes it impossible for Gypsy children to succeed in Czech society. Less extreme versions of school segregation and tracking reinforce racial inequality in the United States.

Racial and Ethnic Inequality in the United States

Racial and ethnic inequality is not new. In this section, we discuss the past, present, and future social positions of selected racial and ethnic groups in the United States.

White Americans

Institutionalized racism occurs when the normal operation of apparently neutral processes systematically produces unequal results for majority and minority groups.

The earliest voluntary immigrants to North America were English, Dutch, French, and Spanish. By 1700, however, English culture dominated the entire Eastern seaboard. The English became the majority group, and everybody else became a minority group. In the 1840s, employers posted signs saying "No Irish need apply." In the 1860s, discrimination focused on Chinese and Japanese, and in the 1890s on Jews and Italians.

This pattern of prejudice and discrimination continues to the present day (if more rarely) against groups as diverse as French Canadians and Arab Americans.

White Ethnicity

Despite and in part because of this history of prejudice and discrimination, many white Americans continue to have a strong sense of connection to their ethnic roots. They are proud to be Italians, Greeks, Norwegians, or Poles, and enjoy eating the foods, celebrating the holidays, and singing the songs of their ethnic group. By the third or fourth generation after immigration, however, ethnic identity is largely symbolic and a matter of choice (McDermott & Samson 2005). This choice carries few risks because white ethnicity rarely presents a barrier to social integration or personal advancement.

Other white Americans no longer can claim an ethnic identity. Some come from families that emigrated to this country generations ago, and others come from families of such mixed heritage that they can no longer identify with a single ethnic group, or even a couple of ethnic groups. These individuals' only ethnic identity is as white Americans.

This shift from Italian-American, Polish-American, and other ethnic identities to "unhyphenated American" identities led some past observers to suggest that America had become a "melting pot," in which (white) ethnic groups had blended together into a new American identity. The reality is more complex. Certainly, our language contains many words borrowed from other languages (*frankfurter, ombudsman, hors d'oeuvre, chutzpah*), and some of us are such mixtures of nationalities that we would be hard-pressed to identify our national heritage. Instead of a blending of all cultures, however, what has occurred is assimilation to the dominant language and culture of the United States. To gain admission into U.S. society and to be eligible for social mobility, one has to learn "correct" English with the "correct" accent, speak without using your hands too much, work on Saturday and worship on Sunday, and, in general, adopt the culture of the northern and western Europeans who dominated the United States for generations.

White Racial Identity

As white Americans' connections to their different ethnic identities have declined, sociologists have begun to focus on whiteness as a *racial* identity (McDermott & Samson 2005). Ironically, one of the most important things to understand about white racial identity is that it typically is invisible. Except in unusual circumstances, such as when they live surrounded by nonwhites, white Americans rarely think of themselves as even having a race. When white people choose to watch football rather than soccer, to listen to rock rather than to salsa music, or to eat apple pie rather than sweet potato pie, they rarely think of these choices as reflections of their white color.

White Privilege

Because white people rarely think of themselves as a racial category, they rarely recognize that the life they enjoy—living in relatively safe neighborhoods, having relatively good jobs, going to relatively good schools—partly resulted from structured racial inequalities built into the system long before they were born. For example, any time an African American is denied a job because of his or her race, the odds of a white person getting hired increase. The term **white privilege** refers to the benefits and opportunities that whites receive simply because they are white (Rothenberg 2002). White privilege benefits all white Americans, whether or not they recognize or want those privileges.

When you go to the shopping mall, do you dress nicely so no one will think you are a shoplifter? If you answered no, you probably are white. The fact that many people assume whites to be law-abiding citizens until proven otherwise is an example of white privilege. White privilege also is typically invisible: Few whites know that law-abiding African Americans are often stopped by security guards, so few whites recognize that their racial identity protects them.

White privilege refers to the benefits whites receive simply because they are white.

African Americans

African Americans now comprise 12.3 percent of the U.S. population. Until very recently, they were the largest minority group in the country, but they were recently passed by Hispanics. Still, their importance goes beyond their numbers: They have made innumerable contributions to U.S. history and culture, and their circumstances have long challenged the United States's view of itself as a moral and principled nation.

The social position of African Americans has its roots in one central fact: Most African Americans are descended from slaves. After slavery ended, both legal barriers (such as patently unfair "literacy tests" that barred African Americans from voting) and illegal barriers (such as the occasional lynching of African Americans who challenged white authority) prevented most African Americans from rising in the American social and economic structure. Real change did not take place until the Civil Rights Act of 1964 and the civil rights activism of the late 1960s.

These days, more African Americans are middle class than ever before. At the same time, however, a troubling fissure has emerged within the African American population: Whereas some African Americans belong to the middle or even upper class and are increasingly integrated into U.S. society, others remain poor and live in segregated neighborhoods with few employment opportunities. In fact, for this second group, the situation has deteriorated (Wilson 1996).

Current Concerns

Since World War II, white attitudes toward African Americans have improved dramatically; most whites now support integration in principle, are comfortable living in neighborhoods where African Americans form a small minority, and no longer disapprove of interracial marriage (Krysan 2000). Similarly, important improvements have been made in many areas of African American life. Nevertheless, neighborhood segregation remains high (Massey 2007), African American infants are more than twice as likely as white infants to die before their first birthday (National Center for Health Statistics 2009), and African American men's life expectancy is still six years less than white men's.

Similarly, although African Americans are rapidly catching up with whites in their educational attainment, they still lag behind (Kao & Thompson 2003; U.S. Bureau of the Census 2009a). Moreover, even when whites and African Americans have the same levels of education, whites have higher incomes. Almost one-quarter of African American families live below the poverty line, and the median income for African American families is only 64 percent that of white families (U.S. Bureau of the Census 2009a). Even more discouragingly, the majority of African American 40-year-olds raised in middle-income families now have *lower* family incomes than did their parents (Isaacs, Sawhill, & Haskins 2008).

These striking economic disadvantages are due to two factors: African American workers earn less than white workers, and African American families are less likely to have two earners.

LOW EARNINGS Even when we look only at people employed full time and year-round, median income for African Americans remains 20 percent lower than for whites (U.S. Bureau of the Census 2009a). In part, this occurs because African Americans more often than whites live in the South, where wages are low for everyone. In addition, African Americans typically have less education than do whites and so disproportionately work in low-paying fields. For example, African Americans make up 11 percent of all employed U.S. civilians but 19 percent of janitors, 26 percent of mail clerks, and 33 percent of nursing aides (U.S. Bureau of the Census 2009a).

Yet these differences account for only part of the earnings gap between African Americans and white people in the United States (Cancio, Evans, & Maume 1996; Pager & Shepherd 2008). The other part is the result of a pervasive pattern of discrimination that produces a very different occupational distribution, pattern of mobility, and earnings picture for African Americans and whites in the United States. Although there are far more African American professionals than in the past, they often are kept outside the true corporate power structure (Collins 1993, 1997): Compared to whites in comparable positions, they receive lower salaries and wield less authority at work (Smith 1997; Wilson 1997).

FEMALE-HEADED FAMILIES About half of the gap between African American and white family incomes is due to the fact that African American families are less likely to include an adult male. Because women earn less than men and because a one-earner family is obviously disadvantaged relative to a two-earner family, these female-headed households have incomes far below those of husband–wife families.

The fact that so many more African American than white families are headed by females—46 percent compared with 14 percent—has led some commentators to conclude that poverty is the result of bad decisions by African American men and women. This type of argument is an example of "blaming the victim," and empirical evidence suggests that it simply isn't true. Rather than *causing* poverty, research indicates that female headship *results* from poverty: African American women are less likely to marry because relatively few men in their community can support a family (Lichter, LeClere, & McLaughlin 1991; Luker 1996; Newman 1999b).

Hispanic Americans

Hispanics (sometimes known as *Latinos*) are an ethnic group rather than a racial category. Hispanics include immigrants and their descendants from Puerto Rico, Mexico, Cuba, and other Central or South American countries. Hispanics constitute 12.5 percent of the U.S. population, making them the largest minority group in the country. About two-thirds of Hispanics in the United States are of Mexican origin, with the rest originating in Central and South America and the Caribbean. Hispanics may also identify as white, black, mixed race, or members of any other racial group.

It is almost impossible to speak of Hispanics as if they were a single group. The experiences of different Hispanic groups in the United States have been and continue to be very different. For example, Cubans who emigrated in the 1960s shortly after the Cuban Revolution typically came from wealthier backgrounds, were lighter-skinned, and were seen as refugees from a hated Communist regime. As a result, they found greater acceptance in the United States than either later waves of Cuban immigrants or immigrants from Mexico or Guatemala.

Figure 8.1 on the next page compares the various Hispanic groups to one another and to the non-Hispanic white, Asian, and African American populations on three measures: education, poverty, and family structure. On two of these measures, a Hispanic group comes out at the very bottom: Mexican Americans are the most poorly educated racial or ethnic group, and Puerto Ricans (many of whom are considered black by other Americans) are the most likely to live in poverty. In addition, Puerto Ricans are second most likely, after African Americans, to live in female-headed households.

Despite the difficulties many Hispanics now face, they remain optimistic about their future prospects in the United States. In a national poll conducted in 2007 by the

FIGURE 8.1 Education, Poverty, and Family Structure, by Race and Hispanic Origin

Compared with other groups, Hispanics—especially Mexicans—are the most likely to lack a high school education, partly because many are recent immigrants. Hispanics and African Americans are more likely than whites and Asians to live below the poverty level, and African Americans are the most likely to live in female-headed households, followed by Puerto Ricans (many of whom are also of African descent).

SOURCE: U.S. Bureau of the Census (2009a).

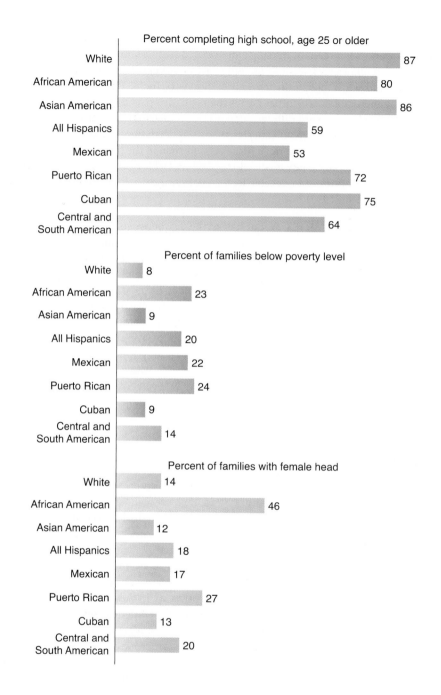

Percent completing high school, age 25 or older

White	87
African American	80
Asian American	86
All Hispanics	59
Mexican	53
Puerto Rican	72
Cuban	75
Central and South American	64

Percent of families below poverty level

White	8
African American	23
Asian American	9
All Hispanics	20
Mexican	22
Puerto Rican	24
Cuban	9
Central and South American	14

Percent of families with female head

White	14
African American	46
Asian American	12
All Hispanics	18
Mexican	17
Puerto Rican	27
Cuban	13
Central and South American	20

New York Times, 74 percent of Hispanics strongly agreed with the statement "If you work hard, you will succeed in the U.S." (Preston 2007).

Current Concerns

The Hispanic population in the United States is growing more rapidly than any other segment, although immigration has slowed considerably due to the economic downturn (U.S. Bureau of the Census 2009a). This rapid growth has raised two concerns among many Americans. First, because most of the new immigrants are young, poorly educated, and (especially if they are undocumented immigrants) willing to accept very low wages, some U.S. citizens fear these immigrants may lower wages for everyone. Second, some fear that Hispanic culture and the Spanish language will "take over" the country. These fears are heightened by the (slightly) increasing residential segregation of Hispanics in the United States (Iceland & Wilkes 2006), which has led some to question whether these new immigrants will ever become socially integrated into U.S. society.

Are new Hispanic immigrants in fact driving down wages? The answer remains unclear. Some researchers have concluded that immigrants stimulate the economy overall and thus benefit all Americans (Card 2005). Other researchers argue that immigration improves the quality of life among affluent Americans (by making cheap labor available), but depresses the wages of Americans who lack high school degrees by as much as 5 percent (Porter 2006; Borjas & Katz 2007).

Will Hispanics become socially integrated into the United States? On the one hand, because of continued immigration from Latin America, U.S.-born Hispanics now can easily enjoy salsa dancing, Mexican fiestas, Guatemalan restaurants, and perhaps romance and marriage with a recent immigrant. As a result, Hispanic ethnic identity is being reinforced even among those whose families emigrated here much earlier (Waters & Jimenez 2005). On the other hand, these earlier generations of Hispanic immigrants nevertheless are relatively socially integrated into the United States (Alba & Nee 2003). Almost all who were born in this country are fluent in English, and those whose parents were also born here often speak little if any Spanish. There is good reason to think the same will be true of new immigrants.

In addition, the caste-like barrier separating races operates much less dramatically for Hispanics. White prejudice against Hispanics is far less strong than against African Americans, and although Hispanic segregation has increased, it remains modest. As a result, the main barrier Hispanics face is class rather than ethnicity—at least if they are white. As a result, by the second and third generation, most Hispanics can translate educational attainment into well-paying jobs and leave the segregated barrios (Iceland & Wilkes 2006).

In sum, there is good reason to be optimistic about the effect of Hispanic immigration on the United States. Nevertheless, concern about rising immigration has fueled recent demands for stricter border controls. In turn, these demands have led to a surge of political activism among Hispanic immigrants and their supporters, calling for more humane treatment of immigrants and perhaps guest worker programs or "amnesty" programs for undocumented immigrants. The results of these efforts remain to be seen.

Asian Americans

The Asian population of the United States (Japanese, Chinese, Filipinos, Koreans, Laotians, and Vietnamese, among others) more than doubled between 1980 and the present but still constitutes only 3.6 percent of the total population. The Asian population can be broken roughly into three segments: descendants of nineteenth-century immigrants (Chinese and Japanese), post-World War II immigrants (Filipinos, Asian Indians, and Koreans), and recent refugees from Southeast Asia (Cambodians, Laotians, and Vietnamese).

Image copyright Andresr, 2009. Used under license from Shutterstock.com

Although life is often difficult for recent Hispanic immigrants, many Hispanic Americans now hold middle- and even upper-class jobs.

A century ago, Asian immigrants encountered sharp and occasionally violent racism. Today, incidents of racial violence directed at Asians are rare but still occasionally make headlines. Despite these handicaps, Asian Americans have experienced high levels of social mobility. A higher percentage of Asian Americans than white Americans have college, doctoral, medical, and law degrees (Le 2006). Educational levels are especially high among Japanese and Chinese Americans, many of whom come from families that have lived in the United States for generations. Education levels are also high among Asian Indians and Filipinos, many of whom came to the United States to get a graduate education or with graduate degrees in hand. Current evidence suggests that the more recent streams of immigrants from Southeast Asia will follow the same path. For example, although many of the Southeast Asian refugees who came to the United States between 1975 and 1984 began their lives here on welfare, almost twice as many Vietnamese youths aged 20 to 24 are enrolled in school as are white youths of the same age.

Current Concerns

The high level of education earned by Asian Americans is a major step in opening doors to high-status occupations, and median income for Asian Americans is very similar to that of whites (U.S. Bureau of the Census 2009a). Yet discrimination is not all in the past. Unofficial policies make it more difficult for Asian American applicants than for white Americans with the same credentials to gain admittance to elite colleges and universities. And highly educated Asian Americans still earn significantly less than whites with the same professional credentials, primarily because they are less likely to move out of professional and technical positions into managerial and executive positions (Le 2006). Asian Americans are often passed over for promotions because white employers assume Asian Americans won't have the personality, social skills, or simply the "look" an executive is assumed to need. In addition, Asian Americans less often even learn about available executive positions because they are less often accepted into the "old boy networks" in which most professional mentoring takes place, and in which individuals gain the contacts that can lead to higher-level jobs. Finally, Asian Americans—even those whose great-great-grandparents emigrated to this country— still are held back by others who assume they aren't "really" Americans. As one third-generation Japanese American said,

> I get real angry when people come up to me and tell me how good my English is. They say: "Oh, you have no accent. Where did you learn English?" Where did I learn English? Right here in America. I was born here like they were. [But] people see me now and they automatically treat me as an immigrant." (quoted in Zhou & Gatewood 2000, 18)

Native Americans

Native Americans (American Indians) are one of the smallest minority groups in the United States (less than 1 percent of the entire population), and nearly half of their members live in just four states: Oklahoma, Arizona, California, and New Mexico. Native Americans are arguably our most disadvantaged minority group. They have the lowest rates of educational achievement and the highest rates of alcoholism and premature death of any U.S. racial or ethnic group (Kao & Thompson 2003). This situation exists despite hopeful new signs of economic vitality on some Indian reservations over the past 20 years, including the development of mineral reserves on the Navajo reservation and the advent of gambling casinos elsewhere.

TABLE 8.2 **Life on the Navajo Reservation**

Percent 65 years and over	9.80%
Percent high school graduate (25 years or older)	63.50%
Percent college graduate (25 years or older)	17.50%
Percent families with female heads	27.80%
Percent families below poverty level	36.20%
Percent using wood to heat home	63.10%
Percent lacking telephones	46.70%
Percent with no vehicle	14.70%
Percent lacking indoor water or toilets	21.20%
Median family income, 2007	$29,846

SOURCE: U.S. Bureau of the Census (2009b).

Table 8.2 summarizes the situation on the Navajo reservation, the largest geographically in the country. More than 250,000 Americans belong to the Navajo (or, in their language, the Diné). Keep in mind, though, that the status of Native Americans is highly diverse. Native Americans represent more than 200 tribal groupings, with different cultures and languages. Some have been successful: fish farmers in the Northwest, ranchers in Wyoming, and bridge builders in Maine. In urban areas, and east of the Mississippi, where the impact of white society has been felt the longest, many Native Americans have blended into the majority culture and entered the economic mainstream. On the other hand, on isolated reservations such as the Navajo reservation, with few economic resources (and little opportunity to draw crowds to casinos), socioeconomic conditions often are quite poor. In addition, in white-dominated towns near large Native American reservations, prejudice and discrimination by whites remain major barriers.

Arab Americans

Although Arab Americans comprise considerably less than 1 percent of the U.S. population, recent world events have given their status in this country special importance.

All Arab Americans are immigrants or children of immigrants from North Africa and the Middle East (including Morocco, Algeria, Saudi Arabia, and Iraq); Iran is not an Arabic country. The largest single group of Arab Americans is from Lebanon ("Arab American Demographics" 2006). Each of these countries has its own traditions, but they share common linguistic, cultural, and historical traditions. Some Arab Americans descend from families that emigrated to the United States in the late 1800s, some emigrated themselves only in the last few years. Two-thirds of Arab Americans are Christians.

Arab Americans are a highly educated population ("Arab American Demographics" 2006). They are as likely as other white Americans to have graduated high school and are slightly more likely to have graduated college. As a result, the majority hold professional jobs, and their median incomes are somewhat above the U.S. average.

Arab Americans, like these Michigan schoolchildren, are an increasingly important minority group in the United States.

© Ed Kashi/Corbis

Current Concerns

The terrorist attacks of 9/11 and other evidence of anti-American sentiment in the Arab world have raised concerns about the status of Arab Americans in the United States. On the positive side, many Americans who had no opinion of Arabs before 9/11 have become more educated since then and have taken pains not to discriminate against all Arabs or Muslims because of the actions of a few. Similarly, a poll conducted in 2007 by the nonprofit Pew Research Center (2007) found that 53 percent reported favorable attitudes toward Muslim Americans—considerably lower than the 76 percent who held favorable views of Catholics and Jews but identical to the percentage with favorable attitudes toward Mormons and much higher than the 35 percent with favorable views of atheists.

On the other hand, Gallup Poll researchers report that about 40 percent of Americans openly admit that they are prejudiced against Muslims in the United States. The same percent would prefer that Muslim Americans be subject to special security requirements, such as carrying special I.D. cards (Saad 2006). American attitudes have grown slightly more negative over the last few years, especially among evangelical Christians and those who rely on the media (rather than personal contact) for their ideas about Arabs and Muslims (Saad 2006; Pew Research Center 2007). Unfortunately, the media often reinforce prejudices in television shows, films, and articles that either ignore Arabs or depict them as anti-American or as terrorists (Semmerling 2006; Byng 2008).

Multiracial Americans

Individuals who identify as more than one race now constitute 1.6 percent of the U.S. population. Although this may seem like a small number, it is a significant change, in two ways. First, the absolute number of multiracial Americans has increased more than ten times in the last half century. Second, for the first time in American history, significant numbers of Americans born to parents of different races now identify as *multiracial* rather than identifying with only one parent's race.

Why have increasing numbers of Americans begun to define themselves as multiracial? In the past, many multiracial children were conceived through rape, leaving them with little desire to identify with their father's race. Similarly, in past decades many who married outside their group were rejected by their families, and so their children only grew up with one set of relatives and one racial identity. Today, most multiracial children are born to loving parents and welcomed by all their relatives. As a result, identifying with only one race can feel like abandoning half of one's family. Yet Americans now must fill out more and more forms that ask them to identify themselves by race. All these pressures led to the rise in individuals who openly identify as multiracial (DaCosta 2007).

Despite the rise of a visible multiracial community, however, many Americans continue to feel uncomfortable when they cannot wedge an individual into a predetermined racial slot. Golf superstar Eldrick "Tiger" Woods has had to fight constantly against journalists and others who want to describe him simply as African American, even though two of his eight great-grandparents were Native American, four were Asian, and one was European American.

The Future of Racial and Ethnic Inequality in the United States

The last few decades have witnessed considerable improvement in the social status of minority groups—as well as the momentous election of the first African American president of the United States (a topic discussed in Focus on American Diversity: The Election of Barack Hussein Obama on the next page). Yet inequality remains. This final section reviews the debate about whether inequality can best be reduced by focusing on race or on class before describing some of the strategies now being used to reduce inequality.

Combating Inequality: Race versus Class

In this chapter and Chapter 7, we have shown how both social class, on the one hand, and race or ethnicity, on the other hand, affect one's life chances. When a person has a lower status on both of these dimensions, we speak of **double jeopardy**. This means that disadvantages snowball. For example, poor African American, Hispanic, and Native American teenagers are more likely than poor white teenagers to be unemployed or to end up in prison.

Sociologists have hotly debated whether race or class is more important for understanding the structure of inequality in the United States today. The question most often asked is, "Is the status of lower-class African Americans due to the color-blind forces of class stratification, or is it due to class-blind racism?" In a series of books and articles, African American sociologist W. J. Wilson (1978, 1987, 1996, 2009) has argued that the problems faced by African Americans stem less from current racism than from the inheritance of poverty and the changing nature of the U.S. economy. As well-paying factory jobs disappeared and as other forms of employment shifted from the inner cities to the suburbs, the position of the poorest third of the African American population has disintegrated. Joblessness is up, the number of female-headed households is up, rates of drug use are up, and so on. For this reason, Wilson argues that African Americans can best be helped through strategies designed to create full employment and better jobs for *all* Americans, such as the movements for fair wages and for increasing educational opportunities described in Chapter 7.

Double jeopardy means having low status on two different dimensions of stratification.

focus on  AMERICAN DIVERSITY

The Election of Barack Hussein Obama

On November 4, 2008, Barack Obama was elected President of the United States. Obama, who identifies as African American, is the son of a white American mother and a black Kenyan father.

Given continuing prejudice against African Americans, how was Obama able to get elected? In part, luck was on his side: The outgoing Republican president, George W. Bush, was the most unpopular president in history; the Republican vice presidential candidate, Sarah Palin, alienated many voters with her highly conservative views and lack of political experience; and both an unpopular war and economic troubles turned voters against the ruling Republican party (Todd & Gawiser 2009).

Obama also won election because he and his campaigners did so many things right. First, they recognized the important growth in urban, suburban, African American, and Hispanic voters and focused on wooing those groups (Todd & Gawiser 2009; Sheldon 2009). Second, they recognized the tremendous potential of new media and took full advantage of email, BlackBerries, blogs, cell phones, Twitter, Facebook, YouTube, and the like (Smith 2009).

Of course, Obama's personal characteristics also contributed to his election. In addition to being highly intelligent and well educated, voters found him articulate, relaxed on stage, funny when appropriate, and amazingly calm—a trait that seemed particularly appealing in such difficult times.

Finally, Obama's election reflects a shift in U.S. attitudes toward race.

No African American candidate, no matter how qualified he or she was or how inept the opposition was, could have been elected president 20 years ago. That said, his odds were certainly improved by the fact that he did not seem "too black" to white voters since he was both light-skinned and obviously upper-middle class.

So far, at least, Obama's election has improved both race relations and *perceptions* of race relations. Just over half of white Americans and almost all African Americans believe his presidency will bring different groups of Americans together. In addition, the percentage who believe that race relations in the United States are generally good increased in less than a year from 55 to 65 percent among whites and from 29 to 59 percent among blacks (*New York Times* 2009).

Most sociologists disagree. They doubt that policies based on social class alone will be enough to resolve the problem of racial inequality in the United States. True, there are middle- and even upper-class minority-group members, and it would be a serious mistake to assume that racism keeps all racial minorities poor and powerless. Nevertheless, race and ethnicity continue to be fundamental dividing lines in U.S. society. Membership in a minority group remains a handicap in social-class attainment and in social relationships. For example, the finding that middle-class African Americans are much more likely than are middle-class whites to live in poor neighborhoods suggests that the issue goes beyond class (Alba, Logan, & Stults. 2000). Any successful strategy for combating inequality in the United States will have to address issues of race and ethnicity as well as social class.

Strategies for Ending Inequality

The major strategies used in the United States to fight against racial and ethnic inequality are antidiscrimination and affirmative action laws. Since 1964, the United States has officially outlawed discrimination on the basis of race, color, religion, sex, and national origin. These laws have had considerable effect. States can no longer declare interracial marriage illegal or refuse to allow African Americans to vote or to attend state schools, and newspapers can no longer advertise that a job is open only to whites.

Whereas antidiscrimination laws make it illegal to discriminate, affirmative action rules require employers, schools, and others to actively work to increase the representation of groups that have historically experienced discrimination.

This means, for example, that a college with very few minority faculty may be required to advertise new jobs through minority faculty organizations, as well as in regular employment bulletins. Affirmative action has proven much more contentious than antidiscrimination laws.

Where This Leaves Us

Racism and interethnic conflicts are problems worldwide, erupting in schoolyards, on street corners, and in courts of law. This does not mean, though, that these conflicts cannot be lessened or even eliminated. Irish people no longer are refused employment, as was common in the nineteenth century, and Jews no longer are prohibited from living in certain neighborhoods or belonging to certain clubs, as was common until the 1960s. Ideas about race and ethnicity are social constructions that change as societies change. To combat prejudice and discrimination, we will need to combat subtle and institutionalized racism, and we will need to address the social class inequalities that support racial and ethnic inequalities. Doing so will be both especially difficult and especially crucial if economic hard times continue in the United States.

Summary

1. A race is a category of people treated as distinct due to physical characteristics that have been given social importance. An ethnic group is a category whose members share a common origin and culture. Both race and ethnicity are socially constructed categories.
2. In the United States, the population is stratified by both race and class. These two factors work together to create greater advantages or disadvantages for different groups.
3. The concepts of majority and minority groups provide a general framework for examining structured inequalities based on ascribed statuses. Interaction between majority- and minority-group members may take the form of pluralism, assimilation, or conflict.
4. Prejudice, discrimination, segregation, and institutionalized racism all help to maintain racial and ethnic inequality. Color-blind racism allows inequality to continue even when majority-group members believe that they are not prejudiced.
5. In the United States, white ethnicity is now largely a symbolic characteristic. Its main consequence is that it has become the "standard" American ethnicity against which other groups are judged. White racial identity is typically invisible and carries considerable if unacknowledged privileges.
6. On many fronts, African Americans have improved their position in U.S. society. Nevertheless, African American families continue to have a median income that is far lower than that of white families. Major areas of continued concern are high rates of female-headed households, unemployment, and housing segregation.
7. Hispanics are the largest and fastest-growing minority group in the United States. They generally have fewer years of education and lower earnings than do other Americans, but they are increasingly assimilating into American culture and life. Hispanic immigration helps the economy overall but may reduce income for the least-educated U.S. citizens.
8. Native Americans are the least-prosperous minority group in the United States. Living conditions and economic prospects are most difficult on geographically isolated reservations.
9. Asian Americans have used education as the road to social mobility. Even the newest immigrant groups outstrip white Americans in their pursuit of higher education. Despite some discrimination, Asian Americans have higher median family incomes than do white Americans and experience low levels of residential segregation.
10. Arab Americans are primarily middle class: well educated, with good jobs. A majority of Americans hold

favorable views toward Arab Americans, but prejudice against them is nonetheless strong.

11. Efforts to reduce racial and ethnic inequality will need to focus not only on reducing prejudice and discrimination but also on tackling broader issues of economic inequality.

Thinking Critically

1. Within the next 50 years or so, non-Hispanic whites will be a *numerical* minority within the United States. In sociological terms, do you think they will be a minority group? What social, economic, or political changes do you expect as a result of changes in the relative size of the different U.S. racial and ethnic groups?

2. In thinking about the relationship between prejudice and discrimination, we generally assume that prejudice leads to discrimination. Can you think of a time or situation when discrimination might have led to prejudice?

3. Consider how people you know talk about Arab Americans and how you have seen them portrayed in the media. Then, using the concepts in this chapter, discuss whether Arab Americans are considered white. (Note: Do not discuss whether they *should* be considered white, just whether they are.)

4. List five things you typically do during the course of the week, such as going shopping or meeting with friends. How would that experience be different if you woke up tomorrow and found that your race had changed to African American or to white?

5. Some scholars contend that the major cause of racial/ethnic inequality in the United States today is institutionalized, not individual, racism. If this is so, what recommendations would you offer to policy makers who wanted to reduce racial or ethnic differences in quality of life?

6. What similarities and what differences do you see between the situation in Darfur and that of African Americans in the United States?

Book Companion Website

Prepare for quizzes and exams with online resources— including tutorial quizzes, a glossary, interactive flash cards, crossword puzzles, essay questions, virtual explorations, and more.

Sex, Gender, and Sexuality

Brand X Pictures/Jupiterimages

Sexual Differentiation

Men and women are different. Biology differentiates their physical structures, and cultural norms in every society differentiate their roles. In this chapter, we describe some of the major differences in men's and women's lives as they are socially structured in the United States. We will be particularly interested in the extent to which the ascribed characteristic of sex has been the basis for structured inequality.

Sex versus Gender

In understanding the social roles of men and women, it is helpful to make a distinction between sex and gender. **Sex** refers to the two biologically differentiated categories, male and female. It also refers to the sexual act that is closely related to this biological differentiation. **Gender**, on the other hand, refers to the normative dispositions, behaviors, and roles that cultures assign to each sex. (See the Concept Summary on Sex versus Gender.)

Although biology provides two distinct and universal sexes, cultures provide almost infinitely varied **gender roles**. Each man is pretty much like every other man in terms of sex—whether he is upper class or lower class, African American or white, Chinese or Apache. Gender, however, is a different matter. The rights, obligations, dispositions, and activities of the male gender are very different for a Chinese man than for an Apache man. Even within a given culture, gender roles vary by class, race, and subculture. In addition, of course, individuals differ in the way they act out their expected roles: Some males model themselves after Brad Pitt and some after Johnny Depp or Will Smith.

Just how much of the difference between men and women in a particular culture is normative and how much is biological is a question of considerable interest to social and biological scientists. This question has led some biologists to investigate whether characteristics we typically think of as male and female also characterize nonhuman species. If they did, that would lend support to the idea that these male/female differences are biological. Results from these studies are decidedly mixed. Among goby fish, females sport bright colors to attract the opposite sex, but among birds it is usually males who do so. Male baboons certainly dominate female baboons, but male marmosets (small monkeys) take care of the young, and male lions depend on the females to do all the hunting. Meanwhile, whales and elephants live in matriarchal families.

For the most part, social scientists are more interested in gender than in sex. They want to know about the variety of roles that have been assigned to women and men and, more particularly, about the causes and consequences of this variation. Under what circumstances does each gender have more or less power and prestige? How does having more or less power affect women's and men's everyday lives? And what accounts for the recent changes that have occurred in gender roles in our society?

Gender Roles across Cultures

A glance through *National Geographic* confirms that gender roles vary widely across cultures. The behaviors we normally associate with being female and male are by no

---

Sex is a biological characteristic, male or female.

Gender refers to the expected dispositions and behaviors that cultures assign to each sex.

Gender roles refer to the rights and obligations that are normative for men and women in a particular culture.

Sex versus Gender

	Sex	Gender
Divides population into:	Male or female (or maybe intersex)	Masculine and feminine
Based on:	Biological characteristics (chromosomes, sex hormones, penises or vaginas, etc.)	Cultural expectations regarding appropriate behaviors and attitudes for each sex
Consequences:	On average, men have more upper body strength than women because of their hormones.	Men also have more upper body strength because women are warned that they will look "too masculine" if they lift weights too much.
	On average, men are taller than women because of their genes.	In poor countries, sex differences in height are amplified because boys receive more food than do girls.

means universal. Among the Wodaabe, a nomadic tribe of western Africa, boys carry mirrors with them from the time they can walk (Bovin 2001). Even when boys spend days alone in the bush herding cows, they begin each day by fixing their hair and putting on their jewelry, lipstick, mascara, and eyeliner. In contrast, because girls are primarily evaluated on their health and ability to work hard, they are expected to pay far less attention to their appearance than do boys. Wodaabe courtship mostly takes place during men's dance competitions, in which women judges select the winners based on the men's physical beauty and charm. Afterward, the women openly approach the men they find most attractive to be their romantic partners.

Despite cross-cultural variations such as these, and despite the fact that women do substantial amounts of work in all societies (often providing more than half of household food), in almost all societies women have less power and less value than men (Kimmel 2000). A simple piece of evidence is parents' almost universal preference for male children (Sohoni 1994), a preference which can be life threatening for girls. Currently, there are about 120 boys for every 100 girls under the age of 5 in China—far higher than the natural ratio of about 105 to 100 (Zhu, Lu, & Hesketh 2009). This difference is primarily due to the use of abortion to kill fetuses identified prenatally as female. Other girls are killed at birth or, more often, die because they receive less food and medical care than their brothers. The preference for boys is less strong in modern industrial nations, but parents in the United States nonetheless prefer their first child to be a boy by a two-to-one margin (Holloway 1994).

Another result of female power disadvantage is widespread violence toward girls and women. According to the respected international organization Human Rights Watch, "Abuses against women are relentless, systematic, and widely tolerated, if not explicitly condoned. Violence and discrimination against women are global social epidemics" (Human Rights Watch 2004). For example:

- Each year, about 1.5 million American women are raped or physically assaulted by intimate partners (Tjaden & Thoennes 2000). Although men are also sometimes assaulted by their partners, they are more likely to be hit in self-defense and less likely to be seriously harmed or killed (Fox & Zawitz 2004). (Violence between intimates is further discussed in Chapter 11.)
- In Uganda, Darfur, Bosnia, and elsewhere, armies have used rape both as a systematic tool to subjugate the population and as a form of "sport" for soldiers.
- Between 100 and 140 million women, mostly in African countries but also in Asia, South America, and Europe, have undergone genital mutilation—removal of some or all of the clitoris and surrounding genitalia (World Health Organization 2008a). Aimed at eliminating sexual desire in women, the practice is dangerous and even deadly.
- In Afghanistan, Islamic fundamentalists have thrown acid onto the faces of girls who dare to go to school, disfiguring and sometimes blinding them (Filkins 2009).

At home and abroad, violence against women results from the lower status accorded to women. In growing numbers, women around the world are demanding equal rights. In some of the least-developed nations, this means changing cultural and legal values that treat women essentially as their husbands' or fathers' property. In the United States and the rest of the developed world, the problems are more subtle. Those problems lead sociologists to ask: How are gendered identities developed? And what are the institutional forces that maintain inequality, with or without overt violence and discrimination?

Theoretical Perspectives on Gender Inequality

Women rather than men bear children because of physical differences between the sexes. Most of the differences in men's and women's life chances, however, are socially structured. Different sociological theories offer different explanations for the persistence of this structured gender inequality.

Structural-Functional Theory: Division of Labor

The structural-functional explanation of gender inequality is based on the premise that a division of labor is often the most efficient way to get a job done. In the traditional sex-based division of labor, the man does the work outside the family and the woman does the work at home. According to this argument, a gendered division of labor is functional because specialization will (1) increase the expertise of each sex in its own tasks, (2) prevent competition between men and women that might damage the family, and (3) strengthen family bonds by forcing men and women to depend on each other.

Of course, as Marx and Engels noted, any division of labor has the potential for domination and control. In this case, the division of labor has a built-in disadvantage for women because by specializing in the family, women have fewer contacts, less information, and fewer independent resources. Because this division of labor contributes to family continuity, however, structural functionalists have seen it as necessary and desirable.

Conflict Theory: Sexism and Discrimination

According to conflict theorists, women's disadvantage is not a historical accident. Instead, it is designed to benefit men and to benefit the capitalist class.

Two major concepts employed by conflict theorists to explain how gender inequality benefits men and capitalists are sexism and discrimination. **Sexism** is the belief that women and men have biologically different capacities and that these differences form a legitimate basis for unequal treatment. Conflict theorists explain sexism as an ideology that is part of the general strategy of stratification. If others can be categorically excluded, the need to compete individually is reduced. Sexism, then, reduces women's access to scarce resources and allows men to keep those resources for themselves.

Discrimination is the natural result of sexism. If we believe that women are better suited to work with children and men are better suited for intellectual work, then we will be more likely to admit men to medical school than women, more likely to hire a man as a doctor than a woman, and more likely to hire a woman as a pediatrician than as a neurosurgeon.

Symbolic Interactionism: Gender Inequality in Everyday Life

Symbolic interactionist theory is particularly useful for understanding the sources and consequences of sexism in everyday interactions. For example, sociologist Karin Martin (1998) was interested in understanding how boys and girls learn gender-normative ways of moving, using physical space, and comporting themselves. To research these questions, she studied 112 preschoolers in 5 different classrooms, at 2 different preschools, with 14 different teachers. She found that teachers routinely structure children's play and impose discipline in ways that reinforce gender differences. Little boys are actively discouraged from playing "dress-up" (even though many of them enjoy doing so). And whereas boys are allowed to have fun shouting, playing rough and tumble games, and moving about wildly, girls are disciplined to raise their hands, lower their voices, and refrain from running, crawling, or lying on the ground. By the end of preschool, then, boys and girls are well on their way to learning the non-verbal behaviors and communication styles that are typical of, and seem so natural for, adult men and women. We will discuss these gendered differences in more detail later in this chapter.

A second study illustrating the symbolic interactionist perspective on gender inequality drew on observations collected at a sleepaway camp during the course of one summer (McGuffey & Rich 1999). At this camp, high-ranking boys attained power and popularity primarily through athletic prowess. They bolstered their positions and won approval from other boys by acting aggressively toward lower-ranking boys and by sexually harassing girls. In addition, and most importantly, high-ranking boys led other boys in teasing, assaulting, or excluding any boys they deemed too "feminine" and any girls they deemed too "masculine." Interestingly, high-ranking boys were able to redefine "feminine" activities they enjoyed (such as hand-clapping games) into masculine activities. In these ways, high-ranking boys maintained their status

Sexism is a belief that men and women have biologically different capacities and that these form a legitimate basis for unequal treatment.

and power over other boys, and almost all boys maintained greater status and power than girls.

Gender as Social Construction and Social Structure

To sociologists, gender is not simply something that individuals have—a biological given—but rather is something that is constantly re-created in individual socialization, in medical and cultural practices, and in social interaction. Similarly, sociologists describe gender as an attribute not only of individuals but also of social structures.

Developing Gendered Identities

From the time they are born, girls are treated in one way and boys in another—wrapped in blue blankets or pink ones, encouraged to take up sports or sewing, described as cute or as strong before they are old enough to truly exhibit individual personalities. In these ways, as symbolic interactionist studies illustrate, children learn their gender and gender roles. By the age of 24 to 30 months, they can correctly identify themselves and others by sex, and they have some ideas about what this means for appropriate behavior (Cahill 1983).

Young children's ideas about gender tend to be quite rigid. They develop strong stereotypes for two reasons. One is that the world they see is highly divided by sex: In their experience, women usually don't build bridges and men usually don't crochet. The other important determinant of stereotyping is how they themselves are treated. Substantial research shows that parents treat boys and girls differently. They give their children "gender-appropriate" toys, they respond negatively when their children play with cross-gender toys, they allow boys to be active and aggressive, and they encourage their daughters to play quietly and visit with adults (Orenstein 1994). When parents do *not* encourage gender-stereotypic behavior, their children are less rigid in their gender stereotypes.

As a result of this learning process, boys and girls develop strong ideas about what is appropriate for girls and what is appropriate for boys. However, boys are punished more than girls for exhibiting cross-gender behavior. Thus, little boys are especially rigid in their ideas of what girls and boys ought to do. Girls are freer to engage in cross-gender behavior, and by the time they enter school, many girls are experimenting with boyish behaviors.

Reinforcing Biological Differences

Because of gender socialization, girls and boys and men and women understand quite well what a "proper" male or female should be like. These ideas can become self-fulfilling prophecies, as the *belief* that males and females are biologically different *keeps* males and females biologically different (Lorber 1994). To understand how this works, Shari Dworkin (2003) spent two years doing participant observation at two gyms.

© Tony Freeman/PhotoEdit

Despite many changes in gender roles in the United States, boys and girls still tend to experience large doses of traditional gender socialization.

She found that trainers at both gyms told women patrons that they could lift weights without fear because only men can "bulk up." Nonetheless, 25 percent of women didn't lift at all because they feared developing "masculine" muscles. Another 65 percent restricted their weight lifting to shorter periods or lighter weights after they *did* develop bigger muscles. By the end of two years training, these women remained relatively unmuscular. They lacked muscles not because they were inherently unable to develop them but because they chose not to do so, based on their beliefs about proper male/female differences.

Biological sex differences can also be reinforced by medical practices. Doctors sometimes prescribe hormones to keep girls from growing "too tall" and boys from being "too short" (Weitz 2010). Doctors also offer plastic surgery to women with small breasts and men with small pectoral muscles. In this way, the very bodies we see around us come to reinforce social ideas about male/female differences.

Our belief in the naturalness of biological differences is also reinforced when we are, in essence, kept from seeing how similar males and females can be. Television offers far more coverage of female cheerleaders and male football players than of male cheerleaders and female football players, reinforcing the idea that it is impossible for women to play strenuous sports and that no "real men" would be interested in cheerleading. Similarly, Olympic games that evaluate female figure skaters on their grace and male skaters on their speed and power force female and male skaters to develop different skills and leave audiences believing that female and male skaters naturally have quite different abilities. The same is true for athletic rules that limit the size of the basketball court on which girls can play or that forbid male and female athletes from competing together.

"Doing Gender"

Gender differences are also reinforced when we "do gender." Sociologists use the term "doing gender" to refer to everyday activities that individuals engage in to affirm their commitment to gender roles (West & Zimmerman 1987). Women who are professional bodybuilders almost always wear long, blonde hair so that no one will question their femininity despite their muscles (Weitz 2004), and male nurses sometimes talk about their athletic interests or heterosexual conquests to keep others from questioning their masculinity. Each of us does gender every day when we (whether male or female) choose to wear skirts or jeans, to speak softly or boldly, to get a butterfly tattoo or shark tattoo, and so on. In these ways we participate in the social construction of gender. Another way to think about this is that gender is not something that we innately have, but rather is something that we *do*.

Because adolescence is a time when individuals are actively creating their self-identities, doing gender is particularly important during those years. Although we often focus on girls and women when we think of gender, boys are also under strong pressure to do gender. In fact, those pressures are so strong that they lead to *compulsive heterosexuality*. **Compulsive heterosexuality** consists of continually demonstrating one's masculinity (which in mainstream culture includes demonstrating one's heterosexuality). In her observations at a California high school, sociologist C. J. Pascoe (2007) found that boys constantly encouraged each other to tell about their female sexual conquests, to physically threaten or assault girls, and to sexually threaten or assault girls. Shockingly, teachers did nothing to stop these behaviors, even when the girls were placed at physical risk. The combination of boys' actions and teachers' *in*action both reinforced the idea that such behaviors were natural aspects of masculinity and helped the boys to prove their masculinity to themselves and others.

Compulsive heterosexuality consists of continually demonstrating one's masculinity and heterosexuality.

Gender as Social Structure

Gender is also a social structure, a property of society (Risman 1998). Gender is built into social structure when workplaces don't provide day care; women don't receive equal pay; fathers don't receive paternity leave; basketballs, executive chairs, and power drills are sized to fit the average man; and husbands who share equally in the housework are subtly ridiculed by their friends. Importantly, this suggests that changing gender roles and attitudes will only produce social change if there are parallel changes in the social structure of gender. Equally important, when social structure changes, gender roles and attitudes change. For example, Barbara Risman (1998) found that fathers whose wives died or deserted them learned quickly how to be good "mothers" who could nurture their children as women would.

Differences in Life Chances by Sex

In terms of race and social class, women and men start out equal. The nurseries of the rich as well as the poor contain about 50 percent girls. After birth, however, different expectations for females and males result in very different life chances. This section examines some of the structural social inequalities that exist between women and men.

Health

Women are at a substantial disadvantage in most areas of conventional achievement; in informal as well as formal interactions, they have less power than men. But men, too, face some disadvantages from their traditional gender roles.

Perhaps the most important difference in life chances involves life itself. Boys born in 2015 can expect to live 76.4 years, whereas girls can expect to live 81.4 years (U.S. Bureau of the Census 2009a). On average, then, women live more than 5 years longer than men. Part of this difference is undoubtedly biological, with women's hormones offering them some protection. But men's gender roles also contribute to their lower life expectancies (Rieker & Bird 2000).

A major way male gender roles endanger men is by encouraging them to "prove" their masculinity through dangerous activities. As a result, compared with young women, young men are twice as likely to die in motor vehicle accidents and six times more likely to be killed by guns (Minino 2002). Similarly, men are far more likely than women to earn their living through dangerous jobs, such as fishing and lumbering.

But risk taking alone cannot explain all the difference between men's and women's life expectancies. For example, research suggests that men are at greater risk of dying from heart disease partly because the male gender role places little emphasis on nurturance and emotional relationships. Maintaining family and social relationships is usually viewed as women's work, and so men who stay single, get divorced, or are widowed often end up alone. Ultimately, this lack of social support leaves men especially vulnerable to stress-related diseases and may explain why their suicide rate is four times higher than women's (Minino 2002; Nardi 1992).

In contrast, in poor countries, women's social positions greatly increase their risk of dying. This topic is explored in Focus on A Global Perspective: Pregnancy and Death in Less Developed Nations on pages 218–219.

Education

Fifty years ago, few young women went to college. Those who did were encouraged to focus not on earning a B.A. but on earning an "MRS" (that is, a marriage certificate). These days, women and men are about equally represented among high school graduates and among those receiving bachelor's and master's degrees. It is not until the level of the PhD or advanced professional degrees (such as in architecture) that women are disadvantaged in quantity of education.

More important than the differences in level of education are the differences in *types* of education. From about the fifth grade on, sex differences emerge in academic aptitudes and interests: Boys take more science and math, whereas girls more often excel in verbal skills and focus their efforts on language and literature. In large part, these sex differences in aptitudes and interests are socially created (Sadker & Sadker 1994). In all subjects, but especially math and science, teachers typically assume that boys have a better chance of succeeding. One result is that teachers more often ask girls simple questions about facts and ask boys questions that require use of analytic skills. When boys have difficulty, teachers help them learn how to solve the problem, whereas when girls have difficulty, teachers often do the problem for them. By the time students arrive at college, girls often lack the necessary prerequisites and skills to major in physical sciences or engineering, even if they should develop an interest in them (Sadker & Sadker 1994). As a result, women college graduates are overrepresented in education and the humanities, and men are overrepresented in engineering and the physical sciences—fields that pay considerably higher salaries.

Table 9.1 shows the proportion of bachelor's degrees earned by women in various fields of study in 1971 and in 2007. You can see from the table that there were changes over this period. Women comprised a far higher proportion of graduates in traditionally male fields in 2007 than in 1971. In fact, women now comprise about half of all graduates in business, pre-law, mathematics, and social sciences and history.

TABLE 9.1 **Percentage of Bachelor's Degrees Earned by Women, by Field, 1971 and 2007**
Between 1971 and 2007, the percentage of college degrees in traditionally male fields that were earned by women increased substantially. Nevertheless, engineering continues to be largely a male preserve, and education—especially home economics education—a female preserve. Because engineers earn roughly three times what teachers earn, this difference in majors is one reason why, on average, women earn less than men.

Field of Study	1971	2007
Business	9	49
Computer and information sciences	14	19
Education	75	79
Engineering	1	18
Health sciences	77	86
Home economics education	97	99
Library and archival sciences	92	88
Pre-law	6	58
Mathematics	38	44
Social sciences and history	37	50

SOURCE: National Center for Education Statistics (2009).

focus on A GLOBAL PERSPECTIVE

Pregnancy and Death in Less-Developed Nations

In the poorer nations of the world, women face a very different set of health risks. Within these nations, the most dangerous thing a woman can do is get pregnant. One of every 75 women in the less-developed nations dies from pregnancy or childbirth—almost *100* times the number that die in the most-developed nations (World Health Organization 2009). Map 9.1 shows how the lifetime risk of dying from pregnancy or childbirth varies around the world.

The high rates of pregnancy-related deaths in the developing nations are a consequence of *social* conditions (World Health Organization 2009). First, in any country, about 10 percent of women may die pregnancy-related deaths if they lack access to medical care—a common situation in the less-developed nations.

The remaining causes of pregnancy-related deaths in the developing nations result from women's low social status. In countries in which women have little value, they rarely get enough to eat. This can be fatal for a pregnant woman, who needs extra nourishment to feed both herself and her developing fetus. Lack of nutrition leaves women more likely to become fatally ill, to hemorrhage during childbirth, or to experience other fatal complications during pregnancy or childbirth.

Similarly, in countries where women's value comes primarily from the children they bear, girls are pressured to marry before their bodies are fully developed. As a result, they may be unable to push a baby out, leaving both baby and mother to die. By the same token, when a woman's worth and a man's power are measured by the number of their sons, women have little choice but to become pregnant repeatedly. For biological reasons, each pregnancy after the third places the mother at greater risk than did the previous one.

Finally, in countries where women's health is valued less than that of their fetuses, the final major cause of pregnancy-related death is unsafe abortion (World Health Organization 2009). Most women who obtain abortions are married mothers who believe they cannot afford to feed another mouth. Deaths typically occur when women swallow toxic chemicals to abort themselves or when others use unsterile instruments or accidentally pierce the uterus during an abortion, leading to infection or hemorrhage (Sedgh et al. 2007). Yet abortion is a technically simple procedure, far safer than childbirth when performed by trained professionals working in sterile conditions with proper tools (World Health Organization 2009). Thus, deaths from abortion typically occur when abortion is illegal or when trained providers are unaffordable.

In sum, the best way to keep girls and women from dying during pregnancy and childbirth is to improve their social position.

Still, striking differences between men and women remain. In 2006, only 19 percent of graduates in engineering and 25 percent of graduates in computer sciences were women. Meanwhile, 79 percent of graduates in education, 86 percent in health sciences (mostly nursing), and almost all graduates in home economics and library sciences were women (U.S. Department of Education 2009). Because engineers and computer scientists earn a great deal more than do home economics teachers, librarians, and nurses, these differences in college majors have implications for future economic well-being. This situation is an example of institutionalized sexism. (Recall that Chapter 8 discussed the parallel concept of institutionalized racism.)

Work and Income

Among Americans ages 16 and over, 68 percent of men compared with 57 percent of women are in the labor force (U.S. Bureau of the Census 2009a). This gap is far smaller than it used to be and will likely continue to shrink (Figure 9.1). Although most young women nowadays still expect to be mothers, they also overwhelmingly expect to work full time after completing their education.

MAP 9.1: **Lifetime Risk of Dying from Pregnancy or Childbirth**

In North America, only 1 out of every 6,000 women eventually dies from pregnancy or childbirth; the risk is similar in Australia, New Zealand, and Europe. in contrast, 1 out of 300 women dies from pregnancy or childbirth in South America, 1 out of 120 dies in Asia, and a stunning 1 out of 26 dies in Africa.

SOURCE: Population Reference Bureau (2008).

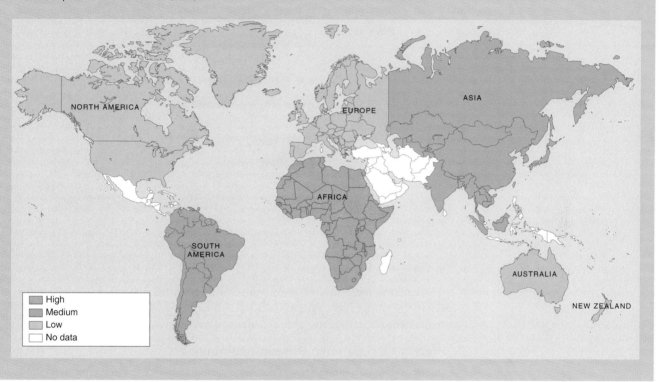

Despite growing equality in labor-force involvement, major inequalities in the rewards of paid employment persist. Women who are full-time workers earn 78 percent as much as men (U.S. Bureau of the Census 2009a). This percentage has not changed much since 1950.

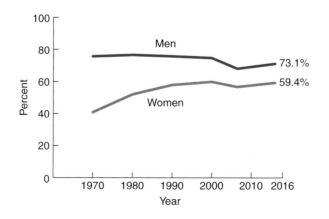

FIGURE 9.1 **Labor-Force Participation Rates of Adult Men and Women, 1970–2016 (estimated)**

SOURCE: U.S. Bureau of the Census (2009a).

Why do women earn less than men? The answers fall into two categories: differences in the types of occupations men and women have and differences in earnings of men and women in the same types of occupations.

Different Occupations, Different Earnings

A major source of women's lower earnings is that women are often employed in different occupations than are men, and women's occupations pay less than men's. The major sex difference as shown in Figure 9.2 is that women dominate sales, office, and service occupations, whereas men dominate blue-collar occupations. The proportion of men and women in professional and managerial occupations is equal. Generally, though, men professionals are doctors and women professionals are nurses; men manage steel plants and women manage dry-cleaning outlets.

There are three major reasons why men and women have different occupations: *gendered occupations*, *different qualifications*, and *discrimination*.

1. *Gendered occupations*. Many occupations in today's segmented labor market are regarded as either "women's work" or "men's work." Construction is almost exclusively men's work; primary school teaching and day care are largely women's work. These occupations are so sex segregated that many men and women would feel uncomfortable working in a job where they were so clearly the "wrong" sex. These stereotypes, combined with the low pay of traditionally female fields, keep most men out of these fields. However, growing numbers of women have moved into jobs that used to be reserved for men, such as insurance adjusting, police work, bus driving, and medicine. This does not, unfortunately, signal that women now have increased access to good jobs. Rather, women by and large move into jobs that men have abandoned because of deteriorating wages and working conditions (Reskin 1989).

focus on  AMERICAN DIVERSITY

Gender Differences in Mathematics

© Tom & Dee Ann McCarthy/Comet/Corbis

Without question, women's position in the work world has improved in recent years. However, they remain underrepresented in the high-paying, high-growth fields of engineering, information technology, and the sciences.

One reason men are overrepresented in these fields is because women less often took advanced math courses or did well on standardized math tests when they were in high school. But are boys and men actually better than girls and women at math, and, if so, is this difference based on nature or nurture?

Neuroscientists interested in this question have begun exploring the relationship between fetal exposure to sex hormones and characteristic differences in the brains of adult men and women. For instance, higher levels of fetal exposure to testosterone (a "male" hormone) are associated with right-brain dominance, while lower exposure levels are associated with left-brain dominance. This association may help explain why, compared with the opposite sex, men more often are left-handed with good visual-spatial skills (a "right-brain" trait) and women more often are right-handed with good verbal skills (a "left-brain" trait).

From findings such as these, some researchers reason that gender differences in mathematical performance are at least partially a result of hormonal differences. But just because hormonal differences are *associated* with mathematical performance does not mean

that the hormonal differences *caused* the differences in performance. For one thing, gender differences in mathematical performance are considerably smaller in countries such as China that less strongly consider mathematics a "male" field (Evans, Schweingruber, & Stevenson 2002). At any rate, the gender differences in performance are small. Because the differences *within* each sex are so much larger than the differences *between* them, critics of the biological perspective argue that hormones can explain only a very small part of the overall variation in mathematical performance. This leaves a great deal of room for the influence of social factors. Evidence for this point of view comes from two lines of research.

First, research suggests that the average test score for girls is lower than that for boys because girls more often respond poorly to the stress of timed tests. Even when girls understand math

as well as boys, they simply don't test as well. In addition, boys typically don't take SAT tests unless they are especially good students, whereas girls often take the tests even if they are only average students. As a result, the mean test score for girls is lower than that for boys (especially on the math section) simply because a broader pool of girls takes the test (Lewin 2006).

Second, research shows that the male advantage in mathematical performance is small, only emerges late in high school, and has declined steadily since the 1960s (Leahey & Guo 2001). One possible explanation for this pattern is that boys and girls are now being socialized more similarly, thereby reducing the traditional male advantage in math.

2. *Different qualifications.* Although the differences are smaller than they used to be, women continue to major in fields of study that prepare them to work in relatively low-paying fields, such as education, whereas men are more likely to choose more lucrative fields. (Some of the reasons for this are discussed in Focus on American Diversity: Gender Differences in Mathematics.) More important than these differences in educational qualifications are disparities in experience and on-the-job training. Believing that women are likely to quit once they marry, have children, or lose interest, employers invest less in training and mentoring them (Tomaskovic-Devey & Skaggs 2002).

TABLE 9.2 Sex Differences in Representation and Median Weekly Earnings, by Occupation*

Women are clustered in lower-paying occupations. But even when women have the same occupation as men, they tend to earn substantially less money. Women tend to be employed in lower-paying firms and subfields and to experience discrimination in hiring, raises, and promotion.

Occupation	Male Income	Female Income	% of Workers Who are Women
Chief executives	$1,903	$1,603	24%
Lawyers	1,751	1,509	38
Computer programmers	1261	1,003	22
Elementary and middle-school teachers	994	871	81
Retail salespersons	623	440	43

SOURCE: U.S. Bureau of Labor Statistics (2009b).
*Full-time, year-round workers only.

As a result, women are less likely to be promoted to management positions—even if they have no intention of having children or marrying.

3. *Discrimination.* Although men and women have somewhat different occupational preparation, a large share of occupational differences is due to discrimination by employers (Hesse-Biber & Carter 2000). Employers reserve some jobs for men and some for women based on their own gender-role stereotypes. As a result, women remain nurses rather than nursing administrators, and salesclerks rather than store managers.

Same Occupation, Different Earnings

Not all occupations are highly sex segregated. Some, such as flight attendant, teacher, and research analyst, contain considerable proportions of both men and women. Within any given occupation, however, men typically earn substantially higher incomes (Table 9.2). There are two main explanations for this: *different titles* and *discrimination*.

1. *Different titles.* Very often, men and women who do the same tasks are given different titles—women will be maids or executive assistants, and men doing the same work will be janitors or assistant executives. Simply because one job category is considered "male" and is occupied by males, it is paid a higher wage.

2. *Discrimination.* Even when women and men have the same job titles, women tend to be paid less. One reason for this is that, within any given occupation, men tend to hold the more prestigious, better-paying positions (Hesse-Biber & Carter 2000; McBrier 2003). Male lawyers tend to be hired in large, high-paying firms to specialize in prestigious fields, whereas women tend to be hired in small, low-paying firms, specializing in less prestigious fields. Male sales staff tend to be hired by stores and departments that offer better salaries or hefty commissions, whereas female sales staff work in less remunerative areas. These differences reflect the segmented labor market (discussed in more detail in Chapter 13).

Even when women and men work in the same occupations and positions; work for the same employers; and have equal education, experience, and other qualifications, women earn less. The absence of any other explanations for this difference has led researchers to conclude that it must be caused by discrimination (Maume 2004).

This discrimination occurs in both female- and male-dominated fields. In female-dominated occupations, women's careers progress gradually. In contrast, men often encounter a "glass escalator" that invisibly helps them to move rapidly into administrative positions and prestigious specialties (Williams 1992; Hultin 2003). In male-dominated occupations, men's careers typically progress gradually, whereas women more often are pressured out of the occupation altogether (Maume 1999). This is often done through subtle discrimination such as exclusion from informal leadership and decision-making networks, sexual harassment, and other forms of hostility from male co-workers (Chetkovich 1998; Jacobs 1989). This informal discrimination creates a "glass ceiling"—an invisible barrier to women's promotions (Freeman 1990).

Gender and Power

As Max Weber pointed out, differences in prestige and power are as important as differences in economic reward. When we turn to these rewards, we again find that women are systematically disadvantaged. In the family, business, the church, and elsewhere, women are less likely to be given positions of authority.

Although sexism continues to have an impact, more and more women are finding employment in fields formerly open only to men.

Unequal Power in Social Institutions

Women's subordinate position is built into most social institutions. In some churches, ministers quote the New Testament command, "Wives, submit yourselves unto your own husbands" (Ephesians 5:22). In colleges, women's basketball coaches are paid less than men's basketball coaches. In politics, prejudice against women leaders remains strong, and women still comprise only a minority of major elected officials in the United States and around the world.

Unequal Power in Interaction

As we noted in Chapter 4, even the informal exchanges of everyday life are governed by norms; that is, they are patterned regularities, occurring in similar ways again and again. Careful attention to the roles men and women play in these informal interactions shows clear differences—all of them associated with childhood socialization and with women's lower prestige and power.

Studies of informal conversations show that men regularly dominate women in verbal interaction (Tannen 1990). Men take up more of the speaking time; they interrupt women more often; and most important, they interrupt more successfully. Finally, women are more placating and less assertive in conversation than men, and women are more likely to state their opinions as questions ("Don't you think the red one is nicer than the blue one?"). This pattern also appears in committee and business meetings, which is one reason women employees are less likely than men to get credit for their ideas (Tannen 1994).

Laboratory and other studies show that this male/female conversational division of labor is largely a result of status differences (Kollock, Blumstein, & Schwartz 1985;

decoding the data

Sexual Harassment on the Job

SOURCE: Calculated from General Social Survey. **http://sda.berkeley.edu**. Accessed June 2009.

	Percentage answering yes	
	Male	Female
College graduates	16%	23%
Not college graduates	20%	29%

By definition, sexual advances or propositions from professors, work supervisors, or others who have power over another individual are considered sexual harassment. Harassment is experienced more often by females and by those who have not graduated from college, regardless of their sex.

Have you ever experienced sexual advances or propositions from supervisors, whether involving physical contact or just sexual conversations?

Explaining the Data: Why would females be more likely than males to experience sexual harassment? Why would those with less education be more at risk of sexual harassment?

Critiquing the Data: Are there any reasons why males would be more likely than females to report these experiences? why females would be more likely than males to do so?

Are there any reasons why college graduates would be more likely than others to report these experiences? why nongraduates might be more likely to do so?

This question was asked of all persons who responded to the national, random General Social Survey, regardless of whether or for how long they had ever held a job. How might the percentage reporting harassment have changed if the question was asked only of full-time workers?

Ridgeway & Smith-Lovin 1999; Tannen 1990). When women clearly have more status than men, such as when a female professor talks with a male student, women do not exhibit low-status interaction styles.

Case Study: Sexual Harassment

The impact of women's relative lack of power becomes clear when we look at the topic of **sexual harassment**—unwelcome sexual advances, requests for sexual favors, and other unwanted verbal or physical conduct of a sexual nature. Although estimates vary widely depending on the definition and sample used, as many as half of all working women probably experience sexual harassment during their lifetime (Welsh 1998). Men also can be sexually harassed—by men as well as by women—but this occurs far less often. The extent and measurement of sexual harassment is discussed further in Decoding the Data: Sexual Harassment on the Job.

There are two forms of sexual harassment (Shapiro 1994). By law, harassment exists when an employer, teacher, or other supervisor expects sexual favors (from inappropriate touching to sexual intercourse) in exchange for something else: keeping one's job, getting a good grade or letter of recommendation, and so on. Sexual harassment ranges from subtle hints about the rewards for being more friendly with the boss or teacher to rape. Sexual harassment also exists when an individual finds it impossible

Sexual harassment consists of unwelcome sexual advances, requests for sexual favors, or other verbal or physical conduct of a sexual nature.

to do his or her job because of a hostile sexual climate, such as when pornographic photographs are posted in an office or co-workers frequently make sexist or sexual jokes.

Sexual harassment exists because women have less power than men. (Similarly, men are only harassed in situations where they have little power.) But sexual harassment not only *reflects* women's relative powerless social position, it also helps to *keep* them in that position. For example, women students in engineering classes or firms who experience sexual harassment are less likely to continue to pursue a career in engineering. They also may lose confidence in their abilities and their judgment and may suffer long-lasting psychological troubles (Sadker & Sadker 1994).

Fighting Back Against Sexism

To fight back against sexual harassment, woman battering, job discrimination, and the other problems discussed in this chapter, women—and men—have united in the feminist movement (Evans 2003; Freedman 2002). At its core, the feminist movement holds that women and men deserve equal rights and that women's lives, culture, and values are as important as are men's. This chapter is an important marker of the success of the feminist movement: Thirty years ago, no sociology textbook would have included a chapter on sex and gender.

This section looks at the history of the feminist movement and at the particular issues involved for nonwhite women.

Sexual harassment remains common—if illegal—in the workplace.

©2009/Jupiterimages

The Feminist Movement

The first wave of the American feminist movement arose in the mid-nineteenth century. At the time, women's legal status was essentially that of property. Like slaves (both male and female), women regardless of race could not own property, vote, make contracts, or testify in a court of law, and only two small colleges admitted women. Many women (both black and white) who were active in the movement to abolish slavery took from their experience a belief in equality and the organizing skills needed to start the feminist movement.

Because of feminist protest, by the end of the nineteenth century, the most egregious legal restrictions on women's lives had been lifted, and a growing (though still small) list of colleges accepted women students. At this point, feminist activity shifted almost entirely to obtaining the vote (suffrage) for women. In 1920, Congress adopted the Nineteenth Amendment, which granted female suffrage.

After the passage of the Nineteenth Amendment, feminist activity declined precipitously. In the 1960s, however, two groups of women began pressing for further change, in what became known as the *second wave* of feminism. The first group (known as *liberal feminists*) came to feminism through involvement in mainstream political and professional organizations and fought for women to gain equal rights within the existing system. They deserve credit for such social changes as requiring selective public high schools and colleges to admit female students and forbidding employers from posting job advertisements "for men only." The second group (known as *radical feminists*) came to feminism through the civil rights and anti-Vietnam War movements and fought for more radical social changes. They deserve credit for bringing public awareness to incest, domestic violence, and date rape. (In fact, the latter two terms didn't even exist before radical feminism.) They also deserve credit for promoting the idea that women can and should enjoy sexual pleasure, which includes the right to birth control.

sociology and you

Did you play on a sports team in high school, or do you play on a team now? If so, and if you are female, you owe your athletic career in part to Title IX of the federal Educational Amendments of 1972. Title IX was a product of liberal feminist activism. It prohibits sex discrimination in any educational institution or activity that receives federal funding. Title IX applies to everything from financial aid and class offerings to athletics and health insurance, from kindergarten through graduate school. It has led to a dramatic rise in women's educational attainment and their athletic participation.

Beginning in the 1990s, a new group of feminists, know as the *third wave*, came to the fore. Because third-wave feminists grew up in a society that had been deeply affected by earlier waves of feminism, they focused on emphasizing how women's position had improved. Similarly, they focused on celebrating women's sexual freedom and pleasure rather than on highlighting sexual dangers. In addition, third-wave feminists emphasized the particular ways that sexism, racism, class inequality, and other forms of inequality differently affect different groups of women (Evans 2003; Freedman 2002).

Fighting Sexism and Racism

Throughout the history of the feminist movement, white and nonwhite women worked together to improve women's lives. Nevertheless, and as third-wave feminists pointed out, the feminist movement has sometimes (if unintentionally) focused on issues that mainly concerned white women.

For nonwhite women, the struggle for equality starts from different places and provokes different questions and answers. Most importantly, nonwhite women face a two-pronged dilemma. First, they have not benefited from the sheltered position of traditional white women's roles. Nonwhite women have always worked outside the home: For example, in 1900 married African American women were six times more likely to be employed than were married white women (Goldin 1992). Although they worked, they still had to face the economic and civic penalties of being women. Consequently, minority women traditionally have had less to lose and more to gain from abandoning conventional gender roles. On the other hand, nonwhite women face a potential conflict of interest: Is racism or sexism their chief oppressor? Should they work for an end to racism or an end to sexism? If they choose to work for women's rights, they may be seen as working against men of their own racial and ethnic group.

Current income figures indicate that sex is more important than race in determining women's earnings: The difference in earnings among Hispanic, African American, and white non-Hispanic women is relatively small compared to the difference between women and men. This suggests that fighting sex discrimination would aid nonwhite women more than fighting racial discrimination would. But this conclusion overlooks the fact that women and their children also need the earnings of their husbands and fathers. For example, because of the low earnings and limited employment opportunities of African American *men*, African American women and children are three times more likely than their white counterparts to live below the poverty level. As a result, nonwhite women have much to gain by fighting racism as well as sexism.

The dilemma remains. The women's rights movement is often seen as a middle-class white social movement; racial and ethnic movements have been seen as men's movements. Nevertheless, minority women have a long history of resistance to both racism and sexism.

The Sociology of Sexuality

Like gender, sexuality is also a product of both biology and culture. Ideas about "proper" sexuality vary cross-culturally, and have varied historically. A hundred years ago, a woman who admitted to enjoying sexual pleasure could have been declared insane and locked in a mental hospital. Now, a woman who does *not* enjoy sexual pleasure may be labeled frigid and referred to a therapist. In ancient Greece, male youths

"Abstinence only" programs now dominate sex education in the United States. Yet research consistently finds that such programs work only in the very short term, if at all.

were expected to engage in homosexual behavior with their adult male mentors; these days, adults (of either sex) who have sexual relations with minors can be imprisoned. In this section, we look at current sexual behavior in the United States.

Sexual Scripts

In few areas of our lives are we free to improvise. Instead, we learn roles and norms scripts that direct us toward accepted behaviors and away from unaccepted ones. Sex is no exception. Cultural expectations regarding who, where, when, why, how, and with whom one should have sex are referred to as **sexual scripts**. Depending on your subculture, you may have learned a sexual script in which sex was something done only between spouses, for the purpose of procreation, at night, behind closed doors. Or you may have learned a script in which sex was something to be celebrated and enjoyed, between any willing partners, in any location and at any time that felt comfortable.

Because no modern culture is fully homogeneous, different sexual scripts are often in conflict. And because we are exposed to sexual scripts from multiple sources—parents, teachers, friends, religious leaders, the mass media—the sexual scripts we adopt often change over time.

Premarital Sexuality

One of the most important sexual scripts has to do with the appropriateness of sexuality outside of (heterosexual) marriage. Premarital intercourse has become increasingly accepted over the last few decades (Ku et al. 1998; Abma et al. 2004). Moreover, whereas in the 1950s couples typically only had sex if they intended to marry, now teens may "hook up" with no intention of even having a relationship.

Similarly, the proportion of never-married teenagers who say that they have had sexual intercourse increased from about 40 percent in the 1950s to about 50 percent

Sexual scripts are cultural expectations regarding who, where, when, why, how, and with whom one should have sex.

for girls and 60 percent for boys by the late 1980s (Abma et al. 2004). Since then, however, rates of sexual intercourse among teens have declined slightly, to about 46 percent among both boys and girls (Abma et al. 2004). What explains this decline?

The answer is definitely *not* the abstinence-only sexual education programs that now dominate in the United States. Research consistently finds no credible evidence that such programs work except in the very short term (Dailard 2003).

More likely, the drop in teenage sexual activity reflects the growing awareness of the threats posed by AIDS and other sexually transmitted diseases. Not surprisingly, the percentage of teenagers who report using condoms the last time they had sexual intercourse has increased steadily since 1988. It is now common for young people to use condoms the first few times they have sexual relations with a new partner. After that, though, most conclude that they know and can trust their partners and so abandon condom use. Women are especially likely to believe that their partner loves them and wouldn't hurt them; men are especially likely to believe that they are invulnerable and don't need to worry. Unfortunately, it is usually impossible to know if someone has a sexually transmitted disease unless they admit it. But many individuals don't know they are infected, while others know but don't tell.

Marital Sexuality

In certain important ways, the sexual scripts followed by married couples have changed little over time. For example, frequency of sexual activity seems to have changed very little among married people over the years (Call, Sprecher, & Schwartz 1995; Laumann et al. 1994). And now, as in the past, most couples find that the frequency of intercourse declines steadily with the length of the marriage. The decline appears to be nearly universal and to occur regardless of the couple's age, education, or situation. After the first year, almost everything that happens—children, jobs, commuting, housework, finances—reduces the frequency of marital intercourse (Call, Sprecher, & Schwartz 1995). Nevertheless, satisfaction with both the quantity and the quality of one's sex life is essential to a good marriage (Blumstein & Schwartz 1983; Laumann et al. 1994).

Despite these historical continuities, the sexual scripts followed by married couples have undergone some important changes in recent decades. First, oral sex, a practice that was limited largely to unmarried sexual partners and the highly educated in earlier decades, is now more common. Second, women and men are now equally likely to have extramarital affairs. The double standard has disappeared in adultery: Studies conducted in the 1990s suggest that as many as 50 percent of both men and women have had an extramarital sexual relationship (Laumann et al. 1994). Unfortunately, more recent data on marital sexuality is not available, because federal funding for sexuality surveys was essentially abandoned under the administration of President Bush.

Sexual Minorities

Although the majority of the population is heterosexual—preferring sex and romance with the opposite sex—significant minorities diverge from this script. This section discusses homosexuals and transgendered persons.

Homosexuality in Society

The largest of the sexual minorities is homosexuals (also known as gays and lesbians). **Homosexuals** are people who prefer sexual and romantic relationships with members

Homosexuals (also known as gays and lesbians) are people who prefer sexual and romantic relationships with members of their own sex.

of their own sex. On well-regarded surveys, somewhere between 2 and 6 percent of Americans admit *recent* homosexual activity or describe themselves as homosexual, with rates about twice as high among men as among women (Binson et al. 1995; Lauman et al. 1994). Considerably more report that they engaged in homosexual activity at some point in their lives. Undoubtedly many others have also done so but have not admitted it to survey researchers.

Attitudes toward homosexuality have fluctuated greatly over time. During the last 50 years, however, American attitudes have become increasingly more positive. In a Gallup Poll conducted in 2008, 55 percent of surveyed Americans agreed that homosexual activity between consenting adults should be legal (Saad 2008). Support for gay rights is highest among persons who are less religious, younger, urban dwellers, non-Southerners, more educated, and more liberal in general.

The Gay and Lesbian Rights Movement

Growing acceptance of homosexuality is a direct outgrowth of the gay and lesbian rights movement. The American gay and lesbian rights movement grew rapidly in the late 1960s and early 1970s, when gays and lesbians who had worked in the civil rights and feminist movements began questioning why they too should not have equal rights (Clendinen & Nagourney 2001; Marcus 2002).

The pivotal moment for the incipient gay rights movement came with the Stonewall Riots, which began June 27, 1969. For many years before that date, the police had routinely raided gay bars in New York City. But something was different that night: This time the bar's patrons fought back. The police responded brutally, but the riot only grew, with about 2,000 people from the heavily gay and lesbian neighborhood joining in over the next few days. By the time the riots ended, the modern gay rights movement had come of age.

The AIDS epidemic also played an important role in the history of the movement. When AIDS was first identified in 1981, many erroneously labeled it a "gay plague," and both prejudice and discrimination increased. As gay men were forced by their illness to reveal their sexual identity or were identified as gay after they died of AIDS, heterosexuals came to realize how many of their friends, relatives, co-workers, neighbors, and favorite film stars (like Rock Hudson) were gay. As a result, stereotypes and prejudices often fell by the wayside.

The gay and lesbian rights movement has achieved some notable successes. The American Psychological Association no longer considers homosexuality per se an illness; at least 21 states and the District of Columbia outlaw discrimination on the basis of sexual orientation (National Gay and Lesbian Task Force 2009); and open homosexuals have been elected to public office, including in the U.S. Senate. Most importantly, in 2003 the U.S. Supreme Court declared that states could no longer criminalize private, consensual, same-sex activities. Currently, the hottest battles are being fought over the right of gays to marry or enter into civil unions.

Transgender in Society

Transgendered persons are individuals whose sex or sexual identity is not definitively male or female. There are two main types of transgendered people: intersex persons and transsexuals.

Intersex persons are individuals who are born with ambiguous genitalia, such as a small penis as well as ovaries. Intersexuality is a naturally occurring, if rare, phenomenon. In the early stage of fetal development, all fetuses are sexually ambiguous. All fetuses (and adult humans) produce both male and female hormones (including estrogen and testosterone), and these hormones lead to sexual differentiation—the

Transgendered persons are individuals whose sex or sexual identity is not definitively male or female.

development of ovaries, penises, and so on—later in fetal development. Intersexuality occurs when that differentiation is incomplete. When such cases are identified, doctors typically use surgery or hormones to transform the individual's body into one that more closely matches our accepted ideas of what males or females should look like. As when plastic surgeons give women larger breasts, these medical interventions serve to reinforce social ideas about proper sexuality. In contrast, some other cultures recognize the existence of more than two sexes (Herdt 1994; Lorber 1994).

Unlike intersex persons, transsexuals' sex is not ambiguous: There are no observable biological differences between them and other heterosexual males or females. Instead, transsexuals are persons who psychologically feel that they are trapped in the body of the wrong sex. As with intersex persons, most doctors consider it appropriate to prescribe hormones or perform surgery (removing penises and constructing vaginas or vice versa) to give transsexuals the bodies they desire. Some observers, however, question the wisdom of these medical interventions (Meyerowitz 2002). They wonder whether, in a society that allowed both men and women more freedom, anyone would feel "trapped" in the wrong body, and they question whether there is really something so wrong with men who enjoy "chick flicks" and taking care of children, or with women who prefer wearing crew cuts and working on cars. To these observers, the medical treatment of transsexuality is another example of the social construction of both gender and sexuality.

Where This Leaves Us

Gender roles have changed dramatically over the last 30 years, in ways that have affected us deeply. As structural functionalists point out, traditional roles had their virtues. Everyone knew what was expected of them, and complementary male/female roles held families together by forcing each sex to depend on the other. In contrast, the decline in traditional gender roles has brought stress to many people—not only to men who lost rights and power but also to women who found themselves caught between changing expectations.

But conflict theorists are also correct: Everyone did not benefit equally from traditional roles, and everyone paid some price for maintaining them. Women endured lower earnings, narrow educational and occupational opportunities, sexual harassment, sexist prejudice and discrimination, and, sometimes, physical violence. Men who held to traditional masculine gender roles experienced more stress, less nurturing relationships, and shorter lives.

Sex is a biological category, something we are born with. But sex, gender, and sexuality are also socially constructed. Doctors can change patients' physical bodies so that individuals' sex and gender better fit social expectations. Society, in general, continually evolves its ideas of what it means to be male and female, masculine and feminine, and all of us contribute to this process when we socialize our children, "do gender," and interact with each other. Creating a more just world will require that we change the social structure of gender and sexuality as well as its interpersonal aspects.

Summary

1. Although there is a universal biological basis for sex differentiation, a great deal of variability exists in the roles and personalities assigned to men and women across societies. In almost all cultures, however, women have less power than men.

2. Structural-functional theorists argue that a division of labor between the sexes builds a stronger family and reduces competition. Conflict theorists stress that men and capitalists benefit from sexism and a segmented labor market that relegates women to lower-status positions. Symbolic interactionism does not address why gender inequality arose but does help us understand how it is perpetuated in interaction.

3. Sex stratification is maintained through socialization. From earliest childhood, females and males learn ideas about sex-appropriate behavior and integrate them into their self-identities. Sex stratification is also maintained by medical and social practices that magnify biological differences between the sexes.

4. Gender is not simply an individual attribute. It is also a property built into social structures and built into our everyday actions. "Doing gender" refers to everyday activities that individuals engage in to affirm that they understand what is expected of them as male or female. Boys' perceived obligation to constantly demonstrate their masculinity and heterosexuality is referred to as *compulsive heterosexuality.*

5. Men as well as women face disadvantages due to their gender roles. For men, these include higher mortality and fewer intimate relationships.

6. Women and men are growing more similar in their educational aspirations and attainments and in the percentage of their lives that they will spend in the workforce.

7. Women who are full-time, full-year workers earn 78 percent as much as men. This is because they have different (poorer-paying) occupations and because they earn less when they hold the same occupations. Causes include different educational preparation and discrimination.

8. Women's subordinate position is built into all social institutions. Although some of this has changed, men disproportionately occupy leadership positions in social institutions. They also dominate women in conversation.

9. For over 150 years, the feminist movement has fought to improve the position of American women. It has had many notable successes.

10. Premarital sexuality is now widely accepted. However, it has declined in frequency since the late 1980s, primarily in response to the AIDS epidemic.

11. Homosexuality is growing more accepted in the United States. Some sociologists question whether the medical treatment of transgendered persons reflects and reinforces traditional ideas about gender roles.

Thinking Critically

1. Suppose you want your daughter to consider science as a future profession. How would you go about encouraging her to consider this career choice? As a member of the PTA at your daughter's school, what changes would you encourage her school to make in order to increase the chances of girls considering science as a profession?

2. Chapter 8 discussed institutionalized racism. Consider the parallels between racism and sexism. Can you think of some specific examples of how institutionalized *sexism* works against women in the workplace? against men? What kinds of programs or policy might help reduce this discrimination against working women?

3. If men have more power, why do they die earlier and have higher rates of heart disease, suicide, and alcoholism? As women gain power, should we expect them to have similar health problems? Why or why not?

4. In TV commercials, males predominate about nine to one as the authority figure, even when the products are aimed at women. Using your sociological knowledge, how would you explain this?

Book Companion Website

www.cengage.com/sociology/brinkerhoff

Prepare for quizzes and exams with online resources—including tutorial quizzes, a glossary, interactive flash cards, crossword puzzles, essay questions, virtual explorations, and more.

CHAPTER 10

Health and Health Care

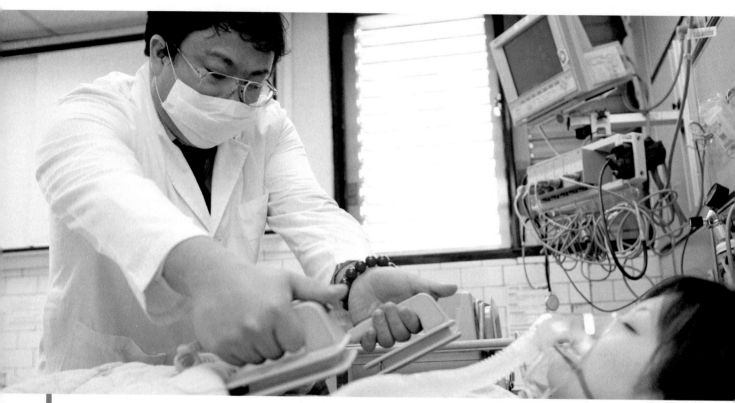

IMAGEMORE Co., Ltd./Getty Images

Health and Health Care as a Social Problem

At first glance, health seems a purely biological state, and health care a purely medical matter. Yet as this chapter will show, health, illness, and health care are deeply affected by social forces and social status.

Although it may seem that health and illness are not issues that need concern college-age students, this is far from true. Illness, disability, and traumatic injury can strike at any age. This is particularly important because the United States is alone among the industrialized nations in not providing access to health care to all citizens. As a result, 45 million Americans under age 65 lacked health insurance in 2007—a number that has surely increased, given current economic conditions—and health-related debt is a major cause of personal bankruptcy (Kaiser Commission on Medicaid and the Uninsured 2008; Newman 1999b; Sullivan, Warren, & Westbrook 2000). Furthermore, health is the single most important factor that influences overall quality of life. Thus, we need to consider not only the social forces that affect health and illness but also why the U.S. health-care system has taken the particular form it has and the consequences of that system. We begin by looking at how sociologists think about illness itself.

Theoretical Perspectives on Illness

Because of their different approaches, each sociological theory of illness focuses on a different set of questions and offers a different set of answers. The classic structural-functionalist theory of illness looks at how (some) illness can help society run smoothly and how society limits illness that can interfere with that smooth flow. Conflict theory illustrates how competing interests lead to different definitions of illness, and symbolic interaction theory has been particularly useful for understanding the experience of illness.

Structural-Functionalist Theory: The Sick Role

The classic sociological theory of illness was first formulated by Talcott Parsons (1951). As a structural-functionalist, Parsons assumed that any smoothly functioning society would have ways to keep illness, like any other potential problem, from damaging it.

Parsons's most important contribution to sociology was the realization that illness is a form of deviance, in that it keeps individuals from performing their normal social roles. The last time you were sick, for example, you might have taken the day off from work, asked your boyfriend or girlfriend to pick up groceries for you, or asked a professor to give you an extension on a paper. You might even have claimed to be sick just to get out of those responsibilities. To Parsons, therefore, illness (or claims of illness) is generally *dysfunctional* because it could threaten social stability.

Parsons also recognized, however, that allowing some illness was good for social stability. If no one could ever "call in sick" or take a "mental health day," no one would have the time needed to recuperate, and resentment would build among workers, students, and spouses who never got a break. In these ways, illness acts as a sort of "pressure valve" for society.

As the sick role describes, when we get sick we are expected to go to the doctor and to follow the doctor's orders.

©2009/Jupiter images

Defining the Sick Role

How does society control illness so that it increases rather than decreases social stability? The answer, according to Parsons, is the **sick role**. The sick role refers to four social norms regarding how sick people should behave and how society should view them. First, sick persons are assumed to have legitimate reasons for not fulfilling their normal social roles. This is why we give sick people time off from work rather than firing them for malingering. Second, cultural norms declare that individuals are not responsible for their illnesses. For this reason, we bring chicken soup to people who have colds rather than jailing them for stupidly exposing themselves to germs. Third, sick persons are expected to consider sickness undesirable and work to get well. This is why we sympathize with those who rest when they are ill and chastise those who don't. Finally, sick persons should seek and follow medical advice.

Critiquing the Sick Role

Parsons's concept of a sick role was a crucial step in beginning to think of illness sociologically. Subsequent research, however, has illuminated the limitations of the sick role model (Weitz 2010). This critique is highlighted in the Concept Summary on Weaknesses of the Sick Role Model.

First, in contrast to Parsons's analysis, ill persons sometimes *are* expected to fulfill their normal social roles. While no one expects persons dying of cancer to continue working, we often expect people with arthritis to do so, as well as those we suspect are malingerers or hypochondriacs because doctors have been unable to diagnose their condition. Similarly, regardless of illness, some professors expect students to turn papers in on time, some husbands expect their wives to cook dinner, and some employers expect their employees to come to work.

Second, sometimes people *are* held responsible for their illnesses. The last time you had a cold, did anyone chastise you for not taking care of yourself well enough? for not taking vitamin C, getting enough sleep, or eating healthy meals? Similarly, newspaper stories and television shows often implicitly blame lung cancer on people who smoke, diabetes on people who eat too much, and so on.

The **sick role** consists of four social norms regarding sick people. They are assumed to have good reasons for not fulfilling their normal social roles and are not held responsible for their illnesses. They are also expected to consider sickness undesirable, to work to get well, and to follow doctor's orders.

Weaknesses of the Sick Role Model

Elements of the Sick Role	Model Fits Well	Model Fits Poorly
Illness is considered a legitimate reason for not fulfilling obligations.	Appendicitis, cancer	Undiagnosed chronic fatigue
Ill persons are not held responsible for illness.	Measles, hemophilia	AIDS, lung cancer
Ill persons should strive to get well.	Tuberculosis, broken leg	Diabetes, epilepsy
Ill persons should seek medical help.	Strep throat, syphilis	Alzheimer's, colds

Third, the sick role's assumption that sick individuals should work to get well simply doesn't fit those who have chronic illnesses that medicine can't cure. In much the same way, the assumption that sick people should follow medical advice ignores those who can't afford or aren't helped by medical care.

Conflict Theory: Medicalization

Like other structural functionalists, Parsons assumed that social ideas about illness (in this case, the sick role) are designed to keep society running smoothly. In contrast, conflict theorists assert that, like other parts of social life, ideas about illness reflect competing interests among different social groups.

One of the major contributions of conflict theory to our understanding of illness is the concept of medicalization. As we saw in Chapter 6, medicalization refers to the process through which a condition or behavior becomes defined as a medical problem requiring a medical solution (Conrad 2007). One hundred years ago, masturbation, homosexuality, and, among young women, the desire to go to college were all considered symptoms of illness. These conditions are no longer considered illnesses not because their biology changed but because social ideas about them did. Similarly, one hundred years ago most women gave birth at home attended by midwives, few boys were circumcised, and plump people were considered attractive and lucky. Nowadays, pregnant women are expected to seek medical care, parents are expected to have their infant sons circumcised by doctors, and overweight people are considered to be at risk for illness or even to have the "illness" of obesity. These are all examples of medicalization.

For medicalization to occur, one or more organized social groups must have both a vested interest in it and sufficient power to convince others to accept their new definition of the situation. The strongest force currently driving medicalization is the pharmaceutical industry, which has a vested financial interest in enlarging the market for its products (Conrad 2007). For example, the pharmaceutical industry was the major force behind defining "male sexual dysfunction" as a disease—to be cured by Viagra (Loe 2004). Pressure for medicalization also can come from doctors who hope to enlarge their markets and from consumer groups who hope to stimulate research on or reduce the stigma of ambiguous conditions such as alcoholism or fibromyalgia (Barker 2005; Conrad 2007).

Mass marketing of Viagra "sold" both the drug and the idea that impotence was a symptom of the disease "erectile dysfunction disorder."

Conversely, doctors sometimes oppose medicalization because they don't want the responsibility for treating a condition (such as wife battering), and consumers sometimes oppose medicalization because they believe a condition is simply a natural part of life (such as menopause). Insurers, too, may support or oppose medicalization, depending on their interests. For example, initially insurers rejected requests for expensive gastric bypass surgery for obese patients, arguing that obesity was not an illness. Now that most insurers have concluded that these surgeries reduce their long-term costs, they support diagnosing obesity as an illness and surgically treating it (Conrad 2007). In each case, the battle over medicalization was won by the group that could bring the most money, influence, and other forms of power to bear.

sociology and you

Social Policy

The next time you are at a doctor's office, keep your eyes open. Do you see pamphlets, posters, pens, mugs, or anything else labeled with names or logos from pharmaceutical companies? Does your doctor offer you free samples of new drugs? Are there any health magazines you don't recognize (likely published by pharmaceutical companies)? All of these are evidence of the pharmaceutical industry's attempts to influence disease diagnosis and treatment.

Symbolic Interaction Theory: The Experience of Illness

The sick role model helps us understand cultural assumptions for how ill people should behave and how they should be treated by others, whereas conflict theory helps us understand how people come to be defined as ill in the first place. In contrast, symbolic interaction theory is particularly useful for understanding what it is like to live with illness on a day-to-day basis and, especially, what happens when doctors and patients have different definitions of the situation. This issue comes to the fore when doctors and patients disagree over treatment.

To doctors, any patient who does not follow their medical orders is engaging in *medical noncompliance*. Doctors typically assume that they know best how a disease should be treated, and therefore assume that any patient who does not follow their orders is either foolish or ignorant. Research by symbolic interactionists, however, suggests that the issue is far more complex. Some patients don't comply because health-care workers offered only brief and confusing explanations of what to do and why. Other patients lack the money, time, or other resources needed to comply. Still others conclude that following medical advice is simply not in their best interests. They may decide, for example, against taking a drug that lowers blood pressure but leaves them unable to achieve erection, that reduces schizophrenic hallucinations but causes obesity, or that brings substantial side effects but seems to have no impact on their symptoms (Lawton 2003). And increasingly, patients reach decisions about treatment based as much on the Internet as on their doctors' advice, a topic discussed in Focus on Media and Culture: The Internet and Health.

In sum, what doctors define as medical noncompliance, patients define as rational decision making. When doctors chastise patients for their noncompliance and fail to understand their perspectives, patients are likely to become even less willing to follow doctors' orders, creating a self-fulfilling prophecy.

The Internet and Health

The rise of the Internet has dramatically affected how doctors, the government, the pharmaceutical industry, and the public deal with illness.

One major change is the shift to online medical records. These records allow multiple doctors, nurses, pharmacists, and others to access the same patient's records, even if they are working at different locations (such as a doctor's office, a hospital, and a drugstore). Such records reduce the chance that a patient will receive prescriptions from different doctors for drugs that interact dangerously and increase the chance that doctors will have a broader understanding of a patient's health problems. However, the use of online medical records raises serious concerns about patient privacy (Alpert 2003; Freudenheim & Pear 2006). For example, if the record indicates that a patient has been treated for alcohol-related problems, many people will *legally* gain access to that information and anyone with good computer hacking skills may do so *illegally*. As a result, patients may experience stigma or even lose their jobs or health insurance. Thus this change has the potential to shift power to any group that has access to the records.

Another major change is the rise in online pharmaceutical sales (Eckholm 2008). These sites benefit consumers by enabling them to purchase needed drugs at reduced costs. On the other hand, these sites enable anyone anywhere to obtain dangerous drugs without prescriptions. In some cases, people may risk their health when they purchase drugs they believe they need without first checking with a doctor. In other cases, people may risk their health by illegally buying addictive drugs such as Valium and Vicodin. Thus these sites have increased the power of individual users and of drug providers while decreasing the power of doctors and the government to control drug use.

Finally, the Internet has affected the entire experience of illness (Barker 2005). These days, many people check the Internet whenever they feel ill—even before calling their doctor. The Internet is in fact a great way to learn, for example, how to tell a simple cold from influenza. The Internet is also especially helpful for those with stigmatized, difficult-to-diagnose, or difficult-to-treat illnesses, such as chronic fatigue syndrome, urinary problems, or multiple sclerosis. Many such individuals have diagnosed themselves (whether accurately or inaccurately), found

Image copyright Konstantin Sutyagin , 2009. Used under license from Shutterstock.com

tremendous emotional support, and garnered practical (if sometimes untested) advice from websites, online message boards, and blogs (Sulik & Eich-Krohm 2008; Seale, Ziebland, & Charteris-Black 2006; Berger, Wagner, & Baker 2005). And many of these have used this information and advice to challenge their doctor's views. Thus the Internet potentially can shift the balance of power between patients and doctors.

The Social Causes of Health and Illness

In a widely cited article titled "A Case for Refocusing Upstream," sociologist John McKinlay (1994) offers the following oft-told tale as a metaphor for the modern doctor's dilemma:

> Sometimes it feels like this. There I am standing by the shore of a swiftly flowing river and I hear the cry of a drowning man. So I jump into the river, put my arms around him, pull him to shore and apply artificial respiration. Just when he begins to breathe, there is another

cry for help. So I jump into the river, reach him, pull him to shore, apply artificial respiration, and then just as he begins to breathe, another cry for help. So back in the river again, reaching, pulling, applying, breathing, and then another yell. Again and again, without end, goes the sequence. You know, I am so busy jumping in, pulling them to shore, applying artificial respiration, that I have *no* time to see who the hell is upstream pushing them all in. (McKinlay 1994, 509–510)

Like the would-be rescuer in this story, doctors have few opportunities to focus upstream and ask why their patients get sick in the first place. Sociologists attempt to answer this question at two levels: the micro-level, in which individuals make choices about adopting behaviors that risk their health, and the macro-level, in which social structures limit the choices available to individuals.

But before we can ask why individuals' health is at risk, we need to know what those risks are. To do so, we need to look at the underlying causes of preventable death.

Underlying Causes of Preventable Death

In a highly influential article published in the *Journal of the American Medical Association*, Mokdad and his colleagues (2004) reviewed all available medical literature to identify the underlying causes of preventable deaths (that is, deaths caused neither by old age nor by genetic disease). Nine factors—tobacco, poor diet and inadequate exercise, alcohol, bacteria and viruses, polluted workplaces and neighborhoods, motor vehicles, firearms, sexual behavior, and illegal drugs—emerged as underlying almost half of all preventable deaths in the United States (Table 10.1).

Of these nine factors, tobacco is clearly the most important—and is far more important than all illegal drugs combined. Whether smoked, chewed, or used as snuff, tobacco can cause an enormous range of disabling and fatal diseases, including heart disease, strokes, emphysema, and numerous cancers (World Health Organization 2008b). About half of all smokers will die because of their tobacco use, with half of these dying in middle age and losing an average of 22 years from their normal life expectancy.

TABLE 10.1 Underlying Causes of Preventable Death in the United States

Cause	Number of Preventable Deaths	Percentage of All Deaths
Tobacco	435,000	18%
Poor diet and inadequate exercise[1]	100–400,000	5–17
Alcohol	85,000	4
Bacteria and viruses[2]	75,000	3
Polluted workplaces and neighborhoods	55,000	2
Motor vehicles[3]	43,000	2
Firearms	29,000	1
Sexual behavior	20,000	1
Illegal drugs	17,000	1

SOURCE: Mokdad et al. 2004.
[1]Estimates vary.
[2]Not including deaths related to HIV, tobacco, alcohol, or illicit drugs.
[3]Includes motor vehicle accidents linked to drug use but not to alcohol use.

The second most common cause of premature deaths is a high-fat diet, sedentary lifestyle, and resulting obesity. Rates of obesity in the United States have skyrocketed since 1980 (Centers for Disease Control and Prevention 2009). The combination of poor diet and insufficient exercise increases the risks of cardiovascular disease, strokes, certain cancers (of the colon, breast, and prostate), and diabetes, among other problems.

The remaining seven factors cause preventable deaths in a variety of ways. Alcohol and illegal drugs make unsafe sex more likely; alcohol, motor vehicles, firearms, and illegal drugs all contribute to deadly accidents; and alcohol, pollution, unprotected sex, and illegal drugs (when injected) can cause cancer, hepatitis, and other illnesses.

As the health belief model suggests, these boys are unlikely to stop smoking because they are unlikely to believe—or even know—that smoking places them at risk for lung cancer and other serious diseases.

Micro-Level Answers: The Health Belief Model

Why do individuals engage in behaviors that endanger their health? Or, to ask the question more positively, why don't individuals adopt behaviors that will *protect* their health? Sociologists have identified four conditions—known collectively as the **health belief model**—that consistently predict whether individuals will adopt healthy behaviors (Becker 1974, 1993). These conditions are:

1. Individuals must believe they are at risk for a particular health problem.
2. They must believe the problem is serious.
3. They must believe that adopting preventive measures will reduce their risks significantly.
4. They must not perceive any significant financial, emotional, physical, or other barriers to adopting the preventive behaviors.

The experience of Pittsburgh Steelers quarterback Ben Roethlisberger illustrates this model (as does the Concept Summary on the Health Belief Model on the next page). In June 2006, Roethlisberger suffered a concussion and numerous other injuries after crashing his motorcycle. He was not wearing a helmet at the time, even though helmets reduce the risk of dying in an accident by at least one-third and reduce the rate of brain injury by two-thirds (National Highway Traffic Safety Administration 2005).

Following his accident, Roethlisberger vowed never to ride a motorcycle without a helmet again. He now realized that the threat of a crash was real, and that the consequences of a crash could be serious or even fatal. Having crashed headfirst into a car's windshield, it now made sense to him that wearing a helmet would significantly reduce his risk of death or brain injury. And when weighed against these potential benefits, the cost and discomfort of a helmet and the potential threat to his "tough guy" image if he wore one no longer seemed like important barriers.

Macro-Level Answers: The Manufacturers of Illness

At first glance, it's easy to conclude that poor individual choices explain most or even all preventable deaths. After all, like Ben Roethlisberger, other people also weigh

According to the **health belief model**, individuals will adopt healthy behaviors if they believe they face a serious health risk, believe that changing their behaviors would help, and face no significant barriers to doing so.

concept summary

Health Belief Model

People Most Likely to Adopt Healthy Behaviors When They	Example: Adopting Healthy Behaviors	
	Likely	Unlikely
Believe they are susceptible:	Forty-year-old smoker with chronic bronchitis who believes he is at risk for lung cancer.	Sixteen-year-old boy who believes he is too healthy and strong to contract a sexually transmitted disease.
Believe risk is serious:	Believes lung cancer would be painful and fatal, and does not want to leave his young children fatherless.	Believes that sexually transmitted diseases can all be easily treated.
Believe compliance will reduce risk:	Believes he can reduce risk by stopping smoking.	Doesn't believe that condoms prevent sexually transmitted diseases.
Have no significant barriers to compliance:	Friends and family urge him to quit smoking, and he can save money by so doing.	Enjoys sexual intercourse more without condoms.

their options and then choose to smoke tobacco, use firearms, engage in risky sex, and so on. But those choices are made in a broader social context. If we look more closely at that social context, we quickly come to what McKinlay (1994) describes as the **manufacturers of illness:** groups that promote deadly behaviors and social conditions. For example, cigarettes, beer, fast cars, good rifles, and sugary foods are inherently appealing to many people. But it is the manufacturers of these goods that largely determine how safe or dangerous their products will be, to whom and how they will be advertised, and where they will be sold. For example, car manufacturers have fought against bumpers that would make SUVs less dangerous to other cars, soda manufacturers have fought for the right to sell their high-calorie products in schools, and tobacco manufacturers have (implicitly) promoted smoking to teens and children through such tactics as the Joe Camel campaign and sponsoring youth-oriented concerts and music festivals (Weitz 2010).

Individual choice is even less a factor for the other underlying causes of death (Weitz 2010). People work with dangerous pesticides, inject illegal drugs that they don't know have been cut with dangerous chemicals, and live in apartments with lead in the water pipes because they lack alternatives. Manufacturers of illness in these circumstances include corporations that expose their workers to dangerous conditions, landlords who don't maintain their buildings, and politicians who oppose legalizing drugs so the drugs can be regulated. Finally, individuals are most likely to engage in unsafe behaviors—from eating doughnuts to shooting crack and having sex without condoms—if they feel they have nothing to look forward to anyway. These feelings are most common among those who are trapped at the bottom of the social class system.

The **manufacturers of illness** are groups that promote and benefit from deadly behaviors and social conditions.

The Social Distribution of Health and Illness

Good health is not simply a matter of good habits and good genes. Although both elements play important parts, health is also strongly linked to social statuses such as gender, social class, and race or ethnicity. In this section, we provide an overview of how these statuses affect health in the United States and then briefly examine how changes in social structure have affected life expectancy in the former Soviet Union.

In the United States, the average newborn can look forward to 78 years of life (National Center for Health Statistics 2009). Although some will die young, the average U.S. resident now lives to be a senior citizen. This is a remarkable achievement given that life expectancy was less than 50 years at the beginning of the twentieth century. Not everyone has benefited equally, however: Men, African Americans, and poorer people on average die younger than women, whites, and more affluent individuals (Table 10.2).

There is much more to health, of course, than just avoiding death. The distribution of illness and disability is at least as important as the distribution of mortality in evaluating a population's overall well-being. For every person who dies in a given year, many more experience serious illness or disability that affects the quality of their lives. In the following sections, we consider why and how gender, social class, and race/ethnicity are related to illness and mortality.

TABLE 10.2 The Impact of Sex, Race, and Family Income on Health
White Americans are healthier than African Americans, and wealthier people are healthier than are poorer people. On average, men have lower life expectancies than do women. Men and women are equally likely to report being in fair or poor health.

	Life Expectancy	Percentage Reporting Fair or Poor Health
Sex		
Male	75	9
Female	80	10
Race		
White	78	9
African Americans	73	14
Hispanic	NA	13
Asian	NA	7
Family income		
Poor	NA	20
Near poor	NA	14
Not poor	NA	6

SOURCE: National Center for Health Statistics 2009.

Gender

On average, U.S. women live about 5 years longer than U.S. men (see Table 10.2). Yet, although women live longer, they also report significantly worse health than men at all ages: more arthritis, asthma, diabetes, cataracts, and so on (Lane & Cibula 2000; Rieker & Bird 2000). These differences mean that women more often than men experience disability and discomfort as they age. In part because of this combination of longer lives and more illnesses, women are considerably more likely than men to eventually enter a nursing home.

How can we explain why, as the saying goes, "Women get sicker but men die quicker"? The answer lies in both biology and society. Probably because of their hormones, females are inherently stronger than males: As long as females receive sufficient food and caring, their chances of survival are greater than for males at every stage of life from conception onward (Rieker & Bird 2000).

Social norms also protect women from fatal disease and injury (Rieker & Bird 2000). Odds are you know a lot more young men than young women who enjoy fast driving, daredevil sports, slugging whiskey, or slugging others. These and other similar behaviors, all of which increase the chances of death, are socially encouraged for males but discouraged for females. Men are also more likely than women to use illegal drugs and to work at dangerous jobs. Finally, women are more likely than men to seek health care when they experience problems, although this has only a small impact on their overall health status.

It is less clear why, despite their lower chances of dying at any given age, women have higher rates of illness than do men (Barker 2005). Most likely, women's higher rates of illness stem from both their hormones (a biological effect) and the fact that, on average, they experience more stress than do men but have less control over the sources of that stress (a social effect). Because stress makes it more difficult for the body's immune system to function, it often leads to ill health.

Social Class

The higher one's income, the longer one's life expectancy and the better one's health (see Table 10.2): Wealthy people live longer on average than do middle-class people, and middle-class people live longer than do poor people (Marmot 2004). This is true even in countries where everyone has access to health care, and even when we compare only people who have similar rates of smoking, obesity, and alcohol use (Banks et al. 2006). Moreover, these differences begin in infancy and childhood. For example, about 50 percent of children in New York City's homeless shelters have asthma, compared with 25 percent of children in the city's poorest neighborhoods and 6 percent of the city's children overall (Pérez-Peña 2004).

The reasons for the link between social class and illness are complex (Robert & House 2000; Marmot 2004; Wilkinson 2005). They are partially attributable to poorer people's inability to afford expensive medical care. However, environmental, economic, and psychosocial factors play even stronger roles in linking poverty with ill health. Lower-income people are more likely to live in unsafe and unhealthy conditions, near air-polluting factories, or in substandard housing. They are more likely to hold dangerous jobs and to lack sufficient, good-quality food. Low-income people also experience more stress than others but have less control over the causes of that stress. As a result, like women at all income levels, they are more likely to experience illness due to stress. In addition, whereas upper-income persons might cope with stress by

Conditions in modern sweatshops, such as this one in New York City's Chinatown, place workers at high risk of injury and illness.

taking a vacation or hiring a maid to help out at home, lower-income people have few such options. Instead, some will try to cope with stress through drinking, smoking, and other calming but health-risking behaviors.

Race and Ethnicity

Although income affects health more than do race and ethnicity (Weitz 2010), the latter nonetheless has a strong and independent effect. Asian Americans of Chinese, Japanese, Filipino, or Indian heritage typically are at least middle class and experience health at least as good as that of whites; the prognosis for recent, poorer immigrant groups from Southeast Asia remains unclear. In contrast, African Americans, Hispanic Americans, and Native Americans are on average poorer than non-Hispanic whites, and primarily as a result suffer disproportionately from the effects of low socioeconomic status on health. Because of lower incomes, these nonwhites are significantly more likely than whites or long-established Asian groups to lack health insurance (Kaiser Commission on Medicaid and the Uninsured 2008). They are also more likely to experience stress and to live or work in areas contaminated by soot, carbon monoxide, ozone, sulfur, pesticides, and even radioactive wastes. For example, *60 percent* of all American children who have dangerously high levels of lead in their blood are African American, and only 17 percent are white non-Hispanic (Meyer et al. 2003).

In addition, regardless of income, the prejudice and discrimination experienced by minorities increases their rates of illness and death (Williams 1998; Williams & Jackson 2005). For example, because of racial segregation, even middle-class African Americans are more likely than whites to live in neighborhoods where violence and pollution threaten their health. Similarly, regardless of patients' symptoms or insurance coverage, doctors are more likely to offer white patients various life-preserving treatments (including angioplasty, bypass surgery, and the most effective drugs for

HIV infection) and more likely to offer minorities various less desirable procedures, such as leg amputations for diabetes (Nelson, Smedley, & Stith 2002).

Taken together, these factors lead African Americans, Hispanics, and Native Americans to have significantly higher rates of illness and higher chances of dying at any given age than do whites.

Age

Not surprisingly, age is the single most important predictor of health, illness, and death. The two groups most at risk are the very young and the very old.

In poor countries, deaths are very common among infants and children younger than age 5. Some die because they are born prematurely, others because they do not get enough food, and still others because their immune systems are unable to fight disease, especially if they are malnourished.

Deaths of young children were also common in the Western world before the twentieth century. These days, such deaths are very rare in the United States, and young people are typically healthy. Compared with other developed nations, however, infant mortality remains shockingly high (Table 10.3). Infant mortality is especially high among African Americans, who (for all the reasons just discussed) are more than twice as likely as white babies to die in infancy (National Center for Health Statistics 2009).

Once past infancy, the chances of dying or developing a disabling illness only begin to rise gradually beginning at about age 40. By age 65, most people will have at least one long-lasting health problem, such as arthritis, hypertension, or hearing loss (Federal Interagency Forum on Aging Related Statistics 2008). Yet even by age 85, the majority report being in good or excellent health. However, the odds of enjoying a healthy old age are significantly lower for racial and ethnic minorities: Among those

TABLE 10.3 Infant Mortality Rates per 1,000 Live Births

Hong Kong	1.6	**U.S., white non-Hispanic**	5.7
Singapore	2.4	Hungary	5.9
Sweden	2.5	Poland	6
Japan	2.8	Slovakia	6.1
France	3.6	**U.S., all races**	6.6
Spain	3.7	Chile	8.8
Germany	3.9	Russia	9
Denmark	4	Bulgaria	9.2
Switzerland	4	Costa Rica	9.7
Italy	4.2	Uruguay	10.5
Netherlands	4.4	Romania	12
Australia	4.7	**U.S., African Americans**	13.7
United Kingdom	4.9	Thailand	16
Cuba	5.3	Mexico	19
Canada	5.4		

SOURCE: Population Reference Bureau 2008.

focus on  AMERICAN DIVERSITY

Changing Populations, Changing Health

With each passing year, fewer babies are born in the United States and more U.S. residents turn 65. At the same time, the white non-Hispanic population is shrinking while the Hispanic and nonwhite populations are growing. What are the combined consequences of these two population trends for health and health care in America?

One obvious result of having more older Americans and more nonwhite Americans is that in the future there will be more Americans who are both older *and* nonwhite. This will have many important consequences, for the experience of old age is substantially different for nonwhite compared with white Americans (Takamura 2002). Most importantly, minority elderly are more likely than others to be poor: Twenty-six percent of African American elderly live below the poverty line, compared with 21 percent of Latino elderly and only 8 percent of white non-Hispanic elderly. This has serious implications for health and health care, because poorer persons are both more likely to need services and less likely to have health insurance and less able to pay for them out of pocket.

But the problems extend beyond those who live in poverty. Even when incomes are equivalent, and after controlling for education, age, sex, marital status, and urban residence, minority elderly are still more likely than white elderly to lack health insurance. As a result, they find it more difficult to get the medical treatment they need, and their health problems are more likely to spiral out of control, making them more difficult (and expensive) to treat in the long run.

Finally, even if they are able to obtain health care, cultural barriers may make that health care less effective than it would otherwise be (Capitman 2002; Hayes-Bautista, Hsu, & Perez 2002; Takamura 2002). When doctors and patients don't speak the same language or come from different cultures or subcultures, doctors may not understand what their patients need and patients may not understand what their doctors want them to do. In these circumstances, patients can become dissatisfied, ignore instructions, or skip follow-up visits. In turn, doctors may come to regard patients as unintelligent or unmotivated. This is a serious problem in the United States, given that most doctors are white and speak only English but many patients are nonwhite

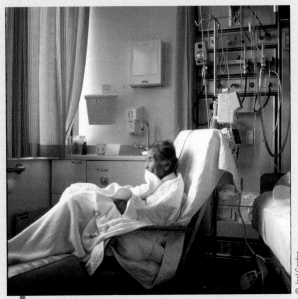

© Joel Gordon

As the number of minority elderly increase, we are likely to see an increased number of elderly people who are poor, who lack health insurance, and who face cultural barriers in interacting with health-care practitioners.

and do not speak English well if at all. Conversely, communication is also a problem for elderly white patients living in nursing homes. Although most doctors and nurses are white, day-to-day care in nursing homes is primarily left to poorly paid nurse's aides. Most of these aides are nonwhite immigrants, many of whom speak English with heavy accents that older people with hearing problems find difficult to understand. For all these reasons, policy makers will need to pay close attention to both these population changes.

aged 68 or older, 80 percent of non-Hispanic whites consider themselves healthy, compared with 65 percent of Hispanics and 63 percent of African Americans (Federal Interagency Forum on Aging Related Statistics 2008).

The health consequences of the shift toward an older and more diverse population are explored more fully in Focus on American Diversity: Changing Populations, Changing Health.

Case Study: Declining Life Expectancy in the Former Soviet Union

The single most important social factor affecting mortality is the standard of living—access to good nutrition, safe drinking water, and adequate housing free from environmental hazards. Differences in living standards help to explain why African American infants in the United States are more than twice as likely as white infants to die in their first year of life and why the average life expectancy of African American men is 6 years less than that of the average white non-Hispanic male. Differences in living standards also help to explain why, on average, Americans can expect to live 30 years longer than citizens of Sierra Leone (Population Reference Bureau 2008).

Throughout the world, improvements in living standards have been accompanied by increased life expectancy. Consequently, the precipitous decline in life expectancy in the former Soviet Union over the last 20 years is one of the most surprising current developments in world health; life expectancy for Russian men is now only 60 years—far lower than before the collapse of the Soviet Union and far lower than in other developed nations (Population Reference Bureau 2008).

What explains this shocking drop in life expectancy? First, during its decades as a dictatorship, the Soviet Union put industrial development above environmental protection. As a result, the countries of the former Soviet Union are now plagued by extensive environmental pollution. This has significantly raised rates of cancer and respiratory diseases, especially in the most industrialized regions (Cockerham 1997; Haub 1994). Second, as we've seen, stress is often an underlying cause of illness. After the collapse of the Soviet Union, incomes plummeted, social services ground to a halt, political uncertainty and corruption increased, and an entire way of life evaporated. The resulting rise in stress levels directly explains much of the increase in deaths. In addition, this stress also fostered sharp increases in smoking and drinking, with resulting deaths from disease, violence, and accidents. Finally, the former Soviet Union had never invested much in health care for the chronic illnesses, such as heart disease, that now cause most deaths, and the health-care system only worsened after the collapse of the Soviet system. Conditions overall have improved in the last few years, but years of environmental damage, social turmoil, and poor living conditions continue to take a large toll.

In sum, in the former Soviet Union as in the United States, social conditions are closely tied to health, illness, and death.

Mental Illness

So far we have talked about health and illness as if the only thing that matters is *physical* health. But mental health is also a crucial issue, affecting millions of people each year.

How Many Mentally Ill?

National random surveys of the U.S. population suggest that during the course of any given year, approximately 11 percent of working-age adults experience a minor but still-diagnosable mental illness, and another 20 percent experience a moderate or severe illness (Kessler et al. 2005). The most common illnesses are major depression and problems with alcohol use. These estimates, however, are probably a bit high, since they are based on reports of symptoms, not medical diagnoses of illnesses (Horwitz 2002). Survey researchers can't know, for example, if someone has lost weight because of depression or because they are getting ready for a wrestling match.

Who Becomes Mentally Ill?

As with physical illness, social factors strongly predict mental illness. We focus here on two important factors: social class and gender.

Social-Class Differences

Since the 1920s, when sociological study of mental disorder began, researchers consistently have found that poorer people experience more mental illness than do wealthier people (Eaton & Muntaner 1999). Researchers disagree, however, on the reasons for this pattern. Some argue that the social stress associated with lower-class life causes mental disorder. Others believe that the onset of mental illness causes people to lose their jobs and drift downward in social class.

Research clearly shows that the lower class does, in fact, experience more of the types of stress (such as job loss or chronic physical disabilities) that can cause mental disorders (Turner, Wheaton, & Lloyd 1995; Turner & Avison 2003; Ali & Avison 1997). The stresses of poverty and economic insecurity appear to be particularly important in understanding the causes of disorders such as major depression.

At the same time, research shows that the onset of disorders such as schizophrenia makes it difficult for people to keep a job. Not only may individuals lose their initial job, but once potential employers discover that an individual has a history of mental disorder, they may be reluctant to hire him or her for anything other than a minimum-wage job (Link et al. 1987, 1997). In these cases, a mental disorder clearly causes people to drift into a lower social class. Social drift, however, explains a lower proportion of mental illness than does the stress of lower social-class life.

In addition to social-class differences in *rates* of mental illness, there are also important differences in the *experience* of mental illness. Lower-class persons diagnosed with mental illness remain in hospitals for longer periods of time and receive less effective types of treatment. In fact, most mental health treatment goes to middle-class persons experiencing short-term emotional problems, rather than to persons (of whatever social class) who are seriously mentally ill. Meanwhile, as funding for hospitals and health care has declined, lower-class mentally ill persons increasingly have been sent to jails or prisons rather than to clinics or hospitals when their behavior becomes socially unacceptable; according to the U.S. Department of Justice, more than half of all jail and prison inmates are mentally ill (James & Glaze 2006).

Gender Differences

Depression is the most common form of mental illness, affecting about 17 percent of all adults living in the United States (Kessler et al. 2005). Because depression is so common and because it is much more commonly diagnosed in women, the overall rates of mental illness are higher for women than for men.

Why are women more likely than men to be diagnosed with depression? Most theorists hypothesize that women have higher rates of depression because they experience more stress *and* have less control over that stress (Horwitz 2002, 173–179). In fact, rates of depression are highest among those women with the least control over their lives: nonworking women and married mothers. A waitress with young children and a husband who expects a hot meal when he gets home, for example, has few means for controlling her life, schedule, or stress levels. By the same token, depression is especially common among men who have less power than their wives, have little control over their work, or lose their jobs.

In contrast, men are more likely than women to report substance abuse and "personality disorders" characterized by chronic maladaptive personality traits, such as compulsive gambling or violence (Kessler et al. 2005). Scholars theorize that because the traditional male role encourages men to respond to stress with aggression or substance abuse, those who experience stress and mental illness are more likely to develop these sorts of symptoms.

Working in Health Care

As in any other area of social life, health care has its own set of roles, statuses, and battles over power. In this section, we look at the two most important health-care occupations, medicine and nursing, and discuss how each has fought to maintain or improve its position in the health-care hierarchy.

Physicians: Fighting to Maintain Professional Autonomy

Less than 5 percent of the medical workforce consists of physicians. Yet they are central to understanding the medical institution. Physicians are responsible both for defining ill health and for treating it. They set the standards for how patients should behave and play a crucial role in setting hospital standards and in directing the behavior of the nurses, technicians, and auxiliary personnel who provide direct care.

As will be described in Chapter 13, a profession is a special kind of occupation that demands specialized skills and permits creative freedom. No occupation better fits this definition than that of physician. Until about 100 years ago, however, almost anyone could claim the title of physician; training and procedures were highly variable and mostly bad (Starr 1982). Some doctors were almost illiterate, many learned to doctor through apprenticeships, and most of the rest learned through brief courses where virtually anyone who could pay the fees could get certified. With the establishment of the American Medical Association in 1848, however, the process of professionalization began; the process was virtually complete by 1910, at which point strict medical training and licensing standards were adopted.

Understanding Physicians' Income and Prestige

The medical profession provides an example for stratification theories. Family practitioners currently earn a median net salary of $156,000, and general surgeons earn an average of $283,000 (U.S. Bureau of Labor Statistics 2009a). Why are physicians among the highest-paid and highest-status professionals in the United States?

According to structural-functionalists, there is a short supply of individuals who have the talent and ability to become physicians and an even shorter supply of those who can be surgeons. Moreover, physicians must undergo long and arduous periods of training. Consequently, high rewards must be offered to motivate the few who can do this work to devote themselves to it. The conflict perspective, on the other hand, argues that the high income and prestige accorded physicians have more to do with physicians' use of power to promote their self-interest than with what is best for society.

In defense of this argument, conflict theorists point to the role played by the American Medical Association (AMA). The AMA sets the standards for admitting physicians to practice, punishes physicians who violate the standards, and lobbies to protect physicians' interests in policy decisions. Although less than half of all physicians belong to the AMA, it nonetheless continues to wield considerable power. It has fought vigorously to ensure the continuance of the free market model of medical care, in which the physician remains an independent provider of medical care on a fee-for-service basis. In pursuit of this objective, the AMA has consistently opposed all legislation designed to create national health insurance, including Medicare, Medicaid, and President Obama's proposals. It also has fought to ban or control a variety of alternative medical practices such as midwifery, osteopathy, and acupuncture (Weitz 2010).

The Changing Status of Physicians

Although physicians have succeeded in maintaining high incomes, they have done less well in maintaining other professional privileges. Until the 1970s, most physicians worked as independent providers with substantial freedom to determine their conditions of work. They also benefited from high public regard; some patients considered them a nearly godlike source of knowledge and help. Much of this is changing. The many signs of changes include the following (Coburn & Willis 2000; Weitz 2010):

- A growing proportion of physicians work in group practices or for corporations, where bureaucrats determine fees, procedures, and working hours. As a result, physicians have lost much of their independence.
- The public has grown increasingly critical of physicians. Getting a second opinion is now standard, and malpractice suits have become much more common.
- Fees and treatments are increasingly regulated by government agencies and insurance companies. Physician autonomy is limited whenever these groups start dictating what treatments will be funded, for which patients, and at what fees.

Doctors have not accepted these changes lying down. Instead, they have fought for legal restrictions on malpractice lawsuits and on insurance company regulations. They have also fought in the court of public opinion to convince patients that physicians continue to have patients' best interests at heart.

Despite these problems, being a physician is still a very good job, offering high income and high prestige. But it is also part of an increasingly regulated industry that is receiving more critical scrutiny than ever before.

Nurses: Fighting for Professional Status

Of the nearly 10 million people employed in health care, the largest single component is that of the 1.8 million registered nurses. No hospital could run without nurses, and no doctor could function without them. Yet despite their great importance to the health-care system, their status remains far lower than we might expect. Why have nurses' attempts to improve their status achieved only modest success?

Nurses' Current Status

Nurses play a critical role in health care, but they have relatively little independence, either in their day-to-day work or in their training and certification. Physicians have a major voice in determining the training standards that nurses must meet and in enforcing these standards through licensing boards. On the job, even the most junior physician can give orders to experienced nurses. Reflecting this status difference, nurses' median income is

The popularity of doctor play sets and the large number of children who aspire to medical professions demonstrate the continuing prestige of doctors in contemporary society.

now $57,000—only one-third the income of doctors in family practice (U.S. Bureau of Labor Statistics 2009a). Even when nurses have PhDs or master's degrees, their salaries remain a fraction of doctors' salaries. Finally, although the general public respects nurses for their dedication, it tends to discount nurses' specialized education. For all these reasons, nursing does not meet the sociological definition of a profession.

Why is nursing's status so low? The primary reason is its history as a traditionally female occupation. Before the twentieth century, most people believed that caring came naturally to women and, therefore, that mothers, daughters, cousins, and sisters should always be willing to help care for any sick family member (Reverby 1987). Nursing did not become a formal occupation until the mid-nineteenth century. Because of its historic roots, from the start it was considered a natural extension of women's character and duty rather than an occupation meriting either respect or rights (Reverby 1987). Nurses were encouraged to enter the field in a spirit of altruism and self-sacrifice and, as proper young women, to accept orders from doctors, hospital administrators, and their nursing superiors. This approach made it difficult for nursing as a field to fight for status, autonomy, or better working conditions. Moreover, the fact that the field was almost solely female in and of itself made it difficult for nursing to obtain the autonomy and public respect for its training and work that define a profession.

Changing the Status of Nurses

To improve the status and position of nurses, nursing's leadership has worked for decades to raise educational levels (Weitz 2010). Until the 1960s, the standard nursing credential was an RN (registered nurse) diploma, obtained through a hospital-based training program. Now, almost all RNs hold 2- or 4-year nursing degrees from community colleges or universities. In addition, a small percentage of nurses obtain graduate degrees and become nurse practitioners or nurse-midwives. These nurses enjoy considerably more autonomy, status, and financial rewards than do other nurses, including the right to prescribe specified medications in most states.

The drive to increase nurses' education and thus their status has succeeded only partially. Because many hospitals believe that associate-degree nurses receive the best practical training and make the best employees, associate-degree programs have remained more popular than higher-level training. Meanwhile, to control costs, hospitals have shifted many services to outpatient clinics where fewer RNs are needed, nurses' salaries are lower, and nursing jobs are less interesting and prestigious (Norrish & Rundall 2001). In addition, hospitals have reduced their nursing staffs and increased the workload of the remaining nurses (Gordon 2005). Finally, although more men now work as nurses, the field is still considered a "woman's profession," and for that reason, salaries and status remain relatively low.

Understanding Health-Care Systems

Ensuring that people have access to health care is one of the most basic tasks of any society. The United States offers many ways through which people can get health care: private and publicly funded insurance, private and public clinics and hospitals, or cash payments. Yet many Americans can obtain only low-quality care, many can obtain care only by making financial sacrifices, and many cannot afford care at all. How does health insurance in the United States work? Why do some people lack insurance, and what are the consequences of being uninsured? How do other countries manage to pay for health care for all their citizens, and why doesn't the United States also have a national health-care program?

Paying for Health Care in the United States

Medical care is the fastest-growing segment of the cost of living. In 1970, Americans spent an average of $372 per person (in 2006 dollars) on medical care. By 2006, they spent an average of $6,561, and by 2013 they will probably spend twice that (U.S. Bureau of the Census 2009a).

There are three primary modes of financing health care in the United States: paying out of pocket, private insurance, and government programs. The cost of health care is so high that only the very rich can afford to pay out of pocket for anything beyond minor problems. As a result, most Americans must rely on private insurance or on government insurance programs. The remainder have no insurance and often are unable to pay for health care. Even those who have insurance often find that it doesn't cover many of their bills. In total, 21 percent of Americans recently surveyed by the Gallup Poll report that they sometimes cannot afford to purchase needed medical care or drugs (Szabo & Appleby 2009). Map 10.1 shows how this varies around the United States.

Private Insurance

Most insured Americans hold private insurance obtained via their employers, their parents' employers, or their spouses' employers. Individuals are most likely to get insurance from their employer if they work for large corporations or government agencies

MAP 10.1: Percent Sometimes Unable to Afford Needed Medical Care or Drugs*
Twenty-one percent of Americans are sometimes unable to afford needed medical care. Rates are highest in southeastern states with many poor, African American residents and lowest in states with state-funded insurance programs, such as Hawaii and Massachusetts.
SOURCE: Szabo & Appleby (2009).
*During previous 12 months.

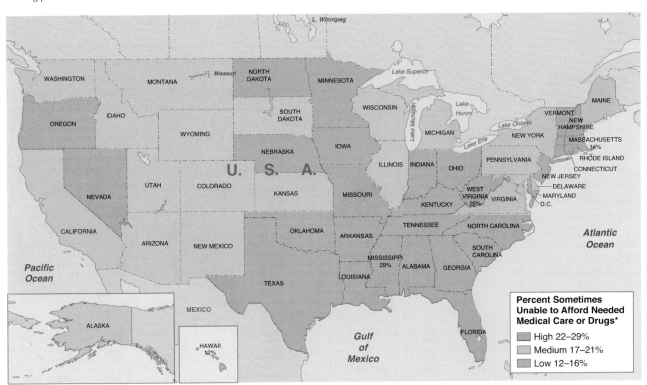

and are least likely to get such insurance if they work for small businesses or in minimum-wage jobs. In the past, the largest private insurers, like Blue Cross and Blue Shield, were nonprofit organizations that at least to some extent tried to keep costs down. These days, however, most private insurance providers are for-profit corporations. This is one reason why individual costs for health care have risen dramatically.

Government Programs

The government has several health insurance programs. The two largest programs are Medicaid and Medicare. In addition, local governments provide medical care through public health agencies and public hospitals.

Medicare is a government-sponsored health insurance program primarily for citizens older than age 65. Because of Medicare, almost all elderly Americans now have health insurance. This is not a cheap program, however: In 2006, the government paid more than $401 billion in Medicare benefits (U.S. Bureau of the Census 2009a). The costs are so high that government officials believe the program could go bankrupt by 2017 unless taxes are raised or costs are somehow reduced.

Unlike Medicare, which is available to almost everyone older than age 65, *Medicaid* provides health insurance based on need. Funds come from both the federal government and state governments. Both eligibility and services are determined by states, some of which offer much more generous medical care than others. Generally speaking, though, you won't get Medicaid unless you are both very poor and either a child or a pregnant woman.

©2009/Jupiter images

The rising cost of health insurance and health care is now one of the top causes of bankruptcy in the United States.

The Uninsured in the United States

As of 2007, about 17 percent of Americans younger than age 65 lacked health insurance. More recent statistics are not yet available, but this percentage has undoubtedly risen since then, given current economic conditions.

Thanks to Medicare, nearly 100 percent of the elderly are insured. Those who fall through the cracks are primarily young or middle-aged, unemployed, working poor,

FIGURE 10.1 America's Uninsured Population

Almost 1 in 5 U.S. residents under age 65 has no health insurance. Most are working-age adults, poor or near poor, and live in a household with at least one full-time worker.

SOURCE: Kaiser Commission on Medicaid and the Uninsured (2008).

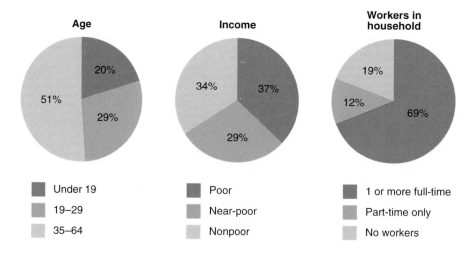

Age	Income	Workers in household
20% / 29% / 51%	37% / 29% / 34%	69% / 12% / 19%

Age:
- Under 19
- 19–29
- 35–64

Income:
- Poor
- Near-poor
- Nonpoor

Workers in household:
- 1 or more full-time
- Part-time only
- No workers

or employees of small businesses (Kaiser Commission on Medicaid and the Uninsured 2008). Figure 10.1 provides a statistical portrait of America's uninsured.

In emergencies, people who lack insurance can get treatment at public hospitals. However, they often must wait several hours before the overworked hospital staff can see them. And once seen, they are more likely than others to receive substandard care. As a result, uninsured Americans are more likely than others with similar conditions to postpone needed medical care, to require hospitalization when they do seek care, and to die whether or not they are hospitalized (Kaiser Commission on Medicaid and the Uninsured 2008; Weitz 2010).

Health Care in Other Countries

The United States is the only industrialized nation that does not guarantee health care to all of its citizens. Instead, health care is sold like any other commodity. Like dry cleaning, you get what you can afford, and if you can't afford it, you may have to go without. In contrast, in the rest of the industrialized world, medical care is like primary education—regarded as something that all citizens should receive regardless of ability to pay. How is health care provided in these countries?

National Health-Care Systems

Different industrialized nations use different systems, but all guarantee that every citizen has affordable access to high-quality health care. Great Britain and Canada provide two useful examples.

In both these countries, health care is provided through a single-payer system. In a **single-payer system**, doctors and hospitals are paid, either directly or indirectly, from a single source: the government. In both countries, doctors who work in hospitals are paid on salary. However, British doctors who work in private offices or clinics usually receive a salary, whereas Canadian doctors who do so are paid a fee for each service they provide (Weitz 2010).

Single-payer systems reduce the cost of care in three ways (Weitz 2010; Physicians for a National Health Program 2009). First, they are very efficient: Whereas a U.S. doctor might have to bill dozens of insurers each week, a Canadian doctor need send a bill only to the government. Second, they are nonprofit: Whereas the Canadian and British health-care systems are motivated solely by the desire to provide health care, the primary aim of U.S. insurance companies is to earn a profit for their stockholders. As a result, U.S. insurance companies prefer to insure only healthy people who will have few medical bills. Finally, single-payer systems are "the only game in town": As the only purchaser of health care (including drugs), single-payer systems can pressure pharmaceutical companies to reduce drug prices, require doctors to keep down their fees, and refuse funds for hospitals to provide new services that government researchers consider unneeded, untested, or ineffective.

The downside of a single-payer system is that it reduces options for doctors and consumers. Doctors can't decide what fees they will charge and consumers can't "shop around" to obtain a treatment that the government does not support.

Good Care at Low Cost

Modern medical technology has enhanced our ability to extend and save lives. It is, however, extraordinarily expensive. So how do some less-developed nations manage to keep their populations healthy?

China provides an interesting example of how this can be done. In China, Western-style medicine has taken a back seat to prevention. The focus has been on

A **single-payer system** (of health care) is one in which doctors and hospitals receive payment solely from the government.

using less-expensive health-care providers such as midwives and nurses; improving sanitation, housing, and food; raising education levels to raise incomes; and using traditional healing practices that Western physicians are only now coming to appreciate. Because of these strategies, life expectancy is now only 5 years less than in the United States, even though China spends several times less on health care (Population Reference Bureau 2008). Other less-developed nations such as Costa Rica, Sri Lanka, Cuba, and Vietnam also have demonstrated the value of preventing illness, increasing education levels, and improving the standard of living, rather than focusing on high-technology health care (Weitz 2010).

Why Doesn't the United States Have National Health Insurance?

Why doesn't the United States have national health care? The answer, sociologists argue, lies in **stakeholder mobilization**: organized political opposition by groups with a vested interest in the outcome (Quadagno 2005).

Opposition to national health care has come from numerous sources (Quadagno 2005; Rothman 1997). In the past, labor unions opposed national health care because health insurance was one of the major benefits they could offer members. Opposition also came from the American Medical Association, which feared doctors might lose income or autonomy under a national health plan, and from middle- and upper-class Americans who had health insurance and saw no reason to pay taxes to support health care for others.

As the health-care crisis has worsened, affecting more and more middle-class Americans, support for national health care has grown among doctors, labor unions, the public, and even some major corporations who are tired of paying high prices for their employees' health insurance. The strongest opposition to national health care now comes from the pharmaceutical and health insurance industries. These industries poured millions into fighting former President Clinton's proposed health plan, outspending those who favored it by a ratio of four to one (Quadagno 2005, 189). In addition, anti-tax sentiment and distrust of "big government" have become powerful forces in U.S. politics since the 1980s, making it difficult to generate support for any governmental programs (Rothman 1997; Skocpol 1996). Nevertheless, polls consistently find that about two-thirds of Americans believe it is the federal government's responsibility to guarantee health care for all members of our society, and they are willing to pay more taxes to fund such services (Everybody In, Nobody Out 2005).

If Americans obtain national health insurance in the future, it will be because the middle class, labor unions, and corporations all find it increasingly difficult to pay their health-care bills and unite to fight against anti-tax lobbies, the health insurance industry, and the pharmaceutical companies.

Where This Leaves Us

Sociological analysis suggests that health and illness are socially structured. To paraphrase C. Wright Mills once again, when one person dies too young from stress or bad habits or inadequate health care, that is a personal trouble, and for its remedy we properly look to the character of the individual. When whole classes, races, or sexes consistently suffer significant disadvantage in health and health care, it is a social problem. The correct statement of the problem and the search for solutions require us to look beyond individuals to consider how social structures and institutions

Stakeholder mobilization refers to organized political opposition by groups with a vested interest in a particular political outcome.

have fostered these patterns. The sociological imagination suggests that significant improvements in the nation's health will require changes in social institutions—increased education, reduced poverty and discrimination, improved access to good-quality housing and food, and so on. Equalizing access to health care will also help but is considerably less important than making these social changes.

Summary

1. A major contribution of structural-functionalist theory to the study of health is the concept of a sick role. This concept explains how (some) illness can help society run smoothly and how society limits illness to keep it from interfering with that smooth flow.

2. From conflict theory we get the concept of manufacturers of illness: groups that benefit from promoting conditions that cause illness and disease. Conflict theory also helps us to understand how definitions of illness develop in the process of medicalization and how competing interest groups battle over different potential definitions of illness.

3. Symbolic interaction theory has been particularly useful for understanding the experience of illness, including why patients sometimes do not follow doctors' orders.

4. Nine factors—tobacco, poor diet and inadequate exercise, alcohol, bacteria and viruses, polluted workplaces and neighborhoods, motor vehicles, firearms, sexual behavior, and illegal drugs—underlie almost half of all preventable deaths in the United States.

5. The sick role consists of four social norms regarding sick people. They are assumed to have good reasons for not fulfilling their normal social roles and are not held responsible for their illnesses. They are also expected to consider sickness undesirable, to work to get well, and to follow their doctors' orders. The sick role model, however, fits some illnesses better than others.

6. The health belief model predicts that individuals will adopt behaviors that will protect their health if they believe they are at risk for a particular health problem, they believe the problem is serious, they believe that changing their behavior will reduce their risks, and no significant barriers keep them from changing their behavior.

7. Gender, social class, race/ethnicity, and age all help explain the patterns of health and illness in the United States. Men, racial and ethnic minorities, those with lower socioeconomic status, and the very young and old have higher mortality rates, largely due to social rather than biological forces.

8. The health disadvantage associated with lower socioeconomic status goes far beyond a simple inability to afford health care. Poorer people experience lower standards of living, more stress, lower education levels, and polluted environments, all of which increase the likelihood that they will experience poor health.

9. Women and lower-class people have higher rates of mental illness than other groups. Although the reasons are complex, differences in exposure to stress appear to be the primary cause. Men have higher rates of substance abuse and personality disorders than women.

10. Physicians are professionals; they have a high degree of control not only over their own work but also over all others in the medical world. Structural functionalists argue that physicians earn so much because of scarce talents and abilities, whereas conflict theorists argue that high salaries are due to an effective union (the AMA). Physicians have less independence than they used to due to increased corporate control, government oversight, and public criticism.

11. Nurses comprise the largest single occupational group in the health-care industry. Nurses earn much less than physicians, have less prestige, and take orders instead of giving them. The reasons for this primarily stem from nursing's position as a traditionally "female" field.

12. Most insured Americans belong to private health insurance plans. Medicare is a government program that insures almost all senior citizens, and Medicaid is a government insurance program primarily for the very poor.

13. The United States is the only industrialized nation that does not make medical care available regardless of the patient's ability to pay. Seventeen percent of U.S. residents under age 65 are uninsured. Uninsured persons are more likely than others to postpone getting needed health care, to become ill, and to die if they become ill.

14. Single-payer systems reduce the costs of health care in other countries because they are efficient, nonprofit, and able to negotiate good prices with health-care providers. However, single-payer systems reduce options for doctors and consumers.

15. The United States lacks national health insurance because of stakeholder mobilization, which currently comes primarily from pharmaceutical and insurance corporations.

Thinking Critically

1. How have "manufacturers of illness" increased deaths caused by tobacco? by alcohol? by toxic agents? by diet?
2. How have social forces and political decisions increased deaths caused by sexual behavior? caused by illicit drugs?
3. Think of the last time you or a close friend or relative was ill. Discuss each of the elements of the sick role, and whether or not it applied in this instance. Then think of someone you know who has a chronic illness, and do the same.
4. Who benefits when "male erectile dysfunction" is defined as an illness? How? Who loses? What do they lose?
5. Think of a friend of yours who smokes or engages in another unhealthy behavior. Use the health belief model to explain what else would have to change before your friend would be likely to change his or her behavior.
6. Why do so few men enter nursing? What could change this gender gap?
7. Who would gain if the United States adopted a national health-care system? Who would lose, and what would they lose? Consider economic, social, political, and psychological costs.

Book Companion Website

www.cengage.com/sociology/brinkerhoff
Prepare for quizzes and exams with online resources—including tutorial quizzes, a glossary, interactive flash cards, crossword puzzles, essay questions, virtual explorations, and more.

CHAPTER 11

Family

©2009/Jupiter images

Marriage and Family: Basic Institutions of Society

Recent decades have seen many changes in American family life. Birth rates have declined sharply, divorce and single-parent families are now common, and the majority of women with small children work in the paid labor force. In addition to these statistical trends, major shifts in attitudes and values have occurred. Homosexuality, premarital sex, and extramarital sex have all become more acceptable. Related to many of these changes are the dramatic changes in the roles of women in our society.

These changes in family life have been felt, either directly or indirectly, by all of us. Is the family a dying institution, or is it simply a changing one? In this chapter, we examine the question from the perspective of sociology. We begin with a broad description of marriage and the family as basic social institutions.

To place the changes in the U.S. family into perspective, it is useful to look at the variety of family forms across the world. What is it that is really essential about the family?

Universal Aspects

In every culture, the family has been assigned major responsibilities, typically including the following (Seccombe & Warner 2004):

- Replacing the population through reproduction
- Regulating sexual behavior
- Caring for dependents—children, the elderly, the ill, and the handicapped
- Socializing the young
- Providing intimacy, belongingness, and emotional support

Because these activities are important for individual development and the continuity of society, every society provides some institutionalized pattern for meeting them. No society leaves them to individual initiative. Although theoretically religious or educational institutions could handle these responsibilities, most societies have found it best to leave them to the family.

Unlike most social structures, the family can be a biological as well as a social group. The **family** is a group of persons linked by blood, adoption, marriage, or quasi-marital commitments. This definition is very broad; it would include a mother living alone with her child as well as a man living with several wives. The important criteria for families are that their members assume responsibility for each other and are bound together—if not by blood, then by some cultural markers such as marriage or adoption.

Marriage is an institutionalized social structure that is meant to provide an enduring framework for regulating sexual behavior and childbearing. Many cultures tolerate other kinds of sexual encounters—premarital, extramarital, or homosexual—but most cultures discourage childbearing outside marriage. In some cultures, the sanctions for nonmarital sexuality and childbearing are severe, but in others they are minimal.

Marriage is also a legal contract, specifying the obligations of each spouse. Until very recently, those obligations were sharply divided by sex: By law, husbands had an obligation to support their wives financially, and wives had an obligation to provide domestic and sexual services to their husbands. These sex-specific obligations only started changing with the rise of the modern feminist movement in the 1970s.

The **family** is a group of persons linked together by blood, adoption, marriage, or quasi-marital commitment.

Marriage is an institutionalized social structure that provides an enduring framework for regulating sexual behavior and childbearing.

Marriage is important for childbearing because it imposes socially sanctioned roles on parents and other relatives. When a child is born, parents, grandparents, and aunts and uncles are automatically assigned certain normative obligations to the child. This network represents a ready-made social structure designed to organize and stabilize the responsibility for children. Children born outside marriage, by contrast, are more vulnerable. The number of people normatively responsible for their care is smaller, and, even in the case of the mother, the norms are less well enforced. One consequence is higher infant mortality for children born outside of marriage in almost all societies, including our own.

Marriage and family are among the most basic and enduring patterns of social relationships. Although blood ties are important, the family is best understood as a social structure defined and enforced by cultural norms.

Cross-Cultural Variations

Families universally are expected to regulate sexual behavior, care for dependents, socialize the young, and offer emotional and financial security. The importance of these tasks, however, varies across societies. Offering economic security is more important in societies without government-provided social services; regulating sexual behavior is more important in cultures without contraception. In our own society, we have seen the priorities assigned to these family responsibilities change substantially over time. In colonial America, economic responsibility and replacement through reproduction were the family's primary functions; the provision of emotional support was a secondary consideration. More recently, however, some of the responsibility for socializing the young has been transferred to schools and day-care centers; financial responsibility for dependent elderly persons has been partially shifted to the government. At the same time, intimacy has taken on increased importance as a dimension of marital relationships.

Although all families share the same basic functions, hundreds of different family forms can satisfy these needs. This section reviews some of the most important ways cultures have fulfilled family functions.

Family Patterns

Throughout history and across cultures, people have typically lived with an assortment of relatives: a husband and one or more wives; their children; and one or more grandparents, uncles, aunts, nephews, nieces, or cousins. This type of family is known as an **extended family**. Extended families have many benefits: There is always someone to hug or to talk with, finding a babysitter is easy, elderly and disabled relatives need not be left alone, and expenses can be shared. In the United States, extended families are particularly common among immigrants, who consider caring for elderly and needy family members both normal and morally required.

Most Americans, however, expect to live in a **nuclear family**. A nuclear family consists of a mother and father and their children. Nuclear families are valued by those who want their independence and who do *not* want parents or in-laws looking over their shoulders.

In reality, less than one-third of U.S. families are nuclear families (U.S. Bureau of the Census 2009d). Moreover, when we look at all U.S. households (rather than just at families), we find that only 22 percent consist of married couples with their own children. Instead, most adults live either alone, with friends or lovers, with children but not a partner, or with relatives. Map 11.1 shows the distribution of nuclear families across the United States.

An **extended family** is a family in which a couple and their children live with other relatives, such as the wife's or husband's parents or siblings.

A **nuclear family** is a family in which a couple and their children form an independent household living apart from other relatives.

MAP 11.1: Nuclear Families as a Percentage of Households
The percentage of households that consist of married couples plus their children is lowest in states with many elderly or poor residents. The percentage is lowest (7 percent) in the District of Columbia and highest (32 percent) in Utah, due to its large Mormon population.
SOURCE: U.S. Bureau of the Census (2009b). GCT1102. Percent of Households That Are Married-Couple Families with Own Children under 18 Years.

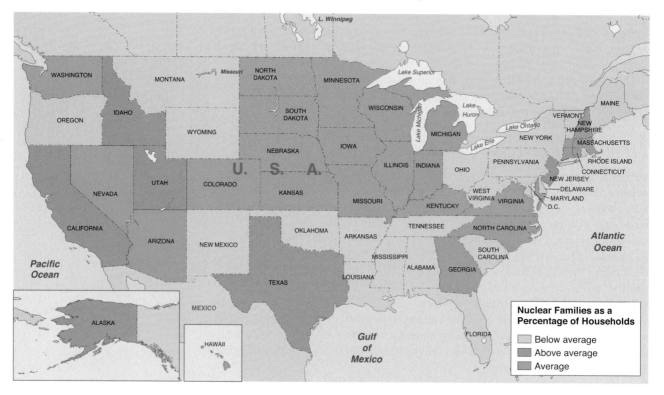

A growing family pattern in the United States is the blended family. A **blended family** is one that includes children born to one parent as well as children born to both parents. Imagine, for example, a marriage between Jim and Jane. Now imagine that Jim has two children from a previous marriage, Jane has one from her first marriage and one from her second marriage, and Jim and Jane together have another child. All of these people belong to one blended family. In addition, each of these children may interact occasionally with Jim and Jane's former spouses and with any other children that those spouses now have.

More recently, the rise in gay and lesbian families has raised further questions about the nature and meaning of family. This topic is explored further in Focus on American Diversity: Gay and Lesbian Families.

Marriage Patterns

In the United States and much of the Western world, a marriage form called **monogamy** is practiced; each man may have only one wife at a time, and each woman may have only one husband at a time. Many cultures, however, practice some form of **polygamy**—marriage in which a person may have more than one spouse at a time. Most often, cultures allow men to have more than one wife, but a small percentage of cultures allow women to have more than one husband.

Even in cultures that allow—or even promote—polygamy, it has limits: Since there are nearly equal numbers of men and women in society, if some men

A **blended family** includes children born to one parent as well as children born to both parents.

Monogamy is a marriage in which there is only one wife and one husband.

Polygamy is any form of marriage in which a person may have more than one spouse at a time.

focus on  AMERICAN DIVERSITY

Gay and Lesbian Families

What does it mean to be a family? As we have seen, a family is a group of persons linked by blood, adoption, marriage, or quasi-marital commitments. By this definition, two men or two women who commit to each other, live together, and, if they have children, parent them together are a family.

Of course, gay or lesbian couples cannot biologically have children together through sexual intercourse. But the same is true of some heterosexual married couples, who also must rely on reproductive technologies or adoption if they want children. Similarly, many lesbians and gay men have children using artificial insemination or the like, and others have children from previous heterosexual relationships whom they raise together.

Issues related to gay families have become matters of fierce public debate in recent years. Should gays be allowed

to adopt children? Is a gay man or lesbian inherently unfit for child custody or visitation rights? Should lesbians be allowed to use artificial insemination? Should gays and lesbians be allowed to marry so their partners can share their health insurance and Social Security benefits?

These are questions that go to the heart of the family. The traditional view is that homosexual unions are both unnatural and sinful. Others define the family by long-term commitment, and they are willing to tolerate or even encourage a variety of family forms—including gay and lesbian families—as long as they contribute to stable and nurturing environments for adults and children. In fact, research consistently finds that growing up with gay or lesbian parents has no measurable effect, other than perhaps increasing children's acceptance of nontraditional gender behavior (Stacey & Biblarz 2001).

There is no question that both homosexual activity and gay marriage are regarded more favorably now than

in the past. A national survey conducted in 2009 found that 42 percent of Americans approve of gay marriages—up from only 22 percent a mere five years earlier—and another 25 percent believe they should be able to form civil unions (New York Times 2009). Despite this growing support, however, few American lesbians and gays have the option of marrying their partners. Only a small number of U.S. jurisdictions allow same-sex couples to register their unions as "domestic partnerships," and even fewer permit same-sex marriages. Moreover, the federal Defense of Marriage Act prohibits any federal recognition of gay marriage. In contrast, a small number of other countries, including Canada, Spain, and South Africa, now recognize same-sex marriage. The question American society must now address is whether gay families should receive the same legal recognition and protection as other families in this country and as gay families in some other nations.

(typically the wealthiest and most powerful) have more than one wife, other men have to do without. Consequently, even in societies where polygamy is accepted, most people actually practice monogamy, and young men may have to go elsewhere to find any wife at all. For example, in recent years hundreds of teenage boys have been banished from U.S. towns controlled by the polygamous Fundamentalist Church of Jesus Christ of Latter Day Saints. (The mainstream Church of Latter Day Saints—commonly known as Mormons—rejected polygamy more than a century ago.) Officially, these boys were banished because of misbehavior. Investigative reporters, however, argue that they were banished because they posed a threat to older men who wanted to take additional, young wives (Krakauer 2003; Eckholm 2007).

As this suggests, polygamy can only exist in societies where men have more power than women and where some men have considerably more power than do other men.

The U.S. Family over the Life Course

Family relationships play an important role in every stage of our lives. As we consider our lives from birth to death, we tend to think of ourselves in family roles. Being a youngster usually means growing up in a family; being an adult usually means having a family; being elderly often means being a grandparent.

sociology and you

The type of family you grew up in likely had a significant effect on your experiences and your future opportunities. If you grew up in a nuclear family, you only needed to share family resources (money, time, food, support for college) with a limited number of people. If you grew up in an extended family, you had to share resources with more people, but may have benefited from having other adults or older kids to care for you. If you grew up in a blended family, or were raised by a single parent or by grandparents, it's more likely that resources were spread thin and that you will need to work harder to support yourself through college.

Some modern American families, like these fundamentalist Mormons, live a polygamous life despite legal and social opposition from most of their fellow citizens and from most other Mormons.

AP Wide World Photos

Because of the close tie between family roles and individual development, we have organized this description of the U.S. family into a life course perspective. This means that we will approach the family by looking at age-related transitions in family roles.

Childhood

U.S. norms specify that childhood should be a sheltered time. Children's only responsibilities are to accomplish developmental tasks such as learning independence and self-control and mastering the school curriculum. Norms also specify that children should be protected from labor, physical abuse, and the cruder, more unpleasant aspects of life.

Childhood, however, is seldom the oasis that our norms specify. A sizable number of children are physically or emotionally abused by their parents. For example, about 10 percent of girls experience rape or attempted rape during childhood (Tjaden & Thoennes 1998). In addition, nearly one-fifth of all American children grow up in poverty—more than in any other Western nation except Russia (Heuveline & Weinshenker 2008).

An important change in the social structure of the child's world is the sharp increase in the proportion of children who grow up in single-parent households: 28 percent of children are now born to single mothers (Childstats.gov 2009). Many more experience the divorce of their parents and sometimes a second divorce between their parents and stepparents (Coleman, Ganong, & Fine 2000). Perhaps because single parents cannot provide as much money or time as married parents, studies show that, on average, children whose parents divorce have poorer self-esteem, academic performance, and social relationships than other children. These differences are slight, however, and stem primarily not from the divorce itself but from the poverty and parental conflicts that precede or follow it (Coontz 1997; Demo & Cox 2000; Lamanna & Riedmann 2000). Consequently, some of these children would not have been any better off if their parents had remained married.

As increasing numbers of U.S. women, including those with infants younger than age 1, have entered the labor force, day-care centers have become much more important aspects of early childhood socialization.

The increasing participation of women in the labor force has added another social structure to the experience of young children: day care. In 2007, about two-thirds of mothers of children younger than age 6 held jobs outside the home (U.S. Bureau of the Census 2009a). About one-third of preschool children with an employed mother attend a day-care center (Smolensky & Gootman 2003).

Research mostly supports the use of day-care centers. Some research suggests that day care can increase children's stress levels and behavioral problems, but the effect is not large (Belsky et al. 2007; Watamura et al. 2003). Other research suggests that day care increases children's math and reading skills and that any negative effects of child care are limited to certain types of children, families, or programs (Love et al. 2003). High-quality programs, which are most often attended by more affluent children, offer especially strong benefits (Kirp 2007). However, because lower-income children come from homes where they are less likely to get intellectual and social stimulation, they benefit considerably from day care, even in lower-quality programs. At any rate, many families cannot afford to have a parent stay home with the children. For these families, day care is far superior to leaving young children alone or in the care of older siblings.

Adolescence

Contemporary social structures make adolescence a difficult period. Because society has little need for the contributions of youth, it encourages young people to become preoccupied with trivial matters—such as eyebrow shaping or loading iPods. Yet, because adolescence is a temporary state, the adolescent is under constant pressure. Questions such as "What are you going to do when you finish school?", "What are you going to major in?", "What went wrong in Friday night's game?", and "How serious are you about that boy [girl]?" can create strain. That strain can be particularly high for gay and lesbian youth, who may find themselves interested in someone of the "wrong" sex, confused about their own feelings, and fearful over how their families might react.

Adolescents are supposed to become independent from their parents, acquiring adult skills and their own values. They are supposed to shift from the family to peer groups as a source of self-esteem. They are supposed to be interested in the opposite sex, but their parents expect this interest to be asexual while their friends may have a very different view. They also must learn how to interact with a broader range of people, and, last but not least, they are supposed to have fun (Gullotta, Adams, & Markstrom 2000). Thus, although society does not appear to expect much from them, adolescents experience a great deal of role strain. Many adults believe adolescence was the worst rather than the best time of their lives.

The Transition to Adulthood

Some societies have **rites of passage**, formal rituals that mark the end of one age status and the beginning of another. In our own society, there is no clear point at which we can say a person has become an adult. However, in the United States adulthood usually means that a person has a job, a place to live other than his or her parents' home, and enough money to support his or her children. Some of these norms are optional, and people may be considered adults who never marry or, in the case of women, hold a paid job. Nevertheless, the exit from adolescence always entails "escaping" from dependence on parents and family.

Making this escape, however, has become a harder and longer process (Settersten, Furstenberg, & Rumbaut 2006). In 1960, more than 80 percent of 30-year-olds (male and female) had left home, finished school, and achieved financial independence. By 2000, the numbers who had done so had dropped to about 70 percent—not a huge drop, but still significant when compared with historical patterns (Furstenberg et al. 2004). With the current economic crisis, significant numbers of young—and not-so-young—people have been forced to move back in with their parents: As of 2008, 17 percent of 25- to 29-year-olds were living with their parents (U.S. Bureau of the Census 2009d).

Why has the transition to adulthood slowed down? First, changing attitudes have allowed women (and thus men) to extend their schooling and delay marriage and parenthood. Second, economic factors have made it hard for many young people to strike out on their own. The cost of living rose rapidly from the 1960s through 2007, and paychecks did not keep pace. Prices have since fallen (along with the economy), but getting a job has become much more difficult. In addition, young people now graduate with more educational debts than was the case a generation ago.

Because of these economic and cultural changes, many young adults continue to live with their parents after leaving school, sometimes leaving home and returning several times before becoming independent. Some live with their parents to make ends meet, some to afford nice cars, cable television, fun vacations, and fast computers. Others can live on their own only because they receive substantial subsidies from their parents; about one-third of young Americans between 18 and 34 receive such subsidies annually (Settersten, Furstenberg, & Rumbaut 2006).

Early Adulthood

Rites of passage are formal rituals that mark the end of one age status and the beginning of another.

Most Americans marry at least once. This strong cultural emphasis on marriage is one of the reasons that so many gays and lesbians want to marry their life partners. Thus one of the key issues in early adulthood is deciding whether and whom to marry.

At first glance, it appears as if all persons are on their own in the search for a suitable spouse; few Americans (outside of certain religious and ethnic communities) rely on matchmakers or arranged marriages. On further reflection, however, it is clear that parents, schools, and churches all try to help young people find suitable partners. Schools hold dances designed to encourage heterosexual relationships, churches have youth groups partly to encourage members to date and marry within their church, parents and friends introduce somebody "we'd like you to meet." Although seeking a marriage partner may be fun, it is also a normative, almost obligatory, social behavior.

Seeking Sexual and Romantic Relationships

In the 1950s, young adults dated in order to find a spouse. Many did so very quickly, and more than 50 percent of U.S. women married before their twenty-first birthday. Times have changed considerably, especially for those who are college educated, live in urban areas, and are not very religious. Nowadays most young people rarely even talk about dates, let alone about finding someone to settle down with. Instead, most prefer to hang out with groups of friends, to "hook up" now and then, and perhaps to find a more serious boyfriend/girlfriend relationship eventually.

By their late twenties, 40 percent of women and 30 percent of men still have never married. Some of them are not interested in marrying, but most are looking for at least a temporary partner. Thus many people continue to seek sexual or romantic relationships into their thirties and later, even if they feel ambivalent about marriage. (This ambivalence is reflected in the growing tradition of bachelorette parties, a topic discussed in Focus on Media and Culture: Understanding Bachelorette Parties on the next page.)

Sorting through the Marriage Market

Over the course of one's single life, one probably meets thousands of potential marriage partners. How do we narrow down the marital field?

Obviously, you are unlikely to meet, much less marry, someone who lives in another community or another state. In the initial stage of attraction, **propinquity**, or spatial nearness, operates in this and a much more subtle fashion, by increasing the opportunity for continued interaction. It is no accident that so many people end up marrying fellow workers or students. The more you interact with others, the more positive your attitudes toward them become—and positive attitudes may ripen into love.

Spatial closeness is also often a sign of similarity. People with common interests and values tend to find themselves in similar places, and research indicates that we are drawn to others like ourselves. Of course, there are exceptions, but faced with a wide range of choices, most people choose a mate who is like them in many ways (Kalmijn 1998). Most marry within their social class, and most also marry within their racial, ethnic, or religious group. Marrying someone who is *similar* to you is called **homogamy**. Marrying *within* one's group—however the group is defined—is called **endogamy**. These two concepts, of course, overlap, since someone from within your group is likely to be somewhat similar to you.

Conversely, marrying someone who is *different* from you is called **heterogamy**, and marrying *outside* one's group is called **exogamy** (or *intermarriage*, in everyday language). Intermarriages can only occur when individuals have contact with persons from other groups and accept those others as more or less equal. Intermarriage is more likely among those with more education: Higher education both brings individuals into contact with others of different backgrounds and exposes individuals to more liberal ideas about whom one could or should marry (Qian & Lichter 2007).

Physical attractiveness may not be as important as advertisers have made it out to be, but studies do show that appearance is important in gaining initial attention (Sullivan 2001). Its importance normally recedes after the first meeting.

sociology and you

Your college education is likely to affect whom you marry. Many people find a spouse in college classrooms or activities (based on propinquity). If you attend a college linked to your religion, race, or ethnic group, you are more likely to marry within your group (endogamy). If college throws you into contact with many others whose backgrounds are different from your own, you will be more likely to marry someone from a different background.

Propinquity is spatial nearness.

Homogamy is the tendency to choose a mate similar in status to oneself.

Endogamy is the practice of choosing a mate from within one's own racial, ethnic, or religious group.

Heterogamy means choosing a mate who is *different* from oneself.

Exogamy means choosing a mate from *outside* one's own racial, ethnic, or religious group.

focus on  MEDIA AND CULTURE

Understanding Bachelorette Parties

A generation ago, most brides-to-be celebrated their upcoming marriages with wedding showers. At these showers, female friends and relatives brought the future bride pots, pans, linens, and the occasional item of lingerie; played humorous games centered on being a wife and mother; and enjoyed light refreshments. These days, many white, middle-class brides (as well as a growing number of others) also celebrate their upcoming weddings at bachelorette parties.

Bachelorette parties are characterized by three things: bonding with female friends, heavy drinking, and a sexualized atmosphere (Montemurro 2006). At a typical party, the bride-to-be spends the night with her female friends, drinking vodka martinis with names like *Sex on the Beach*, watching a male stripper or even getting a lap dance, and playing games that require the bride to do things like kissing male strangers or biting the labels off their briefs.

How did we go from pots, pans, and afternoon tea to lap dances? According to sociologist Beth Montemurro (2006), bachelorette parties reflect the great shifts in women's lives and in cultural attitudes toward gender. First, bachelorette parties celebrate the importance of female friendship, in contrast to earlier norms that expected women to gratefully leave behind their female friends for a man's love. Second, bachelorette parties signal that brides-to-be

have sexual desires and have been sexually active—a major change from earlier ideas about women's sexuality. Finally, the parties signal a defiant belief in gender equality and, specifically, in the idea that women should be free to lead lives independent of their husbands. As one young woman explained when asked why women started having bachelorette parties:

> I think we started it because men always have their bachelor parties...and all we had was a bridal shower, getting stuff for the home. And you never hear of them having a "man shower" where they get hammers and tools...It seemed like theirs was something more about sex and drinking and partying. And it's not fair for the women to miss out on that. (Montemurro 2006, 125)

At the same time, although bachelorette parties are designed partly to celebrate female sexuality, they also signal that the bride-to-be is about to leave her sexual freedom behind. Similarly, the parties celebrate female friendship but also carry a tone of ruefulness when the participants recognize that those friendships will likely weaken after marriage. Finally, the parties

celebrate gender equality and female freedom, but also signal in various ways that the bride-to-be will soon lose some of her control over her life.

In sum, bachelorette parties signal modern women's ambivalence about marriage: Although the women Montemurro interviewed were happy to trade sexual freedom for marriage, they also regretted the losses that they knew marriage would bring. Bachelorette parties serve as a new cultural rite of passage that helps women acknowledge and cope with this ambivalence.

Initial interest is likely to progress toward a more serious relationship if the individuals discover similar interests, aspirations, anxieties, and values (Kalmijn 1998; Seccombe & Warner 2004). When relationships start to get serious, couples begin checking to see if they share values such as the desire for children or commitment to an equal division of household labor. If he wants her to do all the housework and she thinks that idea went out with the hula hoop, they will probably back away from marriage.

Responding to Narrow Marriage Markets

Whether an individual ends up marrying also depends on the local supply of "economically attractive" partners. As early as 1987, William Julius Wilson noted that one of the reasons African American women were much less likely to marry than white women was the shrinking pool of African American men with good educations and jobs. Results from other researchers reinforce this conclusion: A shortage of males employed in good jobs with adequate earnings sharply reduces the likelihood that a woman will marry or even live with a man outside of marriage (Lichter et al. 1992; Raley 1996; Teachman, Tedrow, & Crowder 2000). In fact, differences in the availability of marriageable men account for at least 40 percent of the racial difference in overall marriage rates.

Local marriage markets also affect rates of intermarriage: Minority group members are significantly less likely to intermarry if they can easily find a marriage partner from within their group. So, for example, because of recent immigration from Asia, Asian Americans are now *less* likely to marry non-Asians than they were a decade ago (Qian & Lichter 2007).

As Figure 11.1 shows, Native Americans and Asian Americans are the most likely to marry outside their group (Qian & Lichter 2007). Largely because of gender stereotypes, Asian American women (who are often stereotyped as hyperfeminine) are more likely than Asian American men to find non-Asian spouses (usually white). Similarly, African American men (who are often stereotyped as hypermasculine) are more likely than African American women to find spouses (usually white) from outside their group.

In an interesting sidebar, researchers have found that "economically attractive" women are also more likely to marry. Their greater attractiveness to potential male partners apparently more than makes up for the fact that women with full-time employment and higher earnings tend to be choosier about the men they date and marry (Lichter et al. 1992).

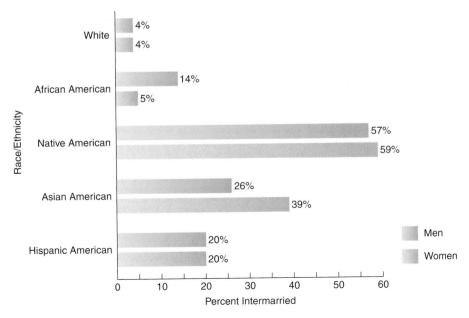

FIGURE 11.1 Intermarriage among Persons Born in the United States Native Americans are more likely than other groups to marry outside of their group (that is to engage in *exogamy*). Exogamy is higher among Asian American women than among Asian American men, and higher among African American men than among African American women.
SOURCE: Qian & Lichter (2007).

Intermarriage has become more common over time, especially between white men and Asian American women.

Middle Age

The busiest part of most adult lives is the time between the ages of 20 and 45. There are often children in the home and marriages and careers to be established. This period of life is frequently marked by role overload simply because so much is going on at one time. Middle age, that period roughly between 45 and 65, is by contrast often a quieter time. Studies show that both men and women tend to greet the empty nest with relief rather than regret (Umberson et al. 2005).

For a growing number of middle-aged couples, however, the nest is far from empty. First, immigrant families often believe that it is proper for adult children to live at home until marriage and for elderly parents to live with their middle-aged children. In these situations, the extended family is accepted and an empty nest may just seem lonely. Second, because of increased life expectancy, many native-born, middle-aged people now find themselves providing care for one or more of their parents. Third, middle-aged people are now more likely to have adult children at home because it has become more difficult for young people to establish themselves financially. Similarly, when young people divorce they are sometimes forced by finances to move back home. Not surprisingly, marital happiness is lower for parents whose adult children live with them (Umberson et al. 2005).

Sadly, the reverse situation—middle-aged adults forced to move in with their adult children—has also grown more common, as increasing numbers of middle-aged people have lost their homes to foreclosures.

Age 65 and Beyond

One of the most important changes in the social structure of old age is that it is now a common stage in the life course—and often a long one. Almost all of us can count on living to age 65. Furthermore, if you live to age 65, you can expect to live an average of 18.7 more years (National Center for Health Statistics 2009). Most of these years will be healthy ones (Federal Interagency Forum on Aging Related Statistics 2008).

Family roles continue to be critical in old age. Having spouses, children, grandchildren, and brothers and sisters all contribute to well-being. Marriage is an especially important relationship, one that provides higher income, live-in help, and companionship. Because of men's shorter life expectancy and their tendency to marry younger women, however, marriage is not equally available: 78 percent of men aged 65 to 74 are still married compared with only 57 percent of women that age.

Whether older people are married or not, relationships with children and grandchildren are typically an important factor in their lives. Most grandparents visit grandchildren every month and report very good relationships with them (American Association of Retired Persons 1999). Many children and families would have great difficulty without the help of grandparents, and many grandparents consider

As people move into the "oldest old" group, most come to rely heavily on their daughters for assistance. This can create considerable strain when the daughters find themselves simultaneously responsible for their parents and their children.

involvement with their grandchildren an important source of personal satisfaction (Allen, Blieszner, & Roberto 2000).

Grandparents are especially important when grandchildren are left parentless or effectively parentless, due to illness, disability, imprisonment, or substance abuse. Currently 1.5 million children live with grandparents rather than with parents (U.S. Bureau of the Census 2009d). These "skipped generation" households of grandparents and grandchildren are sharply at risk for poverty (Newman & Massengill 2006). Usually, though, the alternative is worse: sending the children to foster care, a system rife with problems.

The nature of intergenerational relationships depends substantially on the ages of the generations. When the older generation falls into the "young old" category, they are generally still providing more help to their children than their children are providing to them (Hogan, Eggebeen, & Clogg 1993). They are helping with down payments and grandchildren's college educations or providing temporary living space for adult children who have divorced or lost their jobs.

As the senior generation moves into the "old old" category, however, relationships must be renegotiated (Mutran & Reitzes 1984). Even in the "old old" category, most people continue to be largely self-sufficient, but they eventually will need help of some kind—for shopping, home repairs, and social support. Although these services are available from community agencies, most older people rely heavily on their families, especially their daughters (Kemper 1992; Lye 1996). Understandably, though, both older people and their adult children are happiest when these relationships are free of dependency. Elderly persons much prefer to live alone rather than with their children (Bayer & Harper 2000).

Roles and Relationships in Marriage

Marriage is one of the major role transitions to adulthood, and most people marry at least once. In fiction, the story ends with the wedding, and we are told that the couple lived happily ever after. In real life, though, the work has just begun. Marriage means the acquisition of a whole new set of duties and responsibilities, as well as a few rights. What are they and what is marriage like?

Gender Roles in Marriage

Marriage is a sharply gendered relationship. Both normatively and in actual practice, husbands and wives and mothers and fathers have different responsibilities. Although many things have changed, U.S. norms specify that the husband *ought* to work outside the home; it is still considered his responsibility to be the primary provider for his family—even though in about one-fourth of dual-earner households, wives out-earn their husbands (Winkler, McBride, & Andrews 2005). Similarly, although most Americans now believe that husbands and wives should share in household labor, most still expect that the wife will do the larger share. In fact, women currently perform about two-thirds of household labor (Amato et al. 2007). Interestingly, although husbands are happier when they do *less* housework, and wives are happier when their husbands do *more* housework, the odds that *both* husband and wife will be happy with their marriage is greatest when they evenly *split* the housework (Amato et al. 2007).

Although women who work outside the home typically do less housework than other women, this still leaves many working women (especially those with young children) subject to severe cases of role overload, or role strain. One adaptation women make to this overload is to lower their standards for cleanliness, meals, and other domestic services. They let their family eat at McDonald's and let the iron gather dust.

Another adaptation women make is to hire other women to perform domestic tasks. In this way, domestic labor remains a woman's job, and the idea that women are responsible for this work is reinforced. In addition, since most employers of domestic help are white and middle class and most domestic workers are nonwhite and working class, paid domestic labor also reinforces class and race divisions within society (Hondagneu-Sotelo 2001; Parreñas 2000).

The Parental Role: A Leap of Faith

The decision to become a parent is a momentous one. Children are extremely costly, both financially and in terms of emotional wear and tear—and the costs can continue for decades. It currently costs about $191,000 to raise a child to age 17, and another $42,000 by the time the child reaches age 34 (Settersten, Furstenberg, & Rumbaut 2006).

Parenthood is really the biggest risk most people will ever take. Few other undertakings require such a large commitment on so uncertain a return. The list of disadvantages is long and certain: It costs a lot of money, takes an enormous amount of time, disrupts usual activities, and causes at least occasional stress and worry. Also, once you've started, there is no backing out; it is a lifetime commitment. What are the returns? You hope for love and a sense of family, but you know all around you are parents whose children cause them heartaches and headaches. In fact, the presence of children in the home—especially infants and teenagers—seems particularly likely to reduce marital happiness, and happiness decreases with each additional child (Twenge, Campbell, & Foster 2003). Yet despite all this, most people want and have children.

Mothering versus Fathering

Despite some major changes, the parenting roles assumed by men and women still differ considerably (Cancian & Oliker 2000). Mothers are the ones most likely to drop out of the labor force to care for infants and young children; they are the ones most likely to care for sick children and to go to school conferences (Cancian & Oliker 2000). Fathers, on the other hand, are the ones likely to carry the major burden of providing for their families.

The overwhelming proportion of mothers who are employed—around 80 percent—has exerted pressure for fathers to increase their role in child care. Although research still finds that fathers "help" rather than "take responsibility," and that they are more likely to play with children than to change diapers, fathers have increased their role in child care. A growing proportion of fathers, however, do not live with their children. Among these fathers, contact tends to be low and child care virtually nonexistent.

Stepparenting

Because the U.S. Census collects only limited information on the topic, it is unclear how many U.S. children currently live with a stepparent. Researchers, however, estimate that about one-third will do so before they are 18—most often with a mother and a stepfather (Coleman, Ganong, & Fine 2000). If parenting is difficult, stepparenting is more so (Coleman, Ganong, & Fine 2000). Often stepparents are unsure what role they should take in their stepchildren's lives, and often their spouses and stepchildren are equally ambivalent. Older children, especially, are likely to reject stepparents and to discourage warm relationships, although many eventually develop close relationships with their stepparents. Stepmothers typically face more difficulties than stepfathers because stepmothers typically are more involved in their stepchildren's lives. In addition, stepmothers face more competition for the children's affections, since biological mothers are far more likely to remain involved in their children's lives than are biological fathers.

Although fathers now take more responsibility for child care and household tasks than they did in previous generations, mothers still bear far more of these burdens, leaving many feeling overworked and underappreciated.

Contemporary Family Choices

As discussed in Chapter 2, U.S. norms have changed over time to permit much wider variation in the way that people achieve core values. Although a happy family life remains a central goal for almost all Americans, the ways individuals meet this goal have changed considerably. Increasingly, individuals actively choose whether to marry or just live together, whether to have children (within or outside of marriage), and whether to make work or family their top priority.

Marriage or Cohabitation

Cohabitation means living with a romantic/sexual partner without marrying him or her. During the last 30 years, the chances that an individual will *ever* engage in cohabitation has increased more than 400 percent for men and 1,200 percent for women. Cohabitation is also an increasingly common stage in moving toward marriage: Approximately half of all recently married couples cohabited beforehand (Smock 2000).

But cohabitation is not always a prelude to marriage: Much of the decline in U.S. marriage rates is due to the increasing numbers of individuals who cohabit *instead of* marrying. The proportion of cohabiting couples that married within 3 years declined

Cohabitation means living with a romantic/sexual partner outside of marriage.

by half between the 1970s and the 1990s, and 40 percent of unmarried women who give birth these days are in cohabiting couples (Cherlin 2004). Indeed, Andrew Cherlin, a leading sociologist of the family, argues that we are now witnessing the **deinstitutionalization of marriage**: the gradual disintegration of the social norms regarding the need for marriage and the meaning of marriage. This process has gained ground as cohabiting couples have won legal rights (such as the right to pass on property to each other or to sue for spousal maintenance if they split up). Conversely, the fight for (and against) gay marriage suggests that marriage still means a great deal to most Americans.

Having Children ... or Not

Although most people in the United States plan to have children, increasingly they choose to do so outside of marriage. Others will choose to postpone parenthood, and increasing numbers will choose to remain childless. Still others will conclude that the best way to add children to their family is through adoption.

Nonmarital Births

Almost 40 percent of all births in the United States are to unmarried women. Most of these births (about three-fourths) are to women 20 years of age and older (Hamilton, Martin, & Ventura 2009). Now that most women participate in the labor force, many believe that they have the economic and psychological resources to tackle the tough job of parenting on their own. Some will decide against abortion if they become pregnant accidentally, and others will intentionally become pregnant or adopt even if they are not married (Hertz 2006). For the same reasons, births to unmarried women also have increased in Europe (Figure 11.2).

Nonmarital childbearing among teenagers raises special concern. The rate of teen childbearing declined steadily and considerably from 1991 to 2005, but has risen slightly since then and is higher than in any other industrialized nation (Hamilton, Martin, & Ventura 2009). Because teenage mothers are less likely to complete college or even high school, they are also likely to suffer economic hardship. In many instances, however, teenage pregnancy stems from poverty as well as causing it (Luker 1996; Newman & Massengill 2006; Edin & Kefalas 2006). Girls who face bleak futures sometimes conclude that single motherhood is a reasonable way to seek love and happiness. Other girls become pregnant because they fear that using contraceptives would suggest to a new boyfriend that they are "easy." Still others lack the power to insist that contraception be used.

Having a child outside of marriage, however, does not have to either cause or exacerbate poverty. In Europe, increasing numbers of women are having children outside marriage, but neither the women nor the children fall into poverty as a result. Decoding the Data: Poverty and Single Motherhood explores this apparent paradox.

Nor does having a child outside marriage necessarily mean raising a child alone. About 40 percent of nonmarital childbirths in the United States are to women who live with the fathers of their babies (Smock 2000). Many of these women will eventually marry the father or another man; others will continue to share parenting outside of marriage.

Delayed Childbearing

Many married women are choosing to postpone childbearing until 5 or even 10 years after their first marriages. Today, 28 percent of U.S. women aged 30 to 34 are still

The **deinstitutionalization of marriage** is the gradual disintegration of the social norms that undergird the need for marriage, the meaning of marriage, and expectations regarding marital roles.

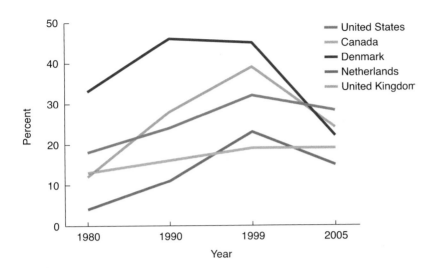

FIGURE 11.2 **Trends in Births to Unmarried Women, 1980 to 2005.** Across nations, births to unmarried women rose between 1980 and 1999, but have fallen or remained stable since then. SOURCE: **Childstats.gov.** Accessed May 2009.

childless as are 19 percent of those aged 40 to 44 (Dye 2005). Many of these intend to have children eventually, but have decided to wait until they are established in a career or in a stable marriage with someone who earns a good income.

decoding the data

Poverty and Single Motherhood

In the United States, single motherhood is closely linked to poverty. In Europe, it's not. The difference lies in the support that different governments give to single mothers.

SOURCE: Gustafsson & Stafford (2009).

	Sweden	Netherlands	United States
Percentage of preschoolers raised by single mothers	11%	6%	28%
Percentage of single mothers who are employed	89	24	66
Percentage of single mothers living in poverty	6	8	53

Explaining the Data: Swedish mothers are more likely to work than are Dutch or American mothers. What kinds of support do you think Sweden offers that allows almost all Swedish mothers to work?

Dutch mothers are much less like likely to work than are Swedish or American mothers. Yet almost no Dutch single mothers are poor. What kind of support do you think they receive from the Dutch government? What resources would American single mothers need to avoid poverty?

How can the ideology of the American Dream help explain the different situations in Sweden, the Netherlands, and the United States?

Choosing Childlessness

While most women and men do eventually want children, increasing numbers have decided that they are uninterested in having children. Of course, this choice depends on access to effective contraception. But it also reflects social changes.

There have always been men who find sufficient satisfaction in their lives that they consider children both unnecessary for happiness and a hindrance to their work and other interests. As more women find satisfaction in their work and other aspects of their lives, they may come to adopt similar attitudes (Park 2005). These decisions are bolstered by the belief—backed by research—that having children *reduces* marital satisfaction and has little impact on happiness during middle age or later life (Umberson et al. 2005). Childlessness is also particularly common among women who were the eldest daughters in large families; these women often feel that they already raised several children and have no interest in doing so again.

Adoption

For those who want children but are single, lesbian, gay, or unable to bear or conceive children, adoption is often the best route to parenthood. In addition, about one-quarter of those who adopt do so simply because they would like to give a needy child a home, while a small percentage adopt stepchildren or the children of relatives (Fisher 2003).

However, it's not easy to find a healthy, white or Asian infant (the preference of most U.S. adoptive families) who is available for adoption. In 1963, when abortion was illegal and single motherhood was highly stigmatized among white Americans, about 40 percent of babies born to unwed white mothers were given up for adoption (Fisher 2003). These days, less than 1 percent are. (Single motherhood has consistently been more common and less stigmatized among African Americans.) As a result, increasing numbers of Americans now seek babies to adopt overseas, and obtaining a baby to adopt is difficult and expensive.

Now that birth control and abortion have significantly reduced the number of unwanted babies, and fewer single mothers give up their babies, it has become increasingly difficult to find babies to adopt. As a result, international adoption has become popular—at least among those who can afford it.

Big Cheese Photo LLC/Alamy

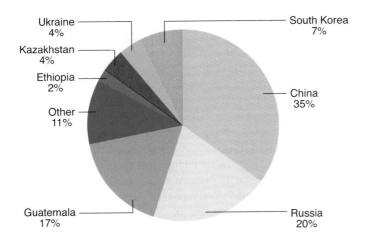

FIGURE 11.3 **Sources of Recent International Adoptions by U.S. Residents**
U.S. residents have adopted more babies from China than from any other country.
SOURCE: Selman (2007).

Adoption can be a wonderful way to create a family, and large, long-term studies find that the overwhelming majority of adoptions are highly successful for both parents and children. This is true even when the children are adopted after spending up to a few years in orphanages (Fisher 2003). But adoption also means the *disruption* of a family: One family lost a baby for another to get a baby. Even when mothers choose to give away a baby, they typically do so because they have no other viable choice: They cannot afford to feed a baby, they and their baby will suffer great stigma if they raise the child out of wedlock, or they lack the basic social support that anyone needs to raise a child. This is why about 50,000 children are adopted yearly from the U.S. child welfare system, whereas in Sweden, where mothers (whether married or not) receive extensive social services and support, fewer than a dozen children are put up for adoption each year (Rothman 2005).

Overseas adoptions also raise serious issues about the **commodification of children**. When couples who want to adopt are willing to pay up to $35,000 for a child, children in poorer countries become *commodities*: goods available for purchase—or theft. The commodification of children refers to the process through which children become treated as goods available for purchase.

There is growing evidence that many children adopted from poorer countries by Westerners have been bought, coerced, or stolen from their birth parents without the knowledge of the adoptive parents (Graff 2008; Smolin 2006). Figure 11.3 shows the nations that have sent the most babies to the United States. The problem is most severe in Guatemala, the source of 17 percent of recent U.S. international adoptions (Figure 11.3).

Work versus Family

These days, among couples with and without children, most spend considerably less time together than couples did twenty years ago (Amato et al. 2007). There are several reasons for this. First, 69 percent of married women aged 25 to 34 are now in the labor force (U.S. Bureau of the Census 2009a). Second, for middle-class Americans, workdays and work weeks are growing longer. Individuals must work early, late, and on weekends and must take work home to demonstrate that they are serious players. Working-class Americans, on the other hand, increasingly can find only part-time employment. Those who have full-time jobs, meanwhile, often must work overtime to

The **commodification of children** refers to the process through which children become treated as goods available for purchase.

earn enough to make ends meet. Still others are pressured to work extra hours off the books and without pay in order to keep their jobs (Ehrenreich 2001). As a result, both working-class and middle-class parents can experience a time bind at home. Family meals are increasingly rare, and time at home becomes rigidly scheduled as parents try to get themselves to work, do the laundry, keep their home reasonably clean, and get their children to school or other activities on time.

This time bind is often explained as the inevitable result of decreasing real wages, global competitiveness in the workplace, and the growing taste for expensive consumer goods. In an influential study, however, sociologist Arlie Hochschild (1997) argues that many middle-class parents are choosing to spend more time at work because they find work more rewarding than being at home with their family. The more hectic it gets at home, the nicer the job looks. Bosses and co-workers hardly ever spill their juice, dirty their diapers, cry, or slam out of the house because they cannot use the car. Compared with home, the workplace tends to be relatively quiet and orderly and the work rewarding. For many, work rather than home is the place where you can put your feet up and drink a quiet cup of coffee, work is the place where you can get advice on your meddlesome mother-in-law or crumbling marriage, and work is the place where employers notice that you're under a lot of stress and provide free professional counseling. Plus, of course, at work there are paychecks, promotion opportunities, and recognition ceremonies.

More recent research, however, suggests that most people work such long hours only because they have no choice (Jacobs & Gerson 2004). On a more positive note, recent research also suggests that, whatever the stresses of long workweeks, parents are finding ways to manage this time bind without cutting back on time spent with children (Bianchi, Robinson, & Milkie 2006). In fact, today's mothers spend as much time with their children as did mothers 40 years ago, and fathers spend considerably more time with children today. The difference is that today's mothers have cut back dramatically on the housework they do, whereas today's fathers do a little more than they used to. In addition, parents preserve the time they can spend with their children by having fewer children and, if they can afford it, by hiring more outside help.

These solutions, however, are simply means to help parents work even longer hours. Real solutions would require a reduction in overtime work, a living wage that enabled individuals to work fewer hours, and a cultural shift that valued raising children as much as careers. For the time being, it seems likely that Americans will continue to be stressed by the competing demands of work and family.

Problems in the American Family

Some couples swear that they never have an argument and never disagree. These people are certainly in the minority, however, for most intimate relationships involve some stress and strain. We become concerned when these stresses and strains affect the mental and physical health of the individuals and when they affect the stability of society. In this section, we cover two problems in the U.S. family: violence and divorce.

Violence

Child abuse is nothing new, nor is wife battering. These forms of family violence, however, didn't receive much attention until recent years. In a celebrated court case in 1871, a social worker had to invoke laws against cruelty to animals in order to remove a child from a violent home. There were laws specifying how to treat your animals, but no restrictions on how wives and children were to be treated. In recent years, however, we have become both more aware and less tolerant of violence in the home.

The incidence of child abuse is particularly hard to measure, since it is difficult to obtain permission to interview children outside of their parents' presence. Surveys of child protective services professionals give us at least a starting point for estimating abuse. These surveys suggest that each year, 1.5 million children are known to be sexually, physically, or emotionally abused by their parents or caregivers, with about one-third of these receiving serious physical injuries (Sedlak & Broadhurst 1996). This figure is obviously an underestimate, as it does not include those whose abuse remains hidden. For this reason, the best data currently available come from a national random survey of 16,000 Americans conducted for the National Institute of Justice (Tjaden & Thoennes 1998). Of women interviewed, 10 percent reported experiencing rape or attempted rape during their childhood, primarily at the hands of family members. These figures, too, are likely to be substantial underestimates, since they do not include those adults who refused to talk about their experiences. In addition, the survey did not include individuals who for whatever reason were in prison, a mental or general hospital, or some other institution at the time of the survey—all settings in which a disproportionate number of residents have experienced childhood abuse.

The same survey gives us our best measure of the extent of violence between adults in families (Tjaden & Thoennes 2000). The survey found that 22 percent of women and 7 percent of men have been physically assaulted by a spouse or cohabitant of the opposite sex (Table 11.1). In addition, women were twice as likely as men to have required medical care after being assaulted. Violence was almost as common in male homosexual couples as in heterosexual couples, but was much rarer among lesbians. In other words, men are more likely than women to batter their partners, whether those partners are male or female. The good news is that violence among married couples has dropped by about half over the last 20 years (Amato et al. 2007).

Recently, concern also has been raised about violence directed at dependent, vulnerable, elderly parents by their adult children. Research has found, however, that most victims of elder abuse have been attacked by their spouses rather than by their children (Bergen 1998).

Although violence among married couples has declined, it remains distressingly common. Men are more likely than women to beat their spouses because men are more likely to believe that it is their right to control their spouses.

Janine Wiedel Photolibrary/Alamy

TABLE 11.1 **Violence between Married and Cohabiting Partners**

	Percentage Who Have Been Assaulted by a Partner
Men with female partners	7%
Women with female partners	11
Men with male partners	23
Women with male partners	22

SOURCE: Tjaden & Thoennes 2000.

Family violence is not restricted to any class or race (Johnson & Ferraro 2000). It occurs in the homes of lawyers as well as the homes of welfare mothers. Violence is most likely to occur when individuals feel they are losing control, whether over their spouse or over other aspects of their lives. One reason men are more likely than women to beat their spouses is because they are more likely to believe that they *should* control their spouses (Johnson & Ferraro 2000).

Ending family violence will not be easy. Nevertheless, new laws against various forms of family violence represent important first steps in this battle.

Divorce

About 10 percent of all U.S. marriages that began in 1890 eventually ended in divorce (Cherlin 1992). Today, it's estimated that 40 to 50 percent of first marriages will eventually end in divorce (Kreider 2005).

Currently, more than 2 million U.S. adults and approximately 1 million children are affected annually by divorce. What factors make a marriage more likely to fail? Table 11.2 displays some of the predictors of divorce within the first 10 years of marriage. Research consistently finds six factors especially important (Bramlett & Mosher 2002; Teachman 2002; Amato et al. 2007):

- *Age at marriage.* Probably the best predictor of divorce is a youthful age at marriage. Marrying as a teenager or even in the early twenties doubles chances for divorce compared with those who marry later (see Table 11.2).
- *Parental divorce.* People whose parents divorced are themselves more likely to divorce.
- *Premarital childbearing.* Having a child before marriage reduces the stability of subsequent marriages. If an unwed woman marries before giving birth, however, that marriage is no more likely than others to end in divorce.
- *Education.* The higher one's education, the less likely one's marriage is to end in divorce. College graduates are only half as likely to divorce as are those without college degrees (Hurley 2005). Partly this is because people with higher educations are more likely to come from two-parent families, avoid premarital childbearing, and marry later. Independent of these other factors, however, higher education does reduce the chances of divorce.
- *Race.* African Americans are substantially more likely than whites, Hispanics, or Asians to get divorced, although the difference has declined over time (Teachman 2002).
- *Religion.* Catholics are significantly less likely than others to get divorced, even after a variety of other demographic variables are taken into account.

Societal-Level Factors

Age at marriage, parental divorce, premarital childbearing, education, race, and religion affect whether a particular marriage succeeds or fails. These personal characteristics, however, cannot explain why between 40 and 50 percent of first marriages begun this year will probably end in divorce, compared to only 10 percent a century ago (Kreider 2005). This huge increase in divorce rates is a social problem, not a personal trouble, and to explain it we need to look at social structure.

Rising divorce rates are not unique to the United States (Figure 11.4 on page 280). Although divorce has always been more prevalent in the United States than

TABLE 11.2 **Factors Predicting Whether First Marriage Will Break Up within the First 10 Years**

The probability that a first marriage will end in divorce is currently between 40 and 50 percent. Divorce is more likely for African Americans; children of divorced parents; and persons who marry young, have limited education, or have a child before marriage.

	% Ending in Divorce
Total	23%
Age at Marriage	
<18	48
18–19	40
20–24	29
25 or over	24
Education	
Less than 12 years	42
12 years	36
13 years or more	29
Children before marriage	
No	31
Yes	50
Race	
White	32
African American	47
Hispanic	34
Asian	20
Children of divorced parents	
Yes	43
No	29

SOURCE: Bramlett & Mosher 2002; U.S. Bureau of the Census 2006.

elsewhere, divorce rates have slowly crept up in other industrialized nations also. These changes are strongly associated with economic changes. In past centuries, individuals' main assets were tools or land. Because divorce meant that one spouse would lose those assets, few could afford to consider it. In today's economy, middle-class individuals' main assets are their education and experience. Because people can walk away from a marriage and take these assets along, divorce no longer seems as risky. At the same time, changes in the economy have made it more difficult for lower-class men and women to support themselves or a family. The resulting economic hardships cause enormous stress within relationships, often resulting in divorce. Finally, now that women have greater opportunities to support themselves outside marriage, divorce can seem a more appealing option. All these

FIGURE 11.4 Trends in Divorce Rates per 1,000 People

Divorce rates have fallen in the United States since 1980, while generally rising in other industrialized nations. Nevertheless, rates remain higher in the United States than in any other nation.

SOURCE: Organization for Economic Cooperation and Development (2009).

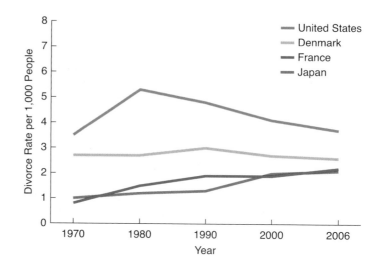

changes leave women and men with fewer reasons to stay married. Nevertheless, it is important to note that most people whose marriages end in divorce eventually remarry, with remarriage especially common among young people, whites, and men (Coleman, Ganong, & Fine 2000).

Despite the very real problems many families face, family relationships continue to be a major source of satisfaction for most Americans throughout their lives.

Where This Leaves Us

Despite all the changes and disruptions in families today, they remain the major source of economic support for children and of social support for people of all ages. Families also provide us with an important arena in which we can develop our self-concept, learn to interact with others, and internalize society's norms. Without the strong bonds of love and affection that characterize family ties, these developmental tasks are difficult if not impossible. Thus, the family is essential for the production of socialized members, people who can fit in and play a productive part in society.

The family is important not just in childhood but throughout the life course. Although we don't always get what we desire, our family members are still usually the ones we turn to when we need love, emotional support, financial assistance, and companionship. If you need an emergency loan to replace your car after an accident, or you need someone who will care for you for weeks on end while you recover from an illness, you are most likely to call on a close family member.

Given these benefits that the family gives to both the individual and society, it makes sense to try both to support the family and to reduce some of its more oppressive features. This goal is not impossible. Despite current rates of divorce, illegitimacy, childlessness, and domestic abuse, there are signs of health in the family: the durability of the mother–child bond, the frequency of remarriage, the number of stepfathers who willingly support other men's biological children, and the frequency with which elderly persons rely on and get help from their children.

There is no doubt that the family is changing. When you ask a young man what his father did when he was growing up, you are increasingly likely to hear, "What father?" or "Which father?" These recent changes must be viewed as at least potentially troublesome. At present we have no institutionalized mechanisms comparable to the family for giving individuals social support or for caring for children. The importance of these tasks suggests that the needs of families and especially children must be moved closer to the top of the national agenda.

Summary

1. Marriage and family are the most basic institutions found in society. In all societies, these institutions meet universal needs such as regulation of sexual behavior, replacement through reproduction, child care, and socialization.

2. Types of families include extended families, nuclear families, and blended families. Nuclear families are no longer very common.

3. High rates of divorce and increases in the participation of women in the labor force have led to major changes in the social structure of childhood. Nowadays, many U.S. children spend some time in a single-parent household before they are 18. About one quarter of preschoolers with employed mothers attend day-care centers.

4. The transition to adulthood occurs later than it used to because young people now attend school longer, marry later, and have more trouble finding work.

5. Mate selection depends on love but also on propinquity, homogamy, and shared values. Intermarriage is now more common, especially among more educated groups.

6. Because children are both leaving home later and returning home for economic reasons, middle-aged couples no longer can count on having an "empty nest." In addition, middle-aged persons may take their parents into their homes, or may find that they need to move in with their children.

7. The increasing participation of wives in the breadwinning role is a major change in family roles. Although

most Americans now believe that husbands and wives should share in household labor, women still perform about two-thirds of household labor.

8. Although fathers' involvement in child care has increased, they are more likely to play with children than to take responsibility for less pleasant, everyday tasks. Stepparenting is particularly difficult because often both adults and children are unclear about stepparents' roles.

9. Cohabitation is now a common choice for couples of all ages. Many cohabiting couples eventually marry, but a sizable minority are content to put off marriage indefinitely. The decline in marriage rates and increase in rates of cohabitation and divorce lead some to suggest that we are now experiencing the deinstitutionalization of marriage.

10. Growing numbers of women now choose to delay childbearing or to forego having children altogether. Forty percent of U.S. births now occur outside of marriage. Teenage pregnancy stems from poverty, a lack of easily available contraception, and a lack of other ways to find meaning and personal satisfaction.

11. As the stigma against single motherhood has declined, it has become more difficult and expensive to find a child to adopt. Babies are most often available for adoption when single mothers are stigmatized and receive few social supports for raising a child on their own. The vast majority of adoptions are successful for both child and adoptive parents.

12. Primarily because of economic pressures, married couples now spend considerably less time together than did couples in the past.

13. Violence against both children and intimate partners is relatively common in U.S. homes. In both homosexual and heterosexual relationships, men are more likely than women to batter their partners, although battering has declined significantly over the last 20 years. Family violence is most likely when individuals feel they have lost control over their lives and their spouses and believe that they have a right to control their spouses.

14. It is estimated that 40 to 50 percent of first marriages will end in divorce. Factors associated with divorce include age at marriage, parental divorce, premarital childbearing, education, race, and religion. Reduced economic dependence on marriage underlies many of these trends.

Thinking Critically

1. What functions are served by nuclear families? What are the major dysfunctions of nuclear families? What are the benefits and problems of extended and blended families?

2. Analyze the mate selection processes that you (or someone close to you) have undergone. Show how propinquity, homogamy, endogamy, and appearance were or were not involved. What role did parents play?

3. Do you know anyone who is taking care of an elderly parent or grandparent? Why do you think that person rather than some other family member has assumed that responsibility? What personal characteristics and what relational characteristics are involved?

4. How many children do you plan to have? What do you think the advantages and disadvantages will be? How and on what basis do you think you and your significant other should divide child-care responsibilities?

Book Companion Website

www.cengage.com/sociology/brinkerhoff
Prepare for quizzes and exams with online resources—including tutorial quizzes, a glossary, interactive flash cards, crossword puzzles, essay questions, virtual explorations, and more.

CHAPTER 12

Education and Religion

John F Clarke/PhotoLibrary

Educational and Religious Institutions

This chapter examines two institutions, education and religion. Both are central components of our cultural heritage and have profound effects on our society and on us as individuals. Most Americans are directly and personally affected by these institutions: almost all people in the United States have attended school, and a strong majority practice a religion. Even those who do not go to school or participate in a religion are affected by the omnipresence, norms, and values of these two institutions.

Theoretical Perspectives on Education

At the broadest level, **education** is the institution within the social structure that is responsible for the formal transmission of knowledge. It is one of our most enduring and familiar institutions. Nearly three of every ten people in the United States participate in education on a daily basis as either students or staff. As former students, parents, or taxpayers, all of us are involved in education in one way or another.

What purposes are served by this institution? Who benefits? Structural-functional and conflict theories offer two different perspectives on these questions.

Structural-Functional Theory: Functions of Education

A structural-functional analysis of education is concerned with the consequences of educational institutions for the maintenance of society. Structural functionalists point out that the educational system has been designed to meet multiple needs. The major manifest (intended) functions of education are to provide training and knowledge, to socialize young people, to sort young people appropriately, and to facilitate positive and gradual change.

Training and Knowledge

The obvious purpose of schools is to transmit knowledge and skills. In schools, we learn how to read, write, and do arithmetic. We also learn the causes of the American War of Independence and the parts of a cell. In this way, schools ensure that each succeeding generation will have the skills needed to keep society running smoothly.

Socialization

In addition to teaching skills and facts, schools help society run more smoothly by socializing young people to conform. They emphasize discipline, obedience, cooperation, and punctuality. At the same time, schools teach students the ideas, customs, and standards of their culture. In American schools, we learn to read and write English, we learn the Pledge of Allegiance, and we learn the version of U.S. history that school boards believe we should learn. By exposing students from different ethnic and social-class backgrounds across the country to more or less the same curriculum, schools help create and maintain a common cultural base.

Education is the institution responsible for the formal transmission of knowledge.

In all societies, education is an important means of reproducing culture. In addition to skills such as reading and writing, children learn many of the dominant cultural values. In Japan, school uniforms emphasize group solidarity over individual achievement.

© Charles Gupton/Stock, Boston Inc.

Sorting

Schools are like gardeners; they sift, weed, sort, and cultivate their products, determining which students will be allowed to go on and which will not. Grades and test scores channel students into different programs—or out of school altogether—on the basis of their measured abilities. Ideally, the school system ensures the best use of each student's particular abilities.

Promoting Change

Schools also act as change agents. Although we do not stop learning after we leave school, new knowledge and technology are usually aimed at schoolchildren rather than at the adult population. In addition, schools can promote change by encouraging critical and analytic skills. Colleges and universities are also expected to produce new knowledge.

Conflict Theory: Education and the Perpetuation of Inequality

Conflict theorists agree with structural functionalists that education reproduces culture, sorts students, and socializes young people, but they view these functions in a very different light. Conflict theorists emphasize how schools reinforce the status quo and perpetuate inequality.

Education as a Capitalist Tool

Some conflict theorists argue that one primary purpose of public schools is to benefit the ruling class. These theorists point to schools' **hidden curriculum**, the underlying cultural messages that schools teach. In public schools, this curriculum includes learning to wait your turn, follow the rules, be punctual, and show respect, as well as learning *not* to ask questions. All of these lessons prepare students for life in the working class (Gatto 2002). A different hidden curriculum in elite private schools trains young people to think creatively and critically and to assume that they are naturally superior and deserving of privilege. Conflict theorists note that both private and public schools

The **hidden curriculum** socializes young people into obedience and conformity.

teach young people to expect unequal rewards on the basis of differential achievement and thus teach young people to accept inequality (Kozol 2005).

Education as a Cultural Tool

Conflict theorists argue that, along with teaching skills such as reading and writing, children learn the cultural and historical perspective of the dominant culture (Spring 2004). For example, U.S. history texts describe the "Indian Wars" but rarely explain why Native American tribes resorted to warfare and give little or no coverage to the waves of anti-Chinese violence in the United States in the late nineteenth century or the removal of Japanese Americans to relocation camps during World War II. Art and music classes typically ignore the cultures of Latin America and Asia and gloss over the many contributions African Americans have made in the United States.

Education as a Status Marker

One supposed outcome of free public education is that merit will triumph over origins, that hard work and ability will be allowed to rise to the top. Conflict theorists, however, argue that evaluating individuals based on their educational credentials is no more egalitarian than evaluating people based on who their parents are (Beaver 2009). Instead of asking about your parents, potential employers may ask where you went to college, and college admissions officers ask how many Advanced Placement (AP) courses you took and whether you graduated high school with an International Baccalaureate (Sacks 2007). (If you came from a poorer family and went to a poorer high school, you may never even have heard of these programs.) Because people from affluent families tend to end up with the best educational credentials—the median family income for Harvard students *who apply for financial aid* is about $150,000 (Leonhardt 2004)—the emphasis on credentials serves to keep "undesirables" out.

Conflict theorists argue that educational credentials are a mere window dressing; apparently based on merit and achievement, credentials are often a surrogate for race, gender, and social class (Brown 2001). In the same way that we use the term *racism* to refer to bias based on race, sociologists use the term **credentialism** to refer to bias based on credentials: *Credentialism* is the assumption that some are better than others simply because they have a particular educational credential.

Unequal Education and Inequality

The use of education as a status marker is reinforced by the very unequal opportunities for education available to different social groups and communities (Kozol 2005; Sacks 2007). In poor communities, students sit in overcrowded classrooms, where undertrained, substitute, or newly graduated teachers focus on training students for standardized tests rather than on developing students' creative thinking skills. Students can choose to take auto mechanics or cosmetology, but their school probably does not offer calculus, creative writing, or AP classes. And regardless of which classes their schools offer, students find it difficult to learn when their classrooms lack proper heating or cooling and they must share outdated textbooks with other students. In contrast, in affluent communities, students sit in state-of-the-art classrooms and science laboratories and can choose from a variety of languages, challenging topics, and AP classes. A staff of advisors will help them gain admission to the most prestigious college that fits their needs and abilities; at the most selective U.S. colleges, 55 percent of freshmen come from families earning in the top 25 percent of income (Leonhardt 2004). Similarly, in mixed-income communities the wealthier students typically receive a far better education, with a very different range of classes, than do the poorer students (Bettie 2003).

Credentialism is the assumption that some are better than others simply because they have a particular educational credential.

It is difficult for children to learn in crowded classrooms that lack proper heating or cooling. It is even more difficult when students are taught by beginning or substitute teachers and must share outdated textbooks with other students. Such conditions are considerably more common in poor and minority communities.

Ethnic differences in access to educational opportunities mirror social-class differences. Public school segregation was outlawed by the U.S. Supreme Court in 1954, and segregation did decline significantly over the next 30 years. Since the mid-1980s, however, judicial support for desegregation programs has declined, and school segregation has steadily increased for both Hispanic and African American students (Frankenberg & Lee 2002). Fewer than 15 percent of students are white in some public schools, from Boston to Birmingham. The higher the percentage of minority students at a school, the lower the chances that the school will offer students the opportunities they need to learn, to graduate from high school, or to go on successfully to college. Within a given school as well, minority students are typically offered far fewer opportunities than are white students (Bettie 2003).

Symbolic Interactionism: The Self-Fulfilling Prophecy

In the modern world, the elite cannot directly ensure that their children remain members of the elite. To pass their status on to their children, they must provide their children with appropriate educational credentials. To an impressive extent, they can indeed do so: Students' educational achievements are very closely related to their parents' social status.

How does this happen? Whereas conflict theorists emphasize how the *structure* of schools leads to these unequal results, symbolic interactionists focus on the processes that produce these results. Perhaps the most important of such processes is the *self-fulfilling prophecy*.

Self-Fulfilling Prophecy

One of the major processes that takes place in schools is, of course, that students learn. When they graduate from high school, many can type, write essays with three-part theses, and even do calculus. In addition to learning specific skills, they also undergo a process of cognitive development in which their mental skills grow and expand. In the

sociology and you

What social-class advantages or disadvantages did you bring with you to college? Did you grow up with parents who read the *New York Times*, or with parents who couldn't read, or couldn't read English? Did your parents pay for you to receive extra tutoring, music lessons, theater tickets, a computer of your own, or a junior year abroad? Or did your parents need you to work to help them pay the household bills? Did your high school have all the latest facilities, or a leaky roof and out-of-date textbooks? These advantages and disadvantages will continue to affect you as you go through college.

ideal case, they learn to think critically, to weigh evidence, and to develop independent judgment.

An impressive set of studies demonstrates that cognitive development during the school years is greatest when teachers set high expectations for their students and, as a result, give their students complex and demanding work. Teachers are most likely to do this when students fit teachers' expectations for how "smart" students should look and behave. This is most likely to happen when students are white and middle or upper class.

One explanation for this is that teachers share the racist and classist stereotypes common in our society. Another explanation is that white, well-off students typically have more cultural capital—attitudes and knowledge common in elite culture (Bourdieu 1984; Bettie 2003). They are more likely to have been introduced at home to the sorts of art, music, and books that middle-class teachers value. They also are more likely to dress and behave in a way that teachers appreciate. This cultural capital helps these students in their interactions with teachers and convinces teachers that they are worth investing time in (Farkas et al. 1990; Kalmijn & Kraaykamp 1996; Bettie 2003).

In contrast, teachers (most of whom are white) are especially likely to assume that African American and Mexican American students are unintelligent and prone to trouble (Ferguson 2000; Bettie 2003). As a result, teachers often focus more on disciplining and controlling minority students than on educating them.

This process is a perfect example of a self-fulfilling prophecy. Those who are now teachers themselves grew up in a society still characterized by racist, sexist, and classist biases. When teachers biases' lead them to assume that certain students cannot succeed, the teachers give those students less opportunity to do so. So girls don't get taught calculus, boys (whether African American or white) don't learn how to cook, and working-class students (whether male or female, white or nonwhite) are encouraged to take cooking or auto mechanics rather than physics. This process helps to keep disadvantaged students from succeeding.

Current Controversies in American Education

In recent years, various proposals have emerged to improve the quality of education in the United States and to give young Americans the tools needed to be more competitive in an increasingly global job market. Three proposals that have been widely adopted are *tracking*, *high-stakes testing*, and *school choice*.

Tracking

Tracking is the use of early evaluations to determine the educational programs a child will be encouraged or allowed to follow. When students enter first grade, they are sorted into reading groups on the basis of ability. By the time they are out of elementary school, some students will be directed into college preparatory tracks, others into general education (sometimes called vocational education), and still others into remedial classes or "special education" programs. At all levels, and regardless of their actual abilities, minority and less affluent students are more likely to be put into lower tracks (Sacks 2007; Bettie 2003; Kao & Thompson 2003; Harry & Klingner 2005).

Ideally, tracking is supposed to benefit both gifted and slow learners. By gearing classes to their levels, both groups should learn faster and should benefit from

Tracking occurs when evaluations made relatively early in a child's career determine the educational programs the child will be encouraged to follow.

increased teacher attention. In addition, classes should run more smoothly and effectively when students are at a similar level. In some ways, this is indeed true. Nevertheless, one of the most consistent findings from educational research is that students are helped modestly by assignment to high-ability groups but hurt significantly if put in low-ability groups (Kao & Thompson 2003).

An important reason students assigned to low-ability groups learn less is because they are taught less. They are exposed to less material, asked to do less homework, and, in general, are not given the same opportunities to learn. Because teachers expect low-track students to do poorly, the students find themselves in a situation where they cannot succeed—a self-fulfilling prophecy (Sacks 2007).

Less formal processes also operate. Students assigned to high-ability groups, for instance, receive strong affirmation of their academic identity and abilities. As a result, they more often find school rewarding, attend school regularly, cooperate with teachers, and develop higher aspirations. The opposite occurs with students placed in low-ability tracks. They receive fewer rewards for their efforts, their parents and teachers have low expectations for them, and there is little incentive to work hard. Many will cut their losses and look for self-esteem through other avenues, such as athletics or delinquency (Bettie 2003). However, these negative effects of tracking diminish in schools where mobility between tracks is encouraged, teachers are optimistic about the potential for student improvement, and schools place academic demands on students who are not in college tracks (Gamoran 1992; Hallinan 1994).

High-Stakes Testing

Both federal and many local laws now require schools to measure student performance using standardized achievement tests. In many school districts, students must now pass these "high stakes" tests before they can move on to a higher grade. In addition, teachers and schools increasingly are evaluated, punished, or rewarded based on results from standardized examinations.

The emphasis on documenting school achievement through standardized test performance has pressed schools to pay more attention to the quality of the education their students receive and has encouraged them to make sure that all students receive good training in basic skills such as reading, writing, and arithmetic.

But high-stakes testing also has had unanticipated negative consequences (Berliner & Biddle 1995; Rothstein 2004). Few schools have received additional resources to meet these new goals. As a result, schools have dropped classes in art, music, physical education, foreign languages, and even history and science so they can use these teachers for classes in reading, writing, and arithmetic—even when the teachers lack the training to teach these subjects (Berliner & Biddle 1995). Furthermore, teachers can afford to spend time only on teaching those aspects of the subjects that appear on the tests. In addition, teachers now must devote

Laurence Gough/Fotolia

The rise of "high-stakes" tests has pressed schools to pay more attention to how their students are doing. It also has forced schools to drop classes in subjects that are not on the tests and pushed teachers to focus on teaching test-taking skills rather than on teaching the subject matter.

time simply to teaching test-taking skills. Meanwhile, the testing process itself costs school districts considerable time, energy, and money.

High-stakes testing also means that some students will be held back a grade and thereby stigmatized as failures. At the end of the 2002/2003 school year, for example, 23 percent of Florida third-graders were held back because they failed to score high enough on the state reading test (Winerip 2003). Yet research suggests that holding students back can *reduce* their long-term academic performance and *increase* their chances of dropping out. Moreover, those who fail are disproportionately lower class and minority, for a variety of reasons. Similarly, when standardized achievement exams are used to determine who should graduate, be admitted to college, or receive financial aid, they typically increase inequality between races and social classes (McDill, Natriello, & Pallas 1986). Finally, there is some evidence that, to artificially improve their schools' rankings on high-stakes tests, schools are encouraging or even forcing low-performing students to leave school before taking the tests—turning potential dropouts into "push-outs" (Nichols & Berliner 2007; Lewin & Medina 2003).

School Choice

Concern about the quality of American public education has led to a variety of proposals and programs for increasing *school choice*. **School choice** refers to a range of options (including tuition vouchers, tax credits, magnet schools, charter schools, and home schooling) that enable families to choose where their children go to school. Tuition vouchers and income tax credits are designed to help families pay for private (and, in some cases, religious) schools. Magnet schools are public schools that try to attract students by offering high-quality special programs or approaches; most commonly these schools emphasize either basic skills, language immersion, arts, or math and science. Charter schools are similar to magnet schools but are privately controlled. Charter schools receive some public funding and are subject to some public oversight, such as requirements that they offer certain courses and that their students meet specified measures of academic performance.

Proponents of school choice argue that when schools compete with each other (and with home schooling) for students, they provide better quality services, in the same way that Ford and Chevrolet compete to provide better cars (Chubb & Moe 1990; Schneider, Teske, & Marschall 2000). The school choice movement reflects the animosity toward "big government" that has been building in the United States for the last quarter century and is part of a broader movement toward *privatization*. **Privatization** refers to the process through which government services are "farmed out" to corporations, redesigned to follow corporate structures and goals, or redefined as matters of individual choice rather than governmental responsibilities.

School choice has found supporters on the left as well as the right: Black separatists, liberal believers in free-form "alternative schools," and, especially, Evangelical Christians all may prefer that their children study at home or attend carefully selected schools where parents' values will be reinforced.

Although there is some merit to the arguments for school choice, it is difficult to document its benefits scientifically. The problem is that students who participate in school choice programs differ from other students from the outset. Their parents are often more educated than other parents. More importantly, by definition their parents value obtaining the best possible education for their children and have the time and other resources needed to do so. As a result, no matter what schools their children attend or whether they study at home, they will likely do well. Currently, the best available research suggests that children sent to charter schools do no better and sometimes

School choice refers to a range of options (vouchers, tax credits, magnet and charter schools, home schooling) that enable families to choose where their children go to school.

Privatization is the process through which government services are "farmed out" to corporations, redesigned to follow corporate structures and goals, or redefined as individual responsibilities.

worse than children in public schools (Renzulli & Roscigno 2007). Children who are home schooled typically perform above national averages on standardized tests, but this may simply reflect selection bias: those with wealthier, more educated parents are more likely to take the tests (Collom 2005; Belfield 2004).

Opponents of school choice identify several unintended negative consequences of these programs. First, the programs reinforce social inequality. Because tuition vouchers and tax credits do not cover the full cost of tuition and transportation, only middle- and upper-income children can afford to use them. The same is often true of magnet and charter schools. Second, school choice programs increase segregation. Many children are home schooled or sent to charter or magnet schools specifically because their parents want to keep them away from children and teachers who are "not like them" (Saporito 2003; Renzulli & Roscigno 2007). Third, school choice programs reduce Americans' commitment to public education and to maintaining high-quality schools in all neighborhoods. Finally, it remains unclear whether children educated in these alternative environments are learning the skills needed to think creatively and to interact with the broad range of people they will meet as adults in ordinary American life (Collom 2005; Belfield 2004).

College and Society

Before World War II, college and even high school graduation were common only among the elite. Since then, however, there has been a tremendous growth in high school and college education, and today almost half of recent high school graduates ages 18 to 21 are enrolled in college. As Figure 12.1 on the next page shows, all segments of the population have been affected by this expansion in education, but significant differences still remain (Kao & Thompson 2003).

Who Goes?

Until recently, non-Hispanic white males were the group most likely to be enrolled in college, but this has changed (Figure 12.2 on the next page). Because young men can earn a good income right out of high school, many decide against going to college—even though in the long run they would earn far more money if they did so (Lewin 2006). Young women, on the other hand, have little chance of earning a good income unless they go to college. As a result, rates of college attendance for women in all ethnic groups have increased steadily, while rates among men have stayed stable. However, white men are still the most likely to receive professional and doctoral degrees and to graduate in the fields that promise the highest incomes.

Overall, though, sex differences in college attendance are fairly small compared to ethnic and social-class differences (Lewin 2006; Mead 2006). Native Americans are the least likely to graduate from high school. African Americans are still slightly less likely than whites or Asians to graduate, and Hispanics are considerably less likely to do so, partly because many emigrated here as adults (U.S. Bureau of the Census 2006).

Why Go?

There is no question that a college education pays off economically. As Table 12.1 on page 293 shows, college graduates are more likely to get satisfying professional jobs with good benefits and are less likely to be unemployed. They also earn nearly double the income of high school graduates.

FIGURE 12.1 **Educational Achievement of Persons 25 and under by Race and Ethnicity**
SOURCE: U.S. Bureau of the Census (2009).

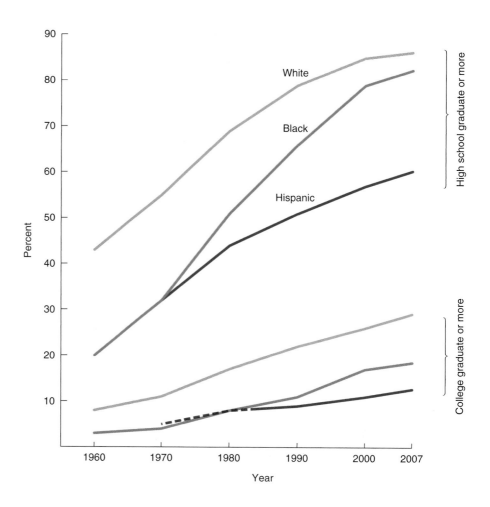

FIGURE 12.2 **Percentage of High School Graduates Ages 18 to 21 Enrolled in College, by Race, Ethnicity, and Sex, 1975 and 2003**
SOURCE: U.S. Bureau of the Census (2006).

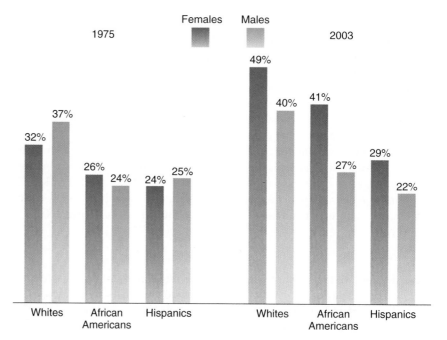

TABLE 12.1 Socioeconomic Consequences of Higher Education

Going to college pays off—literally. Those who graduate college earn nearly twice as much as high school graduates, are more likely to be employed, and are more likely to have a professional job.

Education	Median Annual Income	% with Managerial/ Professional Job	% Unemployed
9–12 years, no degree	$20,873	6.4%	7.1%
High school graduate	31,071	15.5	4.4
Less than 4 years college	32,289	32.1	3.6
College graduate	56,788	72.0	2.0

SOURCE: U.S. Bureau of the Census 2009a.

A college education also offers many less tangible benefits. At its best, college teaches students not only specific skills in math, science, and other fields, but also how to think logically and critically about all aspects of the world. Research shows that students also emerge from college more knowledgeable about the world around them, more active in public and community affairs, and more likely to lead long, healthy lives (Ross & Mirowsky 2002; Hillygus 2005; Thoits & Hewitt 2001).

College conveys psychological and social benefits as well (Kaufman & Feldman 2004). During college, students learn to talk and behave in ways that older adults will interpret as smart and middle class (such as substituting "How are you?" for "Yo, whas up?"). College also teaches students to believe they are intelligent and are entitled to middle-class jobs. As a result, college graduates are more confident and more likely to

At its best, college encourages creative and critical thinking and broadens students' views of the world.

©iStockphoto.com/Joseph C. Justice Jr.

apply for such jobs. At the same time, because American culture stresses that college graduates are more likely than others to have the skills needed for prestigious, high-paying jobs, college graduates are more likely to receive such jobs even if their actual skills are questionable (Brown 2001).

Understanding Religion

Unlike education, which we are forced by law to take part in, we have a choice about participating in religious organizations. Nevertheless, most people in the United States choose to participate, and religion is an important part of social life. It is intertwined with politics and culture, and it is intimately concerned with integration and conflict.

What Is Religion?

How can we define *religion* so that our definition includes the contemplative meditation of the Buddhist monk, the speaking in tongues of a modern Pentecostal, the sacred use of peyote in the Native American Church, and the formal ceremonies of the Catholic Church? Sociologists define **religion** as a system of beliefs and practices related to sacred things that unites believers into a moral community (Durkheim [1915] 1961, 62). Religion includes belief systems (such as native African religions) that invoke supernatural forces as explanations for earthly struggles. It does *not* include belief systems such as Marxism and science that do not emphasize the sacred.

Sociologists who study religion treat it as a set of values. They do not, however, ask whether the values are true or false: whether God exists, whether salvation is really possible, or which is the true religion. Rather sociologists examine the ways in which culture, society, and other social forces affect religion and the ways in which religion affects individuals and social structure.

Why Religion?

Religion is a fundamental feature of all societies; Map 12.1 shows the distribution of religions around the world. Whether premodern or industrialized, every society has forms of religious activity and expressions of religious behavior.

Why is religion universal? One answer is that every individual and every society must struggle to find explanations for, and meaning in, events and experiences that go beyond personal experience. The poor man suffers in a land of plenty and wonders, "Why me?" The woman whose child dies wonders, "Why mine?" The community struck by flood or tornado wonders, "Why us?" Beyond these personal dilemmas, people may wonder why the sun comes up every morning, why there is a rainbow in the sky, and what happens after death.

Religion helps us interpret and cope with events that are beyond our control and understanding; tornadoes, droughts, and plagues become meaningful when attributed to the workings of some greater force. Beliefs and rituals develop as a way to control or appease this greater force, and eventually they become patterned responses to the unknown. Rain dances may not bring rain, and prayers may not lead to good harvests, but both provide a familiar and comforting context in which people can confront otherwise mysterious and inexplicable events. Regardless of whether they are right or wrong, religious beliefs and rituals help people cope with the extraordinary events they experience.

Religion is a system of beliefs and practices related to sacred things that unites believers into a moral community.

MAP 12.1: **Distribution of World Religions**
Christianity is the dominant religion in the Americas, Europe, and Australia, but elsewhere other religions are far more common.
SOURCE: From Warren Matthews, *World of Religions*, 3E, © 1999 Wadsworth, a part of Cengage Learning, Inc.

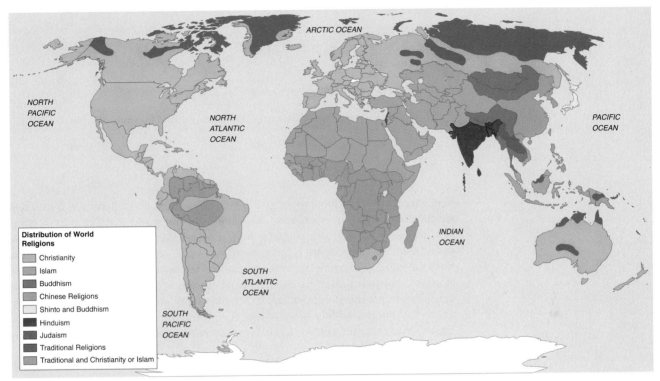

Why Religion Now? The Rise of Fundamentalism

Until the 1970s, many scholars implicitly assumed that religion would decline in importance as science and technology increased society's ability to explain and control previously mysterious events (Emerson & Hartman 2006). As a result, they assumed that **secularization**—the process of transferring things, ideas, or events from sacred authority (the clergy) to nonsacred, or secular, authority (the state, medicine, and so on)—would gradually increase.

Certainly science now explains many phenomena—illness, earthquakes, solar eclipses—that previously had been the territory only of religion. And compared with 40 years ago, many more Americans neither belong to religions, consider religion important in their lives, or even believe in God, as Table 12.2 shows. (In northern and western Europe, especially, the proportion of nonbelievers is exceedingly high.) But despite this evidence of secularization, commitment to fundamentalist religions has increased substantially over the last 30 years, in the United States and elsewhere (Sherkat & Ellison 1999; Stark & Finke 2000; Emerson & Hartman 2006).

Fundamentalism refers to religious movements that believe their most sacred book or books are the literal word of God, accept traditional interpretations of those books, and stress the importance of living in ways that mesh with those traditional interpretations. Fundamentalism exists around the world among Catholics, Protestants, Jews, Muslims, and others. Their beliefs are so strong that a small minority of

Secularization is the process of transferring things, ideas, or events from the sacred realm to the nonsacred, or secular, realm.

Fundamentalism refers to religious movements that believe their most sacred book or books are the literal word of God, accept traditional interpretations of those books, and stress the importance of living in ways that mesh with those traditional interpretations.

TABLE 12.2 Changing Religious Commitment, 1962–2007
During the last 40 years, there has been a small drop in the proportion of Americans who belong to a religion, a bigger drop in those who say religion is very important in their lives, and a *sharp* drop in the proportion who think that the Bible is the actual word of God.

	1962–65	2007
Belong to a religion	98	93
Religion is very important to their own lives	70	56
Believe Bible is actual word of God, to be taken literally word for word	65	31

SOURCE: U.S. **Gallup.com**. Accessed May 2009.

fundamentalists are willing to engage in violence against nonbelievers who they feel are threatening their religion and way of life. Fundamentalist violence is most common in situations in which people believe their religion is being suppressed by the government or their culture is being corrupted by an occupying nation (Emerson & Hartman 2006). Unfortunately, whereas political terrorists aim primarily to get media attention with the goal of promoting social change, religious terrorists (like Christians who attack abortion providers and the Muslims who attacked the World Trade Center) are motivated by a sense of divine duty and often feel that the societies they attack are too morally corrupt to change. As a result, they are willing to kill for their cause (Hoffman 2006).

Rather than modernization reducing religious commitment, as earlier scholars hypothesized, it appears to have *increased* it: As individuals around the world find their basic values about life, the family, gender relations, and society challenged by modernization, they seek out conservative and fundamentalist religions to fight those changes (Emerson & Hartman 2006). Some researchers regard the adamant rejection of modern, Western beliefs about egalitarian gender relations, family structures, and social order to be so important to fundamentalism that they include this rejection in their definition of fundamentalism (e.g., Marsden 2006).

In addition, other theorists argue, commitment to religion remains a *rational* choice for individuals when the time and money costs of commitment are outweighed by its benefits. Those benefits include explanations for otherwise inexplicable events, the promise of supernatural rewards, integration into a community of like-minded individuals, and the lending of supernatural authority to traditional values and practices (Stark & Finke 2000).

Theoretical Perspectives on Religion

As with the study of other social institutions, different sociologists bring different theoretical perspectives to the study of religion. This is the topic of the next section. As we will see, structural functionalists focus on the functions that religion serves for both individuals and societies. Conflict theorists focus on how religion can foster or repress social conflict. A third important perspective, associated with the work of Max Weber, combines elements from the other two perspectives.

Durkheim: Structural-Functional Theory of Religion

The structural-functional study of religion begins, most importantly, with the work of Emile Durkheim. Durkheim began his analysis of religion by identifying the three elements shared by all religions, which he called the *elementary forms of religion* ([1915] 1961).

Elementary Forms of Religion

The first of the three elementary forms is that all religions divide human experience into the *sacred* and the *profane*. The **profane** represents all that is routine and taken for granted in the everyday world—things that are known and familiar, that we can control, understand, and manipulate. The **sacred**, by contrast, consists of the events and things that we hold in awe and reverence—what we can neither understand nor control.

Second, all religions hold beliefs about the supernatural that help people explain and cope with the uncertainties associated with birth, death, creation, success, failure, and crisis. These beliefs form the basis for official religious doctrines.

Third, all religions have rituals. In contemporary Christianity, rituals mark such events as births, deaths, weddings, Jesus's birth, and the resurrection. In earlier eras, many Christian rituals celebrated the planting and harvest seasons—occasions still marked by important ritual occasions in many religions.

Religious rituals help individuals cope with events that are beyond human understanding, such as death, illness, drought, and famine.

The Functions of Religion

Durkheim argued that religion would not be universal if it did not serve important functions for society. At the societal level, the major function of religion is that it gives tradition a moral imperative. Most of the central values and norms of any culture are reinforced through its religions. These values and norms cease to be merely the *usual* way of doing things and become perceived as the *only* moral way of doing them. They become sacred. When a tradition is sacred, it is continually affirmed through ritual and practice and is largely immune to change.

For individuals, Durkheim argued that the beliefs and rituals of religion offer support, consolation, and reconciliation in times of need. On ordinary occasions, many people find satisfaction and a feeling of belongingness in religious participation. This feeling of belongingness creates the moral community, or community of believers, that is part of our definition of religion.

Marx and Beyond: Conflict Theory and Religion

Like Durkheim, Marx saw religion as a supporter of tradition. This support ranges from injunctions that the poor and oppressed should endure rather than revolt (blessed be the poor, blessed be the meek, and so on) and that everyone should pay taxes (give unto Caesar) all the way to the endorsement of inequality implied by a belief in the divine right of kings.

Marx differed from Durkheim by interpreting the support for tradition in a negative light. Marx saw religion as the "opiate of the masses"—a way the elite kept the eyes of the downtrodden happily focused on the afterlife so that the poor would not notice their earthly oppression. This position is hardly value-free, and much more obviously

The **profane** represents all that is routine and taken for granted in the everyday world, things that are known and familiar and that we can control, understand, and manipulate.

The **sacred** consists of events and things that we hold in awe and reverence—what we can neither understand nor control.

than structural-functional theory, it makes a statement about the truth or falsity of religious doctrine.

Modern conflict theory goes beyond Marx's view. Its major contribution is in identifying the role that religions can play in fostering or repressing conflict between social groups. Religion has certainly *contributed* to conflict between Sunni and Shiite Muslims in Iraq and between Protestants and Catholics in Ireland, as well in many other countries. On the other hand, religion has *reduced* conflict when Muslim, Christian, Hindu, and other clergy have taught impoverished people to accept their fate as God's will or have preached that we are all God's children.

Whether it increases or reduces conflict, religion can and has served as a tool for groups to use in their struggles for power. Interestingly, although Marx believed that religion always helps to keep down the oppressed, we now know that oppressed groups can use religion to better their social position. One example of this is the powerful role the African American Church and leaders such as the Reverend Martin Luther King, Jr., played in fighting for civil rights in the United States.

Another contribution of conflict theory to the analysis of religion is the idea of the dialectic, that is, that contradictions build up between existing institutions and that these contradictions lead to change. Specifically, conflict theorists suggest that social change in the surrounding society can foster change in that society's religions. For example, changes in attitudes toward women have led Reform Jews, Methodists, and others to allow women to serve as ministers or rabbis. Conflict theorists also argue that changes in religion can lead to broader social change. For example, the rise of evangelical churches in (traditionally Catholic) U.S. Hispanic communities is playing a substantial role in organizing Hispanics into an effective political lobby. In March 2006, more than 500,000 people, most of them Hispanic and disproportionately evangelical Christians, marched in protest against proposed anti-immigration legislation. Many of these protesters had learned of the march through evangelical ministers.

Weber: Religion as an Independent Force

Max Weber's influential theory of religion combines elements of structural functionalism and conflict theory. Like Durkheim and other structural functionalists, Weber was interested in the forms and functions of religion. But like various conflict theorists, Weber was also interested in the links between social and religious change. However, whereas conflict theorists typically focus on how social conflict can stimulate religious change, Weber focused on how changes in religious ideology can stimulate social change.

For most people, religion is a matter of following tradition; people worship as their parents did before them. To Weber, however, the essence of religion is the search for knowledge about the unknown. In this sense, religion is similar to science: It is a way of coming to understand the world around us. And as with science, the answers religion provides may challenge the status quo as well as support it.

Where do people find the answers to questions of ultimate meaning? Often they turn to a charismatic religious leader. **Charisma** refers to extraordinary personal qualities that set the individual apart from ordinary mortals. Because these extraordinary characteristics are often thought to be supernatural in origin, charismatic leaders can become agents for dramatic social change. Charismatic leaders include Christ, Muhammad, and, more recently, Joseph Smith (Latter Day Saints), David Koresh (Branch Davidians), and the Ayatollah Khomeini (Iranian Islam). Such individuals give answers that often disagree with traditional answers. Thus, Weber sees religious inquiry as a potential source of instability and change in society.

Charisma refers to extraordinary personal qualities that set an individual apart from ordinary mortals.

In viewing religion as a process, Weber gave it a much more active role than did Durkheim. This is most apparent in Weber's analysis of the Protestant Reformation.

The Protestant Ethic and the Spirit of Capitalism

In his classic analysis of the influence of religious ideas on other social institutions, Weber ([1904–1905] 1958) argued that the Protestant Reformation paved the way for capitalism. Early Protestants believed that work, rationalism, and plain living are moral virtues, whereas idleness and indulgence are sinful. Weber labeled these beliefs the **Protestant Ethic**. What happens to a person who follows this ethic—who works hard, makes business decisions based on rational rather than emotional criteria (for example, firing inefficient though needy employees), and is frugal rather than self-indulgent? Such a person is likely to grow wealthier. According to Weber, it was not long before wealth became an end in itself. At this point, the moral values underlying early Protestantism became the moral values underlying early capitalism.

In the century since Weber's analysis, other scholars have explored the same issues, and many have come to somewhat different conclusions. Nevertheless, this research has not changed Weber's major contribution to the sociology of religion: that religious ideas can be the source of tension and change in social institutions.

Tension between Religion and Society

Each religion confronts two contradictory yet complementary tendencies: the tendency to reject the world and the tendency to compromise with the world (Troeltsch 1931). If a religion denounces adultery, homosexuality, and fornication, does it have to categorically exclude adulterers, homosexuals, and fornicators, or can it adjust its expectations to take common human frailties into account? If "it is easier for a camel to go through the eye of a needle than for a rich man to enter the kingdom of God," must a church require that all its members forsake their worldly belongings?

How religions resolve these dilemmas is central to their eventual form and character. Scholars distinguish three general types of religious organizations: *churches*, *sects*, and *new religious movements*.

Churches

In everyday language, we use the term *church* to refer to Christian religious organizations or places of worship. Sociologists, on the other hand, use the term **church** to refer to any religion that accepts the surrounding society and is accepted by it.

In some societies, one church is so interwoven with society that it is strongly supported or even mandated by the government. In these situations, the church is known as a **state church**. For example, in the 1500s in Spain, anyone who wasn't Catholic could be legally sentenced to death by burning. These days in Iran, anyone who doesn't follow strict Islamic rules can be legally sentenced to death by stoning.

In other societies, no church has a monopoly on state power. When a church generally accommodates to the society at large, receives no special state support, and tolerates both the state and other churches, we refer to it as a **denomination**. Most people in the United States belong to denominations, including Conservative Judaism, Roman Catholicism, Lutheranism, and Methodism. Clergy from these groups meet together in ecumenical councils, pray together at commencements, and generally adopt a live-and-let-live policy toward one another.

The **Protestant Ethic** refers to the belief that work, rationalism, and plain living are moral virtues, whereas idleness and indulgence are sinful.

Churches are religious organizations that have become institutionalized. They have endured for generations, are supported by and support society's norms and values, and have become an active part of society.

A **state church** is one that is strongly supported or even mandated by the government.

A **denomination** is a church that accommodates to the state and to the presence of other churches.

Churches' embeddedness in their societies does not necessarily mean that they have compromised essential values. They still retain the ability to protest injustice and immorality. From the abolition movement of the 1850s to the Civil Rights struggle of the 1960s and the demonstrations against torture of prisoners at Guantánamo, churchmen and women have been in the forefront of social protest. Nevertheless, churches are generally committed to working with society. They may wish to improve it, but they have no wish to abandon it.

Structure and Function of Churches

Churches tend to be formal bureaucratic structures with hierarchical positions and official creeds specifying their religious beliefs. Leadership is provided by a professional staff of ministers, rabbis, imams, or priests, who have received formal training at specialized schools. Religious services almost always prescribe formal and detailed rituals, repeated in much the same way from generation to generation. Congregations often function more as audiences than as active participants. They are expected to stand up, sit down, and sing on cue, but the service is guided by ceremony rather than by the emotional interaction of participants.

Generally, people are born into churches rather than converting to them. People who do change churches often do so for practical rather than emotional reasons: They marry somebody of another faith, another church is nearer, or their friends go to another church. Individuals also might change churches when their social status rises above that of most members of their church: Baptists become Methodists, and Methodists become Episcopalians (Sherkat & Ellison 1999). Most individuals who change churches have relatively weak ties to their initial religion. Nevertheless, few make large changes: Orthodox Jews become Conservative Jews and members of one small Baptist church join a different small Baptist church (Stark & Finke 2000).

Churches tend to be large and to have well-established facilities, financial security, and a predominantly middle-class membership. As part of their accommodation to the larger society, churches usually allow scriptures to be interpreted in ways relevant to modern culture. Because of these characteristics, these religions are frequently referred to as *mainline churches.*

Sects

Sects are religious organizations that arise in active rejection of changes they find repugnant in modern society and modern religions (Sherkat & Ellison 1999). Sect members often view themselves as restoring a true faith that had been abandoned by others too eager to compromise with society. Like the Reformation churches of Calvin and Luther, sects want to cleanse religion of secular associations. Most modern sects have emerged as protests against liberal developments in mainstream churches, such as the acceptance of homosexuals, divorce, abortion, or "immodest" dress (for example, short skirts and short hair for women).

Some sects' rejection of society's norms is so great that the relationship between the sect and the larger society becomes fraught with tension and even hostility. Egypt routinely incarcerates members of fundamentalist Muslim sects that it considers too extreme, and the United States in the past jailed Amish men who refused to serve in the military because the Bible says "Thou shalt not kill."

The Amish church is exceptional in that it has managed to maintain its distance from the surrounding social world for generations. In contrast, most sects either dissolve or become increasingly churchlike over time. For example, the Church of Jesus Christ of Latter Day Saints (Mormons) has over time increased its accommodation to

Sects are religious organizations that arise in active rejection of changes they find repugnant in churches.

Although members of the Amish sect reject much of modern life for themselves, they have accommodated to living in the modern world around them.

Martin Thomas Photography / Alamy

the larger society (Arrington & Bitton 1992). Among other things, it officially abandoned polygamy, opened its priesthood to African American men, and left the seclusion of a virtual state church in Utah. The church and its members continue to hold religious and social views that differ from those of many other Americans, but they are now actively involved in the country's political, economic, and educational institutions.

Structure and Function of Sects

The hundreds of sects in the United States exhibit varying degrees of tension with society, but all oppose some basic societal institutions. Not surprisingly, these organizations tend to be particularly attractive to people who are left out of or estranged from society's basic institutions—the poor, the underprivileged, the handicapped, and the alienated. For example, the members of the snake-handling Pentecostal sects of Appalachia are overwhelmingly rural, poor white people who have little chance of succeeding on modern society's terms (Covington 2009). Based on a passage in the New Testament (Mark 16:17–18), church leaders encourage members to speak in tongues, handle poisonous snakes, and drink poisonous potions. Doing so gives individuals a sense that they are close to God and that they control their own lives.

But many who follow sectlike religions are neither poor nor oppressed. Instead, they are seekers of spiritual well-being who find established churches too bureaucratic, or seek a moral community that will offer them a feeling of belongingness and emotional commitment (Saliba 2003; Barker 1986). Others join sects such as Hasidic Judaism or Christian fundamentalist groups because they want to hold on to traditional norms and values that seem to have fallen from favor (Davidman 1991).

Sect membership is often the result of conversion or an emotional experience. Instead of merely following their parents into a sect, individuals actively choose to join. Religious services are more informal than those of churches. Leadership remains largely unspecialized, and there is little, if any, professional training for the calling. The religious doctrines emphasize other worldly rewards, and the scriptures are considered to be of divine origin and therefore subject to literal interpretation.

Sects share many of the characteristics of primary groups: small size, informality, and loyalty. They are closely knit groups that emphasize conformity and maintain significant control over their members.

New Religious Movements

Since the 1960s there has been an explosion of what are known as *new religious movements (NRMs)*. As the term suggests, **new religious movements (NRMs)** are religious or spiritual movements begun in recent decades and *not* connected to a nation's mainstream religious traditions (Clarke 2006; Saliba 2003; Dawson 2006). In common usage, NRMs are often referred to as *cults*, but that term has largely been dropped by sociologists because of its negative connotations. Examples of NRMs are the Church of Scientology, the "neopagan" Wicca religion, various "New Age" spiritual groups that draw on Eastern religions but give them very Western interpretations, and Heaven's Gate, whose members committed mass suicide in 1997 because they believed the Hale-Bopp comet was about to destroy the Earth and believed that their suicides would allow them to survive at a "higher level." Each of these religions stands outside of the Judeo-Christian tradition: They have a different God or Gods or no God at all, and they don't use the Old Testament as a text. NRMs are often led by charismatic leaders who demand strict adherence to specific beliefs and practices that differ from those of the broader society. The Concept Summary on Churches, Sects, and New Religious Movements compares these three types of religious institutions.

NRMs include both groups such as Heaven's Gate that encourage their members to withdraw from mainstream society and "New Age" groups that emphasize using meditation, affirmations, and the like to gain greater success and happiness within mainstream society.

Structure and Function of New Religious Movements

The structure and functions of NRMs strongly resemble those of sects. By definition, since NRMs are new, most members have actively chosen to join rather than simply continuing in their parents' religion. Thus, as with sects, NRMs attract individuals whose spiritual needs are not being met by mainstream religions (Clarke 2006; Saliba 2003; Dawson 2006). Beyond this, however, NRMs differ so greatly from each other that they serve very different purposes for different people. Those NRMs that reject mainstream society best meet the needs of those who are deeply discontented with society or who believe they can never succeed in mainstream society. Those NRMs that emphasize attaining success or happiness in mainstream society obviously are attractive to those who value at least some mainstream cultural norms.

Case Study: Islam

Islam was founded in the seventh century A.D. by an Arab prophet named Muhammad in what is now Saudi Arabia. It is currently the fastest-growing religion in the world, encompassing 21 percent of the world's population, and will likely pass Christianity to become the largest religion in the world within the next 50 years (Ontario Consultants on Religious Tolerance 2009a).

No matter where they live, Muslims (adherents to Islam) share a set of common beliefs. All Muslims believe in a single all-powerful God whose word is revealed to the faithful in the Koran, a book that plays the same role in Islam as the Bible plays for Christians and Jews. All Muslims must follow the Five Pillars of Islam:

New religious movements (NRMs) are religious or spiritual movements begun in recent decades and *not* derived from a nation's mainstream religions.

concept summary

Churches, Sects, and New Religious Movements

Churches, sects, and new religious movements are differentiated based on their attitude toward society, their attitude toward other religions, their position in a given society, and their history.

	Definition	Example	Attitude toward Other Religions
Church	A religion that accepts society as it is and is accepted by society		
State church	A church that is strongly supported or even mandated by the state	Islam in Iran, Roman Catholicism in Medieval Europe	Typically intolerant
Denomination	A church that receives no special state support and tolerates other religions	Methodism, Lutheranism, Roman Catholicism in the United States	Tolerant
Sect	A religion based on rejecting modernizing changes in a given religion	Amish, Ultra-orthodox Judaism, and fundamentalist (polygamous) Mormonism in the United States	Intolerant
New religious movement	A religion that began in recent decades and is *not* the outgrowth of an established religion in a given society	Nation of Islam, Church of Scientology, and "New Age" Buddhist groups	Varies

1. Profess faith in one almighty God and in Muhammad, his prophet,
2. Pray five times daily,
3. Make charitable donations to the Muslim community and the poor,
4. Fast during daylight hours during the month of Ramadan, the time when the Koran was revealed to Muhammad, and
5. Try to make at least one pilgrimage to Mecca.

Muslim prayer usually occurs in a mosque (an Islamic house of worship) and is led by an imam (a religious scholar). Because there is no formal central authority, there is considerable variation across countries in the relationship between Islamic clergy and the government and between Muslims and non-Muslims. In some nations Islam more closely resembles a church, and in others it more closely resembles a sect.

Islam as a Churchlike Religion: Egypt and Iran

Islam is a church in both Iran and Egypt, but in the former it is a *state church* and in the latter, a *denomination*.

In Iran, church and state are intertwined. Because Islam is the state church, Islamic law is used in the courts. Under that law, individuals can be sentenced to death for adultery, armed robbery, homosexuality, or leaving the Islamic faith, among other things. Although many in Iran hope for a more secular society, the clergy continue to hold a great deal of political power, and so there is little tension between religion and the larger society.

In contrast, Islam functions as a *denomination* in Egypt. Egypt's government is more or less secular, even though 90 percent of the nation is Muslim. Tension between Islam and the state is palpable (Rubin 2002). Radical Islamic fundamentalists periodically incite violent anti-government attacks, and the government uses terror and

In Egypt, Islamic moderates and fundamentalists coexist—sometimes peacefully, sometimes not. As a result, Egyptian Muslims have greater freedom than do Iranian Muslims to interpret Islamic rules for themselves, such as rules regarding acceptable clothing.

Bill Lyons / Alamy

repression to keep fundamentalists under control. More moderate Islamic mosques and imams, however, are allowed to function openly, and the government works with them to provide social services to the poor. As a result, Islam remains a highly organized, accepted part of Egypt's culture and society. Thus despite tension between Islam and the government, the religion remains churchlike rather than sectlike.

Islam as a Sectlike Religion: Islamic Fundamentalism

Recent years have seen a worldwide increase in Islamic fundamentalism. Like Christian fundamentalism, Islamic fundamentalism is a *sect*: It emerged in protest against changes occurring within Islam.

In the same way that Christian fundamentalists argue that U.S. society has become corrupted by secularism and turned its back on "true" Christian principles, Islamic fundamentalists call for a rejection of modern secular culture and a return to "true" Islamic principles. Islamic fundamentalism appeals especially to individuals who lack economic and political power in modern society. But it also appeals to educated Muslims who, like those who bombed the World Trade Centers, despair of Western political domination, cultural domination, and, especially, physical occupation of Muslim regions (Amanat 2001; Barber 2001; Jacquard 2002). In the latter case, however, dismay at domination by Westerners usually leads to religious fervor, rather than religious fervor leading to political beliefs and action.

Only the most radical Islamic fundamentalists, however, advocate violence to achieve these goals. Most Muslims, in fact, say the concept of *jihad*—holy war—primarily refers not to actual warfare but rather to the need to defend social justice, first through spiritual, economic, and political means and only if that fails through military means (Lawrence 1998).

Islam as a New Religious Movement: The Nation of Islam

Although Muslims have lived in the United States for centuries, most modern-day U.S. Muslims are recent immigrants or children of recent immigrants (Smith 1999). For most of these immigrants, Islam serves as a denomination: one religion among many co-existing in this country.

In contrast, the Nation of Islam, popularly known as "Black Muslims," is a new religious movement (Clarke 2006). Although it officially began in the 1930s, most of its growth occurred after the 1960s.

The Nation of Islam emerged not in reaction to *Islam* (as a sect would) but in reaction against *Christianity* and *white American society*. Its theology draws on some mainstream Islamic beliefs and practices but adds a belief in the innate superiority of Africans and African Americans (Clarke 2006). Although many former members and leaders have rejected these beliefs and entered mainstream Islam, these beliefs remain strong under the current leadership of the Nation of Islam. The difference between the Nation of Islam and mainstream Islam is so sharp that many mainstream Muslims do not consider members of the Nation of Islam to be Muslim at all.

Membership in this new religious movement is growing most rapidly among poor and disenfranchised African Americans in inner cities and in prisons. For these individuals, Islam can provide a sense of hope, community, identity, and freedom from the white-dominated world around them.

Religion in the United States

The United States is a pluralistic country: Its citizens belong to many different religions and to no religion at all. In this section, we offer a religious portrait of U.S. society.

Trends in U.S. Religious Membership

Throughout its history, members of multiple religions have lived in the United States. Nevertheless, Christians have always been by far the largest group. Although Jews, Buddhists, Muslims, and other non-Christian groups have visible and important presences in the United States, none accounts for much more than 1 percent of the U.S. population. In addition, Judaism is losing population while the other groups are merely holding steady. Muslim Americans are highlighted in Focus on American Diversity: American Muslims on the next page.

That said, Christianity's numeric dominance is now slightly weaker than in the past. The percentage of U.S. residents who identify as Christian dropped from 86 percent in 1990 to 76 percent in 2008 (Kosmin & Keysar 2009). This change primarily reflects a shift away from identifying with any religion, rather than toward identifying with a non-Christian religion. Figure 12.3 on page 307 shows the relative size of various religions both globally and in the United States.

Within Christianity, there has been a steep shift away from mainline Protestant churches such as Lutheranism and Episcopalianism and toward fundamentalist churches such as Pentecostal, Nazarene, and Four Square Gospel churches (Kosmin & Keysar 2009). Similarly, the percentage of U.S. adults who identify as Catholic has held steady over the last 20 years, but only new Hispanic Catholic immigrants have made up for conversions of other Hispanic Catholics to fundamentalist Protestant sects.

The Rise of Emerging Churches

Reflecting these changes, the newest trend in American Christianity is the rise of **emerging churches** (emergingchurch.info 2009, Kimball 2003). This trend reflects rising dissatisfaction with the impersonal, "inauthentic" life of modern Americans and the bureaucratization of religious belief in modern churches. Most who participate in emerging churches are young, white, and urban. Most also consider themselves evangelical Christians, but the appeal of these churches has spread beyond that core base.

Emerging churches are linked by 1) the belief that American life and modern Christian churches are atomized, bureaucratic, and inauthentic and 2) an emphasis on informal rituals, a more open perspective toward scripture and behavior, and living a life of mission, faith, and community.

focus on  AMERICAN DIVERSITY

American Muslims

The history of Muslims in the United States is an old one, going back to the colonial era (Muslim West Facts Project 2009). It is estimated that at least 10 percent of the slaves brought from Africa were Muslim. But the Muslim religion disappeared quickly, as cultural ties to Africa were lost and as slaves were forced to adopt Christianity. However, during the twentieth century many African Americans first joined the Nation of Islam and then joined more mainstream Muslim religious communities.

The first wave of chosen migration of Muslims to the United States occurred after the Civil War. Immigrants from the Arab countries typically became peddlers and factory laborers in the Midwest, while immigrants from India typically entered agricultural labor on the West Coast. Over time, many

of their descendants became well-educated and highly successful.

The second wave of Muslim immigration began in the mid-twentieth century and continues to this day. This wave consists of professionals and university students from across the Muslim world. Finally, in recent years Muslim refugees from war-torn nations such as Somalia, Bosnia, and Ethiopia have settled in communities across the United States.

As this history suggests, American Muslims are a highly diverse population. On average, however, they are well integrated into U.S. society. Their levels of education and income are above the national average and they are almost as likely as other Americans to tell survey researchers that they were treated with respect throughout the day before they were interviewed. Interestingly, considering the stereotype of Muslims as ultra-conservative, a higher percentage

of U.S. Muslims describe themselves as liberal than do members of any religious group other than Jews (Muslim West Facts Project 2009).

On the other hand, Muslims are more likely than other Americans to report feeling stressed, worried, or angered recently. This partly reflects the concerns of lower-income Muslims who are struggling to keep bread on the table and a roof over their heads. But feelings of stress and anger also reflect the changes that have occurred in American society since the attacks of 9/11. As noted in Chapter 4, about 40 percent of Americans now freely admit to prejudice against U.S. Muslims and to concerns about their loyalty (Saad 2006). Nevertheless, almost all U.S. Muslims are citizens, committed to making their homes in the United States.

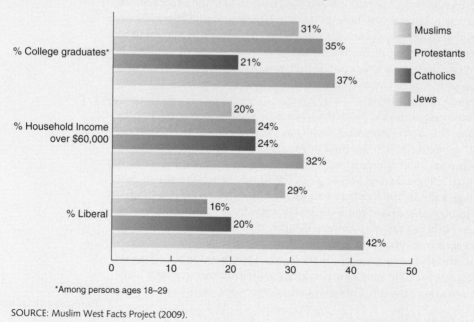

*Among persons ages 18–29

SOURCE: Muslim West Facts Project (2009).

Emerging churches promise an authentic religious experience closely shared with others in an informal space and relying on informal practices. Rather than meeting in formal churches to read prayers and hymns, members meet in homes, talk about their feelings and beliefs, share their questions and tentative answers on matters of

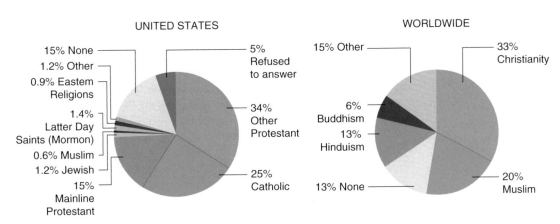

FIGURE 12.3 Religious Affiliation in the United States and Worldwide
Three-quarters of U.S. residents are Christian. Worldwide, Christianity is shrinking and Islam is growing. Mormons are included with other Christians on the world pie chart; Jews do not appear on the chart because they comprise less than 1 percent of the world's population.
SOURCE: Ontario Consultants on Religious Tolerance (2009b); Kosmin & Keysar (2009).

faith, and listen to music straight out of youth popular culture. Emerging churches emphasize how individuals can live a life of mission, faith, and community—qualities many find sorely lacking in a broader culture that emphasizes working, consuming, and individual self-sufficiency. Finally, whereas traditional evangelical churches define themselves partly by their rejection of contemporary American morality and culture, emerging churches have a more open perspective. As a result, they offer a better cultural fit for some young Americans.

The Rise in "No Religion"

Another important trend is the increased number of U.S. residents who claim membership in *no* religion. Currently 15 percent of U.S. residents claim no religion, up from 8 percent in 1990 (Kosmin & Keysar 2009). Some of these individuals nonetheless have strong religious beliefs, but others do not.

Atheists are individuals who believe that there is no God, and *agnostics* are those who do not know whether there is a God. Neither atheism nor agnosticism is a religion, and few atheists or agnostics belong to groups organized around atheism or agnosticism.

When asked about their personal religious identification, less than 2 percent of U.S. adults describe themselves as atheist or agnostic (Kosmin & Keysar 2009). However, when directly asked whether God exists, 2 percent say no and 10 percent say that they don't know. Apparently, many people identify with the religion in which they were raised, even if they now hold atheistic or agnostic beliefs. Thus, data on religious identification underestimate the number of atheists and agnostics. Similarly, although only 15 percent of U.S. adults say they belong to no religion, fully 27 percent do not expect to have a religious funeral when they die. This further indicates that questions about personal religious identification overstate the importance of religion in the United States.

Trends in Religiosity

Religiosity refers to an individual's level of commitment to religious beliefs and to acting on those beliefs. Membership in organized religions is considerably higher in

Religiosity is an individual's level of commitment to religious beliefs and to acting on those beliefs.

the United States than in other developed nations, and reported rates of attendance at religious services have changed very little over the last several decades (although actual rates appear to have declined).

Why is religiosity so strong in the United States? According to sociologists Rodney Stark and Roger Finke (2000), the answer lies in our highly developed, competitive, and unregulated **religious economy**. They argue that because there are so many religious organizations in this country, each must compete with the others to provide better "consumer products," thereby generating greater "market demand" for them.

But although most Americans believe in God, some are more involved in religion than others. Mormons are the most likely to attend religious services at least once per week, and Jews are the least likely; Protestants, Catholics, and Muslims hold similar, medium levels of religious attendance. In addition, across religions, older people, women, Southerners, and African Americans are more likely than others to attend religious services regularly (Sherkat & Ellison 1999; General Social Survey 2009; Muslim West Facts Project 2009).

One interesting topic is the relationship between income, education, and religiosity. In the past, many scholars assumed that religion would appeal disproportionately to the poor, who were in greater need of hope, and to the uneducated, who were more likely to lack "scientific" explanations for natural and human events. It is true that those with a college education are less likely, overall, to say that religion is important to them. However, college graduates are *more* likely to attend church than are nongraduates (General Social Survey 2009). Moreover, among those who consider religion important in their lives, college graduates and nongraduates are equally likely to hold conservative religious beliefs (Sherkat & Ellison 1999; General Social Survey 2009). In general, churchgoing appears to be more strongly associated with being conventional than with being disadvantaged. It is a characteristic of people who are involved in their communities, belong to other voluntary associations, and hold traditional values.

Consequences of Religiosity

Because religion teaches and reinforces values, it has consequences for attitudes and behaviors. People who are more religious tend to be healthier, happier, and more satisfied with their lives (Cotton et al. 2006; Waite & Lehrer 2003; Ferriss 2002). These benefits in large part stem from the social support and sense of belonging that individuals receive from their religious communities.

Persons who are more religious tend to have more conservative attitudes on sexuality and personal honesty; they also may have more conservative attitudes about family life, such as supporting the use of corporal punishment to discipline children (Ellison, Bartkowski, & Segal 1996). Not surprisingly, some conservative religious groups have played significant roles in supporting conservative political movements, such as the antiabortion movement and certain right-wing hate groups.

Yet we should not assume that church members necessarily adopt the attitudes of their churches. For example, although the Pope believes abortion and artificial birth control are sinful, more than three-quarters of U.S. Catholics think abortion is acceptable in some circumstances, and more than half believe teenagers should have access to birth control (General Social Survey 2009).

Moreover, even though religious training generally teaches and reinforces conventional behavior, religion and the church can be forces that promote social change. As noted earlier, African American churches and clergy played a significant role in the civil rights movement of the 1950s and 1960s, and evangelical churches are playing a significant role in the current immigrant rights movement. In Latin America,

---

sociology and you

If you belong to an organized religion, you likely gain certain social benefits from it regardless of its belief system. Your congregation likely affords you a social network to whom you can turn for advice or assistance in bad times. Your network may also help you celebrate your successes and generally give you the sense that you are a valued and worthy person. Finally, your religion's rituals can offer meaning and a sense that things happen for a reason. If you do not belong to an organized religion, you may have sought the same sort of support in other social networks, such as fraternities or friendship circles, and have sought meaning in science, politics, or other belief systems.

Religious economy refers to the competition between religious organizations to provide better "consumer products," thereby creating greater "market demand" for their own religions.

liberation theology aims at the creation of democratic Christian socialism that eliminates poverty, inequality, and political oppression (Smith 1991). Conversely, church members don't always adopt their churches' liberal views: In recent years, some Baptists and Episcopalians, among others, have split from their central churches because they disapprove of growing church support for gay rights and other liberal agendas.

U.S. Civil Religion

As this chapter has demonstrated, Americans are in many ways divided by religion. On the other hand, Americans in general share what has been called a civil religion (Bellah 1974, 29; Bellah et al. 1985). **Civil religion** is a set of institutionalized rituals, beliefs, and symbols sacred to U.S. citizens. These include reciting the Pledge of Allegiance and singing the national anthem, as well as folding and displaying the flag in ways that protect it from desecration. In many U.S. homes, the flag or a picture of the president is displayed along with a crucifix or a picture of the Last Supper.

Civil religion has the same functions as religion in general: It is a source of unity and integration, providing a sacred context for understanding the nation's history and current responsibilities (Wald 1987). For example, shortly after the American colonies declared their independence from Britain, George Washington was declared commander of the U.S. army. With little military experience or charisma, Washington's major qualification for the job was that he didn't want it. Within weeks, he became an object of near worship. Why did this cult of Washington develop? It emerged, in part, because Washington symbolized the fledgling nation's unity and, in part, because his disdain for power made him a hero. In worshipping Washington, the colonists were worshipping their nation and the virtues they believed it embodied (Schwartz 1983).

Since then, we have made liberty, justice, and freedom sacred principles. We believe the American way is not merely the usual way of doing things but also the only moral way of doing them, a way of life blessed by God. The motto on our currency, our Pledge of Allegiance, and our national anthem all bear testimony to the belief that the United States operates "under God" with God's direct blessing.

Within weeks of his appointment as commander of the army, Washington became an object of near worship.

Henry Francis du Pont Winterthur Museum

Where This Leaves Us

Structural-functional theory and conflict theory are both right. On the one hand, schools and churches are preservers of tradition. Both institutions socialize young people to understand and accept traditional cultural values and to find their place in society. Occasionally schools and churches teach people to think for themselves, but more often both stress unquestioning acceptance of authority and of contemporary social arrangements, including social inequalities.

On the other hand, schools and churches are in the forefront of social change. Nowhere are the battles over oppression in the least-developed nations, abortion, or homosexuality fought more bitterly than in the councils of our major churches. Nowhere are the battles over race relations, sex and class equity, and clashing cultural values fought more bitterly than on school boards. Even if you are not religious and even after you finish your education, you cannot afford to ignore the vital roles education and religion play in creating or impeding social change.

Civil religion is the set of institutionalized rituals, beliefs, and symbols sacred to the U.S. nation.

Summary

1. The structural-functional model of education suggests that education meets multiple social needs. It socializes young people to the broader culture, provides knowledge and skills, and can promote social change.

2. Conflict theory suggests that education helps to maintain and reproduce the stratification structure through four mechanisms: training a docile labor force that accepts inequality (the hidden curriculum), using credentialism to save the best jobs for the children of the elite, perpetuating the dominant culture, and ensuring that disadvantaged groups receive inferior educational opportunities.

3. Symbolic interactionists explore some of the processes through which education can reproduce inequality. Key elements of this process are self-fulfilling prophesies and differences in children's cultural capital, both of which keep disadvantaged students from improving their lot.

4. Tracking generally helps students in high-ability groups but hurts those in low-ability groups. High-stakes testing encourages schools to pay more attention to the quality of the education they provide but has forced schools to cut programs and to focus on teaching students how to take tests. School choice gives parents and students options but can reinforce inequality and reduce support for public education.

5. About half of U.S. high school graduates between 16 and 24 are enrolled in college. Women are more likely than men to attend and graduate from college, but class and racial differences are much greater than gender differences. Men from poor, minority families are the least likely to attend college.

6. Education pays off handsomely in terms of increased income, better jobs, and lower unemployment. It also offers nonmonetary benefits such as the likelihood of a longer life.

7. The sociological study of religion concerns itself with the consequences of religious affiliation for individuals and with the interrelationships of religion and other social institutions. It is not concerned with evaluating the truth of particular religious beliefs.

8. Despite earlier predictions, secularization has not increased significantly in the United States. Rather, mainstream religious organizations remain strong, and fundamentalist groups are growing in popularity. Religious membership and attendance remain at stable levels and are far higher than in Europe.

9. Durkheim argued that religion is functional because it provides support for the traditional practices of a society and is a force for continuity and stability. Weber argued that religion generates new ideas and thus can change social institutions. In contrast, Marx argued that religion serves as a conservative force to protect the status quo. More recent conflict theorists have explored the role that religion can play in either fostering or repressing social conflict.

10. All religions are confronted with a dilemma: the tendency to reject the secular world and the tendency to compromise with it. Religions that adapt to the broader world and to other religious groups are called churches. Those that emerge in reaction against modern religions are known as sects. New religions that either promote truly new religious ideas or that draw on religions from outside a given culture are known as new religious movements.

11. Some major developments in U.S. religion are the growth in fundamentalism, in emerging churches, in new religious movements, and in those who identify with no religion.

12. U.S. civil religion is an important source of unity for the U.S. people. It is composed of a set of beliefs (that God guides the country), symbols (the flag), and rituals (the Pledge of Allegiance) that many people of the United States of all faiths hold sacred.

Thinking Critically

1. How have you been helped or harmed by tracking? If you have not experienced it, answer this question based on someone you know.

2. How would you reorganize elementary and secondary classrooms to best meet the needs of all students? What would be the manifest functions of your system? the latent functions? the potential dysfunctions?

3. Given what you now know about the process of secularization and the rise of fundamentalism, do you expect fundamentalism to grow or to recede in coming years? Why? Base your argument on your understanding of sociology, *not* on your religious beliefs.

4. What are the attractions of the emerging churches? Compare the structure (*not* beliefs) of your religion

or the religion of someone you know to the structure of emerging churches. If you belong to an emerging church, compare its structure to that of a friend's religion.

5. If the Religious Right were to gain more power, what changes would you expect to occur in U.S. government? U.S. society? Do you think they would be good for the United States? Why or why not?

Book Companion Website

www.cengage.com/sociology/brinkerhoff

Prepare for quizzes and exams with online resources—including tutorial quizzes, a glossary, interactive flash cards, crossword puzzles, essay questions, virtual explorations, and more.

CHAPTER 13

Politics and the Economy

DAVID NOBLE PHOTOGRAPHY/Alamy

Introducing Politics and the Economy

How do people earn their living? Why are wages so much higher for some types of work than for others, and why are wages so much higher in some countries than in others? How do government leaders get elected—or deposed or assassinated? To answer these questions, we need to look at both politics and the economy. Although we can try to answer the questions separately, they are so interwoven that they are often best treated as one topic. This chapter offers a sociological perspective on politics and the economy that should help you interpret both your own experiences and news headlines.

Power and Politics

Lisa wants to watch *American Idol* and John wants to watch football; fundamentalists want prayer in the schools and the American Civil Liberties Union wants it out; state employees want higher salaries and other citizens want lower taxes. Who decides?

Whether the decision maker is Mom or the Supreme Court, those who can enforce their decisions on others have power. As we discussed in Chapter 7, power is the ability to direct others' behavior, even against their wishes. Here we will describe two kinds of power: *coercion* and *authority.* Although both mothers and courts have power, they differ in the basis of their power, the breadth of their jurisdiction, and the means they use to compel obedience.

Coercion

The exercise of power through force or threats is **coercion**. The force or threat may be physical, financial, or social: We may fear we will be hit, sued, ostracized, fined, killed, or rejected by our friends, among other things. Your parents, for example, may have coerced you into obeying their rules by threatening to spank you, and you may have coerced a younger sibling to follow your rules by refusing to play with him or her otherwise.

Authority

Threats are sometimes effective means of making people follow your orders, but they tend to create conflict and animosity. In some situations, however, threats aren't needed. When power is supported by norms and values that legitimate its use, we call it **authority**. If you have authority, your subordinates agree that, in this matter at least, you have the right to make decisions and they have a duty to obey. This does not mean that the decision will always be obeyed or even that each and every subordinate will agree that the distribution of power is legitimate. Rather, it means that society's norms and values legitimate the inequality in power. For example, if a dad tells his teenagers to be home by midnight, the kids may come in later. They may even argue that he has no right to run their lives. But others in the family likely believe that the father does have this right.

Because authority is supported by shared norms and values, it can usually be exercised without conflict. Ultimately, however, authority rests on the ability to back up commands with coercion. Parents may back up their authority over teenagers with

Coercion is the exercise of power through force or the threat of force.

Authority is power supported by norms and values that legitimate its use.

threats to ground them or take the car keys away. Employers can fire or demote workers. Thus, authority rests on a legitimization of coercion (Wrong 1979).

In a classic analysis of power, Weber distinguished three bases on which individuals or groups gain acceptance as legitimate authorities: *tradition, extraordinary personal qualities* (known as *charisma*), and *legal rules*.

Traditional Authority

When power is based on the sanctity of time-honored routines, it is called **traditional authority** (Weber [1910] 1970b, 296). Monarchies and patriarchies are classic examples of this type of authority. For example, a half century ago, the majority of Americans believed that husbands ought to make all the major decisions in the family. In other words, husbands had authority. Today, much of that authority has disappeared.

Charismatic Authority

When an individual gains the right to make decisions because of perceived extraordinary personal characteristics, he or she holds **charismatic authority** (Weber [1910] 1970b, 295). In many cases, an individual holds charismatic authority because his or her followers believe the individual has been chosen by God. But charismatic authority does not have to be linked to religion. Mahatma Gandhi, for example, was neither an elected politician nor a religious leader, yet he led a political revolution in India. More recently and less positively, Osama bin Laden's followers also grant him charismatic authority.

Rational-Legal Authority

When individuals hold power based on rationally established rules, we say they hold **rational-legal authority**. An essential element of rational-legal authority is that it is impersonal. You do not need to like or admire or even agree with the person in authority; you simply follow the rules.

Our government runs on rational-legal authority. When we want to know whether Congress has the right to make certain decisions, we check our rule book: the Constitution. As long as Congress follows the rules, most of us agree that it has the right to make decisions and we have a duty to obey.

Combining Bases of Authority

Analytically, we can make clear distinctions among these three types of authority. In practice, the successful exercise of authority usually combines two or more types. An elected official who adds charisma to his rational-legal authority will increase his power; depending on your politics, Ronald Reagan or Barack Obama could serve as examples. Similarly, a charismatic leader who establishes a rational-legal system to manage her followers will also increase her power; Mary Baker Eddy, who founded the Christian Science religion and turned it into a large, bureaucratic organization, is an example. All types of authority, however, depend on subordinates agreeing that the person in charge has the right to make decisions and that they have a duty to obey.

The Concept Summary on Power and Authority illustrates the differences between power, coercion, and authority, as well as between the different types of authority.

AP Images

Traditional authority, like that enjoyed by King Mohamed VI of Morocco, exists when an individual's right to make decisions for others is widely accepted based on time-honored beliefs.

Traditional authority is the right to make decisions for others that is based on the sanctity of time-honored routines.

Charismatic authority is the right to make decisions that is based on perceived extraordinary personal characteristics.

Rational-legal authority is the right to make decisions that is based on rationally established rules.

concept summary

Power and Authority

Concept	Definition	Example
Power	Ability to get others to act as one wishes despite their resistance; includes coercion and authority	Someone gets you to mow the lawn even though you don't want to.
Coercion	Exercise of power through force or threat of force	"Mow the lawn or I'll spank you."
Authority	Power supported by norms and values	"It's your duty to mow the lawn."
Traditional authority	Authority based on sanctity of time-honored routines	"As your father, I'm ordering you to mow the lawn."
Charismatic authority	Authority based on extraordinary personal characteristics of a leader	You are so moved by President Obama's call for service that you volunteer to mow an elderly neighbor's lawn.
Rational-legal authority	Authority based on submission to a set of rationally established rules	"You know the rules: Your sister mowed the lawn last week so it's your turn now."

Politics

Power inequalities are built into all social institutions. In institutions as varied as the school and the family, roles such as student–teacher and parent–child specify unequal power relationships as normal.

In a very general sense, **politics** refers to all institutions concerned with the social structure of power, including the family, the workplace, the school, and even the church or synagogue. The most prominent political institution, however, is the state.

Power and the State

The **state** is the social structure that holds a monopoly on the legitimate use of coercion and physical force within a territory. It is distinguished from other political institutions by two characteristics: (1) Its jurisdiction for legitimate decision making is broader than that of other institutions, and (2) it controls the use of legalized coercion in a society.

Jurisdiction

Whereas the other political institutions of society have rather narrow jurisdictions (over church members or over family members, for example), the state exercises power over the society as a whole.

Generally, states are responsible for arbitrating relationships among the parts of society, maintaining relationships with other societies, and gathering resources (taxes, draftees, oil) to meet collective goals. As societies have become larger and more

Politics is the social structure of power within a society.

The **state** is the social structure that successfully claims a monopoly on the legitimate use of coercion and physical force within a territory.

decoding the data

Attitudes toward Government Responsibilities

SOURCE: General Social Survey (2009).

Percentage Who Agree That:	Low Income	Middle Income	High Income
The government in Washington should do everything possible to improve the standard of living of all poor Americans.	41%	30%	23%
It is the responsibility of the government in Washington to see to it that people have help in paying for doctors and hospital bills.	59%	53%	46%

Explaining the Data: It's easy to see that those who can afford to pay their own bills are less likely to think the government should help people with their bills. But what other reasons might explain why those with lower incomes are more likely to favor government helping the poor and the sick? How do the life experiences of low-, middle-, and upper-income people differ, and how might this affect their views?

Why would high-income Americans be more likely to believe that the government should help people pay their medical bills than to help improve the standard of living of the poor?

Critiquing the Data: How could you reword the survey statements so that *more* people would agree with them? How could you reword them so that *fewer* people would agree?

---

complex, the state's responsibilities have grown to include things such as providing sex education to children and providing subsidies to families at risk of losing their homes. Decoding the Data: Attitudes toward Government Responsibilities explores Americans' attitudes toward government responsibilities.

State Coercion

The state claims a monopoly on the legitimate use of coercion. To the extent that other institutions use coercion (for example, the family or the school), they do so only with the approval of the state. For example, state laws now forbid husbands from beating their wives and parents from beating their children.

The state uses three primary types of coercion. First, the state can legally arrest, attack, imprison, and even kill citizens in certain circumstances. Second, the state can legally take money from citizens through taxes and fines. Finally, the state legally can negotiate with other countries and can use its armed forces to attack and kill in other countries.

Different states, however, obtain power and use coercion in very different ways. The most basic distinction is between authoritarian systems and democracies.

Authoritarian Systems

Authoritarian systems are political systems in which the leadership is not selected by the people and legally cannot be changed by them.

Most people in most times have lived under **authoritarian systems**. Authoritarian governments go by a lot of other names: totalitarianism, dictatorships, military juntas, despotisms, monarchies, theocracies, and so on. In all cases, however, the leadership

All South Africans regardless of race now have the right to vote. Democracy triumphed when the financial and political power of the white minority was finally counterbalanced by the sheer numbers and political determination of the black majority.

1994 Tom Muscionico/Contact Press Images

was not selected by the people and cannot be changed by them (except through revolution). Even if the state allows elections, those elections will be rigged so that only certain individuals can win. Afghanistan under the Taliban was an authoritarian system, as is Libya under Muammar al-Gaddafi.

Authoritarian structures vary in the extent to which they attempt to control people's lives and the extent to which they use terror and coercion to maintain power. Some authoritarian governments, such as monarchies, govern through traditional authority; others have no legitimate authority and rest their power almost exclusively on coercion.

Democracies

Democracies come in many forms. All, however, share two characteristics: They have regular, legal procedures for changing leaders, and these leadership changes reflect the will of the majority.

In a democracy, two basic groups exist: the group in power and one or more legal opposition groups that are trying to get into power. The rules of the game call for sportsmanship on all sides. The winners can't punish or kill the losers, the losers must accept their loss and wait until the next legal opportunity to try again, and both sides must let the public participate in deciding who wins.

Why are some societies governed by democracies and others by authoritarian systems? The answer appears to have less to do with virtue than with economics. Democracy occurs primarily in the wealthier nations of the world, especially those with large middle classes. Middle-class citizens usually have sufficient social and economic resources to organize effectively and to hold the government accountable. However, democracy also exists in poorer nations with relatively little income inequality, such as Costa Rica and Sri Lanka. But democracy can exist even in the absence of these conditions: The largest democracy in the world, India, has a relatively small middle class and tremendous income inequality.

Democracies are political systems that provide regular, constitutional opportunities for a change in leadership according to the will of the majority.

Democracy also flourishes in societies with many competing groups, each of which comprises less than a majority. In such a situation, no single group can win a majority of voters without negotiating with other groups; because each group is a minority, safeguarding minority political groups protects everybody (Weil 1989). However, if competing interest groups don't share basic values and interests, they likely won't abide by the rules of the game. The repeated failures of peace talks and eruptions of violence between Israelis and Palestinians demonstrate how fundamental differences can make it difficult for democracy to flourish.

Globalization and State Power

As the Israeli and Palestinian governments have fought for land and autonomy, each has been both helped and hindered by organizations outside their borders. The United Nations and the European Union send diplomats and peace-keeping forces, the World Court judges whether either government has broken international laws, the World Bank decides whether to extend low-interest loans to build the economy, and multinational oil companies pressure politicians in the United States, the Middle East, and elsewhere to safeguard the companies' interests. Each of these is an example of globalization—in this case, the globalization of the economy and law.

Because of globalization, some argue, multinational corporations and international organizations now hold much of the power once held by states (Sassen 2006; Appelbaum 2005). For example, corporations have fought successfully against minimum-wage laws in the United States and against price controls on tortillas in Mexico. Similarly, international regulatory organizations and associations such as the European Union and the International Monetary Fund also have imposed new rules on states.

In contrast, others argue that globalization has been going on since the days of the great sailing ships without threatening state power. Indeed, these scholars argue, the power of the state over the economy and citizenry is greater than ever (Wolf 2005). Moreover, with the current global economic crisis, many nations have decided to protect themselves first. Consequently, they have withdrawn their support from

Concern over globalization has led to protests around the world, such as this one in Brazil. "Guerra" means "war" in Portuguese.

REUTERS/Mariana Bazo/Landov

concept summary

Two Models of American Political Power

	Pluralist Model	Power-Elite Model
Basic units of analysis	Interest groups	Power elites
Source of power	Situational: Depends on issue	Inherited and positional; top positions in key economic and social institutions
Distribution of power	Dispersed among competing diverse groups	Concentrated in relatively homogeneous elite
Limits of power	Limited by shifting and cross-cutting loyalties	Limited when other groups unite in opposition
Role of the state	Arena where interest groups compete	One of several sources of power

agreements that fostered globalization, such as treaties requiring states to drop taxes on imported goods (Erlanger 2009).

Who Governs? Models of U.S. Democracy

Almost everyone agrees that the United States is a democracy. Political parties with different economic and social agendas vie for public support, and every 4 years the voters can replace the president if they want to. Many, however, question whether the decisions made by U.S. leaders really reflect the will of the majority. This section outlines the two major sociological models of how these decisions are made: the pluralist model and the power-elite model. The Concept Summary on Two Models of American Political Power summarizes the differences between these models.

Structural-Functional Theory: The Pluralist Model

Like all structural-functionalist models, the pluralist model of political power assumes that the various parts of our political process typically run smoothly and harmoniously, for the good of all. The pluralist model focuses on the processes of checks and balances within the U.S. government and on coalition and competition among governmental and nongovernmental groups. This model argues that the system of checks and balances built into the U.S. Constitution makes it nearly impossible for either the judicial, legislative, or executive branch of government to force its will on the other branches. Similarly, the model argues that different groups with competing vested interests hold power in different sectors of American life. Some groups have economic power, some have political power, and some have cultural power. Because each group has some power, all are reasonably content and no extreme group can force its views on the others.

Research suggests the limits of the pluralist model. Typically, the power elite stick together, while other groups lack the resources to successfully challenge the elite (Burris & Salt 1990; Clawson & Su 1990; Korpi 1989). In the United States, programs designed to share wealth or award opportunities more equitably—such as civil rights laws or the Social Security system—have succeeded only when (1) a crisis caused the elite to favor at least some change, and (2) the elite disagreed among themselves (Jenkins & Brent 1989).

Conflict Theory: The Power-Elite Model

In contrast to the pluralist model, the power-elite model, which is based on conflict theory, contends that a relatively unified elite group makes all major decisions, based on its own interests (Domhoff 2009). In his classic work, *The Power Elite*, C. Wright Mills (1956) defined the **power elite** as the people who occupy the top positions in three bureaucracies: the military, industry, and the executive branch of government. Through a complex set of overlapping cliques, these people share decisions on national and international issues (Mills 1956, 18). Consequently, creating meaningful social change is difficult unless the non-elite organize together in unions, social movements, and the like.

Without question the power elite has become more diverse since Mills's day. The independent power of the military has declined, whereas that of the cultural elite—which includes both movie stars and religious leaders—has grown. Increasing numbers of African Americans, Hispanics, and women hold high corporate positions and elected office, especially at local levels. On the other hand, white males still greatly outnumber women and minorities in positions of power. Moreover, most "outsiders" who become part of the power elite come from at least middle-class homes, attend elite schools, and are willing and able to fit in: light-skinned minorities, Jews who marry Christians, and women who learn to play golf and even to smoke cigars, for example (Zweigenhaft & Domhoff 1998).

Individual Participation in U.S. Government

So far, we've focused on the role of leaders, elites, and other organized interests. But by definition democracy requires the participation of individual citizens as well. This section describes how and why citizens do—or do not—participate as voters in U.S. politics.

Who Votes?

Although the United States is a democracy, about one-third of its voting-age population does not even register to vote, and almost half (44 percent in 2008) do not vote even in presidential elections (U.S. Bureau of the Census 2009a). An astonishing 75 to 80 percent do not vote in typical local elections.

This low level of political participation poses a crucial question about power in U.S. democracy. Who participates? If they are not a random sample of citizens, then some groups must have more influence than others.

Social Class

One of the firmest findings in social science is that political participation (indeed, social participation of any sort) is strongly related to social class. Whether we define

The **power elite** comprises the people who occupy the top positions in three bureaucracies—the military, industry, and the executive branch of government—and who are thought to act together to run the United States in their own interests.

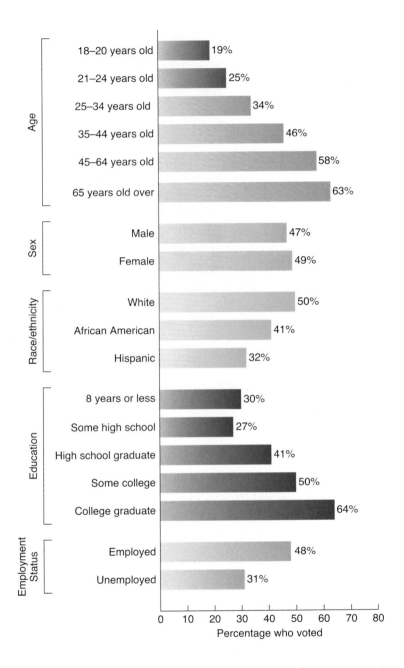

FIGURE 13.1 **Percentage Who Voted in 2006 (among Voting-Age Population)**
Older, better-educated, employed, and non-Hispanic Americans are more likely to vote.
SOURCE: U.S. Bureau of the Census (2008c).

participation as voting or letter writing, people with more education, more income, and more prestigious jobs are more likely to be politically active. They know more about the issues, have stronger opinions, more often believe they can influence political decisions, and thus more often try to do so. Data on voting support and illustrate this conclusion. As Figure 13.1 shows, those who have graduated from college are more than twice as likely to vote as those who have not completed high school.

Age

Age also affects political participation: Older persons are considerably more likely than younger persons to vote (Figure 13.1). Even in the turbulent years of the Vietnam War, when young antiwar demonstrators were so visible, young adults were significantly less

Although all U.S. citizens over the age of 18 have the right to vote, white, middle-aged, better-off, and better-educated citizens are most likely to do so.

likely to vote than were middle-aged individuals. In that period, many young adults engaged in other forms of political participation that did, in fact, influence political decisions. In most time periods, however, the low participation of younger people at the polls is a fair measure of their overall participation.

Race and Ethnicity

Race and ethnicity also affect the likelihood of voting. Whites are more likely than African Americans to vote, and Hispanics are less likely to vote than either whites or African Americans. The low rates of voting among Hispanics reflect both their lower average socioeconomic status and the fact that many lack U.S. citizenship and therefore can't vote.

Which Party?

Unlike the United States, most European nations have parliamentary governments. In these nations, parties are awarded seats in Parliament based on the percentage of the votes they won: If 10 percent of citizens voted for the Green Party, for example, the Green Party would get 10 percent of seats in Parliament. As a result, many different parties can have members in Parliament.

In contrast, seats in the U.S. Congress (and other U.S. political offices) are won through a "winner take all" process: In each election, whoever receives the most votes wins. As a result, only candidates from the two largest parties—the Democratic Party and the Republican Party—have much chance of winning elections. Consequently, citizens rarely bother to support candidates from smaller parties such as the Green Party.

Although both major political parties in the United States are basically centrist, there are philosophical distinctions between them. For the last century, the Democratic Party has been more associated with liberal morality; support for social services; and the interests of the poor, the working class, and minorities. The Republican Party has been more associated with conservative morality, tax cuts, and the interests of industry

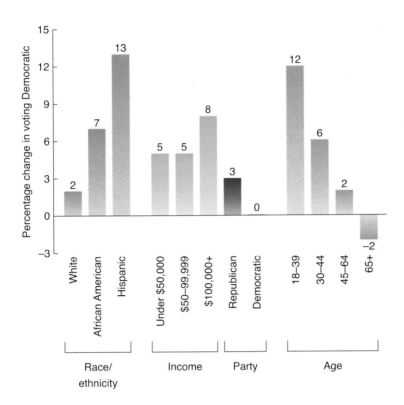

FIGURE 13.2 **Percentage Increase or Decrease in Voting Democratic, 2004–2008 Presidential Elections** Across the board, Americans were more likely to vote for the 2008 Democratic candidate (Barack Obama) than for the 2004 Democratic candidate (John Kerry). SOURCE: Pew Research Center (2008).

and the affluent. As a result, voters who are female, younger, minority, or less educated tend to favor the Democratic Party. However, in 2008 Democrats gained votes across all segments of the population (Figure 13.2). It remains to be seen, however, whether this shift will continue or whether it merely reflected the unusual circumstances of the 2008 election (a highly unpopular Republican president in office and a highly unusual African American candidate running against him).

A growing proportion of voters align themselves with neither party but vote based more on issues than on party loyalty. When the 10 percent (or more) of voters who call themselves independent go to the polls, however, they usually have to choose between a Democratic and a Republican candidate.

Why So Few Voters?

The United States prides itself on its democratic traditions. Yet U.S. citizens are only half as likely to vote as are citizens of other Western nations. Moreover, although studies consistently find that those with more education and higher income are more likely to vote, voting rates in the United States have declined steadily for the last century, even though both income and educational levels have increased. Why are voting rates in the United States so low?

Some scholars argue that political participation has declined because more and more Americans believe that the political process is corrupt, that the Democrats and Republicans are more similar than different, and that it makes little difference who gets elected (Southwell & Everest 1998). Others argue that voting rates are so low because politicians have made it so difficult for people to vote (Piven & Cloward 1988, 2000). Until only a few years ago, both registering to vote and voting were more cumbersome in the United States than in any other Western democracy. In many

KYNDELL HARKNESS/MCT/Landov

Voting rates have increased among African Americans and other minorities when social and political movements (such as the civil rights movement and the Obama campaign) have reached out to them and convinced them that they can make a difference.

states, individuals had to register annually, pass literacy tests, or pay special taxes. They also had to both register and vote in specific locations during specific limited hours, which was especially difficult for persons who held strictly scheduled, working-class jobs.

Voter registration has increased significantly since passage of the National Voter Registration Act of 1993. However, barriers to voting remain. In the 2008 presidential election, for example, potential voters (especially in poor, minority, and Democratic districts) were hampered by broken voting machines, polling places that closed too early for some working people to vote, and new legal requirements that removed people from voting rolls (People for the American Way 2008).

Still others argue that relatively few Americans vote because no major political party has sought to involve poor, minority, and disenchanted Americans or to address their concerns. In contrast, voting rates have increased when social and political movements (such as the civil rights movement and the Obama campaign) have reached out to such Americans and convinced them that they can make a difference (Winders 1999).

Case Study: Ex-Felon Disenfranchisement

As we've seen, a surprising number of Americans choose not to participate in the democratic process. An even more surprising number of Americans *cannot* legally vote. An estimated 5.4 *million* Americans are barred from voting—disenfranchised—because they were once convicted of a felony (Manza & Uggen 2006). In some states, only those still in prison are forbidden from voting; in other states, a felony conviction brings lifelong **ex-felon disenfranchisement**. Because the United States has both a high rate of felony convictions (primarily for drug-related crimes) and unusually restrictive laws on the voting rights of ex-felons, the United States has a higher rate of ex-felon disenfranchisement than almost any other country (Hull 2005; Manza & Uggen 2006). In essence, the very possibility of rehabilitation is ignored: Someone convicted at age 20 of selling marijuana, for example, might be ineligible to vote for the rest of his or her life, even if he or she never again commits a crime and becomes a successful worker, parent, and community citizen.

Importantly, because poverty sometimes pushes individuals to commit crimes, and because the criminal justice system more often convicts poor criminals than equally guilty wealthy criminals, those subject to ex-felon disenfranchisement overwhelmingly are poor. The number of disenfranchised poor people is high enough to significantly decrease the chances of electing politicians who favor helping the poor (Uggen & Manza 2002; Hull 2005; Manza & Uggen 2006).

Ex-felon disenfranchisement is the loss of voting privileges suffered by those who have been convicted of a felony. In some states, ex-felon disenfranchisement applies only to those in prison; in other states, it is lifelong.

Modern Economic Systems

As we've seen, from the role of the working class to the role of the power elite, understanding politics requires understanding underlying economic issues. In this

Joel Stettenheim/Corbis

Under capitalism, those who experience financial hardship for whatever reason typically must rely on themselves or on charity.

section we look at modern economic systems in general, before turning to the U.S. economy.

The **economy** consists of all social structures concerned with the production and distribution of goods and services. Production includes issues such as how much or how little to build, whether to invest in manufacturing more weapons or growing more food, and whether to encourage large factories or smaller enterprises. Distribution includes issues such as how money is divided between workers and owners, who should support those who can't work, and whether individuals should receive income based on need, effort, or ability. The distribution aspect of the economy intimately touches the family, stratification systems, education, and government.

In the modern world, there are basically two types of economic systems: *capitalism* and *socialism*. Because economic systems must adapt to different political and natural environments, however, we find few instances of pure capitalism or pure socialism. Most modern economic systems represent some variation on the two and often combine elements of both.

Capitalism

Capitalism is the economic system in which most wealth (land, capital, and labor) is private property, bought and sold on the open market and used by its owners for their own gain. Capitalism is based on market competition. Each of us seeks to maximize our own profits by working harder or devising more efficient ways to produce goods. Such a system encourages hard work, technical innovation, and a sharp eye for trends in consumer demand. Because self-interest is a powerful spur, such economies can be very productive.

Even when it is very productive, though, a capitalist economy has drawbacks. These drawbacks all center around problems in the *distribution* of resources. First, the capitalist system represents a competitive bargain between labor (workers) and capital (owners of industries), both of whom control a necessary resource. But this is not a

The **economy** is everything involved in the production and distribution of goods and services.

Capitalism is the economic system, based on competition, in which most wealth (land, capital, and labor) is private property, to be used by its owners to maximize their own gain.

bargain between equals: Almost always, capital has more bargaining power than does labor. As a result, workers earn only a fraction of what capitalists earn. Second, those who have neither labor nor capital to bargain with (children, stay-at-home moms, the elderly, the disabled, and workers whose jobs have disappeared) always lose out. They rely on aid from others—if they are lucky. Third, because public services such as paved streets, parks, sanitary water systems, or national armies offer no profit, pure capitalism has no interest in providing them. Yet society cannot function without these services. Thus capitalist systems must have some means of distribution other than the market.

Socialism

If capitalism is an economic system that maximizes production at the expense of distribution, socialism is a system that stresses distribution at the expense of production. In its ideal form, **socialism** is an economic structure in which the workers own the means of production and use them for the collective good.

In theory, socialism has several major advantages over capitalism. First, societal resources can be used for the benefit of society as a whole rather than for individuals. For example, theoretically a socialist system could protect the environment for everyone's sake, rather than allowing corporations to pollute it for private profit. Similarly, a socialist economy could divert resources from profitable industries such as television production to industries more likely to benefit everyone in the long run, such as education, agriculture, or steel. The major advantage claimed for socialism, however, is that it produces equitable (although not necessarily equal) distribution.

The creed of pure socialism is "From each according to his or her ability, to each according to his or her need." Under socialism, everyone should receive what they need to survive, and everyone should work their best to achieve that common goal. Workers are expected to be motivated by loyalty to their community and their comrades. In reality, the hard-working woman with no children is not likely to work her hardest when the lazy worker next to her takes home a larger paycheck simply because she has more children and thus greater need. Nor is the farmer as likely to make the extra effort to save the harvest from rain or drought if his rewards are unrelated to either effort or productivity. Because of this factor, production is usually lower in socialist economies than in capitalist economies.

Mixed Economies

Most Western societies today represent a mixture of both capitalist and socialist economic structures. In many nations, services such as the mail and the railroads and key industries such as steel and energy are socialized. This socialism rarely results from pure idealism. Rather, public ownership is often seen as the best way to ensure continuation of vital but unprofitable services. Other services—for example, health care and education—have been partially socialized because societies have judged it unethical to deny these services to the poor and too inefficient to provide them on the open market.

In several nations, socializing services has gone far toward meeting the maxim "from each according to his or her ability, to each according to his or her need." There are still inequalities in education and health care, but far fewer than there would be if these services were available on a strictly cash basis. The United States has done the least among major Western powers toward creating a mixed economy, and our future

Socialism is an economic structure in which productive tools (land, labor, and capital) are owned and managed by the workers and used for the collective good.

direction is unclear. By and large, the Republican Party has pushed to reduce government provision of social services and the Democratic Party has pushed to increase such services. The future mix of socialist and capitalist principles will reflect political rather than strictly economic conditions.

The Political Economy

Political economy refers to the interaction of political and economic forms within a nation. Both capitalism and socialism can coexist with either authoritarian or democratic political systems. Many Western European nations, such as the United Kingdom and Sweden, combine socialism and democracy. (The Swedish system is discussed more fully in Focus on a Global Perspective: Democratic Socialism in Sweden on the next page.) Other nations, such as China and Cuba, combine socialism with an authoritarian political system. We often use (and misuse) the term *communist* to refer to societies in which a socialist economy is guided by a political elite and enforced by a military elite. The goals of socialism (equality and efficiency) are still there, but the political form is authoritarian rather than democratic.

Likewise, some capitalist nations are democratic and some are authoritarian. Both the United States and Japan have capitalist economies and democratic political systems. Singapore and Saudi Arabia, on the other hand, have capitalist economies but autocratic political systems, in which elections are either nonexistent or virtually meaningless. These examples remind us that both capitalism and socialism can coexist with authoritarian and democratic regimes.

Privatization and the U.S. Political Economy

As we've seen, the United States is a democracy based in capitalism. This capitalistic basis of our system is reflected in the recent trend toward the privatization of government services. As described in the previous chapter, privatization is the process of "farming out" government services to corporations, redesigning those services to fit a corporate mold, or redefining them as private choices rather than government responsibilities (Hacker 2006).

Privatization has affected many types of government services (Jurik 2004). Some cities and states contract out water testing and delivery to private bidders. Others deliver public water very cheaply to private bottlers, who earn extraordinarily high profits by filtering it and selling it as a luxury good. Yet public water supplies are both more heavily regulated and safer than are Perrier, Calistoga Springs, or other private waters (Public Citizen 2006). Similarly, health care in U.S. prisons is now primarily offered by doctors working under contract for private firms. Some states have gone even farther and have hired private companies to run their prisons, welfare systems, and other government services (Hallett 2002). Meanwhile, the federal government now encourages citizens to create their own pension savings accounts and health savings accounts, rather than relying on Social Security or government-funded health insurance programs.

Major public universities illustrate the second form of privatization: redesigning public services to mimic corporate processes. These universities are still owned and run by state and city governments but, like corporations, they increasingly focus on the bottom line (Washburn 2006). They now hire and fire professors less on the quality of their teaching, or even the quality of their research, and more on whether their research will bring grant dollars or remunerative patents to their university.

Political economy refers to the interaction of political and economic forms within a nation.

focus on A GLOBAL PERSPECTIVE

Democratic Socialism in Sweden

What would it be like to live and work in Sweden? You would have guaranteed access to quality public transportation; guaranteed income if you were ill, disabled, or elderly; guaranteed access to comfortable housing; and free education all the way through college, graduate school, or professional training. If you or your partner gave birth or adopted a baby, you'd receive a full year of paid parental leave. When you returned to work, you could use a free, high-quality, state-funded daycare center. In exchange for these benefits, you would pay about 25 percent of your paycheck in federal income taxes and almost as much in local taxes.

Sweden is a democratic socialist society with an economy that mixes corporate capitalism with substantial welfare benefits for everyone. Because Sweden is a democracy, the majority of Swedes have voted to receive these benefits and to pay high taxes for them. But Sweden's economy wasn't always arranged this way.

Sweden owes its political economy in part to the rise of a strong labor movement (Koblik 1975). As industrialization began in Sweden in the 1870s, labor union members worked to create the Social Democratic Party, a political party dedicated to equitable wages, job security, and welfare programs for the entire society. While Communists in Russia were fighting and winning the Russian Revolution in 1914–1917, members of Sweden's Social Democratic Party were fighting for seats in Parliament. After

All children in Sweden have access to free, high-quality day care.

holding power on and off during the 1920s, the Social Democratic Party won an important election in 1932. It has dominated Parliament in most elections since then, giving its social welfare politics time to develop deep roots.

The welfare state's emergence and success also reflect the deeply held Swedish belief that the community should look out for all its members. This attitude, in turn, has been fostered by the cultural homogeneity of the Swedish population. Until about 1980, the population of Sweden was overwhelmingly ethnically Swedish. Currently, however, foreign immigrants and their children comprise close to 20 percent of the Swedish population. Many now question whether support for the welfare state will decline either if ethnic Swedes become unwilling to extend their social

welfare system to immigrants or if immigrants reject the philosophies underlying the social welfare system.

Other forces also are putting pressure on Sweden's social welfare system. The globalization of the economy has made it more difficult for the Swedish government to keep transnational corporations based in Sweden from exporting jobs (Olsen 1996). In addition, economists point out that Sweden's system is based on an inherent irony: Strong and profitable capitalist businesses are necessary so that workers can be employed and taxes for welfare benefits can be collected. But insisting on generous worker benefits and full employment eats into capitalist profits (Olsen 1996).

Supporters of privatization argue that it brings greater efficiency to water supplies, prisons, universities, and other government services by motivating individuals to work hard and keep a sharp eye on cost-benefit ratios. Opponents argue that professors, scientists who test our water supply, guards who staff our prisons, and the like should make decisions based on what is best for our society, rather than on what will generate the greatest profit.

Primary production involves direct contact with natural resources—fishing, hunting, farming, forestry, and mining.

© Mira/Alamy

The U.S. Economic System

Why are lawyers paid more than schoolteachers? Why are so many small grocery stores in New York and Los Angeles run by Korean immigrants? Why do so few farm kids stay on the farm? And how much can you expect to earn after you graduate? To answer these questions, we first need to understand the economic "big picture." To do so, we need to address three topics: the *postindustrial economy,* the *corporate economy,* and the *"Wal-Mart Economy."*

The Postindustrial Economy

In a *preindustrial* economy, the vast majority of the labor force works in the primary sector. The **primary sector** is that part of the economy involved in extracting raw materials from the environment. The primary sector includes farming, herding, fishing, hunting, and mining. Such activities characterized Europe until 500 years ago and are still common in the least-developed societies.

The Industrial Revolution brought a shift from the primary sector to the secondary sector. The **secondary sector** is that part of the economy involved in processing raw materials. For example, the steel, textile, and lumber industries process raw materials into ore, cotton, and wood, respectively. Other industries in the secondary sector then turn these materials into automobiles, clothing, and furniture.

Postindustrial economies rest on the **tertiary sector** of the economy, the sector involved in the production of *services*. The tertiary sector includes a wide variety of occupations: physicians, schoolteachers, hotel maids, short-order cooks, and police officers. It includes everyone who works for hospitals, governments, airlines, banks, hotels, schools, or grocery stores. Rather than producing tangible goods, these organizations provide services to others. They count their production not in barrels or tons but in numbers of customers.

The tertiary sector has grown very rapidly in the last half century and is projected to grow still more. As Figure 13.3 illustrates, the tertiary sector grew from only

sociology and you

During the course of the day, you interact in some way with each sector of the economy. The *primary* sector provides any food that you eat "as is," such as fruits and vegetables. The *secondary* sector processes raw foods into other foods (turning grain into flour and then into bread or cupcakes, for example). The *tertiary* sector ships those foods to a grocery store or restaurant and provides the salesclerks or waiters who sell the food to you.

The **primary sector** extracts raw materials from the environment.

The **secondary sector** processes raw materials for sale.

The **tertiary sector** produces services for sale.

FIGURE 13.3 Changing Labor Force in the United States

Over the last 200 years, there has been a drastic shift in the U.S. labor force from the primary sector (agriculture, fishing, and so on), to manufacturing and commerce, and then to service provision.

SOURCE: U.S. Bureau of Labor Statistics (2002); Figueroa &Woods (2007).

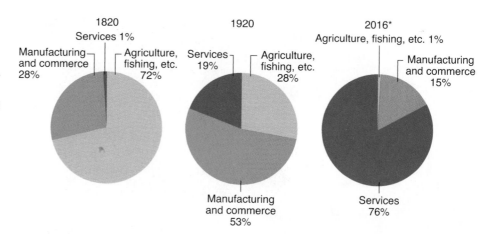

1 percent of the labor force in 1820 to 19 percent in 1920. By 2016 it is expected to include 76 percent of the labor force. These days, almost no one is employed in the primary sector, and jobs in the secondary sector have fallen dramatically. This does not mean that the primary and secondary sectors no longer matter, however. The nation's service sector has been able to grow so large because the other sectors are now so efficiently productive and because we can draw on the primary and secondary sectors in other nations. This reflects the globalization of the economy that we first explored in Chapter 7.

The Corporate Economy

More than 250,000 businesses operate in the United States, ranging from hot-dog pushcart vendors to Microsoft. However, most of the nation's capital and labor are tied up in a few giant, transnational corporations. The top 20 U.S. companies are huge bureaucracies that control billions of dollars of assets and employ thousands of individuals. These giants loom large on both the national and international scene.

At the local level, you may know of one major employer who holds city and county government hostage and bargains for tax advantages and favorable zoning regulations in exchange for increasing or retaining jobs. Because of the growing size and interdependence of corporations, this scene is now reenacted at the federal and even international level.

Wealthy capitalists link to each other through shared ownership of large firms; large firms link to one another through the members on their boards of directors, the businesses they invest in, and the businesses that invest in them. As a result of this interdependence, relations among large firms have become more cooperative than competitive. Although decreased competition reduces productivity and efficiency, it increases joint political influence (Mizruchi 1989, 1990). For example, as the proportion of the nation's assets held by the top 100 firms increased, their political power increased and the taxes they were required to pay decreased—even while individual income taxes rose (Jacobs 1988).

That political power can extend to influencing U.S. foreign policy. A desire to protect the interests of transnational companies like Dole and United Fruit certainly played a role in the U.S. decision to support dictatorships in Guatemala, Honduras, and other Latin American countries during the twentieth century. Similarly, to protect transnational oil companies, the United States covertly orchestrated the 1951 coup against elected Iranian Prime Minister Mohammad Mossadegh and subsequently propped

Although individual consumers benefit from Wal-Mart's low prices, the low pay and limited benefits it offers employees, coupled with its ability to drive competitors out of business, hurts communities in many ways. These factors have led to protests around the country.

up the authoritarian regime of the Shah of Iran (Kinzer 2003). (Popular resentment of the Shah's repressive regime eventually led to the Islamic Iranian revolution in 1979, which stimulated Islamic fundamentalism worldwide.) More recently, some observers argued that the United States invaded and occupied Iraq more to protect U.S. oil interests than to fight terrorism.

The "Wal-Mart Economy"

So far we have talked about entire segments of the economy at a time—large corporations, informal businesses, and so on. The implicit message is that no one corporation or organization is that important on its own. In fact, however, one corporation—Wal-Mart—is now so large and so powerful that it affects the entire U.S. economy.

Until the 1980s, federal laws prohibited any corporation from becoming a monopoly. A monopoly is a corporation that holds so large a market share for a given good or service that it could drive any competitors out of business and then set any prices it wanted for its goods and services. These laws were substantially weakened by elected officials, beginning with Ronald Reagan, who were opposed to "big government" of all sorts. The Wal-Mart economy is the result.

Wal-Mart earns its profit not by setting prices high, but by setting prices low and selling in vast quantities. Without question Wal-Mart's low prices benefit individual consumers. But everyone pays for these cheap goods in other ways. Because Wal-Mart holds such a large share of any given market (for toys, for tires, for clothing), any manufacturer that doesn't sell its products through Wal-Mart risks being driven out of business. Meanwhile, manufacturers that do work with Wal-Mart *also* can be driven out of business when Wal-Mart requires them to price their goods so low that the manufacturers no longer earn a profit. To avoid this fate, manufacturers have either cut wages to the bone or have moved jobs overseas in search of cheaper labor.

Many stores that used to compete with Wal-Mart also have been driven out of business by the company's predatory pricing; in towns across the United States, the arrival of Wal-Mart has quickly led entire downtowns to virtually shut down and has led to significant drops in wages at stores that continue to compete with Wal-Mart.

MAP 13.1: **Number of Foreclosed Homes per 10,000 Homes on Market**
Foreclosures have recently skyrocketed across the United States. The hardest hit states are those that experienced explosive growth in housing and housing costs over the last decade. In April 2009, 1 home was foreclosed in Nevada for every 67 homes on the market. In contrast, 1 home was foreclosed in South Dakota for every *21,000* homes on the market.
SOURCE: Calculated from data at **realtytrac.com**. Accessed May 2009.

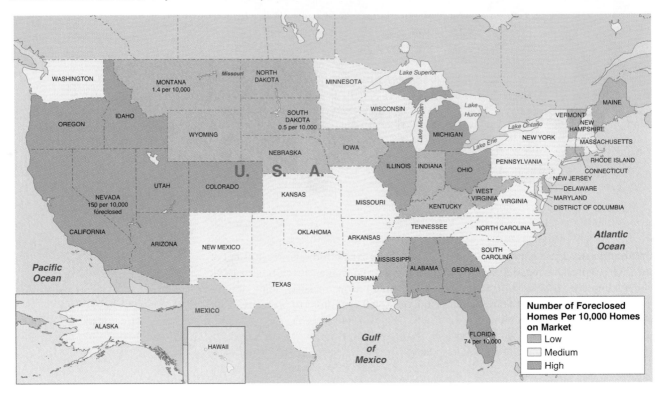

When small businesses go under, not only do individuals lose jobs, but towns lose a stable middle class with a vested interest in civic affairs. The resulting loss in jobs leaves many workers with no option other than to seek employment at Wal-Mart, where average salaries are below the poverty level (United Food and Commercial Workers 2006). In sum, in the new U.S. economy, Wal-Mart not only sets its own prices and employees' wages, but also effectively sets the prices for goods from its suppliers and for wages at both its suppliers and its competitors (Fishman 2006; Lynn 2006).

The Economy in Crisis

The U.S. economy is now a system in crisis. As of April 2009, one-quarter of U.S. residents can no longer pay their bills and more than half fear someone in their household will lose his or her job (New York Times 2009). Meanwhile, more than a million homes are in foreclosure. (Map 13.1 shows the distribution of foreclosed homes around the nation.) Finally, unemployment has soared and the stock market has plummeted, taking many Americans' pensions and savings accounts along with it. What has caused this crisis?

According to most observers, the crisis was caused primarily by soaring levels of debt, made possible by cutbacks in government regulation. These factors in turn led housing and stock prices to rise sharply and then to plummet, taking much of the economy with them (Bernanke 2009; Posner 2009; Phillips 2008).

Over the last two decades, the U.S. government increasingly relaxed financial regulations to make it easier for individuals to get credit—whether or not anyone believed those individuals could pay their debts. If individuals eventually fell behind on payments, their banks and credit card companies charged high penalty fees and raised the interest rates they charged—both actions previously forbidden by government regulations. As a result, banks and credit card companies grew richer while individuals grew poorer.

Over time, however, as more and more Americans fell behind on their mortgages, the banks that had loaned money for those mortgages found it increasingly hard to pay their own bills. This triggered a crisis in public confidence in the entire financial system, leading stock investors to begin selling off their stocks. With many people selling and few buying, stock prices fell. Similarly, more and more people put their homes on the market, leading housing values and sales to drop as well. Because so much of the U.S. economy is linked to home construction, home sales, home furnishings, and the like, the drop in housing rippled throughout the economy.

Meanwhile, the same trend toward less regulation and more financial risk-taking put other parts of the economy at risk. In addition, banks became increasingly concerned about their own financial losses and increasingly afraid of loaning money to anyone. Because most businesses rely on constant loans to buy the goods they sell or the raw materials they need to produce those goods, when banks stopped giving out loans, many businesses failed, taking jobs with them.

Because of globalization, the economic crisis has spread around the world, with devastating results. For example, when Americans stop replacing their old computers with new ones, people in India lose jobs answering Americans' phoned-in computer questions, people in Thailand can no longer earn a living scavenging scrap metal from used computers that Americans throw out, and children in Mexico no longer receive money from immigrant parents who used to work in U.S. computer factories or stores.

The current economic crisis strongly suggests that the free-market, capitalist system only works when balanced by government regulation (Posner 2009).

Work in the United States

From the individual's point of view, the economy often boils down to jobs. For some, jobs are just jobs; for others, they are careers. Either way, work plays a central role in most people's lives. This section looks at the different types of work, the experience of work, the nature of unemployment and underemployment, and the future of work in the United States.

Occupations

Your occupation affects your life in many different ways. Here we look at the important differences between professional occupations, nonprofessional occupations, and occupations in the "underground economy."

Professional Occupations

The most prestigious occupations are the **professions**. Sociologists generally define an occupation as a profession when it meets three characteristics: autonomy, highly

Professions are occupations that demand specialized skills and creative freedom.

specialized training, and public trust. First, a profession must have the autonomy to set its own educational and licensing standards and to police its members for incompetence or malfeasance. For example, doctors, rather than consumers, make up the licensing boards that judge doctors accused of incompetence. Second, a profession must have its own technical, specialized knowledge, learned through extended, systematic training. For example, both lawyers and car mechanics have specialized knowledge, but lawyers must study for years before entering the field, whereas mechanics need study only for months. As a result, sociologists consider lawyers to be professionals, but not car mechanics. Third, a profession must be believed by the public to follow a code of ethics and to work more from a sense of service than from a desire for profit. So, for example, even though the public realizes that some individual ministers, doctors, and lawyers place personal profits above public service, it believes that most members of these professions do not.

The rewards that professionals receive vary considerably: Physicians certainly earn higher income and prestige than do schoolteachers. All professionals, however, enjoy greater-than-average freedom from supervision. Because their work is nonroutine and requires personal judgments, professionals can demand the right to do their work more or less their own way.

That said, professionals' freedom from supervision has declined over time. Increasingly, professionals work for others in bureaucratic structures that reduce their autonomy. Teachers must now spend considerable class time prepping students for standardized exams, and doctors must limit their prescriptions to drugs approved by health insurance companies.

Nonprofessional Occupations

Most U.S. workers hold nonprofessional occupations. Nonprofessional occupations do not require long years of education, do not have the autonomy to set their own educational and licensing standards, and do not have the public's confidence that they are motivated primarily by a code of ethics and a sense of service. To label these jobs *nonprofessional*, however, does *not* imply that these workers do not try their best to do high-quality work. As in any occupation, some individuals are skilled, reliable, and caring and some just skate by.

Nonprofessional jobs vary enormously, from store managers, small business owners, and auto mechanics to janitors, typists, and call-center operators. Some work with their hands, some with their minds; some work on their own, some under heavy supervision. Some (such as electricians) can earn more than some professionals (such as public defense lawyers). It is thus difficult to draw generalizations about these jobs.

Nevertheless, nonprofessional occupations typically offer lower incomes, lower status, lower security, closer supervision, and more routine than do professional occupations. In addition, and reflecting the changes in the U.S. economy shown in Figure 13.3, nonprofessional jobs increasingly are located in the service (tertiary) sector: Compared to past years, far fewer U.S. residents now work in factories and far more fry hamburgers, collect bad debts for credit card companies, or work as nurses' aides.

"Underground" Work

An important type of work that often goes unnoticed is employment in the **underground economy**. This is the part of the economy associated with workers who are trying to hide from state regulation. It includes illegal activities such as prostitution; selling fake Gucci bags; and smuggling immigrants, drugs, or cigarettes. It also includes a large variety of activities that would be legal if the workers or employers

The **underground economy** is associated with workers who are trying to hide from state regulation.

The fastest-growing jobs in the United States today are minimum-wage service jobs that offer few benefits and fewer prospects for advancement.

© Seth Resnick

met government standards. For example, native-born citizens may work as contractors without proper licenses, undocumented immigrants may work on construction jobs without necessary visas, and legal immigrants may work in sweatshops that don't meet government health and safety standards.

Underground work can occur in professional as well as nonprofessional occupations. A doctor from Russia might sell his services to fellow immigrants in the United States even though he lacks a license to practice here, or a graduate student from Mexico might work as a computer programmer even though her visa forbids her from earning an income here.

Often referred to disparagingly as "fly-by-night" businesses, underground enterprises are nevertheless an important source of employment. This is especially true both for poor communities that lack the services and jobs available in other communities and for individuals who want to avoid government notice: undocumented immigrants, disabled people who don't want their earnings to reduce their disability benefits, adolescents too young to meet work requirements, and many others.

The Experience of Work

For most of us, work is a necessary means of earning a living. In addition, as noted in Chapter 7, work also gives us our position in the stratification structure and affects our health, happiness, and lifestyle.

Work also structures our lives. It determines when we wake up, what we do all day, who we do it with, and how much time we have left for leisure. If we ourselves do not hold a job, our parents' or spouses' jobs may structure our lives: There's a big difference—one that goes beyond mere income—between being a preacher's kid or an army brat, a doctor's wife or a janitor's wife. Thus, the nature of our work and our attitude toward it can have a tremendous impact on whether we view our lives as fulfilling or painful. If we are good at our work, if it gives us a chance to demonstrate competence, and if it is meaningful and socially valued, then it can significantly increase our life satisfaction.

Work Satisfaction

U.S. surveys consistently find that most workers (80 percent) report satisfaction with their work. This statistic may represent acceptance of one's lot more than real enthusiasm, but it's nonetheless remarkable.

Generally, professionals report the most job satisfaction. Professionals have considerable freedom to plan their own work, to express their talents and creativity, and to work with others. They also enjoy both public respect and good incomes. The least satisfied workers are those who work on factory production lines. Although some earn good incomes, their work offers little emotional satisfaction, they have little control over the pace or content of their work, and they have few opportunities to interact with co-workers. Skilled and semiskilled workers fall between these two extremes. Nevertheless, even those who hold highly routine, physically demanding jobs such as cashiers and cooks at fast-food restaurants often enjoy the satisfactions of doing a job well, earning a steady paycheck, and socializing with fellow workers (Newman 1999b).

Alienation

Another dimension of the quality of work life is alienation. **Alienation** occurs when workers have no control over their labor. Workers feel alienated when they do work that they think is immoral (build bombs) or meaningless (push papers or brooms, or put together small pieces without understanding how those small pieces will form a larger whole). Work is also alienating when it takes physical and emotional energy without providing emotional satisfaction in return. Alienated workers feel *used*.

The concept of alienation was first developed by Karl Marx to describe the factory system of the mid-nineteenth century. In 1863, a mother gave the following testimony to a committee investigating child labor:

> When he was seven years old I used to carry him [to work] on my back to and fro through the snow, and he used to work 16 hours a day.... I have often knelt down to feed him, as he stood by the machine, for he could not leave it or stop. (as quoted in Hochschild 1985, 3)

This child was used as a tool, just like a hammer or a shovel, to create a product for someone else.

Although few Americans work on assembly lines anymore, modern work nonetheless can be alienating. Service work, in fact, has its own forms of alienation, known as **emotional labor**. In occupations from nursing to teaching to working as flight attendants, both our bodies and our emotions become tools. To satisfy customers, we must smile and act cheerful even when customers are mean, rude, or abusive.

Performing emotional labor can be very stressful. After smiling for 8 hours a day for pay, we may feel that our smiles have no meaning at home. We may lose touch with our emotions and feel alienated from ourselves, especially if we have no control over our job conditions (Hochschild 1985; Bulan, Erickson, & Wharton 1997).

Unemployment and Underemployment

But even the worst job is typically better than no job at all. Between December 2007 and May 2009, the United States lost *6 million* jobs. Currently, the government estimates that 10 percent of Americans are unemployed (Haugen 2009).

Alienation occurs when workers have no control over the work process or the product of their labor.

Emotional labor refers to the work of smiling, appearing happy, or in other ways suggesting that one enjoys providing a service.

Many waitresses are required to engage in emotional labor—smiling, laughing, even allowing patrons to hug or pinch them—as part of their job. These Hooters waitresses are expected to help customers celebrate their birthdays even if the customers are drunk or rude.

Enigma/Alamy

According to the federal government, an individual is **unemployed** if she lacks a job, is available for work, *and* has actively sought work during the last 4 weeks (gone on a job interview, sent out a resume, or the like). This is the definition typically used whenever politicians, researchers, or newscasters talk about unemployment. This definition, however, leaves out anyone who has not looked for work in more than 4 months because it seems hopeless.

The official definition of unemployment leaves out those who are *underemployed*. People who used to work full time but now can find only part-time work and people who used to work as managers or carpenters but now can find work only as sales-clerks are considered **underemployed**. Because the official unemployment rate leaves out both those who have become discouraged about job seeking and those who are underemployed, many argue that it underestimates unemployment levels. When we combine unemployed workers, discouraged workers, and underemployed workers, the unemployment rate increases by about 50 percent, to 16 percent of Americans currently (Haugen 2009).

Unemployment and underemployment rise and fall together: Whenever people lose their jobs, more and more people must chase fewer and fewer jobs. Figure 13.4 on the next page shows the change over time in the number of people seeking work compared to the number of job openings. As of March 2009, there were 4.8 workers for each job opening.

The Future of Work

What will the world of work look like for Americans in coming years? To answer this question, we need to look at the shifting nature of the U.S. economy, the growing impact of technology, the potential impact of globalization, and the policies the U.S. government may employ to protect jobs.

Unemployed people are those who lack a job, are available for work, *and* are actively seeking work.

Underemployed people hold jobs more appropriate for someone with fewer skills or hold part-time jobs only because they can't find full-time jobs.

FIGURE 13.4 The Loss of Jobs for U.S. Workers

Beginning in December 2006, the number of individuals seeking work rose sharply compared to the number of jobs available. As of March 2009, there were 4.8 workers for each job opening.

SOURCE: Shierholz (2009). Permission of Heidi Shierholz and Economic Policy Institute. From Nearly Five Unemployed Workers for Every Available Job. http://www.epi.org/publications/entry/jolts_20090512/.

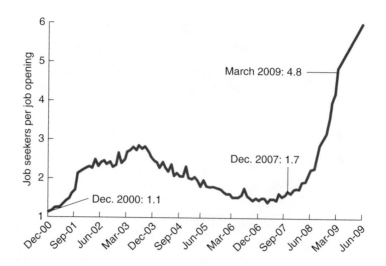

Occupational Outlook

As Figure 13.3 indicated, U.S. jobs increasingly cluster in the tertiary (service) sector. In fact, as Figure 13.5 shows, the ten fastest growing occupations are all service jobs, and seven of these ten are low-wage and low-skill jobs.

In contrast, other occupations are expected to suffer major declines. Job losses are expected to be highest in fields that rely on older technologies (such as

FIGURE 13.5 The Ten Fastest-Growing Jobs

The occupations listed will likely experience the greatest increase in new job openings between 2006 and 2016. The increase in jobs will be greatest in personal services, information technology, and health care.

SOURCE: U.S. Bureau of Labor Statistics (2007).

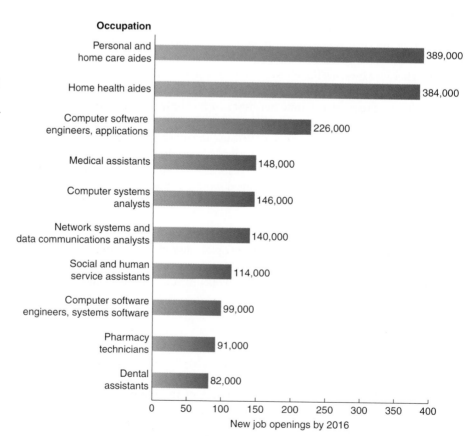

newspaper reporters) and fields that can be shifted overseas (such as printing and farming).

The big question is what kind of new jobs the economy will offer. Optimistic observers note that many executive and professional jobs are growing faster than average and suggest that the high quality and good pay of these new jobs indicate what awaits today's college graduates. Others focus on the rapid increase in what one critic has called "McJobs" (Ritzer 1996). Although not all these jobs entail selling hamburgers, many—such as health aides, personal and home-care aides, and cashiers—offer little status, low pay, and no benefits.

Both the optimists and the critics are correct in part. Although a 4-year college degree will not guarantee a secure, well-paying job, good jobs for college graduates and those with technical training—computer engineers and scientists, registered nurses, and systems analysts—nonetheless are growing rapidly. So, too, are low-paying, low-status jobs—such as nurses' aides, child-care workers, and waiters—often done by women (James, Grant, & Cranford 2000). Thus, the fastest-growing occupations require either years of advanced education or almost no skill at all, with the latter offering very little reward. The traditional working class stands to lose the most. Unlike their parents, who could find good, unionized jobs, young working-class people who do not obtain a 4-year college degree will find a hard road ahead (Blau 2001; Perrucci & Wysong 2002).

Technology and the Future

In our modern world, the experience of work is intimately linked to the nature of technology. But does technology help or hurt workers? Critics of technology argue that it harms workers by deskilling them, displacing them, and increasing supervision over them.

DESKILLING Because of technology, many occupations now require so little skill that workers find it hard to take pride in a job well done. Such *deskilling* can occur either when a job is mechanized or when workers must perform their job in ways set by others.

Deskilling affects both professional and nonprofessional workers (Burris 1998). For example, nurses and doctors these days often are required to follow set protocols for treating patients and have little freedom to make independent decisions (Weitz 2010). Similarly, in modern sawmills, computers now calculate how to cut each log to get the most usable lumber from it—a task that used to be performed by highly skilled and valued workers.

DISPLACEMENT OF THE LABOR FORCE As the sawmill example suggests, in many industries technology has replaced people with machines. In the automobile industry—or what's left of it—robots have replaced thousands of workers. In grocery stores, computerization has largely eliminated inventory clerks and pricing personnel and is increasingly replacing cashiers with "self-check-out" aisles. Meanwhile, in industry after industry, sophisticated technology has sharply reduced the time—and the number of workers—needed to produce goods and services. Thus one reason workers seldom complain about deskilling is that they are happy to still have a job.

GREATER SUPERVISION Computerization and automation give management more control over both production and workers. For example, the scanner machines used at grocery store checkouts do not simply total your grocery bill. They also keep tabs on the checker by producing statistics such as number of corrections made per hour, number of items run through per hour, and average length of time per customer.

THE IMPACT OF TECHNOLOGY Whether new technologies are an enemy of labor may depend on which laborer we ask. Those persons whose jobs are being replaced by new technologies are unlikely to see anything good about them. This is true not only for working-class people whose jobs have been mechanized but also for professionals whose jobs can now be outsourced via the Internet to people in other states or nations. On the other hand, those who have good jobs in new industries made possible by new technologies obviously benefit from these technologies. But even they may occasionally wonder how much they benefit when technologies such as BlackBerries, Twitter, and blogs allow—or even require—that they work at home, expanding work into a 24-hour-a-day job.

Technology by itself is a neutral force: It can aid management or it can aid the workers. Which technologies are implemented and the way they are implemented reflect a struggle between workers and management, and this struggle, not the technology itself, will determine the outcome.

Globalization and the Future

The globalization of the economy has led to the loss of many jobs in the United States. For example, during 2007 and 2008 Honeywell International (based in New Jersey) closed factories and fired hundreds of workers in the United States. In 2009, it announced plans to hire 3,000 workers for a new research and development center in India.

In some ways, globalization is leading our national economy through a process of reverse development: Like a least-developed country, we export raw materials such as logs and wheat and import manufactured products such as DVD players and automobiles. People in Mexico, Japan, and Korea have jobs manufacturing products for the U.S. market while U.S. workers are making hamburgers.

But factory workers are not the only ones affected by the loss of American jobs. Increasingly, even professional work like computer programming and scientific research has moved overseas in search of cheaper workers. Countries like India now offer highly skilled workers, fluent in English and better at math than most Americans, who are willing to work for far less than will U.S. workers. As a result, the safest bet for Americans are jobs that require customer and worker to be in the same geographic location, such as dog walker, grocery clerk, or doctor.

Protecting U.S. Jobs

How can policy makers protect jobs in the United States? There are three general policy options: the *conservative approach,* the *liberal approach,* and the *social investment approach.*

THE CONSERVATIVE APPROACH: FREE MARKETS Generally, business leaders and conservatives argue that the way to keep jobs in the United States is to reduce government oversight and leave wages and benefits up to market forces. In effect, this means reducing wages and benefits so businesses will have less incentive to automate jobs or move them overseas.

This approach has been adopted across the nation. In response to threats from businesses to close plants and eliminate jobs, workers have accepted lower wages and reduced benefits, and cities have agreed to reduce business taxes.

THE LIBERAL APPROACH: GOVERNMENT POLICIES Liberals argue that private profit should not be the only goal of economic activity. Instead, they suggest that governments should strive to protect workers. Among other things, liberals recommend that governments should (1) invest in industries that will provide decent jobs, (2) oversee corporate mergers and plant closings to protect workers' interests, and (3) enact subsidies and surcharges to make U.S.-made goods more competitive and to reduce the advantage that foreign-made products have in the United States.

In addition, liberals favor social welfare policies to protect those who lose their jobs. These policies include offering generous unemployment benefits, developing programs to retrain laid-off workers, and requiring corporations to give workers advance notice of plant closings. Such policies are common in Western Europe, especially in the Scandinavian countries.

THE SOCIAL INVESTMENT APPROACH Finally, some observers (both liberal and conservative) note that low-tech jobs move overseas solely to save money, but high-tech jobs move overseas both to save money *and* to seek educated workers (Lohr 2006). These observers argue that the United States can best protect high-income jobs in information technology and scientific research by ensuring that American students receive quality education in reading, writing, science, and mathematics, from grade school through graduate school.

Where This Leaves Us

As you study to prepare for a career, the economy is changing all around you. As a result, you likely will need to seek new jobs and new job skills several times before you eventually retire. New technologies, globalization, the "Wal-Mart Economy" and continuing economic troubles will further change the job situation in coming years.

In Western Europe, citizens have long expected their governments to help workers when times are tough. The current crisis has led increasing numbers of Americans to feel the same way. On the other hand, so long as relatively few Americans vote, and voting rates remain especially low among those who most need government help, politicians have little incentive to protect ordinary workers. If this economic crisis deepens, however, these patterns could change.

Summary

1. Power may be exercised through coercion or through authority. Authority may be traditional, charismatic, or rational-legal

2. Any ongoing social structure with institutionalized power relationships can be referred to as a form of politics. The most prominent political institution is the state. It is distinguished from other political institutions because it claims a monopoly on the legitimate use of coercion and it has power over a broader array of issues. Globalization, however, may limit this power.

3. Democracy is most likely to flourish in societies that have vibrant, competing interest groups, large middle classes, and relatively little income inequality.

4. The two major models used to describe the U.S. political process are the pluralist model and the power-elite model. Although they disagree on whether power is centered in an elite or more broadly distributed, they agree that organized groups have far more power to influence events than do individuals.

5. Voting and other forms of political participation are especially low in the United States. Political participation is greater among older persons, whites, and those with more income or education.

6. Although the Democratic Party tends to attract working-class and minority voters and the Republican Party tends to attract white and better-off voters, both U.S. political parties tend to have middle-of-the-road platforms with broad appeal. Public sentiment may be shifting toward the Democratic Party.

7. Some sociologists believe that voting rates are lower in the United States than in other industrialized countries because potential voters are politically alienated. Others believe that rates are low because government policies make it difficult or impossible for Americans to vote and because no major political party has actively sought to involve marginalized groups.

8. Capitalism is an economic system that maximizes productivity but pays little attention to the equitable distribution of resources to the people; socialism emphasizes distribution of resources but neglects aspects of production. Most societies mix capitalist and socialist elements in their economies.

9. Each nation has its own political economy: the particular combination and interaction of political and economic forms within a nation. Both capitalist and socialist nations can be either democratic or dictatorial.

10. Changes from preindustrial to industrial to postindustrial economies profoundly affect social organization. The tertiary sector of the economy provides employment for about three-quarters of the U.S. labor force; it includes doctors and lawyers as well as truck drivers and waitresses.

11. A small number of a few giant, transnational corporations now hold considerable power over local, national, and even international political and economic matters. These corporations often work together for their shared interests.

12. Currently, one corporation, Wal-Mart, has the power to affect organizations and individuals across the economic spectrum. Because of its great market share, it can affect prices for workers' labor, for its suppliers, and for its competitors.

13. The current economic crisis was caused primarily by soaring levels of debt, made possible by cutbacks in government regulation. Because of globalization, this economic crisis has spread around the world.

14. Professional occupations are those that (1) can set their own educational and licensing standards; (2) have their own specialized knowledge, learned through years of training; and (3) are believed by the public to be motivated by ethics and a sense of service.

15. The underground economy consists of all income-generating activities that are hidden from government regulation, including prostitution and working without proper licenses or visas.

16. Although most U.S. workers report satisfaction with their work, many nevertheless feel alienated because they are estranged from the products of their labor or from their emotions.

17. Unemployed persons are those who lack a job, are available for work, *and* have actively sought work during the last 4 weeks. Underemployed persons are those who can find only part-time work or work that does not fit their credentials and experience. Both unemployment and underemployment have soared recently.

18. For the near future, the largest number of new jobs will likely be in service work, much of which offers little status and low wages.

19. Critics argue that technology has had three ill effects on labor: deskilling jobs, reducing the number of jobs, and increasing control over workers.

Thinking Critically

1. The family and the classroom are more often authoritarian than democratic. Give examples of how this works, and explain the pros and cons of autocratic versus democratic approaches.

2. Keeping in mind what you just read about the factors associated with voting, how (if at all) do you think the 2008 presidential election will affect future voting rates? Why?

3. As an *employee*, what would you like about working in Sweden? What would you dislike? As an *employer*, what would you like and dislike about doing business in Sweden? How can Sweden's democratic socialist

government continue to resolve the differences between the interests of workers and those of business?

4. How has technology affected your schoolwork in the last 10 years? How has it made your work easier? harder? How has it made it easier or harder for teachers to monitor or control your behavior?

5. How do you think a postindustrial economy will affect your work life? How will a globalized economy affect you?

6. Which of the three general policy options outlined in the text do you think the United States should follow? Why?

Book Companion Website

www.cengage.com/sociology/brinkerhoff
Prepare for quizzes and exams with online resources—including tutorial quizzes, a glossary, interactive flash cards, crossword puzzles, essay questions, virtual explorations, and more.

CHAPTER 14

Population and Urbanization

Image copyright Jeremy Richards, 2009. Used under license from Shutterstock.com.

Populations, Large and Small

Birth and death—nothing in our lives quite matches the importance of these two events. Naturally, each of us is most intimately concerned with our own birth and death, but to an important extent, our lives are also influenced by the births and deaths of those around us. Do we live in large or small families, large or small communities? Is life predictably long or are families, relationships, and communities periodically and unpredictably shattered by death?

In this chapter we take a historical and cross-cultural perspective on the relationship between social structures and population. The study of population is known as **demography**, and those who study it are known as demographers. Demographers focus primarily on three issues: births, deaths, and migration patterns. Here we will look at these three issues and also at the effect of population size on social relationships within communities. We are interested in questions such as how births, deaths, and community size affect social structures and, conversely, how changing social structures affect births, deaths, and community size.

Currently, the world population is 6.7 billion people, give or take a couple hundred million. This is two and a half times as many people as lived in 1950. World population has grown for two basic reasons. First, the **mortality rate** (or *death rate*)—the number of *deaths* per every 1,000 persons in a given population in a given time period—has declined rapidly. Most babies now survive until adulthood, and many adults live into old age. Meanwhile, the **fertility rate**—the number of *births* per every 1,000 *women* in a population—has decreased only slowly. Similarly, the **birth rate**—the number of *births* per every *person* (male or female) in the population—has decreased slowly. In other words, births are now outpacing deaths, and so each year there are more and more people. In part because of this population growth, millions are poor, underfed, and undereducated; pollution is widespread; and the planet's natural resources have been ransacked.

These problems are among the causes of **migration**, the movement of people from one geographic area to another. We use the term **internal migration** to refer to migration to find new homes *within* a country and the term **immigration** to refer to migration *between* countries to find new homes.

Migration, in turn, leads to another set of social concerns, as nations wrestle with how to respond to the newcomers in their midst. Some nations, like the United States, allow immigrants to eventually become citizens. Other nations refuse citizenship not only to almost all immigrants but also to their children and grandchildren. For example, Germany generally will not grant citizenship to the children of Turkish immigrants, even if these children are born, raised, and educated in Germany. Immigration has substantial consequences, then, not only for population growth and economic development but also for issues such as the meaning of citizenship and nationality.

In sum, population size and population change are vitally linked to many important social issues. The next section examines how the world's population reached its current size.

Understanding Population Growth

The human population continues to grow each day, as Table 14.1 shows. Worldwide, the birth rate in 2008 was 21 births per 1,000 population; the mortality rate was a much lower 8 per 1,000. Because the number of births exceeded the number of deaths

---

Demography is the study of population—its size, growth, and composition.

The **mortality rate** is the number of deaths per every 1,000 people in a given population during a given time period.

The **fertility rate** is the number of births per every 1,000 women in a population during a given time period.

The **birth rate** is the number of births per every 1,000 persons in a population during a given time period.

Migration is the movement of people from one geographic area to another.

Internal migration is the movement of people to new homes within a country.

Immigration is the movement of people to find new homes in a different country.

TABLE 14.1 World Population Picture, 2008

In 2008, the world population was 6.7 billion and growing at a rate of 1.3 percent per year. Growth was uneven, however; the less developed areas of the world were growing much more rapidly than the more developed areas. As a result, most of the additions to the world's population were in poor nations.

Area	Birth Rate per 1,000 Persons	Mortality Rate per 1,000 Persons	Annual Percentage Increase in Population*	Projected Population Increase, 2006–2025
World	21	8	1.3%	1,294,652,000
More-developed nations	12	10	0.2%	41,853,000
Less-developed nations	23	8	1.5%	1,252,798,000

SOURCE: Population Reference Bureau (2008).
*Rate of natural increase.

by 13 per 1,000, the world's population grew at 1.3 per hundred, or 1.3 percent. If your savings were growing at the rate of 1.3 percent per year, you would undoubtedly think that the growth rate was low. A growth rate of 1.3 percent in population, however, means that the planet will hold an extra 2.6 *billion* people by the year 2050.

Importantly, all those new people will not be spread equally around the world. Instead, as Table 14.1 shows, populations are growing more rapidly in some nations than in others. Less developed nations in Africa, for example, may double population size in less than 30 years, whereas the developed nations of Europe will have shrinking populations.

Because most population growth is occurring in poor nations, the world will likely be poorer in 2025 than it is now. How did these different population patterns evolve?

Population in Former Times

For most of human history, both birth rates and mortality rates were about 40 per 1,000. Because both rates were similar, populations grew slowly if at all. Translated into personal terms, this means that the average woman spent most of the years between the ages of 20 and 45 either pregnant or nursing. If both she and her husband survived until they were 45, she would produce an average of 6 to 10 children. The average life expectancy was perhaps 30 or 35 years. Such a low life expectancy was largely due to very high *infant mortality rates*. The **infant mortality rate** is the number of babies who die during or shortly after childbirth per every 1,000 live births in a given population. Throughout much of human history, perhaps one-quarter to one-third of all babies died before they reached their first birthday. Both birth and death were frequent occurrences in most preindustrial households.

The Demographic Transition in the West

Beginning in the eighteenth century, a series of events occurred that revolutionized population in the West. First, death rates fell substantially while birth rates remained high. As a result, the population grew rapidly. Then birth rates, too, dropped. Once birth and death rates reached similar levels, they balanced each other out

The **infant mortality rate** is the number of babies who die during or shortly after childbirth per every 1,000 live births in a given population.

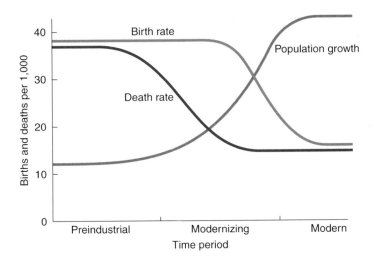

FIGURE 14.1 **The Demographic Transition in the West**
In the preindustrial West, both birth and death rates were high. As living conditions improved and death rates began to fall, the population grew. Eventually, however, birth rates also fell and population size stabilized. This process is known as the *demographic transition*.

and population size stabilized, as Figure 14.1 illustrates. Because studies of population are called demography, this change from a population characterized by high birth and death rates to one characterized by low birth and death rates is called the **demographic transition**. It results in longer life expectancies. Although this transition occurred at different times in different countries, the process was more or less similar across Europe and in the United States. More recently, birth rates have fallen still further, and populations in some nations are shrinking.

The Decline in Mortality Rates

Prior to the demographic transition, widespread malnutrition was an important factor underlying high mortality rates. Although few died of outright starvation, poor nutrition increased the susceptibility of the population to disease. Improvements in nutrition were the first major cause of the demographic transition's decline in mortality rates, beginning in the 1700s and continuing into the early twentieth century. New crop varieties from the Americas (especially corn and potatoes), new agricultural methods and equipment, and increased trade all helped improve nutrition in Europe and the United States. The second major cause of the decline in mortality rates was a general increase in the standard of living, as improved shelter and clothing left people healthier and better able to ward off disease. Changes in hygiene were vital in reducing communicable diseases, especially those affecting young children, such as typhoid fever and diarrhea (Kiple 1993).

In the late nineteenth century, public-health engineering led to further reductions in communicable disease by providing clean drinking water and adequate treatment of sewage. For example, between 1900 and 1970, the life expectancy of white Americans increased from 47 to 72, and the life expectancy of African Americans increased from 33 to 64 (U.S. Bureau of the Census 1975, 2006). Thus, although life expectancy has been increasing gradually since about 1600, the fastest increases occurred in the first few decades of the twentieth century. Medical advances probably account for no more than one-sixth of this overall rise in life expectancy (Bunker, Frazier, & Mosteller 1994). Instead, public-health initiatives, better nutrition, and an increased standard of living are largely responsible for rising life expectancies (McKinlay & McKinlay 1977; Weitz 2010). Interestingly, once the standard of living in a nation reaches a certain point—approximately $6,400 per capita income—further increases in life expectancy depend less on increasing income than on reducing the income gap

Demographic transition is the process through which a population shifts from high birth and death rates to low birth and death rates.

Have you ever traveled to a less-developed country? If you did, the odds are that you got a nasty stomach virus for a day or two, but otherwise suffered no health problems. Yet malaria, cholera, dysentery, and the like kill millions in these countries each year. Why are American tourists virtually immune? Vaccinations, antibiotics, and access to soap and water help. But the most important reason is that, unlike many residents of less-developed countries, tourists start out healthy, well nourished, well sheltered, and well clothed. As a result, even if they come in contact with dangerous germs, their bodies most likely will be able to fight against infection.

between rich and poor (Wilkinson 1996). This is one major reason why, on average, Cubans live almost as long as do Americans, and Swedes live longer than Americans.

The Decline in Fertility Rates

The Industrial Revolution also affected fertility rates, although less directly. Industrialization meant increasing urbanization, greater education, and the real possibility of getting ahead in an expanding economy. Pensions and other social benefits became more common with industrialization, so people no longer needed to have many children to care for them in their old age (Friedlander & Okun 1996). In addition, as mortality rates dropped, parents no longer needed to have eight children to count on two surviving. Perhaps even more important, industrialization created an awareness of the possibility of doing things differently than they had been done by previous generations. As a result, the idea of controlling family size to satisfy individual goals spread even to areas that had not experienced industrialization, so that by the end of the nineteenth century, the idea of family limitation had gained widespread popularity (van de Walle & Knodel 1980). Currently in Europe and North America, birth and death rates are about even, and there is little population growth.

The Demographic Transition in the Non-West

In the less-developed nations of the non-West, birth and death rates remained at roughly preindustrial levels until the first decades of the twentieth century. After that, in some areas such as Latin America, Taiwan, Singapore, and South Korea, economic development and improvements in the standard of living caused both death and birth rates to plummet, much as they had previously done in the West.

The poorest nations of the world followed a somewhat different path. Death rates in these nations only began to fall in the second half of the twentieth century, following basic improvements in sanitation and in health care (especially the adoption of childhood vaccinations and of new treatments for childhood diarrhea). Because death rates fell while birth rates remained stable, initially this shift led to population growth. More recently, however, birth rates also have declined, and population growth has slowed considerably. In addition, in the countries hardest hit by AIDS, such as Botswana, Swaziland, and Lesotho, death rates have soared, and population growth has dropped dramatically (UNAIDS/WHO 2007).

Population and Social Structure: Two Examples

In this section we explore contemporary relationships between social structure and population in two societies: Ghana, where the fertility rate is high, and Italy, where the fertility rate is low.

Figure 14.2 illustrates the differences between the populations in these two countries through the use of "population pyramids." A population pyramid shows the number of people in a nation's population, broken down by age group. Males are shown on the left-hand side and females on the right-hand side.

Ghana's population pyramid actually looks like a pyramid because many Ghanaians are very young and relatively few Ghanaians survive into old age. In contrast, Italy's pyramid bulges out in the middle because there are so many middle-aged Italians. Moreover, its pyramid shows that Italy has almost as many old people as young people.

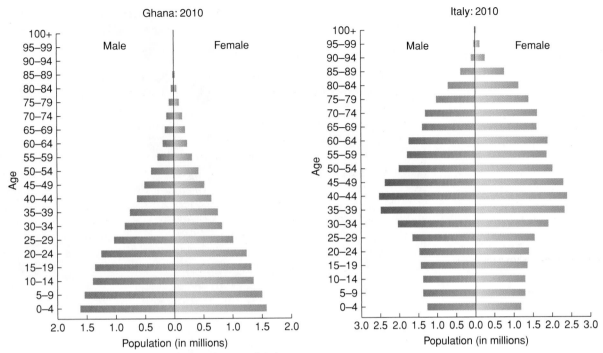

FIGURE 14.2 Population Pyramids for Ghana and Italy

Ghana's population pyramid looks like a pyramid because it includes many young people and few old people. Italy's pyramid bulges out in the middle with middle-aged people and is balanced top and bottom with reasonably similar numbers of young and old people.

SOURCE: U.S. Census Bureau, International Data Base, Accessed June 2009.

Ghana: Is the Fertility Rate Too High?

Ghana is an example of a society in which traditional social structures encourage a high fertility rate. It is also an example of a society in which high fertility may ensure continuing traditionalism—and poverty.

The Effects of Social Roles on the Fertility Rate

Fertility rates have declined in Ghana in recent years but remain high. Ghana still has a crude birth rate of 32 per 1,000 population. The mortality rate, however, is down to 10 per 1,000. This means that the rate of natural increase in Ghana is 2.2 percent per year (Population Reference Bureau 2008). If that rate continues, the population could double in less than 30 years.

One of the most important reasons for this high fertility rate is the nature of women's roles in Ghanaian society. In Ghana, children are an important—or even the only—source of esteem and power open to many women. Women who cannot bear children risk divorce or abandonment. This is especially true for the 22 percent of Ghanaian women who live in polygamous unions (Social Institutions & Gender Index 2009). The number of children a woman has—especially the number of sons—strongly affects her position relative to that of her co-wives. Moreover, because infant mortality rates remain relatively high, Ghanaian women believe they must have four or more children to ensure that two survive to adulthood.

Another important cause of the high fertility rate is the need for economic security. Most Ghanaians work in subsistence agriculture. To survive, families need

Many families in Africa, especially polygamous families, have numerous children, and overpopulation is a cause for concern.

AP Images

children as well as adults to work in the fields. In addition, when children grow up and marry, they can add to the family's economic and political security by creating political and social allegiances to other families. Finally, children are the only form of old-age insurance available to Ghanaians: Parents who grow old or ill must rely on their children to support them. Conversely, having children is relatively inexpensive: No expensive medical treatment is available for children, schooling is either inexpensive or unaffordable, and children don't expect to own designer jeans or $150 tennis shoes. With a cost/benefit ratio of this sort, it is not surprising that Ghanaians desire many children.

The Effects of High Fertility Rates on Society

Although individual women may benefit from Ghana's high fertility rate, Ghanaian society as a whole has suffered. If its population continues to explode, Ghana will have to increase its governmental expenditures dramatically just to maintain current levels of support for education, highways, agriculture, and the like. Thus, a decision that is rational on the individual level turns out to be less wise on the societal level.

This problem sometimes leads people in the West to ask: "Are they stupid? Can't they figure out they would be better off if they had fewer children?" Unfortunately for the argument, nations don't have children; people do. A high fertility rate continues to be a rational choice for individual Ghanaians.

Policy Responses

To reduce its population growth, Ghana has established an excellent family-planning program that makes contraception available, convenient, and affordable to women who want it. When women want several children, however, access to contraception has limited impact. Currently, only 17 percent of all married women in Ghana use modern contraceptive methods (Population Reference Bureau 2008). Contraceptive use is considerably higher among younger, better-educated, urban women. Study after study has found that the best way to reduce the fertility rate is to combine access to

TABLE 14.2 **Population Change in Europe**
Overall, deaths are now slightly exceeding births in Europe. Thus, some nations are already experiencing population decline. The last column in the table shows the combined impact of births, deaths, and migration into and out of a country.

Country	Birth Rate per 1,000 Persons	Mortality Rate per 1,000 Persons	Annual Percentage Change in Population*	Average No. of Children per Woman	Projected Population Change, 2000–2050** (percent)
Denmark	12	10	0.2%	1.8	0%
Germany	8	10	−0.2	1.3	−13
Hungary	10	13	−0.4	1.3	−11
Italy	9	10	0.0	1.3	3
Romania	10	12	−0.2	1.3	−20
Spain	11	9	0.2	1.4	−6
United Kingdom	13	9	0.3	1.9	26

SOURCE: Population Reference Bureau (2008).
*Rate of natural increase.
**Reflects the impact of immigration as well as birth and death rates. (United Kingdom receives more immigrants and Romania loses more to immigration than do the other countries in this table.)

contraception with educational and economic development and higher status levels for women (Poston 2000).

Italy: Is the Fertility Rate Too Low?

In a world reeling from the impact of doubling populations in the less-developed world, it is ironic that many developed countries worry that their fertility rates are too low. Yet low fertility also can cause serious problems.

The Effects of Social Roles on Fertility Rates

With modern mortality rates, fertility rates must average 2.1 children per woman if the population is to replace itself: two children so that the woman and her partner are replaced and a little extra to cover unavoidable childhood deaths. Such a fertility rate is called **zero population growth**. If the fertility rate is less than this, the next generation will be smaller than the current one.

Currently in Italy, the average woman is having only 1.3 children (Table 14.2). This means that the next generation of Italians will be much smaller than previous ones, unless the country absorbs many new immigrants. The same scenario holds true across most of Europe, as Table 14.2 shows.

Why is the fertility rate so low in Italy? In essence, the situation in Italy is the reverse of that in Ghana. Most Italian women are educated, and many hold paying jobs outside of the home. Women's social status is close to that of men, so women do not need to have children to have a purpose in life or to assure their social standing. Because few Italians work in agriculture, and all children are expected to be in school,

Zero population growth exists when the fertility rate is about 2.1 births per woman, the rate needed to maintain the population at a steady size.

In Europe, many families have only one child, and underpopulation is increasingly a cause for concern.

having children doesn't add to a family's labor pool. Finally, the Italian government provides a good safety net in the form of disability insurance, health care, old-age pensions, and the like, which means that couples do not need to have children to take care of them in sickness or old age.

The Effects of Low Fertility Rates on Society

Given the serious worldwide dilemmas posed by population growth and Italy's very high density, why should we consider a low fertility rate a problem? There are two main concerns: the large numbers of old people compared with young people, and rising nationalistic fears resulting from the importing of immigrant labor.

A very low fertility rate creates an age structure in which the older generation is as large as or larger than the younger generation on whom it relies for support. As a result, it is increasingly difficult for Italy to fill all the occupations—from taxi drivers to doctors—needed to keep the nation running. At the same time, the cost of paying for old-age pensions and health care is growing rapidly. (The same is true of Social Security in the United States.) As a result, the most-industrialized nations must spend more and more of their national net income on pensions and health care for older citizens.

To counteract this problem, Italy has imported workers from other countries, primarily neighboring Albania. This has led to nationalist fears of cultural dilution. A survey conducted in 2003 found that an astounding 80 percent of Italians believed that Albanian immigrants were bad for Italy (Pew Research Center 2003). These feelings have provoked anti-immigrant violence in Italy and have led Italy to clamp down on immigration. In turn, the isolation and discrimination experienced by immigrants in Italy have also led to outbreaks of violence by immigrants themselves. Similar conditions elsewhere in western Europe have produced similar results, such as the riots that blazed across France's immigrant neighborhoods for 3 weeks in late 2005.

Another consequence of low fertility was tragically illustrated on May 12, 2008 when a devastating earthquake struck China's Sichuan province. To relieve overpopulation and protect its declining environment, China's authoritarian government refuses to allow most couples to have more than one child and punishes severely those who ignore this rule. When the earthquake struck, almost 7,000 poorly built classrooms collapsed, killing thousands of children—most of whom were their parents' only child.

The loss of a child is always a tragedy. Losing an only child, however, is particularly devastating, since parents lose both the sense of a future for their family and the sense of security that children can bring to aging parents. China has agreed to relax its one-child policy for couples who lost their only child in the earthquake, but some couples may be too old to take advantage of this.

Policy Responses

In response to the various concerns raised by low fertility, Italy and other European nations have established incentives to encourage a higher fertility rate. Among them are paid, months-long maternity leave; cash bonuses and housing subsidies for having more children; and monthly subsidies for children until age 3 (Oleksyn 2006).

TABLE 14.3 **Average Number of Births per Woman, 1950–2008**
In the last half century, the average number of children per woman has declined worldwide.

Region	Average Number of Births per Woman	
	1950	2008
Africa	6.6	4.9
Asia	5.9	2.4
Europe	2.6	1.5
Latin America	5.9	2.5
North America	3.5	2.1
Oceania	3.8	2.4

SOURCE: Gelbard, Haub, & Kent (1999); Population Reference Bureau (2008).

Nevertheless, the costs of raising children far outstrip these benefits. As a result, while these incentive plans have kept birth rates from falling drastically, they have not helped to raise birth rates in Italy or other countries where women have attractive alternatives outside the home (Gautier & Hatzius 1997).

Population and Social Problems: Two Examples

Analysis of world population growth reveals a good news/bad news situation. The good news is that the average number of births per woman has declined in every part of the world (Table 14.3). The bad news is that the population of the world will nonetheless increase dramatically over the next 50 years. The reason for this gloomy prediction lies in the age structure of the current population. The next generation of mothers is already born—and there are a lot of them. Thus, we must plan for a world that will soon hold 8 or 9 billion people.

Population pressures can contribute to numerous social problems. In this section, we address two of them—environmental devastation and poverty.

Environmental Devastation: A Population Problem?

All around the world, there are signs of enormous environmental destruction: In the developed world, we have acid rain and oil spills; in Africa, desert environments are spreading rapidly due to deforestation and overgrazing. Both of these pose serious threats to the environment, but only the latter is truly a population problem.

The United States is responsible for far more than its share of environmental destruction. Our affluent, throwaway lifestyle requires large amounts of petroleum and other natural resources. Obtaining these resources results in the destruction

© Harold Castro

Deforestation is devastating tropical rainforests in Brazil, the Philippines, and elsewhere.

of wilderness, the loss of agricultural lands, and the pollution of oceans. Using these resources causes illness-inducing air pollution, acid rain, and smog that are killing our forests. Although these problems would be less severe if there were half as many of us (and hence half as many cars, factories, and Styrofoam cups), they are not really population problems. They stem from our way of life rather than our numbers.

In sub-Saharan Africa, however, population pressure is a major culprit in environmental destruction. In rural areas, the typical scenario runs like this: Population pressure forces farmers to plow marginal land and to plant high-yielding crops in quick succession without soil-enhancing rotations or fallow periods. The marginal lands and the overworked soils produce less and less food, forcing farmers to push the land even harder. They cut down forests and windbreaks to free more land for production. Soon, water and wind erosion becomes so pervasive that the topsoil is borne off entirely, and the tillable land is replaced by desert or barren rock. This cycle of environmental destruction—which destroys forests, topsoil, and the plant and animal species that depend upon them—is characteristic of high population growth in combination with poverty. When one's children are starving, it is hard to make long-term decisions that will protect the environment for future generations.

In sum, reducing population growth would reduce future pressure on natural resources, but it would not solve the current problem. The solution rests in an international moral and financial commitment to reducing rural poverty, improving farming practices, reducing the foreign debt of the less- and least-developed nations, *and* curbing wasteful and destructive practices in the developed nations.

Poverty in the Least-Developed World

Perhaps 500 million people around the world are seriously undernourished, and each year outbreaks of famine and starvation occur in Africa and Asia. A billion more are poorly nourished, poorly educated, and poorly sheltered. These people live in the same nations that have high population growth.

Some observers blame poverty in the developing nations on the high fertility rates in these nations. Yet high fertility rates are not the only or even the primary cause of this poverty. Poverty and malnutrition result primarily from war, corruption, and inequality in nondemocratic countries and from a world economic system that extracts raw goods and profits from poorer countries (Chase-Dunn 1989; Dreze & Sen 1989; Sen 1999). It is a terrible irony that most poor countries export more food than they import (Lappé, Collins, & Rosset 1998). Cuba, for example, became poorer in the 1990s not because of population growth but because its authoritarian government failed to develop a strong economy and instead relied heavily on subsidies from the now-defunct Soviet Union. People in the Democratic Republic of the Congo, meanwhile, are dying of starvation because of war rather than because of a high fertility rate.

Policy Responses

Although many factors contribute to poverty, almost all world leaders agree that reducing the fertility rate is an important step toward increasing the standard of living in the poorer nations of the world. The most successful programs to reduce fertility rates have combined an aggressive family-planning program, economic and educational development, and improvements in the status of women (Poston 2000).

FAMILY-PLANNING PROGRAMS Family-planning programs are designed to make modern contraceptives and sterilization available inexpensively and conveniently to individuals who desire to limit the number of their children. For example, between 1975 and 1991, an aggressive family-planning program increased contraceptive use in Bangladesh by 500 percent and decreased the average number of children per woman from 7 to 5 in just 16 years (Kalish 1994).

ECONOMIC AND EDUCATIONAL DEVELOPMENT Experience all over the world shows that fertility rates decline as education increases and the country undergoes economic development. For example, South Korea's fertility rate has plummeted from 6.0 children per woman in 1960 to only 1.3 currently in the wake of its dramatic economic development (Population Reference Bureau 2008).

IMPROVING THE STATUS OF WOMEN In countries where women have low status, they can only increase their social value and guarantee support in their old age by having many children—especially sons. When women have greater education and can earn even a small income on their own, they gain greater power within the family. As a result, they typically marry later and have fewer children. In addition, they are better able to protect their daughters from being married off while still children. Consequently, the countries that have proven most successful in family planning and in economic growth are those, such as South Korea and Singapore, that have made particular efforts to increase education, economic options, and legal rights for women (United Nations Population Fund 2000).

Efforts to reduce birth rates in poor nations like Afghanistan have been most successful when they have combined family-planning programs with increases in access to jobs and education, especially for girls and women.

Photo by Tim Graham/Getty Images

Population in the United States

The U.S. population picture is similar to that in Italy, with low mortality and fertility rates, but there are also several differences. First, although the fertility rate is close to the zero population growth level, it has not dropped significantly below this level as has happened in Italy. Second, immigration continues to add substantially to the size of our population. Third, and partly because of this immigration, our population is younger than Italy's. In this section, we briefly describe fertility rates, mortality rates, and migration issues in the United States.

Fertility Rates

For nearly 20 years, the number of children per woman in the United States has remained just around or just under 2.1—the zero population growth level. This low fertility rate has been accompanied by sharp reductions in social-class, racial, and religious differences in fertility rates. Some women will give birth when they are teenagers and some when they are 40, but increasingly they will stop at 2 children.

Mortality Rates

Death is almost a stranger to U.S. families. The average age at death is now in the late seventies, and many people who survive to age 65 live another 20 years. Parents can feel relatively secure that their infants will survive. If they don't divorce, young newlyweds can safely plan on a golden wedding anniversary.

Since 1970, we have added about 7 years to the average life expectancy. This increase is primarily due to better diagnosis and treatment of the degenerative diseases (such as heart disease and cancer) that strike elderly people. In addition, increases in life expectancy have been made possible by reducing (although not eliminating) racial and social-class differentials in mortality rates. In the early 1940s, African American women lived a full 12 years less than white women; today, the gap is down to a bit over 4 years (U.S. Bureau of the Census 2009a).

On the other hand, the AIDS epidemic, first recognized in 1981, has given death a new face. Although death rates from AIDS have fallen in recent years, AIDS remains a leading cause of death for all persons ages 25 to 44, but especially for African Americans and Hispanics. Often spread through intravenous drug use (which has the most appeal for those who have the least to look forward to), AIDS is becoming a disease of the poor and disadvantaged.

Migration Patterns

Although it can safely be ignored as a factor in world population growth, migration often has dramatic effects on the growth of individual nations. The United States is one of the nations for which immigration has had an important impact, particularly in Sunbelt states such as California, Arizona, and Florida.

Most U.S. citizens are descended from people who emigrated to the United States to improve their economic prospects, such as many recent migrants from Mexico. Other immigrants, such as those from Iraq, Bosnia, and the Sudan, are primarily refugees driven from their homes by warfare or the economic destruction that often follows in its wake (see Focus on a Global Perspective: International Migration on pages 358–359). Patterns of both internal migration and international immigration have created a unique set of problems in the United States and have dramatically changed our political landscape.

Immigration

The United States has always been a country of immigrants. Immigration peaked between 1880 and 1920, and then fell with the passage of restrictive immigration laws. Immigration then rose steadily until 2008, when the sharp loss in U.S. jobs made immigration to this country much less appealing (Preston 2009). Future immigration will depend on how the U.S. economy compares to that in Mexico and other countries.

An estimated 1 million immigrants enter the United States each year. Almost all recent immigrants come from Latin America or Asia. Perhaps as many as half are illegal immigrants, most from Mexico or Central America.

Immigrants to the United States divide roughly into two very different groups. The first is skilled, well educated, able to speak English, and here legally, such as doctors and computer scientists from India. The second group is made up of low-skilled workers with little education or ability to speak English, many of whom are here illegally. Most of these workers come from Latin America. The experiences of these two groups, and their impact on the United States, differ markedly.

Because of immigration, the United States does not need to fear population decline. The racial and ethnic composition of the nation will change substantially, however. By 2050, it is estimated that the combination of Hispanic immigration and a low fertility rate among whites will reduce the proportion of our population that is white non-Hispanic from 69 percent in 2001 to 50 percent (Figure 14.3 on the next page).

Most immigrants to the United States, both legal and illegal, are pushed from their native lands by poor local economies and are pulled by an unmet demand in

The Hispanic population in the United States has grown considerably in recent years. This family is waiting to apply for legal residency.

AP Images

International Migration

During 2008, approximately 11.3 million refugees fled their homes involuntarily, and several million more (who had fled earlier) remained outside their home countries as either stateless persons or asylum seekers (United Nations High Commissioner for Refugees 2009). Millions more chose voluntarily to seek new lives and new opportunities in other countries.

We often hear debate about immigrants and refugees in the United States, but what do we know about international migration? Map 14.1 shows recent migration patterns around the world. Most refugees flee from one developing nation to a neighboring developing nation, whereas many voluntary migrants move to industrialized nations in search of a better life.

Demographers believe that the economic and political turmoil of the last two decades, coupled with the opportunities presented by globalization, have substantially increased international migration. At least 191 million people lived outside their country of birth or

citizenship in 2005, almost twice the number 50 years ago (United Nations Population Fund 2006). Although push factors such as war and famine account for much of this international migration, some migrants are also pulled by the economic growth and employment opportunities in newly industrializing nations, such as South Korea, Singapore, and Malaysia. Pull factors also account for much of the immigration from less-developed to more-developed countries. Strong European economies provide increasing numbers of jobs to a growing non-Western labor force. Migrants traditionally have been young men, but women and girls now comprise about half of those leaving their home countries (United Nations Population Fund 2009). Many of these are mothers who, in growing numbers, seek employment opportunities in more affluent neighboring countries in order to send money to the family, friends, or neighbors who are raising their children.

The money sent back home by migrants—both men and women—is an important source of revenue for many nations. During 2005 alone, migrants sent more than $232 billion

to their home nations (United Nations Population Fund 2006). These funds have helped millions of people in poorer nations to rise out of poverty (United Nations Population Fund 2006). With the current economic crisis, however, these funds have fallen dramatically, and whole communities in nations such as Mexico have suffered as a result.

It is not yet clear who profits most from the international migrant stream. Although countries such as Germany, France, and Italy face new challenges stemming from an ethnically diverse population, workers from developing nations help to sustain the continued expansion of these nations' economies. Low birth rates have led to smaller labor forces and aging populations in Europe, Japan, and elsewhere. Thus, migrants from countries such as Turkey and Pakistan fill the demand for more workers, particularly at the low end of the labor hierarchy. Whether the money that migrants send home will significantly improve the quality of life in less-developed nations remains an open question.

FIGURE 14.3 The Changing U.S. Population
If immigration and fertility rates remain stable, the proportion of Hispanic and Asian Americans will likely increase and the proportion of non-Hispanic whites will likely decrease.

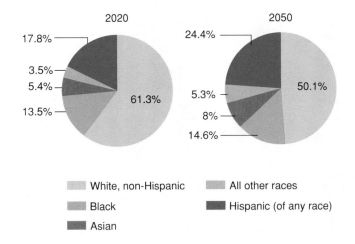

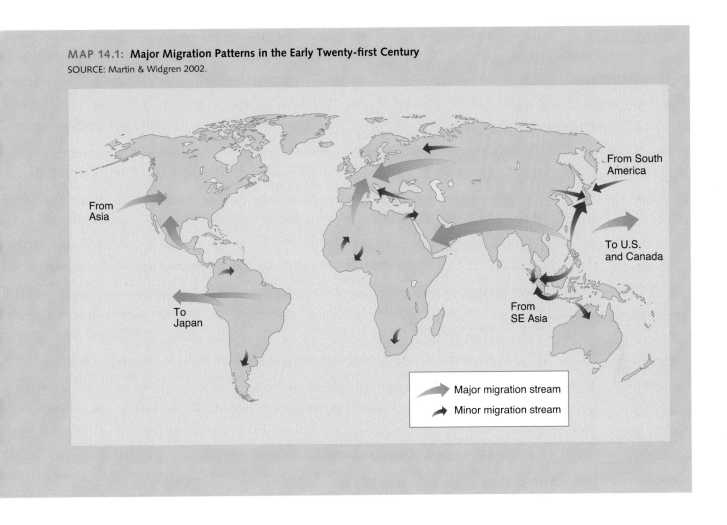

MAP 14.1: **Major Migration Patterns in the Early Twenty-first Century**
SOURCE: Martin & Widgren 2002.

the United States for low-skilled, low-paid labor. In the past, many immigrants (especially from Mexico) would come to the United States to work briefly and then return to their home countries, a cycle they repeated whenever they needed to earn extra money. Ironically, the clampdown on border crossings has made it too dangerous to cross the border repeatedly, and so many of these migrants instead have settled in the United States (Massey 2006). Meanwhile, that clampdown has had no impact on the number who cross the border: One study of 1,000 persons who chose to migrate from southern Mexico found that *all* eventually made it across (Preston 2009). However, the current economic downturn has made the United States a less attractive destination, and for the moment immigration has decreased.

The consequences of current immigration trends are likely to be both economic and cultural. From the standpoint of economics, research suggests that (1) immigrants are not taking jobs away from U.S. citizens, but (2) the availability of low-wage illegal immigrants may depress wages for the least educated American citizens. Some economists believe immigrants have no effect on wages; others believe they reduce wages for high school dropouts by as much as 5 percent (Borjas & Katz 2007; Card 2005). From the standpoint of culture, it is likely that the United States will become a more pluralistic society in which salsa and soccer are as popular as hot dogs and baseball, but that the new immigrants will integrate into American society as did earlier waves of Hispanic and other immigrants (Alba & Nee 2003).

Internal Migration

Until recently, internal migration—movement from one part of the country to another—has been higher in the United States than in most of the developed world. However, the recent economic downturn has dramatically changed this: The percentage of Americans who moved homes in 2008 was the lowest in *60 years* (Edwards 2009). Many people cannot move because no one can afford to buy their homes, whether because potential buyers have lost their jobs or savings or because banks have tightened the rules for giving mortgages. Others cannot move because the value of their homes has dropped substantially and they fear financial catastrophe if they sell at current low prices.

The most striking and largest example of internal migration in U.S. history was the exodus of about 1 million people triggered by 2005's Hurricane Katrina. New Orleans had already been a city in decline for many years before 2005. But things got much worse when the hurricane and the flooding that followed it destroyed homes, businesses, and basic services, such as sewer systems and electric lines. As a result, many fled the city. Although living conditions have improved and some have returned to the city, its population remains about 40 percent lower than it was before the hurricane. Many of the hurricane refugees—both those who returned to the city and those who have relocated elsewhere—remain mired in deeper poverty than before the disaster, and many are still living in "temporary" mobile homes designed only to serve as emergency shelters.

New Orleans, of course, is a unique case. More generally, the history of internal migration in the United States has been a story of **urbanization**—the increasing movement of people from towns and farms into cities. For most of our history, urban areas grew faster than rural areas, with the largest urban areas growing the most. Since about 1970, however, this has all changed. Currently, the three major trends in internal migration are Sunbelt growth, migration from central cities to suburbs, and the resurgence of some nonurban areas.

Since 1970, there has been consistent movement of people from the Midwest and the northern states to the Sunbelt states of the Southeast and Southwest. Working people have followed jobs, and retirees have followed the sunshine. Most urban growth has occurred in these areas as well. However, the crash in jobs and housing prices has hit hardest in these areas, and many are now losing population.

In the rest of the country, central cities have declined while suburbs surrounding them have grown. **Suburbs** are communities that develop outside of cities and that, historically, primarily provided housing rather than services or employment. Importantly, the middle class has disproportionately left the cities, so that increasingly cities are home to only the wealthy and the poor (Scott 2006). Urban poverty has sharply increased as jobs have moved to the suburbs, public transportation to the suburbs remains minimal, and the wealthy have driven up the cost of urban housing ("Out of Sight" 2000).

At the same time that central cities have been shrinking, nonurban areas have experienced some modest growth (Johnson 2003). Most of this growth, again, has occurred in Sunbelt states, especially in retirement destinations and in areas within a few hours' drive of a big city. But rising home prices, rising numbers of retirees,

Urbanization is the process of concentrating populations in cities.

Suburbs are communities (primarily residential) that develop outside of cities.

Since Hurricane Katrina in 2005, many New Orleans neighborhoods have lost homes. As a result, many families have moved from New Orleans to Houston and elsewhere, in a process known as *internal migration*.

and rising numbers of workers who can live anywhere there is an Internet connection have led to a small but growing migration to more distant towns in more varied places, like northern Michigan and Archer County, Texas (Fessenden 2006).

Because suburban and nonurban life is so dependent on automobile transportation, migration to the suburbs and beyond has increased air pollution. In addition, much of the geographic relocation of the U.S. population since 1970 has been to those regions of the country that are least able to withstand the ecological impact of a large population. In many areas of Florida, California, and the Southwest, the demand for water already outstrips the supply. As states argue over water rights, political tensions are likely to increase; within states, competition for access to water may increase conflict between agricultural and urban interests.

Fertility rates, mortality rates, and migration patterns in the United States provide clear examples of the interrelationships between population and social institutions. Social class, women's roles, and racial and ethnic relationships are all intimately connected to changes in population. One additional element of population that is especially important for social relationships is community size, an issue to which we now turn.

Urbanization

Most of our social institutions evolved in agrarian societies, where the vast bulk of the population lived and worked in the countryside. As late as 1850, only 2 percent of the world's population lived in cities of 100,000 or more (Davis 1973). Today, nearly a quarter of the world's population and more than two-thirds of the U.S. population live in cities larger than 100,000. (This population shift is illustrated in less-developed and more-developed countries in Figure 14.4 on the next page.) How did these cities develop, and what are they like?

Theories of Urban Growth and Decline

Structural-functionalist theorists and conflict theorists hold very different views of the sources, nature, and consequences of urban life. Structural functionalists emphasize

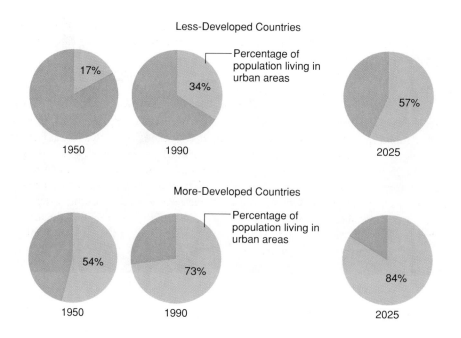

the benefits of urban growth and decline, while conflict theorists emphasize the political struggles that undergird these changes.

Structural-Functional Theory: Urban Ecology

Early structural-functional sociologists, many of whom lived in the booming Chicago of the 1920s and 1930s, assumed that cities grew in predictable ways. Some argued that (like Chicago), cities naturally grew outward in concentric circles from central business districts (Burgess 1925). Others believed that cities grew in wedge-shaped sectors, along transit routes, or in other patterns (Hoyt 1939). All structural functionalists, however, agreed that healthy and natural competition between economic rivals would lead cities to grow in whatever ways offered the most efficient means for producing and distributing goods and services. More recently, structural functionalists have assumed that urban decline and the growth of suburbs similarly reflect natural progress toward superior and more efficient ways of organizing economic and social life.

Conflict Perspectives: White Flight and Government Subsidies

In contrast, conflict theorists note that no patterns of urban growth have yet been discovered that hold across time and across different locations. Thus they conclude that there is nothing natural about urban growth or decline. Rather, they argue, each city grows or declines in its own unique way, depending on the relative power of competing economic and political forces (Feagin & Parker 1990).

These competing forces appear to have played an important role in drawing middle-class Americans from cities during the last half century. Western culture has long held an anti-urban bias, assuming that rural life is "purer" than city life. This view gained strength during the early decades of the twentieth century, as first foreign immigrants and later African Americans moved in large numbers from the South to the cities of the Northeast and Midwest. These changes contributed greatly to white Americans' sense that the city was a dangerous place and encouraged middle- and upper-class Americans to flee the cities, a process known as "white flight." In contrast, throughout most of the world, the upper classes live in central cities, and the poor are relegated to city outskirts and rural areas.

The abandonment of American cities was greatly assisted by government subsidies for **suburbanization**, the growth of suburbs (Goddard 1994; Moe & Wilkie 1997). Since the 1930s, federal and local governments have responded to pressure from auto manufacturers and suburban developers by steadily reducing financial support for public transit while tremendously expanding subsidies for auto manufacturing, highways, road maintenance, and the like. As a result, people found it increasingly difficult to live, work, shop, or travel in dense cities with limited parking and decaying transit systems. In addition, since the 1950s the government has provided inexpensive home mortgages (along with tax breaks) to suburbanites while routinely denying mortgages to city dwellers. During the 1960s and 1970s, the government implemented a catastrophic "urban renewal" program that placed highways in the middle of stable, urban neighborhoods (most of which were minority and poor or working class) and moved dislocated residents to poorly constructed, public, high-rise housing. Finally, in the last two decades, local suburban governments have used tax subsidies to entice corporations to relocate to the suburbs.

All these changes pressured middle-class Americans to move to the suburbs, further contributing to the decay of our cities (Moe & Wilkie 1997). Of course, many people gratefully left their urban homes for suburbia and relished the freedom automobiles promised. But many others only reluctantly exchanged their close-knit urban neighborhoods, where they could read the newspaper while riding the bus to work, for sprawling suburbs where high walls separate neighbor from neighbor and long, nerve-wracking drives to work are the norm.

The Nature of Modern Cities

From the Industrial Revolution to the present, the modern city has grown in size and changed considerably in character. We look here at the development of *industrial* and *postindustrial* cities.

The Industrial City

With the advent of the Industrial Revolution, production moved from the countryside to the urban factory, and industrial cities, such as Boston, Detroit, and Pittsburgh, were born. These cities were mill towns, steel towns, shipbuilding towns, and, later, automobile-building towns; they were home to slaughterers, packagers, millers, processors, and fabricators. They were the product of new technologies, new forms of transportation, and vastly increased agricultural productivity that freed most workers from the land.

Fired by a tremendous growth in technology, the new industrial cities grew rapidly during the nineteenth century. In the United States, the urban population grew from 2 to 22 million in the half century between 1840 and 1890. In 1860, New York was the first U.S. city to reach 1 million in population. The industrial base that provided the impetus for city growth also gave the industrial city its character: tremendous density and a central business district.

DENSITY Until the middle of the twentieth century, most Americans walked to work—and everywhere else, for that matter. The result was dense crowding of working-class housing around manufacturing plants. Even in 1910, the average New Yorker

Suburbanization is the growth of suburbs.

As this 1950s photo of Yorkshire, England shows, industrial cities are characterized by dense crowding of working-class housing around often-polluting manufacturing plants.

Photo by Bentley Archive/Popperfoto/Getty Images

commuted only two blocks to work. Entire families shared a single room, and in major cities such as New York and London, dozens of people crowded into a single cellar or attic. The crowded conditions, accompanied by a lack of sewage treatment and clean water, fostered tuberculosis, epidemic diseases, and generally high mortality rates.

CENTRAL BUSINESS DISTRICT The lack of transportation and communication facilities also contributed to another characteristic of the industrial city, the central business district. The central business district is a dense concentration of retail trade, banking and finance, and government offices, all clustered close together so messengers could run between offices and businesspeople could walk to meet one another. By 1880, most major cities had electric streetcars or railway systems to take traffic into and out of the city. Because most transit routes offered service only into and out of the central business district rather than providing crosstown routes, the earliest improvements over walking enhanced rather than decreased the importance of this district.

The Postindustrial City

The industrial city was a product of a manufacturing economy plus a relatively immobile labor force. Beginning about 1950, these conditions changed, and a new type of city began to grow. Among the factors prominent in shaping the character of the postindustrial city are the change from secondary to tertiary production and the greater ease of communication and transportation. These changes have led to the rise of urban sprawl and edge cities.

CHANGE FROM SECONDARY TO TERTIARY PRODUCTION As we noted in Chapter 13, the last decades have seen a tremendous expansion of jobs in tertiary production and the subsequent decline of jobs in secondary production. The manufacturing plants that shaped the industrial city are disappearing. Many of those that remain have moved to the suburbs, where land is cheaper, and have taken working-class jobs, housing, and trade with them.

Instead of manufacturing, the contemporary central city is dominated by medical and educational complexes, information-processing industries, convention and entertainment centers, and administrative offices. These are the growth industries. They are also white-collar industries. These same industries, plus retail trade, also dominate the suburban economy.

EASIER COMMUNICATION AND TRANSPORTATION Development of telecommunications and good highways has greatly reduced the importance of physical location. The central business district of the industrial city was held together by the need for physical proximity. Once this need was eliminated, high land values and commuting costs led more and more businesses to locate on the periphery, where land was cheaper and housing more desirable. Many corporate headquarters moved from New York or Chicago all the way to Arizona or Texas.

A key factor in increasing individual mobility was the automobile. Without the automobile, workers and businesses could not have moved to the city periphery, and space-gobbling single-family homes would not have been built. In this sense, the automobile and the automotive industry have been the chief architect of U.S. cities since 1950.

URBAN SPRAWL AND EDGE CITIES These changes have led to the collapse of many central business districts. In their stead, urban sprawl and edge cities have emerged. Postindustrial cities, such as Atlanta, Las Vegas, and Miami, are much larger in geographical area than the industrial cities were. The average city in 1940 was probably less than 15 miles across; now many metropolitan areas are 50 to 75 miles across. No longer are the majority of people bound by subway and railway lines that only go back and forth to downtown. Retail trade is dominated by huge, climate-controlled, suburban malls. A great proportion of the retail and service labor force has also moved out to these suburban centers, and many of the people who live in the suburbs also work in them. Suburban areas that now have an existence largely separate from the cities that spawned them are known as **edge cities** (Garreau 1991).

Urbanization in the United States

What is considered urban in one century or nation is often rural in another. To impose some consistency in usage, the U.S. Bureau of the Census has replaced the common words *urban* and *rural* with two technical terms: *metropolitan* and *nonmetropolitan*.

A **metropolitan statistical area** is a term used by federal researchers to refer to a county that has a city of 50,000 or more in it *plus* any neighboring counties that are significantly linked, economically or socially, with the core county. Some metropolitan areas have only one county; others, such as New York, San Francisco, or Detroit, include half a dozen neighboring counties. In each case, the metropolitan area goes beyond the city limits and includes what is frequently referred to as, for example, the Greater New York area. A **nonmetropolitan statistical area** is a county that has neither a major city in it nor close ties to such a city.

Currently, 78 percent of the U.S. population lives in metropolitan areas. This metropolitan population is divided between those who live in the central city (within the actual city limits) and those who live in the surrounding suburban ring. More than half

Edge cities are suburban areas that now have an existence largely separate from the cities that spawned them.

A **metropolitan statistical area** is a county that has a city of 50,000 or more in it plus associated neighboring counties.

A **nonmetropolitan statistical area** is a county that has no major city in it and is not closely tied to such a city.

The suburbs are growing faster than rural or urban areas in the United States, and edge cities are growing faster than other cities.

Peter Menzel

of the metropolitan population live in the suburbs rather than in the central city itself. Although these people have access to a metropolitan way of life, they may live as far as 50 miles from the city center.

The nonmetropolitan population of the United States has shrunk to 22 percent of the U.S. population. Although there are nonmetropolitan counties in every state of the Union except New Jersey, the majority of the nonmetropolitan population lives in either the Midwest or the South. Only 5 percent are farmers, and many live in small towns rather than in purely rural areas.

Urbanization in the Less-Developed World

The growth of large cities and an urban way of life has occurred everywhere very recently; in the less- and least-developed nations, this growth is happening almost overnight. Mexico City, São Paulo, Bogotá, Seoul, Kinshasa, Karachi, Calcutta, and other cities in developing nations continue to grow rapidly. Their populations are likely to double in about a decade. The roads, the schools, and the sewers that used to be sufficient no longer are; neighborhoods triple their populations and change their character from year to year. These problems are similar to the problems that plagued Western societies at the onset of the Industrial Revolution, but on a much larger scale.

Urbanization in the less-developed world differs from that in the developed world, not only in pace but also in causes. First, more than half of the growth in developing cities is due to a high excess of births over deaths, rather than to migration from the countryside. Second, many of the large and growing cities in the less-developed world have never been industrial cities. They are government, trade, and administrative centers. More than one-third of the regular full-time jobs in Mexico City are government jobs. These cities offer few working-class jobs, and the growing populations of unskilled men and women become part of the informal economy—artisans, peddlers, bicycle renters, laundrywomen, and beggars.

Place of Residence and Social Relationships

Every year, new films and television shows depict the evils of city life, the boredom of suburbs, and the intolerance of small towns. How realistic are such images? This section explores the pleasures and perils of modern urban, suburban, and small-town life.

Urban Living

One of the primary questions raised by sociologists who study cities is the extent to which social relationships and the norms that govern them differ between rural and urban places. Here we look at sociological theories of urban life and research on the realities of urban living.

Theoretical Views

As we saw earlier, the Western world as a whole has an antiurban bias. Big cities are seen as haunts of iniquity and vice, corruptors of youth and health, and destroyers of family and community ties. City dwellers are characterized as sophisticated but artificial; rural people are characterized as unsophisticated but warm and sincere. This general antiurban bias (which has been around at least since the time of ancient Rome), coupled with the very real problems of the industrial city, had considerable influence on early sociologists.

The classic statement of the negative consequences of urban life for the individual and for social order was made by Louis Wirth in 1938. Wirth argued that the greater size, heterogeneity, and density of urban living necessarily led to a breakdown of the normative and moral fabric of everyday life.

Greater size means that many members of the community will be strangers to us. Wirth postulated that urban dwellers would still have primary ties but would keep their emotional distance from, for example, store clerks or strangers in a crowded elevator by developing a cool and calculating interpersonal style.

Wirth also believed that when faced with a welter of differing norms, the city dweller was apt to conclude that anything goes. Such an attitude, coupled with the lack of informal social control brought on by size, would lead to greater crime and deviance and a greater emphasis on formal controls.

Later theorists have had a more benign view of the city. Sociologists now suggest that individuals experience the city as a mosaic of small worlds that are manageable and knowable. Thus, the person who lives in New York City does not try to cope with 9 million people and 500 square miles of city; rather the individual's private world and primary ties are made up of family, a small neighborhood, and a small work group. In addition, sociologists point out, urban life provides the "critical mass" required for the development of tight-knit subcultures, from gays to symphony orchestra aficionados to rugby fans. Wirth might interpret some of these subcultures as evidence of a lack of moral integration of the community, but they can also be seen as private worlds within which individuals find cohesion and primary group support.

Realities of Urban Living

Does urban living offer more disadvantages or advantages? This section reviews the evidence about the effects of urban living on *social networks, neighborhood integration*, and *quality of life*.

Many people enjoy the excitement of city life.

Alan Chandler/Photo Library

SOCIAL NETWORKS The effects of urban living on personal integration are rather slight. Surveys asking about social networks show that urban people have as many intimate ties as rural people. There is a slight tendency for urban people to name fewer kin and more friends than rural people. The kin omitted from the urban lists are not parents, children, and siblings, however, but more distant relatives (Amato 1993). There is no evidence that urban people are disproportionately lonely, alienated, or estranged from family and friends.

NEIGHBORHOOD INTEGRATION Empirical research generally reveals the neighborhood to be a very weak group. Most city dwellers, whether central city or suburban, find that city living has freed them from the necessity of liking the people they live next to and has given them the opportunity to select intimates on a basis other than physical proximity. This freedom is something that people in rural areas do not have. There is growing consensus among urban researchers that physical proximity is no longer a primary basis of intimacy (Flanagan 1993). Rather, people form intimate networks on the basis of kin, friendship, and work groups, and they keep in touch by telephone, e-mail, or instant messaging rather than by face-to-face communication. In short, urban people do have intimates, but they are unlikely to live near each other. When in trouble, they call on their good friends, parents, or adult children for help. In fact, one study of neighborhood interaction in Albany-Schenectady-Troy, New York, found that a substantial share—15 to 25 percent—of all interaction with neighbors was with *family* neighbors—parents or adult children who lived in the same neighborhood (Logan & Spitze 1994).

Neighbors are seldom strangers, however, and there are instances in which being nearby is more important than being emotionally close. When we are locked out of the house, need a teaspoon of vanilla, or want someone to accept a United Parcel Service package, we still rely on our neighbors (Wellman & Wortley 1990). Although we generally do not ask large favors of our neighbors and don't want them to rely heavily on us, most of us expect our neighbors to be good people who are willing to help in a pinch. This has much to do with the fact that neighborhoods are often segregated by social class and stage in the family life cycle. We trust our neighbors because they are people pretty much like us.

QUALITY OF LIFE Big cities are exciting places to live. People can choose from a wide variety of activities, 24 hours a day, 7 days a week. The bigger the city, the more it offers in the way of entertainment: libraries, museums, zoos, parks, concerts, and galleries. The quality of medical services and police and fire protection also increases with city size. These advantages offer important incentives for big-city living.

On the other hand, there are also disadvantages: more noise, more crowds, more expensive housing, and more crime. The rates of both violent crimes and personal crimes are considerably lower in rural areas than in either suburban areas or cities, and the largest cities have higher crime rates than do smaller cities (Federal Bureau of Investigation 2009). (On the other hand, methamphetamine use and associated crimes are now more common in rural areas than in suburban or urban areas, as this chapter's Focus on American Diversity box discusses.)

Because of these disadvantages, many people would rather live close to a big city than actually in it. For most Americans, the ideal is a large house on a spacious lot in

focus on  AMERICAN DIVERSITY

Methamphetamines in Rural America

Since the 1980s, poverty has increased in rural America. Small family farms have been bought by large corporations, and rural mines and manufacturing plants have closed. As a result, rural areas have lost stores, services, and population, with better-educated and younger people especially likely to move to cities.

This shift has contributed to a stunning rise in the use of illegal methamphetamine ("meth") in rural America, especially among working-class white youths (Grant et al. 2007; Van Gundy 2006; NIDA Research Report 2006). Like other amphetamines, meth is a highly addictive stimulant. Users experience very pleasurable sensations that last only briefly, leading some to take the drug repeatedly—sometimes without stopping to eat or sleep. Long-term use can result in anxiety, insomnia, violence, paranoia, hallucinations, and possibly brain damage (NIDA Research Report 2006).

Meth use is now considerably more common in rural America than in other areas, even when we look only at poor people in these different areas. Meth labs, too, are most common in rural areas, which offer both the basic ingredients (such as fertilizer) needed to produce the drug and abandoned buildings on deserted roads to serve as labs. Meth production raises the risks for rural areas, since producing one pound of meth releases five pounds of toxic chemicals into the environment (NIDA Research Report 2006). Moreover, in untrained hands meth production can easily lead to dangerous explosions that can harm anyone in the vicinity.

Unfortunately, it's particularly difficult for rural methamphetamine users to obtain treatment (NIDA Research Report 2006). Many rural communities have neither substance abuse treatment facilities nor support groups like Narcotics Anonymous. At any rate, many rural dwellers cannot afford to pay for treatment or even to pay for transportation to treatment facilities. In addition, many steer away from treatment due to fears of stigma—a realistic concern in small, conservative communities where everyone knows everyone else's business.

To deal with the rural meth epidemic, we will need to address both the underlying social problems that lead to drug use and the lack of social and medical services in rural America.

the suburbs, but close enough to a big city that they can spend an evening or afternoon there. Some groups, however, prefer big-city living, in particular, childless people who work downtown. Many of these people are decidedly pro-urban and relish the entertainment and diversity that the city offers. Because of their affluence and childlessness, they can afford to ignore many of the disadvantages of city living.

Sociological attention has been captured by cities such as Manhattan and San Francisco with their bright lights, ethnic diversity, and crowding. Nevertheless, only one-quarter of our population actually lives in these big-city centers. The rest live in suburbs and small towns. How does their experience differ?

Suburban Living

The classic picture of a suburb is a development of very similar single-family detached homes on individual lots. This low-density housing pattern is the lifestyle to which a majority of people in the United States aspire; it provides room for dogs, children, and barbecues. This is the classic picture of suburbia. How has it changed?

The Growth of the Suburbs

The suburbs are no longer bedroom communities that daily send all their adults elsewhere to work. They are increasingly major manufacturing and retail trade centers. Most people who live in the suburbs work in the suburbs. Thus, many close-in suburban areas have become densely populated and substantially interlaced with retail trade centers, highways, and manufacturing plants.

These changes have altered the character of the suburbs. Suburban lots have become smaller, and neighborhoods of townhouses, duplexes, and apartment buildings have begun to appear. Childless couples, single people, and retired couples are seen in greater numbers. Suburbia has become more crowded and less dominated by the minivan set.

With expansion, suburbia has become more diverse. Although each suburban neighborhood tends to have its own style, stemming in large part from each development including houses of similar size and price, there are a wide variety of styles. In addition to classic suburban neighborhoods, there are now areas of spacious mini-estate suburbs where people ride horses and lawn mowers, as well as dense suburbs of duplexes, townhouses, and apartment buildings. Some of the first suburbs, which were built after World War II, are now more than 50 years old. Because people tend to age in place, these suburbs are often characterized by retirees living on declining incomes (Lambert & Santos 2006). Many houses are becoming run-down, and renting is becoming increasingly common.

Suburban Problems

Many of the people who moved to suburbia did so to escape urban problems: They were looking for lower crime rates, less traffic, less crowding, and lower tax rates. The growth of the suburbs, however, has brought its own problems (Langdon 1994). Three of the most important are weak governments, car dependence, and social isolation and alienation.

The county and municipal governments of suburban towns and cities are fragmented and relatively powerless. One result of this is the very haphazard suburban growth associated with weak and inadequate zoning authority. In addition, because there is rarely any governmental body that has the power to make decisions for a city and its suburbs as a whole, it is nearly impossible to coordinate decisions across a metropolitan region. This means, for example, that if one suburb or city decides to ban smoking in restaurants, business will simply move to the next suburb.

The lack of regional planning is particularly important when it comes to transportation. Without effective regional decision making, it is difficult to develop effective mass transit systems or even highways. This leaves suburban dwellers in the lurch, since most commute from suburb to suburb or suburb to city. It also makes suburban dwellers even more dependent than others on automobiles. People who don't have cars are basically excluded from the suburban lifestyle and from jobs in either suburb or city. If you can't afford a car or can't drive one due to disability, aging, or youth, your quality of life in suburbia plummets.

Long commutes leave individuals with little time to socialize with co-workers after work or with neighbors and family once they arrive home. In addition, suburban zoning laws that forbid businesses such as cafes, beauty parlors, and taverns in residential neighborhoods deprive people of the natural gathering places that foster social relationships and a sense of community. Similarly, suburban houses with high fences and no front porches make it nearly impossible for neighbors to meet informally (Oldenburg 1997). When people live in one community and work in another, they may end up feeling alienated from both.

Small-Town and Rural Living

Approximately 25 percent of the nation's population lives in rural areas or small towns (less than 2,500 people). Some live within the orbit of a major metropolitan area, but most live in nonmetropolitan areas, from Maine to Alabama, California to Florida. These areas vary greatly and include everything from millionaire second-home towns like Telluride, Colorado, to dying farm or mill towns in Kansas or Maine, flourishing

Suburbs are intensely car dependent. As suburbs have grown, so have traffic jams and long commutes.

Amish communities in Pennsylvania, and booming Nebraska poultry processing towns. Some rural areas are overwhelmingly white, some are overwhelmingly African American, and a growing number have substantial Hispanic populations.

Across the board, people find rural and small-town living attractive for a number of reasons (Brown & Swanson 2003). It offers lots of open space, low property taxes, and affordable housing (except in vacation areas). There's much less worry about crime and drugs, although alcohol and methamphetamine abuse are actually most common in rural areas. Many also appreciate the more conservative views on politics, premarital sex, religion, and the like that typify nonmetropolitan areas. In addition, community ties remain strong in the small minority of rural towns still characterized by deep family roots, family-run farms, civically engaged churches, and small rather than large manufacturing plants, and both children and adults benefit from the neighborliness and community sentiment. In a city, you might find a bar like Cheers "where everybody knows your name." In a rural area, practically everyone does.

Although young people who grow up in nonmetropolitan areas often must leave to get an education or a job, these areas continue to grow (Johnson 2003). Most of this growth, however, is in "recreational" areas that attract second-home owners and retirees, areas near large cities that attract long-distance commuters, and areas with large-scale food manufacturing plants (meat packing, canning, and so on).

The major problem with rural life is the dearth of jobs, especially well-paying jobs with benefits (Jensen, McLaughlin, & Slack 2003). Family farms have all but disappeared, driven out of business by global competition or bought out by huge agribusinesses (the only ones with the money and power to compete in this global market). Only 5 percent of nonmetropolitan dwellers still work in agriculture, while the majority now work in low-wage service jobs in prisons, casinos, fast-food restaurants, and the like (McGranahan 2003). Because of these problems, many nonmetropolitan

In the most desirable rural areas, well-paying jobs are scarce and housing is expensive. As a result, many rural families must live in inexpensive manufactured homes.

Jeff Morgan tourism and leisure/Alamy

dwellers must endure long commutes to jobs in distant metropolitan areas. Stress over low wages, underemployment, and unemployment coupled with the physical stresses of the available work, lack of social resources, and limited access to health care combine to leave nonmetropolitan residents, on average, in somewhat poorer physical and mental health than urban or suburban residents (Morton 2003).

In addition, nonmetropolitan areas that have experienced inflows of "city people" are experiencing new strains due to growing stratification: The economic and cultural differences between the upper and lower ends of the population are far greater than in the past (Brown & Swanson 2003). Forty years ago, ski resort owners and ski resort workers all lived in Telluride, if in different conditions. Now, resort owners and their clients live in luxury homes in or near the center of town, while most workers can only afford to live far from town in "rural ghettoes" of mobile homes and concentrated poverty. This stratification is particularly hard on schoolchildren, who find themselves increasingly marginalized and stigmatized by teachers and wealthier children whose expectations for clothing, vacations, and academic preparation cannot be met by poorer children. In sum, although life in small towns and rural areas still brings benefits, it can bring high costs as well.

Where This Leaves Us

There's no question about it: Numbers matter. As the world's population grows—and, in places, shrinks—all of us are affected. Population growth in the United States has enormous consequences for the environment because of the huge amounts of natural resources Americans use. Population growth in the less-developed nations is especially important because it not only stems from poverty but also produces even more poverty. Meanwhile, population loss in Europe leaves nations grappling with problems brought on by having too few young people compared with the number of old people.

The problems of population growth are intimately connected to the problems of urbanization—and suburbanization. Cities emerged with the rise in industrialization, a process that is still continuing in the developing nations. In turn, problems with urban life, accentuated by various social policies, have stimulated the growth of suburbs and "edge cities." Each of these environments offers its own dangers and its own rewards.

Summary

1. For most of human history, fertility rates and mortality rates were about equal, and the population grew slowly or not at all. Childbearing was a lifelong task for most women, and death was a frequent visitor to most households, claiming one-quarter to one-third of all infants in the first year of life.

2. The demographic transition—the decline in mortality and fertility rates—developed over a long period in the West. Mortality rates declined because of better nutrition, an improved standard of living, improved public sanitation, and to a much more limited extent, modern medicine. Somewhat later, changes in social structure associated with industrialization caused fertility rates to decline. In the developing nations, mortality rates have declined rapidly, and fertility rates are only slowly declining in response.

3. Social structure, fertility rates, and mortality rates are interdependent; changes in one affect the others. Among the most important causes and consequences of high fertility rates is the low status of women.

4. The fertility rate in a society is directly linked to the costs and rewards of childbearing. In traditional societies, such as Ghana, most social structures (the economy and women's roles, for example) support high fertility rates. In many modern societies, such as Italy and the United States, social structure imposes many costs on parents.

5. When a nation's fertility rate declines, the nation faces several problems. Among these are labor-force shortages, difficulties in funding health and pension benefits for a burgeoning number of older people, and nationalistic fears over growing numbers of foreign workers.

6. Population growth is an important cause of environmental devastation in the less-developed world but not in the developed world (where most environmental resources are used). Although population growth does contribute to poverty in the less-developed world, other factors are much stronger causes of poverty.

7. In the United States, life expectancy is high and continues to increase. Childlessness is increasing and fertility is near the zero population growth level. Because of high immigration rates, however, the U.S. population is unlikely to decline. Because many of the new Americans are Asian and Latino, the racial and ethnic composition of the U.S. population is likely to change substantially. Immigration has not taken jobs from U.S. citizens but may have reduced wages among the least-educated native-born Americans.

8. Since the 1970s, central cities in most of the nation have shrunk, and urban poverty has increased. Meanwhile, suburban towns and cities have grown significantly, and nonurban areas have experienced modest growth. This movement to suburbia and to nonurban locations raises serious environmental questions. Across categories—urban, suburban, and nonurban—the Sunbelt states have seen the most growth.

9. Structural functionalists argue that cities grow and decline in predictable and natural ways, reflecting the most efficient means for producing and distributing goods and services. Conflict theorists, on the other hand, argue that city growth and decline reflect the outcomes of economic and political struggles between competing groups. Government subsidies played a major role in the twentieth-century growth of suburbs and decline of central cities.

10. The industrial city has high density and a central-city business district. The postindustrial city reflects the shift to tertiary production and increased ease in communication and transportation and is characterized by lower density and urban sprawl.

11. Urbanization is continuing rapidly in the less-developed world; many of its large cities will double in size in a decade. This urban growth is less the result of industrialization than of high urban fertility.

12. There are competing theories about the consequences of urban living. Wirth's theory suggests that urban living

will lead to nonconformity and indifference to others. Other theorists suggest that the size of the city is managed through small groups and allows for the development of unconventional subcultures.

13. Urban living is associated with less reliance on neighbors and kin and more reliance on friends, with greater risk of crime.

14. Suburban living has become more diverse. Retail trade and manufacturing have moved to the suburbs, and the suburbs are now more densely populated, more congested, and less dominated by the minivan set. Suburban living has its own problems, including weak governments, transportation problems, and social isolation and alienation.

15. Among the benefits of small-town and rural living are less crime, stronger community ties, more open spaces, and more affordable housing (except in vacation areas). The most serious problem is the dearth of well-paying jobs with benefits, which results in somewhat poorer physical and mental health.

Thinking Critically

1. Unless you are much older than most college students, your generation is considerably smaller than your parents' generation. How will this affect you? Consider the impact on you now, as your parents and their generation retire, and as you approach retirement. Think about both personal finances and resources and government programs and spending.

2. How is dormitory life similar to urban living? similar to small-town living?

3. Make a list of the environmental resources you use in a day. Consider "natural" products, such as oranges, as well as manufactured products, such as computers. How would your list compare with that of someone in a developing nation?

4. What would the United States be like if all immigration ceased? What would be the benefits? the disadvantages?

Book Companion Website

www.cengage.com/sociology/brinkerhoff
Prepare for quizzes and exams with online resources—including tutorial quizzes, a glossary, interactive flash cards, crossword puzzles, essay questions, virtual explorations, and more.

CHAPTER 15

Social Change

Ellen McKnight/Alamy

How Societies Change

Social institutions do not stand still. Often, things change without our knowing how or why. Immediately after the terrorist attacks of September 11, 2001, thousands of Europeans held candlelight vigils to express their solidarity with the United States. These days, Europeans are far more often found demonstrating against the continued U.S. presence in Iraq. Meanwhile, nations in Eastern Europe, Africa, and Latin America lurch toward new economic and political forms, while the fortunes of all nations increasingly depend upon an international political and economic system.

What is going on? Many Americans shake their heads in confusion. The last decade has brought great changes, such as new drugs to treat cancer, ever-smaller laptop and notebook computers, and the election of our first African American president. Balanced against these positive changes, however, are civil war and malnutrition in many developing nations, the destruction of the Amazon rainforest, and an epidemic of repetitive stress disorders linked to computer use.

All of these changes—both positive and negative—are referred to by sociologists as social change. **Social change** is defined as any significant modification or transformation of social structures or institutions over time. The rapid pace of social change and the complexity of twenty-first-century problems lead many individuals to feel a sense of both urgency and helplessness. In this chapter, we describe three potential sources of social change: *collective behavior*, *social movements*, and *technology*.

Collective Behavior

- After the film *Twilight* opened, teenage girls around the country gathered in large numbers to scream, hug, and cry wherever the film's stars appeared in public.
- In March 2009, hundreds of Bolivians wielding sticks and whips looted the house and attacked the family of an unpopular politician.

Despite the differences between these actions, both are examples of collective behavior. **Collective behavior** is spontaneous action by groups in situations where cultural rules for behavior are vague, inadequate, or contested (Marx & McAdam 1994). It includes such diverse actions as mob violence and spontaneous candlelight vigils to protest mob violence, as well as the behavior of crowds surging into Wal-Mart for a sale or carousing in the streets during Mardi Gras. These are unplanned, more or less spur-of-the-moment actions, where individuals and groups improvise a joint response to an unusual or problematic situation. Collective behavior differs from social movements (discussed below) in that it is usually short-lived, at least in part because participants lack a clearly defined social agenda and the resources needed to affect public policy. (Some sociologists include social movements as collective behavior, but others, including this textbook's authors, prefer to separate these topics.)

As noted, collective behavior occurs when cultural rules are (1) vague, (2) inadequate, or (3) contested. Cultural rules are *vague* in many areas: Should a woman tattoo her whole arm? Should someone take a year off between high school and college? Cultural rules are often *inadequate* during crises or periods of rapid social change: Who should be rescued first during a disaster? What is appropriate—or safe—to post on a Facebook page? Cultural rules are *contested* when some social groups feel that the normal rules of the society work against them and decide to subvert or protest those rules.

Social change is any significant modification or transformation of social structures and sociocultural processes over time.

Collective behavior is spontaneous action by groups in situations where cultural rules for behavior are unclear.

Collective behavior, such as mosh pits and crowd surfing at rock concerts, differs from social movement in being more spontaneous and relatively unplanned.

© Neal Preston/Corbis

Collective behavior can occur anywhere there is a group, from sidewalks, to prisons, to corporations (Marx & McAdam 1994). A rumor can lead illegal street vendors to quickly pack up their goods, and a prison may erupt in violence over squalid conditions. Within a corporation, a particular Windows desktop wallpaper may suddenly become popular on a floor, employees might help each other escape a disaster (as when the World Trade Center was attacked), or they might begin an informal work slowdown as a silent protest against low pay.

Even when collective behavior is not designed as protest, however, it can have the effect of challenging the status quo. For example, if enough college students post descriptions of drinking binges or wild sexual activity on Facebook or MySpace, then that behavior will likely come to seem more acceptable. The difference between collective behavior and social movements, however, is that social movements are organized, relatively broad based, long term, and intended to foster social change.

Social Movements

Social movements are individuals, groups, and organizations united by a common desire to change social institutions, attitudes, or ways of life (Tilly 2004). Examples include the immigrant rights and environmental movements, as well as the grassroots struggle against drunk driving. A social movement is extraordinarily complex. It may include sit-ins, demonstrations, and even riots, but it also includes meetings, fundraisers, legislative lobbying, and letter-writing campaigns.

Both collective behavior and social movements challenge the status quo. As a result, they are related in at least two ways. First, social movements need and encourage some instances of collective behavior simply to keep issues in the public eye (Marx & McAdam 1994). There is nothing like a riot or police breaking up an illegal demonstration to get people's attention. Second, even though collective behavior is usually

A **social movement** is an ongoing, goal-directed effort to fundamentally challenge social institutions, attitudes, or ways of life.

concept summary

Theories of Social Movements

Theory	Major Assumption	Causes of Social Movements
Structural-Functional Theory: Relative Deprivation	Social movements are an abnormal part of society	Social change produces disorganization and discontent
Conflict Theory: Resource Mobilization	Social movements are the normal outgrowth of competition between groups	Competition between organized groups
Symbolic Interaction Theory: Political Process	People join social movements because they have developed an "insurgent consciousness"	Political opportunities combine with an individual sense that change is needed and possible

sociology and you

Sooner or later, most people experience relative deprivation. Any time you have felt unhappy because someone you knew had a bigger allowance, nicer clothes, or a newer car than you did, you experienced relative deprivation. This is true even if most observers would consider both of you to be poor or consider both of you to be wealthy.

limited to a particular place and time, it can be a repeated mass response to problematic conditions. When this happens, collective behavior at a grassroots level may be a driving force in mobilizing social movements (Tilly 2004).

As we documented in Chapter 13, most people in the United States have relatively little interest in politics. Why, then, do some people shake off this lethargy and try to change the system? And under what circumstances do social movements succeed or fail?

Theoretical Perspectives on Social Movements

Three major theories explain the circumstances in which social movements arise: *relative-deprivation theory*, *resource mobilization theory*, and *political process theory*. All three theories suggest that social movements arise out of inequalities and cleavages in society, but they offer somewhat different assessments of the meaning, sources, and tactics of social movements. These differences are described in the Concept Summary on Theories of Social Movements.

Structural-Functional Theory: Relative Deprivation

Poverty and injustice are universal phenomena. Why is it that they so seldom lead to social movements? According to **relative-deprivation theory**, social movements arise when we *believe* we should have more than we *actually* have—especially if we feel this deprivation is a result of unfair treatment (Walker & Smith 2002). Our expectations, in turn, are usually determined by comparing ourselves with others or with past situations. Because the theory refers to deprivation relative to other groups or times rather than to absolute deprivation, it is called *relative-deprivation theory*.

Figure 15.1 diagrams three conditions for which relative-deprivation theory would predict the development of a social movement. In Condition A, disaster or taxation suddenly reduces the standard of living (or "rewards") for everyone. Unless people's expectations also drop, they will resent their new deprivation. In Conditions B and C, the standard of living has risen, but expectations have risen even further. Consequently, people feel deprived relative to what they had anticipated. Relative-deprivation theory has the merit of providing a plausible explanation for many social movements occurring in times when objective conditions are either improving (Condition C) or at least are better than in the past (Condition B).

Relative-deprivation theory argues that social movements arise when people experience an intolerable gap between their expectations and the rewards they actually receive.

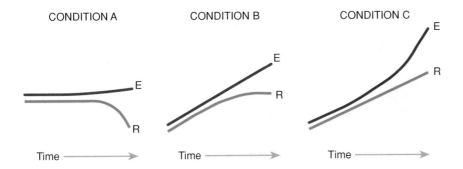

CONDITION A   CONDITION B   CONDITION C

Time ⟶   Time ⟶   Time ⟶

FIGURE 15.1 Expectations, Rewards, and Relative Deprivation
Relative-deprivation theory suggests that relative deprivation exists whenever there is a gap between expectations (E) and rewards (R). It may occur when the rewards available to individuals decline (Condition A), when the rewards level off (Condition B), or even when rewards steadily increase (Condition C).

Relative-deprivation theory is a structural-functional theory. Like other structural-functional theories, it assumes that in normal circumstances society functions smoothly. According to this theory, then, social movements arise only when social change occurs unevenly across social or cultural institutions or when the pace of change is simply too rapid.

There are two major criticisms of relative-deprivation theory. First, empirical evidence does not bear out the prediction that those who are most deprived, absolutely or relatively, will be the ones most likely to participate in social movements. Often, social movement participants are the best off in their groups rather than the worst off. For example, almost all of the 19 terrorists who destroyed the World Trade Center and attacked the Pentagon, as well as Osama bin Laden, were well educated and middle class or wealthy. In many other situations, individuals participate in and lead social movements on behalf of groups to which they do not belong, such as South African whites who fought against apartheid and people who fight for animal rights. Second, the theory fails to specify the conditions under which relative deprivation will lead to social movements. Why do some relatively deprived groups form social movements and others don't? Relative deprivation can play a role, but by itself it is not a good predictor of the development of social movements (Gurney & Tierney 1982).

Conflict Theory: Resource Mobilization

While structural functionalists assume that society generally works harmoniously, conflict theorists assume that conflict, competition, and, as a result, deprivation are common. If deprivation were all it took to spark a social movement, we would have active social movements all the time. Yet social movements only arise sporadically. Consequently, conflict theorists argue, relative deprivation by itself cannot explain why social movements emerge when they do. Rather, they argue, social movements emerge when individuals who experience deprivation can garner the resources they need to mobilize effectively for action. This theory is known as **resource mobilization theory**, and it is the most commonly used theory among American sociologists (McAdam & Snow 1997).

According to resource mobilization theory, then, the spark for turning deprivation into a movement is not anger and resentment but rather organization and resources. As a result, social movements will be more common in affluent societies than in poorer ones, since in affluent societies even the least well-off may have access to the minimum resources needed for protest. Similarly, the building blocks of social movements are organized groups whose leaders are relatively well provided with resources, rather than discontented individuals from the lower classes.

Resource mobilization theory suggests that social movements develop when individuals who experience deprivation can garner the resources they need to mobilize for action.

The "immigrants' rights" movement in the United States reflects "insurgent consciousness"—the belief that change in the system is both needed and possible—among both immigrants and their supporters.

At the same time, the rise of information technologies has made resource mobilization easier and faster for people around the globe. For example, Barack Obama's 2008 campaign for the Presidency effectively used YouTube videos, Facebook, e-mail, blogs, and other Internet resources to spread his message, raise funds, and attract people to campaign events. Similarly, in April 2009, anti-government activists in the eastern European nation of Moldova used cell phones, e-mail, Facebook, and Twitter to draw more than 10,000 young people to a political protest on short notice.

Symbolic Interaction Theory: Political Process

Resource mobilization theory remains very important within sociology, but it has been criticized for two reasons. First, it downplays the importance of grievances and spontaneity as triggers for social movements (Klandermas 1984; Morris & Mueller 1992). Second, it overlooks the crucial process through which vague individual grievances lead to new collective identities and organized political agendas (Jasper & Poulsen 1995; Williams 1995). **Political process theory** has arisen to fill this gap. According to political process theory, a social movement needs two things: political opportunities and an "insurgent consciousness." **Political opportunities** include preexisting organizations that can provide the new movement with leaders, members, phone lines, copying machines, and other resources. Whether or not political opportunities will exist depends on a number of factors, including the level of industrialization in a society, whether a war is going on, and whether other cultural changes are underway (Meyer 2004).

Insurgent consciousness is the individual sense that change is both needed and possible. In the same way that symbolic interactionism argues that individuals develop their identities and understanding of the social world through interactions with significant others, political process theory argues that individuals develop their sense of identity and of the possibility of change through interaction with others. For example, until the 1970s, newspapers regularly listed job ads in separate columns for men and for women, top universities refused to admit women as students, and some ministers told battered wives that they must have done something to cause their husbands to

Political process theory suggests that social movements develop when political opportunities are available and when individuals have developed a sense that change is both needed and possible.

Political opportunities are resources that allow a social movement to grow; they include preexisting organizations that can provide the new movement with leaders, members, phone lines, copying machines, and other resources.

Insurgent consciousness is the individual sense that change is both needed and possible.

TABLE 15.1 **Social Movement Outcomes**
According to William Gamson, the outcomes of social movements take four possible forms. These outcomes depend on whether the movement achieves its goals and whether it gains acceptance from society at large.

Level of Goal Achievement	Level of Social Acceptance	
	Considerable social acceptance	**Little social acceptance**
Many goals achieved	*Outcome*: Success *Example*: The U.S. abolitionist movement. After the Civil War, slavery was abolished. Eventually, most Americans supported this change.	*Outcome*: Preemption *Example*: Feminism Most Americans now agree that women deserve equal rights but still equate feminism with man-hating.
Few goals achieved	*Outcome*: Cooptation *Example*: The "green housing" movement. Most Americans agree we should use less energy at home. Builders now use the "green" label to sell huge, energy-sucking homes with a few "green" details like insulated windows.	*Outcome*: Collapse *Example*: The U.S. movement to legalize prostitution. Earned little social acceptance, achieved no goals (except in a few counties in Nevada), and essentially disappeared.

beat them. The growth of the women's movement depended upon convincing women that these were not merely personal problems but rather were problems they shared with other women *simply because they were women*. This point is neatly summed up in the feminist slogan "The personal is political."

Why Movements Succeed or Fail

Why do some movements succeed while others disappear? Based on a historical review of 53 diverse social movement organizations (SMOs), sociologist William Gamson (1990) identified four possible outcomes of social movement activities. A fully successful SMO is one that both *achieves its goals* and *wins acceptance* as a legitimate, reputable organization. Nelson Mandela's African National Congress, for example, now controls the government of the Republic of South Africa and has improved the situation of South Africa's black population enormously. Other SMOs, however, have not been as successful. Some SMOs are *co-opted* when their rhetoric and ideology gain nominal public approval, but the real social changes they had advocated have not occurred. Other SMOs are *preempted* when those in power adopt their goals and programs but continue to denigrate the organization and its ideology; many politicians, for example, now support the idea of equal pay for equal work but continue to belittle the feminists who brought the issue to public attention. Still other SMOs have little lasting effect on society. Table 15.1 outlines the four movement outcomes discussed by Gamson.

Empirical analysis of social movements in the United States and around the world suggests that a number of factors are important for movement success. Movements are most likely to succeed if they contain diverse organizations using diverse tactics, if they can garner sufficient resources, and if they can frame their goals and ideology in a way that attracts and keeps members.

Diverse Organizations and Tactics

A social movement is the product of the activities of dozens and even hundreds of groups and organizations, all pursuing, in their own way, the same general goals. For example, there are probably dozens of different SMOs within the environmental movement, ranging from the relatively conventional Audubon Society and Sierra Club to the radical Greenpeace organization and the ecoterrorists of the Earth Liberation Front (ELF). The organizations within a movement may be highly divergent and may compete with each other for participants and supporters. Because this assortment of organizations provides avenues of participation for people with a variety of goals and styles, however, the existence of diverse SMOs is functional for the social movement.

SMOs can be organized in one of two basic ways: as professional or as volunteer organizations. On the one hand, we have organizations such as the American Civil Liberties Union or the National Rifle Association, which have offices in Washington, D.C., and a relatively large paid staff, some of whom are professional fund-raisers or lobbyists who develop an interest in an issue only after being hired. At the other extreme is the SMO staffed on a volunteer basis by people who are personally involved—for example, neighbors who organize in the church basement to prevent a nuclear power plant from being built in their neighborhood. These two types of SMO are referred to, respectively, as the *professional* SMO and the *indigenous* SMO.

Evidence suggests that the existence of both types of organizations facilitates a social movement. The professional SMO is usually more effective at soliciting resources from foundations, corporations, and government agencies. It appeals to individuals who are ideologically or morally committed to the group's cause. On the other hand, because employees of professional SMOs are not themselves underprivileged and because they work daily with the establishment, professional SMOs sometimes lose the sense of grievance that is necessary to motivate continued, imaginative efforts for change. As a result, a social movement also requires sustained indigenous organizations (Jenkins & Eckert 1986). Indigenous organizations perform two vital functions. First, by keeping the aggrieved group actively supportive of the social movement, they help to maintain the sense of urgency necessary for sustained effort. Second, their anger and grievance propel them to more direct-action tactics (sit-ins, demonstrations, and the like) that publicize the cause and keep it on the national agenda.

The feminist movement is an excellent example of a social movement that combines both professional and indigenous SMOs. Informal networks continue to keep the discussion of equal rights and equal opportunities alive, even in periods when professional SMOs are nonexistent or marginalized. The most successful periods of feminist activism have been when professional SMOs, such as the National Organization for Women (NOW), worked in close cooperation with indigenous SMOs made up of informal networks and passionate individuals (Buechler 1993). In the absence of direct actions—candlelight vigils for victims of wife abuse, boycotts of pornography stores, or equal rights rallies—pressure from both professional and indigenous SMOs can produce only modest results, at best.

Mobilizing Resources

Mobilization is the process through which a social movement gains needed resources, of many types. These resources may be weapons, technologies, goods, money, or members. The resources available to a social movement depend on two factors: the amount of personal resources controlled by movement members and

Mobilization is the process by which a social movement gains control of new resources.

For a social movement to succeed, it needs to mobilize many resources—sometimes including weapons.

the proportion of those resources that members will contribute to the movement. Thus, mobilization can proceed by increasing the size of the membership, increasing the proportion of assets that members are willing to give to the group, or recruiting richer members. Mobilization can also mean getting other organizations to work with a social movement. For example, the civil rights movement relied on aid from African American churches, and the anti-pornography movement has garnered support from both fundamentalist churches and feminist organizations.

Organizational factors also affect the odds that an SMO will succeed. Most importantly, SMOs must be able to mobilize sufficient resources to achieve their ends. Those resources can take many forms. During the spring of 2006, tens of thousands of high school students across the country walked out of their schools in protest against proposed anti-immigration legislation. These students were mobilized virtually overnight through text messaging and cell phone calls—movement resources that Karl Marx never envisioned. In addition, SMOs are more likely to be successful when individuals must actively participate in the movement to derive any of the benefits from its victories. SMOs are also more likely to succeed if they have a centralized, bureaucratic structure; are able to avoid infighting; and cultivate alliances with other organizations (Gamson 1990).

Frame Alignment

Political process theory has pointed to the importance of frame alignment for attracting and mobilizing new members. **Frame alignment** is the process that movements use to convince individuals that their interests, values, and beliefs are complementary to those of the SMO (Benford & Snow 2000; Snow et al. 1986). The Sierra Club, for example, might mail pamphlets to members of the Audubon Society in hopes of convincing them to join. It also might hold public meetings in a town plagued by pollution in hopes of convincing parents that their children's illnesses are caused by pollution, not by bad luck or bad genes. Other organizations, like cults and extremist

Frame alignment is the process used by a social movement to convince individuals that their personal interests, values, and beliefs are complementary to those of the movement.

If a missionary has ever tried to convince you to save your soul by joining his religion, that missionary was engaging in frame alignment: trying to convince you that your interests and those of his movement overlapped. The same is true whenever someone running for student government or someone hoping you will join Greenpeace tries to convince you that their movement's interests, values, and beliefs mesh well with yours.

groups, try to gain new members by convincing individuals that the way they have seen things is entirely wrong.

Who is most likely to be recruited through frame alignment? Studies of social movement activists show that, although ideology and grievances are important in bringing in new participants, the key factor is personal ties and networks. No matter how deeply committed individuals might be to a movement's ideology, they are not likely to become active members unless they belong to a network of like-minded others. Conversely, they also are unlikely to become active if their friends, relatives, and acquaintances oppose the movement (McAdam 1986; McAdam & Paulsen 1993).

Countermovements

Countermovements are social movements that seek to reverse or resist changes advocated by an opposing movement (Lo 1982; Meyer & Staggenborg 1996). Countermovements can arise in response to any movement and can be either left-wing or right-wing.

Countermovements are most likely to develop if three conditions are met (Meyer & Staggenborg 1996). First, the original movement must have achieved moderate success. If the movement appears unsuccessful, then few will feel it worth their while to oppose it. Conversely, if the movement appears totally successful, then opposition will seem futile. Most tobacco smokers, for example, simply accepted new restrictions on smoking in the workplace rather than trying to resist them. On the other hand, when cities passed laws banning smoking in restaurants and bars, smokers realized that they had new allies: bar and restaurant owners who feared loss of customers. As a result, a countermovement has appeared to fight these laws.

Second, countermovements only arise when individuals feel that their status, power, or social values are threatened. This is most likely to happen if the original movement frames its goals broadly. The nineteenth-century temperance movement, which opposed all alcohol use, generated a strong countermovement. In contrast, the current movement against drunk driving, which identifies individual drunk drivers as the problem rather than alcohol consumption per se, has met almost no opposition.

Third, countermovements emerge when individuals who feel threatened by a new movement can find powerful allies. Those allies can come from within political parties, unions, churches, or any other important social group. Again, the alliance between smokers and bar owners is an example.

The conflict over abortion provides an excellent example of the interrelationship between movements and countermovements. The abortion rights movement of the 1960s was a quiet campaign, largely run by political elites—doctors, lawyers, and women active in mainstream political groups. For this reason, perhaps, it received little media coverage (Luker 1985). Its victory in the 1973 *Roe v. Wade* Supreme Court decision caught the country by surprise and galvanized the antiabortion movement (Meyer & Staggenborg 1996). That countermovement drew its supporters from women and men who believed that the legalization of abortion threatened religion, the stability of the family, and traditional ideas regarding women's nature and role. The antiabortion movement gained further support through highly visible, "newsworthy" actions that won media coverage for its views. In the years since *Roe v. Wade*, both the movement and the countermovement have sought political allies—the pro-choice movement primarily within the Democratic Party and the antiabortion movement primarily within the Republican Party. Neither group, however, has yet achieved a decisive legal victory.

A **countermovement** seeks to reverse or resist change advocated by an opposing social movement.

As these antiabortion and pro-choice protesters illustrate, whenever a social movement succeeds in creating social change, a countermovement is likely to develop.

© Sylvia Johnson/Woodfin Camp & Associates

Case Study: How the Environmental Movement Works

Being in favor of protecting the environment sounds like an innocuous position to take. After all, who is in favor of polluted air, dirty water, and disappearing species? Yet by default, nearly all of us are.

Our modern lifestyle depends on ruining the environment. The average American produces 35 pounds of garbage each week but recycles only a tiny fraction of this. Environmental protection, on the other hand, carries significant costs that few care to bear: higher-priced goods, more bother over recycling, more regulation, fewer consumer goods, and the loss of some jobs. Despite this apparent ill fit between environmentalism and modern life, the environmental movement continues to fight for its cause.

The Battle over Environmental Policy

This battle is being fought on many fronts—nuclear power, oil exploration in protected areas, hazardous wastes, forests, and suburban sprawl. Sometimes the battle takes extreme forms. "Mink liberators" in Utah have released animals from fur farms, bombed the fur breeder's cooperative that provides most of the food for the state's $20-million-a-year mink industry, and even set fire to a leather store. The Earth Liberation Front (ELF) announced that it firebombed and destroyed a $12-million mountaintop restaurant and ski-lift facility in 1998 to protect the last, best lynx habitat in Colorado (Glick 2001). Elsewhere, groups protesting suburban sprawl have set fire to sport utility vehicles and luxury home construction sites. Although many environmentalists disagree with this illegal sabotage, the spokesperson for one ELF cell says, "We know that the real 'ecoterrorists' are the white male industrial and corporate elite. They must be stopped" (Murr & Morganthau 2001).

Although militants do much to publicize and galvanize the environmental movement, they cannot succeed on their own. Arson, freeing animals, and bombing may

AP Images

Ecoterrorists who oppose suburban sprawl and the sale of gas-guzzling vehicles have taken actions such as spray-painting sport-utility vehicles and burning dealerships where SUVs are sold.

buy time, but permanent victory in protecting forests, wildlife, and the rest of the environment involves court orders, legal battles, and other strategies. Thus, both professional and indigenous, conservative and radical SMOs help to push the movement forward.

The professional SMOs of the environmental movement—the Sierra Club, the Environmental Defense Fund, the National Audubon Society, and others—write letters to congressional representatives to urge support for clean-air laws or to lobby against dam projects or unrestrained suburban growth. They pay a battery of lawyers to get court injunctions when needed and to push for change in government policies. And, increasingly, they work with corporations to develop corporate policies that will protect the environment without hurting those corporations' bottom lines. For example, the Environmental Defense Fund prodded FedEx to use delivery trucks with hybrid fuel systems. This shift reduced air pollution, gasoline consumption, and FedEx's costs while burnishing the company's public image (Deutsch 2006; FedEx 2006).

The Environmental Movement Assessed

One reason corporations and federal agencies have adopted more environmentally friendly policies is that concern for the environment has increased markedly over the last two decades; most Americans now say they are willing to pay more taxes to clean up the environment.

Reflecting this growing public support, the environmental movement has had some notable successes. These include the rise in recycling, the establishment of new wilderness areas, and the passage of the Endangered Species Act. However, since the 1980s, increased anti-government, anti-tax, and pro-business sentiment has dramatically limited economic and political support for environmental protection. Moreover, as the economy has faltered, Americans have become less willing to sacrifice economic growth for environmental benefits: In 2009, for the first time in 25 years, Americans surveyed by Gallup Poll researchers rated protecting the economy as more important than protecting the environment (Figure 15.2). Similarly, another large, random poll conducted in 2009 found that 85 percent of Americans rated the economy a top priority, but only 41 percent rated the environment a top priority (Pew Research Center for the People and the Press 2009). If Americans continue to believe that environmental protectionism threatens their livelihoods, then the environment and the environmental movement are likely to suffer.

Technology

In social movements, individuals consciously aim to change their society. In other cases, people's intentions are more modest but may lead to great social change nonetheless. Such is the case with technology.

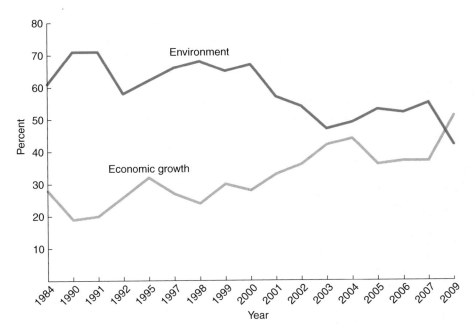

FIGURE 15.2 Environmental versus Economic Concerns
In 2009, for the first time in 25 years, Americans surveyed by Gallup Poll researchers rated protecting the economy as more important than protecting the environment.
SOURCE: Gallup.com (2009).

Technology is more pervasive than ever in our daily lives. Perhaps you woke up to an alarm this morning to find coffee already brewed in your preset electric coffee-maker, checked your cell phone for messages, and listened to MP3 files on your laptop, all before you made it to your first class. Technology is also more powerful and danger-ous than ever before: The lethal power of a car or nuclear bomb is far greater than that of a horse-drawn cart or sword. It is vitally important, then, that we think about the social changes that technology can bring.

Technology is defined as the human application of knowledge to the making of tools and to the use of natural resources. It is important to note that the term *technology* refers not only to the tools themselves (material culture) but also to our beliefs, values, and attitudes toward them (nonmaterial culture). While we may be inclined to think of technology in terms of today's high-tech advances, it also includes relatively simple tools such as pottery and woven baskets. Thus, technology has been a component of culture from the beginning of human society.

Because technology defines the limits of what a society can do, technological in-novation is a major impetus to social change. As we saw in Chapter 4, technology helped to transform hunting, fishing, and gathering societies to horticultural, then agricultural, and then industrial societies. Currently, new technologies are developing to meet new needs created by a changing culture and society. The result is a never-ending cycle in which social change both causes and results from new technology. In this section, we briefly review two theories of technologically induced social change and present a case study of how information technology may change society. We then discuss the benefits and costs of two new technologies: information technology and reproductive technology.

Theoretical Perspectives on Technology and Social Change

Since the nineteenth century, sociologists have been interested in the link between technology and social change; as we saw in Chapter 1, many early scholars entered

Technology involves the human application of knowledge to the making of tools and to the use of natural resources.

sociology because of their interest in the sources and consequences of the Industrial Revolution, an event that triggered dramatic social change. This section explores how structural functionalism and conflict theory explain the connections between technology and social change.

Structural-Functional Theory: Technology and Evolutionary Social Change

While structural-functional theory primarily asks how social organization is maintained in an orderly way, the theory does not ignore the fact that societies and cultures change. As pointed out in Chapter 1, according to the structural-functional perspective, change occurs through evolution: Social structures adapt to new needs and demands in an orderly way, while outdated patterns, ideas, and values gradually disappear. Often, the new needs and demands that prompt this evolution are technological advances.

But even if change is evolutionary, it does not always occur smoothly. One reason for this is that changes in one aspect of a culture invariably affect other aspects. Structural-functionalists believe that typically cultures will adapt to these changes, but recognize that adaptation may take a while. As a result, societies can experience a "cultural lag" during which some aspects of a culture haven't kept up with changes in other aspects. For example, the rise of factories led to skyrocketing rates of industrial accidents beginning in the 1870s, but laws providing compensation to injured workers were not passed until the 1920s—a cultural lag of about 50 years. Cultural lag is the temporary period of maladjustment during which the social structure adapts to new technologies.

Conflict Theory: Technology, Power, and Social Change

While structural functionalism sees social change as orderly and generally consensual, conflict theorists contend that change—including the adoption of new technologies—results from conflict between competing interests. Furthermore, conflict theorists assert that those with greater power can direct technological and social change to their own advantage. In a process characterized by conflict and disruption, social structure changes (or does not change) as powerful groups act either to alter or to maintain the status quo.

According to Thorstein Veblen (1919), those for whom the status quo is profitable are said to have a vested interest in maintaining it. **Vested interests** represent stakes in either maintaining or transforming the status quo; people or groups who would suffer from social change have a vested interest in maintaining the status quo, while those who would profit from social change have a vested interest in transforming it. Electric companies have a vested interest in promoting electric cars; gas companies have a vested interest in impeding this. College students have a vested interest in downloading textbooks for free from the Internet; publishers have a vested interest in preventing this.

Just as the benefits of a particular technology are unevenly distributed, so also are the costs. Conflict theorists argue that costs tend to go to the less powerful. Pollution-producing factories, which can earn great profits for corporations, are typically located in poor neighborhoods and never located in places like Beverly Hills or Scarsdale.

Like evolutionary theories, the conflict perspective on social change makes intuitive sense to many, and there is empirical evidence to support it. A general assumption of the conflict perspective is that those with a disproportionate share of society's wealth, status, and power have a vested interest in preserving the status quo. In today's

Vested interests are stakes in either maintaining or transforming the status quo.

rapidly changing society, this may no longer be the case, as powerful factions may be just as likely to support as to oppose technological innovations. Microsoft, for example, is fully in favor of developing new technologies that it can profit from, like Windows Vista, even while it works to impede innovations that others control, like Linux and Apple software and computers. Furthermore, some scholars have argued that technology is virtually "autonomous." That is, once the necessary supporting knowledge is developed, a particular invention—like the personal computer or the atomic bomb—will be created by someone. And once created, it will be used. In other words, technological changes may be put in motion by social forces beyond our effective control.

The Costs and Benefits of New Technologies

Almost all of us are glad that personal computers now exist: Their benefits are obvious, and the problems they create seem small by comparison. Far fewer of us are happy that the atomic bomb exists, although most Americans were happy that our government was able to use it during World War II.

As these examples suggest, new technologies always offer both benefits and costs, many of which are not immediately obvious. Focus on a Global Perspective: India Meets the Cell Phone on the next page explores the many ways that cell phones are affecting Indian society.

In the following paragraphs, we discuss two examples of new technologies: *new reproductive technologies* and *information technology*. We then explore two general problems that can arise along with any new technologies : the *technological imperative* and *normal accidents*.

New Reproductive Technologies

New reproductive technologies—some simple, some complex—have substantially expanded the options of women and men who want children who are genetically related to them. Men whose wives are infertile can have their sperm inseminated into another woman who agrees to serve as a "surrogate mother" (usually for a fee). Women whose husbands are infertile can be inseminated with another man's sperm. Women who cannot conceive can have their eggs surgically removed, fertilized by sperm in a test tube, and then surgically implanted in their uteruses. The same technology enables women who lack viable eggs (including post-menopausal women) to bear children using another woman's eggs. These technologies are available not only to childless couples but also to single men and women and to gay and lesbian couples. Currently, about 50,000 babies are born each year as a result of these technologies (U.S. Bureau of the Census 2009a). An unknown additional number of babies were born when women were inseminated vaginally after taking prescription hormones or undergoing surgical procedures to restore their fertility

Although these reproductive technologies have increased childbearing options, some sociologists have raised concerns about their health, social, and ethical implications (Rothman 2000). The potential health problems are numerous. Women who take prescription hormones to increase their chances of conceiving risk breast cancer or ovarian cancer in the future. Other women face long, difficult, and potentially life-threatening pregnancies when these hormones leave them carrying twins, triplets ... or even septuplets. The children they give birth to are disproportionately likely to be born prematurely, and as a result to have greater risks of a wide variety of lifelong cognitive and health problems. Finally, women who undergo surgical procedures for infertility face all the dangers inherent in any surgery.

focus on A GLOBAL PERSPECTIVE

India Meets the Cell Phone

For people in the United States, having a cell phone makes life more convenient. For people in India, a cell phone can change everything.

Cell phone usage is growing more rapidly in India than anywhere else in the world, with about one-third of all Indians now owning one (Giridharadas 2009). Moreover, cell phone ownership has grown among the poor as well as the wealthy and in small towns and villages as well as in cities.

Cell phones have proven so popular because they serve so many different needs (Giridharadas 2009). Until recently, few Indians could afford or obtain land-line telephones. Meanwhile, those who did have land-line telephones could rarely use them with any privacy, since phones typically were placed in central locations for easy sharing by all members of the family. Other electronic equipment, such as cameras, DVD players, and stereos, also remain relatively rare, while even the small percentage who own computers or laptops rarely have Internet connections.

In this context, the cell phone has proven revolutionary. As in the United States, Indians now use their cell phones as flashlights and as a means of connecting to the Internet, keeping a daily calendar, taking photos, and so on. The difference is that in India, most have no other tools available for these tasks.

The cell phone has also dramatically increased access to privacy for Indians—especially young people. In a society in which arranged marriages remain the norm and social contact between unmarried men and women is viewed with suspicion, Indian young people happily use their cell phones to surreptitiously text or call members of the opposite sex. As a result, cell phones are changing ideas about both privacy and romance.

Finally, cell phones are changing political life in India. Activist groups now use cell phones to broadcast information about political candidates, journalists use them to poll viewers on current events and politics, and citizens use them to send political comments to television stations that run these comments as an on-screen "crawl." These actions have already had an impact on some local elections and court cases (Giridharadas 2009).

In other ways, however, the cell phone has become a new way of reinforcing old cultural and social divisions in India. As one observer wrote, cell phones

announc[e] who outranks whom. Small people have small phones, and big people have big ones. Small people have numerical-soup numbers, and big people have numbers that end in 77777 or something equally important-sounding or easy to remember. Small people have one phone, and big people have two. Small people set their phones merely to ring, and big people make Bollywood songs play when you call them. (Giridharadas 2009, WK3)

It seems, then, that like other technologies in other cultures, cell phone usage in India both reflects the existing culture and has considerable power to change that culture.

The social and ethical problems implicit in new reproductive technologies are more subtle. Perhaps most important, some of these techniques have low success rates, especially with older women. Even those who eventually give birth typically have to endure several cycles of treatment costing around $12,000 each before they have a baby. Yet the constant development of new techniques makes it difficult for childless individuals and couples to decide to adopt or to accept their childlessness. Finally, these technologies raise the question of whether we are turning children into commodities available to the highest bidder; they also may encourage a narrow definition of parenthood as having genetic ties to a child rather than a broader definition of parenthood as loving and raising a child.

Information Technology

Consider the college student in 1970 who is assigned the task of writing a term paper on the consequences of parental divorce. She goes to the library and walks through the periodicals section until she stumbles on the *Journal of Marriage and the Family*, in which she eventually finds five articles—the number her professor requires—on her topic. She takes notes on three-by-five-inch cards (there are no photocopying machines) and goes home to draft her paper on her new electric typewriter. She cuts

Because today's computers are better and cheaper than those of even 10 years ago, a very large portion of all college students bring their own to campus with them. In fact, some colleges now require students to have their own notebook computer.

and tapes together her draft copy, moving sections around until it looks good, checks words of dubious spelling in her dictionary, and then retypes a final copy. She uses carbon paper to make a copy for herself. (Ask your mom or dad to explain this to you.) When she makes a mistake, she erases it carefully and tries to type the correction in the original space.

Now consider a student today. This student starts her paper by logging onto *Sociological Abstracts*, an online bibliography of more than 100,000 sociology articles. When she enters the keywords *divorce* and *parental*, the program responds with full citations and summaries for 41 articles. After identifying and downloading the 5 articles she wants, the student drafts a report on her laptop, edits it to her satisfaction, runs it through her spelling checker, and adjusts the vocabulary a bit by using her laptop's built-in thesaurus. She also runs the report through her grammar checker, which will catch errors in punctuation, capitalization, and so forth. Finally, she sends the whole thing to her mother (who lives 2,000 miles away) by e-mail and asks her to read it for logic and organization. She receives the edited version from her mother in an hour, prints two copies, and hands in the report. Or she may send the paper to her instructor via e-mail.

Information technology—computers and telecommunication tools for storing, using, and sending information—has changed many aspects of our daily lives. Over the past few decades, the United States has become an "information society." More and more people work in information acquisition, processing, and communication. Aside from enabling us to write term papers more easily, how will information technology change our lives in the future? Will it reduce or increase social-class inequality? Will it make life safer and better? Or will it make life more stressful and isolated?

The answer is likely to be some of each. As shown in Map 15.1, access to the Internet has spread rapidly—if unevenly—around the world. This means we can link via computer to distant family and friends, to doctors and medical information, to libraries and databanks, and to world events. For example, U.S. soldiers in Iraq can stay in touch with their families via e-mail and web cams, and U.S. residents can follow the

Information technology comprises computers and telecommunication tools for storing, using, and sending information.

MAP 15.1: **Percent of Residents with Internet Access**

SOURCE: International Telecommunications Union. World Telecommunication/ICT Indicators 2008. **http://www.itu.int/ITU-D/ict/statistics.** Accessed June 2009.

situation in Iraq on Internet chat rooms, blogs, RSS feeds, and tweets. Both soldiers and citizens also can follow events on 24-hour satellite and cable news stations (some broadcast from Europe or the Arab world). Iraqi citizens, meanwhile, can use their computers to find out both where the most recent bombs exploded and who won the Academy Awards. Similarly, during the highly contested 2009 presidential election in Iran, Iranian citizens used Twitter, Facebook, blogs, and other Internet sites to obtain and share information that countered that available via government-controlled newspapers and television

Information technology also allows us to participate more fully in the political process by making it possible to communicate more effectively and directly with our elected officials. By linking us to distant work sites, computers and e-mail allow us to work from home, reducing the time spent commuting to work and increasing the time we have for friends and families.

On the downside, advances in information technology have introduced new forms of crime (hacking and electronic theft), new defense worries (breaches of defense data systems and faulty software programs that may inadvertently launch World War III), new health problems (eyestrain and repetitive stress injuries), and new inefficiencies ("I'm sorry, the system is down"). They also have introduced new forms of social control. Information technology has given corporations, the police, lawyers, and government bureaucrats, among others, greater ability to build databases about you, combining information on the cars you buy, the websites you visit, and the type of music you like with whether or not you have recently married, moved, had a child, or received a speeding ticket. Similarly, others now can obtain access to your computer files, deleted e-mail messages, and phone logs. One survey of 1,000 major corporations showed that almost two-thirds of these corporations engage in some form of "electronic surveillance" of their employees (Rosen 2000). Finally, new technologies have lengthened the number of working hours in a day, as notebook computers, e-mail, faxes, BlackBerries, and cell phones increasingly invade our homes and even vacations.

The long-term effect of information technology on society will depend as much on social institutions as it does on the technological capacities of computers and telecommunications. Information technology offers us more freedom of residence and more input into local and federal legislative bodies, but we simultaneously lose some privacy and autonomy. Whether the blessings or costs will predominate will depend on how these technologies are implemented in schools, workplaces, and government bureaucracies. To the extent that they affect relationships among work, class, neighborhood, and family, the new technologies are of vital interest to those concerned with social institutions.

Making the best use of advancing technology and helping to ensure that advances prompt desirable social changes require social planning—the conscious and deliberate process of investigating, discussing, and coming to agreement about desirable actions based on common values.

The Technological Imperative

As we've already noted, once the knowledge needed to devise a certain technology is available, that new technology is likely to appear and to gain adherents. But we can make an even stronger statement: Once that technology is available, it becomes more and more difficult for anyone to decide against using it.

At home and even on vacation, new communication technologies keep us connected to our offices and the world of work.

Consider the automobile. In 1925, any city dweller who had enough money could choose to commute to work by car. But if he chose not to do so, he could rely on a broad network of trolleys running on a frequent schedule to get him to his destination. He almost certainly lived fairly close to where he worked and could also choose to enjoy the walk instead. These days, the automobile has become completely enmeshed in our way of life. Billions of dollars in public subsidies pay for road building and parking lots and keep down the price of oil and gas for consumers. Meanwhile, public transportation has been cut to the bone. In many cities, walking or bicycling is dangerous or unpleasant because of high-speed traffic or freeways that divide neighborhoods.

This situation is an example of the **technological imperative**: the idea that once a technology is available, it becomes difficult to avoid using it. Think how annoyed people sometimes feel when their friends don't use cell phones, e-mail, or instant messaging and the pressures on holdouts to get these technologies.

Normal Accidents

As our lives come increasingly to depend on highly complex and interconnected technologies, our vulnerability to technological problems increases exponentially. In the nineteenth century, most people got water from wells and used candles for lighting. If a well dried up or a house burned down, the disaster was limited to no more than a few households. Now we get our water from municipal water systems and our electricity from electric companies. When things go wrong, they go wrong big time.

The blackout of August 2003 provides a perfect example of this vulnerability. Electricity is provided to American households by a network of cooperating utility

Technological imperative refers to the idea that once a technology becomes available, it becomes difficult to avoid using it.

Air travel is exceedingly safe. But because both the jets themselves and the air traffic system are so complex, tragedies are bound to happen sooner or later. Such tragedies are known as *normal accidents*.

Ellen McKnight/Alamy

Normal accidents are accidents that can be expected to happen sooner or later, no matter how many safeguards are built into a system, simply because the system is so complex.

companies sharing a vast grid of electric cables. This grid depends on complex computerized technologies, designed to spread the demand over a broad region and reduce the chance of overloading the system in any one region. But because of its complexity and interconnectedness, a small problem can quickly mushroom to a huge problem for a huge area. In 2003, for example, overloaded circuits in the Midwest caused 50 million people in the Midwest, Northeast, and Canada to lose electric power for as much as several hours.

The 2003 blackout highlighted how dependent we have become on technology, and how vulnerable we are when that technology fails. Because the system for distributing water to consumers runs on electricity, the blackout left thousands without water. Flashlight batteries ran down, leaving people with only candles for lighting. Many found themselves with no way of communicating with friends and relatives. Laptops and PDAs quickly ran out of power, while cell phone networks either lost power or became overloaded, so people could neither phone nor e-mail. Even those who had working phones could not telephone others if they kept their phone directories on computers.

This process is an example of a *normal accident.* **Normal accidents** are accidents that can be expected to happen sooner or later, no matter how many safeguards are built into a system, simply because the system is so complex (Perrow 1984). Normal accidents such as space shuttle crashes, accidental releases of radiation from nuclear power plants, and electrical blackouts are the price we pay for modern technology.

Where This Leaves Us

Whether it originates in a social movement or in a new technology, any social change will have opponents. Every winner potentially produces a loser. This means that change creates a situation of competition and conflict.

In Chapter 1, we discussed the appropriate role of sociologists in studying social issues. Should they be value free, or should they take a stand? Issues of social change and conflict bring this question into sharp focus. Although most sociologists restrict their work to teaching and research, a vocal minority argue that sociologists should take a more active role in monitoring and even creating social change. They believe that sociologists should be actively involved in helping individuals understand and resolve the conflicts that arise from competition, inequality, and social change.

What can sociologists contribute to ensure that social changes enhance social justice? Three particularly useful things sociologists can do are:

- *Study conflict resolution.* A growing number of universities have special courses or programs on conflict resolution. These courses are concerned with the development of techniques for handling disputes and negotiating peaceful settlements that can lead to positive social changes. Sociological research on topics such as small-group decision making and organizational culture are relevant here.
- *Develop social justice perspectives.* At its core, sociology is concerned with the interaction of social groups and the role that power plays in those interactions. In their research and teaching, sociologists can explore how individuals, groups, and nations obtain and use power and how that power can be distributed and used more equitably.
- *Model social change strategies.* Sociological research may lead to the development of more effective programs for improving the well-being of individuals and social groups, from Head Start programs to transnational investments.

The involvement of sociologists in issues of conflict resolution, social justice, and social change is not likely to be the crucial factor that creates a better world. We can be sure, however, that scholarly neglect of these issues is both shortsighted and immoral. To the extent that developing knowledge of the principles of human behavior will help us reduce social conflict, we have an obligation—as scholars, students, and citizens—to seek out knowledge and to apply it. Our future depends on this.

Summary

1. Collective behavior and social movements, although related, are distinct activities. Collective behavior is spontaneous and unplanned; a social movement is organized, goal oriented, and long term.

2. According to relative-deprivation theory, social movements arise when individuals experience an unacceptable gap between what they have and what they expect to have. Expectations are derived from comparisons with other groups and other points in time.

3. Resource mobilization theory argues that social movements emerge when individuals are able to bring together the resources needed to create social change.

4. Political process theory builds on resource mobilization theory by recognizing that in addition to access to political opportunities and resources, successful movements must build a sense among participants that change is both needed and possible.

5. A successful movement needs a diverse range of organizations to accomplish different goals. It also must be able to mobilize needed resources of all sorts. To get new members, it must frame its ideology in ways that convince individuals that a problem is serious, that taking action on a problem is both proper and effective, and that individuals' interests, values, and beliefs mesh well with those of the movement. Regardless of ideology, however, individuals are most likely to be recruited when they have social ties to movement members and lack ties to movement opponents. Finally, successful movements need innovative tactics that will garner media attention.

6. Countermovements are social movements that seek to resist or reverse changes advocated by other social movements. A countermovement is most likely to develop if the original movement achieves modest success, if some individuals feel that their social position or values are

threatened by changes achieved by the original movement, and if potential countermovement participants believe that they will have powerful allies.

7. In its effort to affect public policy, the environmental movement uses a variety of tactics, ranging from courtroom battles to sabotage. Among the reasons for the movement's growing successes are the wide variety of SMOs within the movement.

8. Technology is the human application of knowledge to the making of tools and hence to humans' use of natural resources. The term refers not only to the tools themselves (aspects of material culture) but also to people's beliefs, values, and attitudes regarding those tools (aspects of nonmaterial culture).

9. Social change is any significant modification or transformation of social structures or institutions over time. Technology is one important type and cause of social change.

10. Structural-functional theory primarily asks how technology contributes to orderly and positive social change.

Cultural lag can be a serious problem when a technology enters a society too quickly for the culture to adapt to the changes it brings.

11. Conflict theorists contend that technological change results from and reflects conflict between competing interests. People or groups who would either suffer or profit from social change have vested interests—stakes in either maintaining or transforming the status quo.

12. Information technology has changed many aspects of our daily lives. It links us to people and information but has also created new defense worries, new inefficiencies, new forms of social control, and new illnesses and injuries. Similarly, new reproductive technologies have expanded the options of those who want children genetically related to them. At the same time, they have raised serious health, social, and ethical questions, such as whether we are turning children into commodities available to the highest bidder.

Thinking Critically

1. What social structural conditions in the larger society do you think helped spark the environmental movement? What countermovements do you know of that may impact the movement's success?

2. Suppose you were interested in mobilizing public opinion against the death penalty. What kind of activity or event would you try to use to get the media's attention?

3. How would you analyze the current debate over affirmative action policies and programs in terms of various groups' vested interests?

4. Europeans have opposed genetically modified plants and food much more vigorously than have Americans. How

would you explain this difference based on your understanding of the factors that make societies more or less likely to adopt new technologies and attitudes (see Chapter 2) and on your understanding of how social movements are able to successfully mobilize?

5. If you were to run for office, how would you use e-mail in your campaign? Which groups of your constituents would you be more likely to hear from via the Internet? How would you know whether they were actually U.S. citizens with the legal right to vote—or would it matter? How might you make sure that other voices, those without high-speed data ports and modems, were heard as well?

Book Companion Website

www.cengage.com/sociology/brinkerhoff
Prepare for quizzes and exams with online resources—including tutorial quizzes, a glossary, interactive flash cards, crossword puzzles, essay questions, virtual explorations, and more.

Glossary

A

Accounts are explanations of unexpected or untoward behavior. They are of two sorts: excuses and justifications.

An **achieved status** is a status that an individual earns, such as being a criminal or a college graduate.

The **agents of socialization** are all the individuals, groups, and media that teach social norms.

Agricultural societies are based on growing food using plows and large beasts of burden.

Alienation occurs when workers have no control over the work process or the product of their labor.

Anomie is a situation in which the norms of society are unclear or no longer applicable to current conditions.

Anticipatory socialization is the process that prepares us for roles we are likely to assume in the future.

An **ascribed status** is fixed by birth and inheritance and is unalterable in a person's lifetime.

Assimilation is the process through which individuals learn and adopt the values and social practices of the dominant group, more or less giving up their own values in the process.

An **authoritarian personality** is submissive to those in authority and antagonistic toward those lower in status.

Authoritarian systems are political systems in which the leadership is not selected by the people and legally cannot be changed by them.

Authoritarianism is the tendency to be submissive to those in authority, coupled with an aggressive and negative attitude toward those lower in status.

Authority is power supported by norms and values that legitimate its use.

B

A **blended family** includes children born to one parent as well as children born to both parents.

The **bourgeoisie** is the class that owns the tools and materials for their work—the means of production.

Bureaucracy is a special type of complex organization characterized by explicit rules and hierarchical authority structure, all designed to maximize efficiency.

C

Capitalism is the economic system, based on competition, in which most wealth (land, capital, and labor) is private property, to be used by its owners to maximize their own gain.

Caste systems rely largely on ascribed statuses as the basis for distributing scarce resources.

Charisma refers to extraordinary personal qualities that set an individual apart from ordinary mortals.

Charismatic authority is the right to make decisions based on perceived extraordinary personal characteristics.

Churches are religious organizations that have become institutionalized. They have endured for generations, are supported by and support society's norms and values, and have become an active part of society.

Civil religion is the set of institutionalized rituals, beliefs, and symbols sacred to the U.S. nation.

Class, in Marxist theory, refers to a person's relationship to the means of production.

Class consciousness occurs when people understand their relationship to the means of production and recognize their true class identity.

Class systems rely largely on achieved statuses as the basis for distributing scarce resources.

Coercion is the exercise of power through force or the threat of force.

Cohabitation means living with a romantic/sexual partner outside of marriage.

Cohesion in a group is characterized by high levels of interaction and by strong feelings of attachment and dependency.

Collective behavior is spontaneous action by groups in situations where cultural rules for behavior are vague, inadequate, or debated.

Collective efficacy refers to the extent to which individuals in a neighborhood share the expectation that neighbors will intervene and work together to maintain social order.

Color-blind racism refers to the belief that all races are created equal, that racial equality has already been achieved, and that therefore any minorities who do not succeed have only themselves to blame.

The **commodification of children** is the process of turning children into goods available for purchase.

A **community** is a collection of individuals characterized by dense, cross-cutting social networks.

Competition is a struggle over scarce resources that is regulated by shared rules.

Complex organizations are large formal organizations with elaborate status networks.

Compulsive heterosexuality consists of continually demonstrating one's masculinity and heterosexuality.

Concentrated poverty refers to areas in which very high proportions of the population live in poverty.

Conflict is a struggle over scarce resources that is not regulated by shared rules; it may include attempts to destroy, injure, or neutralize one's rivals.

Conflict theory addresses the points of stress and conflict in society and the ways in which they contribute to social change.

Consumerism is the philosophy that says "buying is good" because "we are what we buy."

Content analysis refers to the systematic examination of documents of any sort.

A **control group** is the group in an experiment that does not receive the independent variable.

Cooperation is interaction that occurs when people work together to achieve shared goals.

Core societies are rich, powerful nations that are economically diversified and relatively free from outside control.

Correlation exists when there is an empirical relationship between two variables (for example, income increases when education increases).

Countercultures are groups that have values, interests, beliefs, and lifestyles that are opposed to those of the larger culture.

A **countermovement** seeks to reverse or resist change advocated by another social movement.

Credentialism is the assumption that some are better than others simply because they have a particular educational credential.

Crime is behavior that is subject to legal or civil penalties.

Cross-sectional design uses a sample (or cross section) of the population at a single point in time.

Crude birth rate refers to the number of live births per 1,000 persons in a given population.

Crude death rate refers to the number of deaths per 1,000 persons in a given population.

Cultural capital refers to having the attitudes and knowledge that characterize the upper social classes.

Cultural diffusion is the process by which aspects of one culture or subculture are incorporated into another.

Cultural lag occurs when one part of a culture changes more rapidly than another.

Cultural relativity requires that each cultural trait be evaluated in the context of its own culture.

Culture is the total way of life shared by members of a community. It includes not only language, values, and symbolic meanings but also technology and material objects.

The **culture of poverty** is a set of values that emphasizes living for the moment rather than thrift, investment in the future, or hard work.

Culture shock refers to the discomfort that arises from exposure to a different culture.

D

Deduction is the process of moving from theory to data by testing hypotheses drawn from theory.

The **deinstitutionalization of marriage** refers to the gradual erosion of social norms that stress the need for marriage and dictate how spouses should behave.

Democracies are political systems that provide regular, constitutional opportunities for a change in leadership according to the will of the majority.

Demographic transition refers to the shift from a society characterized by high birth rates and low life expectancies to one characterized by low birth rates and high life expectancies.

Demography is the study of population—its size, growth, and composition.

A **denomination** is a church that accommodates both to the society at large and to the presence of other churches.

The **dependent variable** is the effect in cause-and-effect relationships. It is dependent on the actions of the independent variable.

Deterrence theories suggest that deviance results when social sanctions, formal and informal, provide insufficient rewards for conformity.

Development refers to the process of increasing the productivity and standard of living of a society—longer life expectancies, more adequate diets, better education, better housing, and more consumer goods.

Deviance refers to norm violations that exceed the tolerance level of the community and result in negative sanctions.

Dialectic philosophy views change as a product of contradictions and conflict between the parts of society.

A **differential** is a difference in the incidence of a phenomenon across social groups.

Differential association theory argues that people learn to be deviant when more of their associates favor deviance than favor conformity.

A **disclaimer** is a verbal device employed in advance to ward off doubts and negative reactions that might result from one's conduct.

Discrimination is the unequal treatment of individuals on the basis of their membership in categories.

Double jeopardy means having low status on two different dimensions of stratification.

Dramaturgy is a version of symbolic interaction that views social situations as scenes manipulated by the actors to convey the desired impression to the audience.

Dysfunctions are consequences of social structures that have negative effects on the stability of society.

E

Economic determinism means that economic relationships provide the foundation on which all other social and political arrangements are built.

The **economy** consists of all social structures involved in the production and distribution of goods and services.

Edge cities are suburban areas that now have an existence largely separate from the cities that spawned them.

Education is the institution responsible for the formal transmission of knowledge.

Emerging churches are characterized by (1) the belief that American life and modern Christian churches are atomized, bureaucratic, and inauthentic and (2) an emphasis on informal rituals, a more open perspective toward scripture and behavior, and living a life of mission, faith, and community.

Emotional labor refers to the work of smiling, appearing happy, or in other ways suggesting that one enjoys providing a service.

Empirical research is research based on systematic, unbiased examination of evidence.

Endogamy is the practice of choosing a mate from within one's own racial, ethnic, or religious group.

Environmental racism refers to the disproportionately large number of health and environmental risks faced by minorities.

An **ethnic group** is a category whose members are thought to share a common origin and important elements of a common culture.

Ethnocentrism is the tendency to judge other cultures according to the norms and values of one's own culture.

Ex-felon disenfranchisement is the loss of voting privileges suffered by those who have been convicted of a felony. In some states, ex-felon disenfranchisement applies only to those in prison; in other states, it is lifelong.

Exchange is a voluntary interaction from which all parties expect some reward.

Excuses are accounts in which one admits that the act in question is wrong or inappropriate but claims one couldn't help it.

The **experiment** is a method in which the researcher manipulates independent variables to test theories of cause and effect.

An **experimental group** is the group in an experiment that experiences the independent variable. Results for this group are compared with those for the control group.

An **extended family** is a family in which a couple and their children live with other kin, such as the wife's or husband's parents or siblings.

Exogamy is the practice of choosing a mate from *outside* one's own racial, ethnic, or religious group.

F

False consciousness is a lack of awareness of one's real position in the class structure.

The **family** is a group of persons linked together by blood, adoption, marriage or quasi-marital commitment

The **fertility rate** is the number of births per every 1,000 women in a population during a given time period.

Folkways are norms that are the customary, normal, habitual ways a group does things.

Formal social controls are administrative sanctions such as fines, expulsion, or imprisonment.

A **frame** is an answer to the question, What is going on here? It is roughly identical to a definition of the situation.

Frame alignment is the process used by a social movement to convince individuals that their personal interests, values, and beliefs are complementary to those of the movement.

Functions are consequences of social structures that have positive effects on the stability of society.

Fundamentalism refers to religious movements that stress traditional interpretations of religion and the importance of living in ways that mesh with those traditional interpretations.

G

Gemeinschaft refers to societies in which most people share close personal bonds.

Gender refers to the expected dispositions and behaviors that cultures assign to each sex.

Gender roles refer to the rights and obligations that are normative for men and women in a particular culture.

The **generalized other** is the composite expectations of all the other role players with whom we interact; it is Mead's term for our awareness of social norms.

Genocide refers to mass killings aimed at destroying a population.

Gesellschaft refers to societies in which people are tied primarily by impersonal, practical bonds.

Globalization refers to the process through which ideas, resources, practices, and people increasingly operate in a worldwide rather than local framework.

Globalization of culture is the process through which cultural elements (including musical styles, fashion trends, and cultural values) spread around the globe.

A **group** is two or more people who interact on the basis of shared social structure and recognize mutual dependency.

Groupthink exists when pressures to agree are strong enough to stifle critical thinking.

H

The **health belief model** proposes that individuals will be most likely to adopt healthy behaviors if (1) they believe their health is at risk, (2) they believe the risk is a serious one, (3) they believe that changing their behaviors would significantly reduce those risks, and (4) they face no significant barriers that would make changing their behaviors difficult.

Heterogamy means choosing a mate who is *different* from oneself.

The **hidden curriculum** consists of the underlying cultural messages taught by schools. Both public and private schools teach young people to accept inequality.

High culture refers to the cultural preferences associated with the upper class.

Homogamy means choosing a mate who is similar to oneself.

Homosexuals (also known as gays and lesbians) are people who prefer sexual and romantic relationships with members of their own sex.

Horticultural societies are characterized by small-scale, simple farming, without plows or large beasts of burden.

Hunting-and-gathering societies are those in which most food must be obtained by killing wild animals or finding edible plants.

A **hypothesis** is a statement about relationships that we expect to find if our theory is correct.

I

The **id** is the natural, unsocialized, biological portion of self, including hunger and sexual urges.

An **ideology** is a set of beliefs that strengthen or support a social, political, economic, or cultural system.

Immigration is the movement of people to find new homes in a different country.

Impression management consists of actions and statements made to control how others view us.

Incidence is the frequency with which an attitude or behavior occurs.

Income refers to money received in a given time period by an individual, household, or organization.

Income inequality refers to the extent to which incomes vary within a given population.

The **independent variable** is the cause in cause-and-effect relationships.

The **indirect inheritance model** argues that children have occupations of a status similar to that of their parents because the family's status and income determine children's aspirations and opportunities.

Induction is the process of moving from data to theory by devising theories that account for empirically observed patterns.

Industrial societies are characterized by mass production of nonagricultural goods.

The **infant mortality rate** is the number of babies who die during or shortly after childbirth per every 1,000 live births in a given population.

Informal social control is self-restraint exercised because of fear of what others will think.

Information technology comprises computers and telecommunication tools for storing, using, and sending information.

Institutionalized racism occurs when the normal operation of apparently neutral processes systematically produces unequal results for majority and minority groups.

An **institution** is an enduring social structure that meets basic human needs.

Internal migration is the movement of people to new homes within a country.

Insurgent consciousness is the individual sense that change is both needed and possible.

J

Justifications are accounts that explain the good reasons the violator had for choosing to break the rule; often they are appeals to some alternate rule.

L

Labeling theory is concerned with the processes by which labels such as *deviant* come to be attached to specific people and specific behaviors.

Language is the ability to communicate in symbols—orally, by manual sign, or in writing.

Latent functions or dysfunctions are consequences of social structures that are neither intended nor recognized.

Laws are rules that are enforced and sanctioned by the authority of government. They may or may not be norms.

Least-developed countries are characterized by poverty and political weakness and rank low on most or all measures of development.

Less-developed countries are those nations whose living standards are worse than those in the most-developed countries but better than in the least-developed nations.

The **linguistic relativity hypothesis** argues that the grammar, structure, and categories embodied in each language affect how its speakers see reality. Also known as the *Sapir-Whorf hypothesis.*

Longitudinal research is any research in which data are collected over a long period of time.

The **looking-glass self** is the process of learning to view ourselves as we think others view us.

M

Macrosociology focuses on social structures and organizations and the relationships between them.

A **majority group** is a group that is culturally, economically, and politically dominant.

Manifest functions or dysfunctions are consequences of social structures that are intended or recognized.

The **manufacturers of illness** are groups that promote and benefit from deadly behaviors and social conditions.

Marriage is an institutionalized social structure that provides an enduring framework for regulating sexual behavior and childbearing.

The **mass media** are all forms of communication designed to reach broad audiences.

McDonaldization is the process by which the principles of the fast-food restaurant—efficiency, calculability, predictability, and control—are coming to dominate more sectors of American society.

Medicalization refers to the process through which a condition or behavior becomes defined as a medical problem requiring a medical solution.

A **metropolitan statistical area** is a county that has a city of 50,000 or more in it plus any neighboring counties that are significantly linked, economically or socially, with the core county.

Microsociology focuses on interactions among individuals.

Migration is the movement of people from one geographic area to another.

A **minority group** is a group that is culturally, economically, and politically subordinate.

Mobilization is the process by which a social movement gains control of new resources.

Modernization theory sees development as the natural unfolding of an evolutionary process in which societies go from simple to complex economies and institutional structures.

Monogamy is the term for marriages in which there is only one wife and one husband.

Moral entrepreneurs are people who attempt to create and enforce new definitions of morality.

Mores are norms associated with fairly strong ideas of right or wrong; they carry a moral connotation.

The **mortality rate** is the number of deaths per every 1,000 people in a given population during a given time period.

Most-developed countries are those rich nations that have relatively high degrees of economic and political autonomy.

Multiculturalism is the belief that the different cultural strands within a culture should be valued and nourished.

N

The **near poor** live in households earning from just above the federal poverty level to twice the federal poverty level.

New religious movements (NRMs) are religious or spiritual movements begun in recent decades and not derived from a nation's mainstream religions.

A **nonmetropolitan statistical area** is a county that has no major city in it and is not closely tied to such a city.

The **norm of reciprocity** is the expectation that people will return favors and strive to maintain a balance of obligation in social relationships.

Normal accidents are accidents that can be expected to happen sooner or later, no matter how many safeguards are built into a system, simply because the system is so complex.

Norms are shared rules of conduct that specify how people ought to think and act.

A **nuclear family** is a family in which a couple and their children form an independent household living apart from other kin.

O

An **operational definition** describes the exact procedure by which a variable is measured.

Operationalizing refers to the process of deciding exactly how to measure a given variable.

Organizational culture refers to the pattern of norms and values that structures how business is actually carried out in an organization.

P

Participant observation refers to conducting research by participating, interviewing, and observing "in the field."

The **peer group** refers to all individuals who share a similar age and social status.

A **peer** is a member of a peer group.

Peripheral societies are poor and weak, with highly specialized economies over which they have relatively little control.

Pluralism is the peaceful coexistence of separate and equal cultures in the same society.

Political economy refers to the interaction of political and economic forms within a nation.

Politics is the social structure of power within a society.

Political opportunities are resources that allow a social movement to grow; they include preexisting organizations that can provide the new movement with leaders, members, phone lines, copying machines, and other resources.

Political process theory suggests that social movements develop when political opportunities are available and when individuals have developed a sense that change is both needed and possible.

Polygamy is any form of marriage in which a person may have more than one spouse at a time.

Popular culture refers to aspects of culture that are widely accessible and commonly shared by most members of a society, especially those in the middle, working, and lower classes.

Postindustrial societies focus on producing either information or services.

Power is the ability to direct others' behavior even against their wishes.

The **power elite** comprises the people who occupy the top positions in three bureaucracies—the military, industry, and the executive branch of government—and who are thought to act together to run the United States in their own interests.

Prejudice is an irrational, negative attitude toward a category of people.

Prestige refers to the amount of social honor or value afforded one individual or group relative to another. Also referred to as *status*.

Primary groups are groups characterized by intimate, face-to-face interaction.

The **primary sector** is that part of the economy concerned with extracting raw materials from the environment.

Primary socialization is personality development and role learning that occurs during early childhood.

Privatization is the process through which government services are "farmed out" to corporations, redesigned to follow corporate structures and goals, or redefined as individual responsibilities.

The **profane** represents all that is routine and taken for granted in the everyday world, things that are known and familiar and that we can control, understand, and manipulate.

Professional socialization is the process of learning the knowledge, skills, and cultural values of a profession.

Professions are occupations that demand specialized skills and creative freedom.

The **proletariat** is the class that does not own the means of production. They must support themselves by selling their labor to those who own the means of production.

Propinquity is spatial nearness.

The **Protestant Ethic** refers to the belief that work, rationalism, and plain living are moral virtues, whereas idleness and indulgence are sinful.

R

A **race** is a category of people treated as distinct because of *physical* characteristics to which *social* importance has been assigned.

Racism is the belief that inherited physical characteristics associated with racial groups determine individuals' abilities and personalities and provide a legitimate basis for unequal treatment.

Random samples are samples chosen through a random procedure, so that each individual in a given population has an equal chance of being selected.

Rational-legal authority is the right to make decisions based on rationally established rules.

Reference groups are groups that individuals compare themselves to regularly, either because they identify with the group or aspire to it.

Relative deprivation exists when we compare ourselves to others who are better off than we are.

Relative-deprivation theory argues that social movements arise when people experience an intolerable gap between their expectations and the rewards they actually receive.

Religion is a system of beliefs and practices related to sacred things that unites believers into a moral community.

Religiosity is an individual's level of commitment to religious beliefs and to acting on those beliefs.

Religious economy refers to the competition between religious organizations to provide better "consumer products," thereby creating greater "market demand" for their own religions.

Replication is the repetition of empirical studies by another researcher or with different samples to see if the same results occur.

Reproductive labor refers to traditionally female tasks such as cooking, cleaning, and nurturing that make it possible for a society to continue and for others to work and play.

Resocialization is the process of learning a new self-concept and a radically different way of life (often against our will).

Resource mobilization theory suggests that social movements emerge when individuals who experience deprivation can garner the resources they need to mobilize for action.

Rites of passage are formal rituals that mark the end of one age status and the beginning of another.

A **role** is a set of norms specifying the rights and obligations associated with a status.

Role conflict is when incompatible role demands develop because of multiple statuses.

Role strain is when incompatible role demands develop within a single status.

Role taking involves imagining ourselves in the role of others in order to determine the criteria they will use to judge our behavior.

S

The **sacred** consists of events and things that we hold in awe and reverence—what we can neither understand nor control.

Sampling is the process of systematically selecting representative cases from the larger population.

Sanctions are rewards for conformity and punishments for nonconformity.

The **Sapir-Whorf hypothesis** argues that the grammar, structure, and categories embodied in each language affect how its speakers see reality. Also known as the *linguistic relativity hypothesis.*

Scapegoating occurs when people or groups blame others for their failures.

School choice refers to a range of options (vouchers, tax credits, magnet and charter schools, home schooling) that enable families to choose where their children go to school.

Secondary groups are groups that are formal, large, and impersonal.

The **secondary sector** is that part of the economy concerned with the processing of raw materials.

Sects are religious organizations that arise in active rejection of changes they find repugnant in churches.

Secularization is the process of transferring things, ideas, or events from the sacred realm to the nonsacred, or secular, realm.

Segregation refers to the physical separation of minority- and majority-group members.

The **self-concept** is our sense of who we are as individuals.

Self-fulfilling prophecies occur when something is *defined* as real and therefore *becomes* real in its consequences.

Sex is a biological characteristic, male or female.

Sexism is a belief that men and women have biologically different capacities and that these form a legitimate basis for unequal treatment.

Sexual harassment consists of unwelcome sexual advances, requests for sexual favors, or other verbal or physical conduct of a sexual nature.

Sexual scripts are cultural expectations regarding who, where, when, why, how, and with whom one should have sex.

The **sick role** consists of four social norms regarding sick people. They are assumed to have good reasons for not fulfilling their normal social roles and are not held responsible for their illnesses. They are also expected to consider sickness undesirable, to work to get well, and to follow doctor's orders.

Significant others are the role players with whom we have close personal relationships.

A **single-payer system** is a health care system in which doctors and hospitals are paid solely by the government.

Social change is any significant modification or transformation of social structures and sociocultural processes over time.

Social class is a category of people who share roughly the same class, status, and power and who have a sense of identification with each other.

The **social construction of race and ethnicity** is the process through which a culture (based more on social ideas than on biological facts) defines what constitutes a race or an ethnic group.

Social control consists of the forces and processes that encourage conformity, including self-control, informal control, and formal control.

Social-desirability bias is the tendency of people to color the truth so that they sound more desirable and socially acceptable than they really are.

Social interaction refers to the ways individuals interact with others in everyday, face-to-face situations.

Social mobility is the process of changing one's social class.

A **social movement** is an ongoing, goal-directed effort to fundamentally challenge social institutions, attitudes, or ways of life.

A **social network** is an individual's total set of relationships.

Social processes are the forms of interaction through which people relate to one another; they are the dynamic aspects of society.

A **social structure** is a recurrent pattern of relationships among groups.

Socialism is an economic structure in which productive tools (land, labor, and capital) are owned and managed by the workers and used for the collective good.

Socialization is the process of learning the roles, statuses, and values necessary for participation in social institutions.

A **society** is the population that shares the same territory and is bound together by economic and political ties.

Sociobiology is the study of the biological basis of all forms of human (and nonhuman) behavior.

Socioeconomic status (SES) is a measure of social class that ranks individuals on income, education, occupation, or some combination of these.

The **sociological imagination** is the ability to see the intimate realities of our own lives in the context of common social structures; it is the ability to see personal troubles as public issues.

Sociology is the systematic study of human society, social groups, and social interactions.

The **sociology of everyday life** focuses on the social processes that structure our experience in ordinary, face-to-face situations.

A **spurious relationship** exists when one variable *seems* to cause changes in a second variable, but a third variable is the *real* cause of the change.

Stakeholder mobilization refers to organized political opposition by groups with a vested interest in a particular political outcome.

The **state** is the social structure that successfully claims a monopoly on the legitimate use of coercion and physical force within a territory.

A **state church** is one that is strongly supported or even mandated by the government.

Status is an individual's position within a group relative to other group members; also social honor, expressed in lifestyle.

Status set refers to the combination of all statuses held by an individual.

A **stereotype** is a preconceived, simplistic idea about the members of a group.

Strain theory suggests that deviance occurs when culturally approved goals cannot be reached by culturally approved means.

Stratification is an institutionalized pattern of inequality in which social statuses are ranked on the basis of their access to scarce resources.

Strong ties are relationships characterized by intimacy, emotional intensity, and sharing.

Structural-functional theory addresses the question of social organization (structure) and how it is maintained (function).

Subcultures are groups that share in the overall culture of society but also maintain a distinctive set of values, norms, and lifestyles and even a distinctive language.

Suburbanization is the growth of suburbs.

Suburbs are communities that develop outside of cities and that, historically, primarily provided housing rather than services or employment.

The **superego** is composed of internalized social ideas about right and wrong.

Survey research is a method that involves asking a relatively large number of people the same set of standardized questions.

Symbolic interaction theory addresses the subjective meanings of human acts and the processes through which people come to develop and communicate shared meanings.

T

Technological imperative refers to the idea that once a technology becomes available, it becomes difficult to avoid using it.

Technology involves the human application of knowledge to the making of tools and to the use of natural resources.

Terrorism is the deliberate and unlawful use of violence against civilians for political purposes.

The **tertiary sector** is that part of the economy concerned with the production of services.

A **theory** is an interrelated set of assumptions that explains observed patterns.

Total institutions are facilities in which all aspects of life are strictly controlled for the purpose of radical resocialization.

Tracking occurs when evaluations made relatively early in a child's career determine the educational programs the child will be encouraged to follow.

Traditional authority is the right to make decisions for others that is based on the sanctity of time-honored routines.

Transgendered persons are individuals whose sex or sexual identity is not definitively male or female. Some are hermaphrodites, some are transsexuals.

Transnational corporations are large corporations that produce and distribute goods internationally.

A **trend** is a change in a variable over time.

U

Underemployed people hold jobs more appropriate for someone with fewer skills or hold part-time jobs only because they can't find full-time jobs.

The **underground economy** is economic activity associated with workers who are trying to hide from state regulation such as prostitutes, unlicensed contractors, and work by undocumented laborers.

Unemployed people are those who lack a job, are available for work, *and* are actively seeking work.

Urbanization is the process of concentrating populations in cities.

V

Value-free sociology concerns itself with establishing what is, not what ought to be.

Values are shared ideas about desirable goals.

Variables are measured characteristics that vary from one individual or group to the next.

Vested interests are stakes in either maintaining or transforming the status quo.

Victimless crimes such as drug use, prostitution, gambling, and pornography are voluntary exchanges between persons who desire illegal goods or services from each other.

Voluntary associations are nonprofit organizations designed to allow individuals an opportunity to pursue their shared interests collectively.

W

A **war** is an armed conflict between a national army and some other group.

Weak ties are relationships characterized by low intensity and little intimacy.

Wealth refers to the sum value of money and goods owned by an individual or household.

White-collar crime refers to crimes committed by respectable people of high status in the course of their occupation.

White privilege refers to the benefits whites receive simply because they are white.

World-systems theory is a conflict perspective of the economic relationships between developed and developing countries, the core and peripheral societies.

Z

Zero population growth exists when the fertility rate is about 2.1 births per woman, the rate needed to maintain the population at a steady size.

References

Abma J. C., G. M. Martinez, W. D. Mosher, and B. S. Dawson. 2004. "Teenagers in the United States: Sexual activity, Contraceptive Use, and Childbearing, 2002." National Center for Health Statistics. Vital and Health Statistics 23(24):1–48.

Agadjanian, Victor, and Cecilia Menjívar. 2008. "Talking about the 'Epidemic of the Millennium': Religion, Informal Communication, and HIV/AIDS in Sub-Saharan Africa." Social Problems 55:301–321.

Alba, Richard D., John R. Logan, and Brian J. Stults. 2000. "How Segregated Are Middle-Class African Americans?" Social Problems 47:543–558.

Alba, Richard, and Victor Nee. 2003. Remaking the American Mainstream: Assimilation and Contemporary Immigration. Cambridge, MA: Harvard University Press.

Alcock, John. 2001. The Triumph of Sociobiology. New York: Oxford University Press.

Ali, Jennifer, and William R. Avison. 1997. "Employment Transitions and Psychological Distress: The Contrasting Experiences of Single and Married Mothers." Journal of Health and Social Behavior 38:345–362.

Allan, Emilie, and Darrell Steffensmeier. 1989. "Youth Unemployment and Property Crime." American Sociological Review 54:107–123.

Allen, Katherine R., Rosemary Blieszner, and Karen A. Roberto. 2000. "Families in the Middle and Later Years: A Review and Critique of Research in the 1990s." Journal of Marriage and the Family 62:911–926.

Alpert, Sheri A. 2003. "Protecting Medical Privacy: Challenges in the Age of Genetic Information." Journal of Social Issues 59(2):301–322.

Altheide, David. 2002. Creating Fear: News and the Construction of Crisis. Piscataway, NJ: Aldine Transaction.

———. 2006. Terrorism and the Politics of Fear. Lanham, MD: AltaMira Press.

Amanat, Abbas. 2001. "Empowered through Violence: The Reinventing of Islamic Extremism." In Strobe Talbott and Nayan Chanda (eds.), The Age of Terror. New York: Basic Books.

Amaro, Nelson, Christopher Chase-Dunn, and Susanne Jonas. 2001. Globalization on the Ground: Postbellum Guatemalan Democracy and Development. Lanham, MD: Rowman & Littlefield Publishers.

Amato, Paul R. 1993. "Urban-Rural Differences in Helping Friends and Family Members." Social Psychology Quarterly 56:249–262.

Amato, Paul R., Alan Booth, David R. Johnson, and Stacy J. Rogers. 2007. Alone Together: How Marriage in America Is Changing. Cambridge, MA: Harvard University Press.

American Association of Retired Persons. 1999. The AARP Grandparenting Survey: Sharing and Caring between Mature Grandparents and Their Grandchildren. Washington, D.C.: American Association of Retired Persons.

Anderson, David C. 1998. Sensible Justice: Alternatives to Prison. New York: Norton.

Anne E. Casey Foundation. 2009. Juvenile Detention Alternatives Initiative. http://www.aecf.org/MajorInitiatives/JuvenileDetentionAlternativesInitiative/JDAIResults.aspx. Accessed April 2009.

Appelbaum, Richard. 2005. Critical Globalization Studies. New York: Routledge.

"Arab American Demographics." 2006. Arab American Institute. http://www.aaiusa.org/arab-americans/22/demographics. Accessed May 2006.

Arana-Ward, Marie. 1997. "As Technology Advances, a Bitter Debate Divides the Deaf." The Washington Post. May 11:A11.

Arrigo, Bruce A. (ed.). 1998. Social Justice/Criminal Justice. Belmont, CA: Wadsworth.

Arrington, Leonard J., and Davis Bitton. 1992. The Mormon Experience: A History of the Latter-Day Saints (2nd ed.). Urbana, IL: University of Illinois Press.

Asch, Solomon. 1955. "Opinions and Social Pressure." Scientific American 193:31–35.

Atkinson, Michael. 2003. Tattooed: The Sociogenesis of a Body Art. Toronto, Ontario: University of Toronto Press.

Atkinson, Michael, and Kevin Young. 2001. "Flesh Journeys: Neo Primitives and the Contemporary Rediscovery of Radical Body Modification." Deviant Behavior 22:117–146.

Austin, Roy, and Mark Allen. 2000. "Racial Disparity in Arrest Rates as an Explanation of Racial Disparity in Commitment to Pennsylvania's Prisons." Journal of Research in Crime and Delinquency 37:200–220.

Babbie, Earl. 2010. The Practice of Social Research (12th ed.). Belmont, CA: Wadsworth.

Bai, Ruoyun. 2003. "Chicken Wings." Studies in Symbolic Interaction 26:263–265.

Banks, James, Michael Marmot, Zoe Oldfield, and James P. Smith. 2006. "Disease and Disadvantage in the United States and in England." Journal of the American Medical Association 295:2037–2045.

Barber, Benjamin R. 2001. Jihad vs. McWorld: How Globalism and Tribalism Are Reshaping the World (rev. ed.). New York: Ballantine.

Barker, Eileen. 1986. "Religious Movements: Cult and Anticult since Jonestown." Annual Review of Sociology 12:329–346.

Barker, Kristin. 2005. The Fibromyalgia Story: Biomedical Authority and Women's Worlds of Pain. Philadelphia: Temple University Press.

Bayer, Ada-Helen, and Leon Harper. 2000. Fixing to Stay: A National Study on Housing and Home Modification Issues. Washington, D.C.: American Association of Retired Persons.

Bearman, Peter S., and James Moody. 2004. "Adolescent Suicidality." American Journal of Public Health 4:89–95.

Beaver, William. 2009. "A Matter of Degrees." Contexts 8:22–26.

Becker, Howard S. 1963. Outsiders: Studies in the Sociology of Deviance. New York: Free Press.

Becker, Marshall H. (ed.). 1974. The Health Belief Model and Personal Health Behavior. San Francisco: Society for Public Health Education.

———. 1993. "A Medical Sociologist Looks at Health Promotion." Journal of Health and Social Behavior 34:1–6.

Beckwith, Carol. 1983. "Niger's Wodaabe: 'People of the Taboo.'" National Geographic 164(10):482–509.

Belfield, Clive R. 2004. "Home-Schooling in the U.S." Occasional Paper No. 88. National Center for the Study of Privatization in Education, Teachers College, Columbia University.

Bellah, Robert N. 1974. "Civil Religion in America." In Russell B. Richey and Donald G. Jones (eds.), American Civil Religion. New York: Harper & Row.

Bellah, Robert N., Richard Madsen, William M. Sullivan, Ann Swidler, and Steven M. Tipton. 1985. Habits of the Heart: Individualism and Commitment in American Life. Berkeley: University of California Press.

Belsky, Jay, Deborah L. Vandell, Margaret Burchinal, Alison K. Clarke-Stewart, Kathleen McCartney, and Margaret T. Owen. 2007. "Are There Long-Term Effects of Early Child Care?" Child Development 78:681–701.

Benford, Robert D., and David A Snow. 2000. "Framing Processes and Social Movements: An Overview and Assessment." Annual Review of Sociology 26:611–639.

Bergen, Doris. 2003. War & Genocide: A Concise History of the Holocaust. Lanham, MD: Rowman & Littlefield.

Bergen, Raquel Kennedy. 1998. Issues in Intimate Violence. Thousand Oaks, CA: Sage.

Berger, Magdalena, Todd H. Wagner, and Laurence C. Baker. 2005. "Internet Use and Stigmatized Illness." Social Science & Medicine 61(8):1281–1287.

Berliner, David C., and Bruce J. Biddle. 1995. The Manufactured Crisis: Myths, Fraud, and the Attack on America's Public Schools. New York: Longman.

Bernanke, Ben S. 2009. "The Crisis and the Policy Response. Board of Governors of the Federal Reserve System." http://www.federalreserve.gov/newsevents/speech/bernanke20090113a.htm. Accessed May 2009.

Bertrand, Marianne, and Sendhil Mullainathan. 2004. "Are Emily and Greg More Employable Than Lakisha and Jamal? A Field Experiment on Labor Market Discrimination." American Economic Review 94:991–1013.

Bettie, Julie. 2003. Women without Class: Girls, Race, and Identity. Berkeley: University of California Press.

Bianchi, Suzanne M., John P. Robinson, and Melissa A. Milkie. 2006. Changing Rhythms of American Family Life. ASA Rose Monograph Series. New York: Russell Sage.

Binson, D., S. Michaels, R. Stall, T. J. Coates, J. H. Gagnon, and J. A. Catania. 1995. "Prevalence and Social Distribution of Men Who Have Sex with Men: United States and Its Urban Centers." Journal of Sex Research 32:245–254.

Blass, Thomas. 1999. "The Milgram Paradigm after 35 Years: Some Things We Now Know about Obedience to Authority." Journal of Applied Social Psychology 25:955–978.

Blau, Joel. 2001. Illusions of Prosperity: America's Working Families in an Age of Economic Insecurity. New York: Oxford University Press.

Blau, Peter M. 1987. "Contrasting Theoretical Perspectives." In J. Alexander, B. Giesen, R. Munch, and N. Smelser (eds.), The Micro-Macro Link. Berkeley: University of California Press.

Blendon, Robert, and John T. Young. 1998. "The Public and the War on Illicit Drugs." Journal of the American Medical Association 279:827–832.

Blum, Deborah. 2002. Love at Goon Park: Harry Harlow and the Science of Affection. Cambridge, MA: Perseus Publishing.

Blumstein, Phillip, and Pepper Schwartz. 1983. American Couples. New York: William Morrow.

Bobo, Lawrence, and Vincent Hutchings. 1996. "Perceptions of Racial Group Competition: Extending Blumer's Theory of Group Position to a Multiracial Social Context." American Sociological Review 61:951–972.

Bobo, Lawrence, and James R. Kluegel. 1993. "Opposition to Race Targeting: Self-Interest, Stratification Ideology, or Racial Attitudes?" American Sociological Review 58:443–464.

Bonacich, Edna, Lucie Cheng, Norma Chinchilla, Nora Hamilton, and Paul Ong. 1994. Global Production. The Apparel Industry in the Pacific Rim. Philadelphia: Temple University Press.

Bonilla-Silva, Eduardo. 2006. Racism without Racists: Color-Blind Racism and the Persistence of Racial Inequality in the United States. Lanham, MD: Rowman & Littlefield.

Booth, Alan, and D. Wayne Osgood. 1993. "The Influence of Testosterone on Deviance in Adulthood: Assessing and Explaining the Relationship." Criminology 31: 93–117.

Bordo, Michael D., Alan M. Taylor, and Jeffrey G. Williamson. 2003. Globalization in Historical Perspective. National Bureau of Economic Research. Chicago: University of Chicago Press.

Borgonovi, Francesca. 2008. "Doing Well by Doing Good. The Relationship between Formal Volunteering and Self-Reported Health and Happiness." Social Science & Medicine 66:2321–2334.

Borjas, George J., and Lawrence F. Katz. 2007. "The Evolution of the Mexican-Born Workforce in the United States." In George J. Borjas (ed.), Mexican Immigration. Chicago: University of Chicago Press.

Bourdieu, Pierre. 1984. Distinction: A Social Critique of the Stratification of Taste. Cambridge, MA.: Harvard University Press.

Bourgois, Philippe. 1995. In Search of Respect: Selling Crack in El Barrio. New York: Cambridge University Press.

Bovin, Mette. 2001. Nomads Who Cultivate Beauty: Wodaabe Dances and Visual Arts in Niger. Uppsala, Sweden: Nordiska Afrikainstitutet.

Boynton Health Service. 2007. Health and Academic Performance: Minnesota Undergraduate Students. Minneapolis, MN: Boynton Health Service.

Braithwaite, John. 1985. "White Collar Crime." Annual Review of Sociology 11:1–25.

Bramlett, M. D., and W. D. Mosher. 2002. "Cohabitation, Marriage, Divorce, and Remarriage in the United States." Vital Health Statistics 23(22):1–87.

Brinkman, Richard L., and June E. Brinkman. 1997. "Cultural Lag: Conception and Theory." International Journal of Social Economics 24:609–627.

Brissett, Dennis, and Charles Edgley. 2005. Life as Theater: A Dramaturgical Sourcebook (2nd ed.). New Brunswick, N. J.: Transaction Publishers.

Brooks, David. 2000. Bobos in Paradise: The New Upper Class and How They Got There. New York: Simon & Schuster.

Brown, David K. 2001. "The Social Sources of Educational Credentialism: Status Cultures, Labor Markets, and Organizations," Sociology of Education, Extra Issue: 19–34.

Brown, David L., and Louis E. Swanson (eds.). 2003. Challenges for Rural America in the Twenty-First Century. University Park, PA: Pennsylvania State University Press.

Brulle, Robert J., and David Naguib Pellow. 2006. "Environmental Justice: Human Health and Environmental Inequalities." Annual Review of Public Health 27:103–124.

Buechler, Steven M. 1993. "Beyond Resource Mobilization? Emerging Trends in Social Movement Theory." Sociological Quarterly 34:217–235.

Bulan, Heather Ferguson, Rebecca J. Erickson, and Amy S. Wharton. 1997. "Doing for Others on the Job: The Affective Requirements of Service Work, Gender, and Emotional Well-Being." Social Problems 44:235–255.

Bullard, Robert D., Rueben C. Warren, and Glenn S. Johnson. 2001. "The Quest for Environmental Justice." In Ronald L. Braithwaite and Sandra E. Taylor (eds.), Health Issues in the Black Community (2nd ed.). San Francisco: Jossey-Bass: 471–488.

Bunker, John P., Howard S. Frazier, and Frederick Mosteller. 1994. "Improving Health: Measuring Effects of Medical Care." Milbank Quarterly 72:225–258.

Burger, Jerry M. 2009. "Replicating Milgram: Would People Still Obey Today?" American Psychologist 64:1–11.

Burgess, Ernest W. 1925. "The Growth of the City: An Introduction to a Research Project." In Robert E. Park, Ernest W. Burgess, and Roderick D. McKenzie (eds.), The City. Chicago: University of Chicago Press: 47–62.

Burris, Beverly H. 1998. "Computerization of the Workplace." Annual Review of Sociology 24:141–157.

Burris, Val, and James Salt. 1990. "The Politics of Capitalist Class Segments: A Test of Corporate Liberalism Theory." Social Problems 37:341–359.

Byng, Michelle D. 2008. "Complex Inequalities: The Case of Muslim Americans after 9/11." American Behavioral Scientist 51:659–674.

Cahill, Spencer E. 1983. "Reexamining the Acquisition of Sex Roles: A Social Interactionist Perspective." Sex Roles 9:1–15.

Call, Vaughn, Susan Sprecher, and Pepper Schwartz. 1995. "The Incidence and Frequency of Marital Sex in a National Sample." Journal of Marriage and the Family 57:639–652.

Camacho, David E. (ed.). 1998. Environmental Injustices, Political Struggles: Race, Class, and the Environment. Durham, NC: Duke University Press.

Cancian, Francesca M., and Stacey J. Oliker. 2000. Caring and Gender. Thousand Oaks, CA: Pine Forge Press.

Cancio, A. Silvia, T. Davic Evans, and David J. Maume, Jr. 1996. "Reconsidering the Declining Significance of Race: Racial Differences in Early Career Wages." American Sociological Review 61:541–556.

Capitman, John. 2002. Defining Diversity: A Primer and a Review." Generations 26(3):8–14.

Card, David. 2005. "Is the New Immigration Really So Bad?" NBER Working Paper No. 11547. http://www.phil.frb.org/econ/conf/ immigration/card.pdf. Accessed June 2006.

Carroll, Joseph. 2005. "Who Supports Marijuana Legalization?" http://www.gallup.com/poll/19561/who-supports-marijuana-legalization.aspx. Accessed June 2009.

Centers for Disease Control and Prevention. 2009. "Overweight and Obesity: Home." http://www.cdc.gov/nccdphp/dnpa/obesity. Accessed June 2009.

Central Intelligence Agency. 2008. World Factbook 2008. Washington, D.C.: Central Intelligence Agency.

Chapkis, Wendy. 1997. Live Sex Acts: Women Performing Erotic Labor. New York: Routledge.

Charon, Joel M. 2006. Symbolic Interactionism: An Introduction, An Interpretation (9th ed.). Englewood Cliffs, NJ: Prentice-Hall.

Chase-Dunn, Christopher. 1989. Global Formation: Structure of the World Economy. London: Basil Blackwell.

Cherlin, Andrew. 1992. Marriage, Divorce, and Remarriage (rev. ed.). Cambridge, MA: Harvard University Press.

Cherlin, Andrew J. 2004. "The Deinstitutionalization of American Marriage." Journal of Marriage and Family 66:848–861.

Chesney-Lind, Meda, and Randall G. Shelden. 2004. Girls, Delinquency, and Juvenile Justice (3rd ed.). Belmont, CA: Wadsworth.

Chetkovich, Carol. 1998. Real Heat. New Brunswick, NJ: Rutgers University Press.

Childstats.gov. 2009. Single-Parent Households. Accessed May 2009.

Chirot, Daniel. 1977. Social Change in the Twentieth Century. San Francisco: Harcourt Brace Jovanovich.

———. 1986. Social Change in the Modern Era. San Diego: Harcourt Brace Jovanovich.

Chodak, Symon. 1973. Societal Development: Five Approaches with Conclusions from Comparative Analysis. New York: Oxford University Press.

Chodorow, Nancy. 1999. The Power of Feelings: Personal Meaning in Psychoanalysis, Gender, and Culture. New Haven, CT: Yale University Press.

Chubb, John E., and Terry M. Moe. 1990. Politics, Markets, and America's Schools. Washington, D.C.: Brookings Institute.

Clarke, Peter B. (ed.). 2006. Encyclopedia of New Religious Movements. New York: Routledge.

Clawson, Dan, and Tie-ting Su. 1990. "Was 1980 Special? A Comparison of 1980 and 1986 Corporate PAC Contributions." Sociological Quarterly 31:371–387.

Clendinen, Dudley, and Adam Nagourney. 2001. Out for Good: The Struggle to Build a Gay Rights Movement in America. New York: Simon & Schuster.

Coburn, David, and Evan Willis. 2000. "The Medical Profession: Knowledge, Power, and Autonomy." In Gary L. Albrecht, Ray Fitzpatrick, and Susan C. Scrimshaw (eds.), Handbook of Social Studies in Health and Medicine. Thousand Oaks, CA: Sage: 377–393.

Cockerham, William C. 1997. "The Social Determinants of the Decline of Life Expectancy in Russian and Eastern Europe: A Lifestyle Explanation." Journal of Health and Social Behavior 38:117–130.

Coleman, Marilyn, Lawrence Ganong, and Mark Fine. 2000. "Reinvestigating Remarriage: Another Decade of Progress." Journal of Marriage and the Family 62:1288–1307.

Collins, Sharon M. 1993. "Blacks on the Bubble: The Vulnerability of Black Executives in White Corporations." Sociological Quarterly 34:429–447.

———. 1997. "Black Mobility in White Corporations: Up the Corporate Ladder but Out on a Limb." Social Problems 44:55–67.

Collom, Ed. 2005. "The Ins And Outs of Homeschooling: The Determinants of Parental Motivations and Student Achievement." Education and Urban Society 37:307–335.

Conley, Dalton. 2004. The Pecking Order: Which Siblings Succeed and Why. New York: Pantheon.

Conrad, Peter. 2007. The Medicalization of Society: On the Transformation of Human Conditions into Treatable Disorders. Baltimore, MD: Johns Hopkins University Press

Cook, Philip J., and John H. Laub. 1998. "The Unprecedented Epidemic in Youth Violence." In Michael Tonry and Mark H. Moore (eds.), Youth Violence. Chicago: University of Chicago Press.

Cooley, Charles Horton. 1902. Human Nature and the Social Order. New York: Scribner's.

———. 1967. "Primary Groups." In A. Paul Hare, Edgar F. Borgotta, and Robert F. Bales (eds.), Small Groups: Studies in Social Interaction (rev. ed.). New York: Knopf. (Originally published 1909.)

Cooney, Mark. 1997. "The Decline of Elite Homicide." Criminology 35:381–407.

Coontz, Stephanie. 1997. The Way We Really Are: Coming to Terms with America's Changing Families. New York: Basic Books.

Corcoran, M. 1995. "Rags to Rags: Poverty and Mobility in the United. States." Annual Review of Sociology 21:237–267.

Correspondents of the New York Times. 2005. Class Matters. New York: Times Books.

Corsaro, William. 2003. "We're Friends, Right?": Inside Kids' Culture. Washington, D.C.: National Academies Press.

———. 2004. Sociology of Childhood (2nd ed.). Thousand Oaks, CA: Sage Press.

Coser, Lewis A. 1956. The Functions of Conflict. New York: Free Press.

Cotton, Sian, Kathy Zebracki, Susan L. Rosenthal, Joel Tsevat, and Dennis Drotar. 2006. "Religion/Spirituality and Adolescent Health Outcomes: A Review." Journal of Adolescent Health 38:472–480.

Covington, Dennis. 2009. Salvation on Sand Mountain: Snake Handling and Redemption in Southern Appalachia. New York: Da Capo Press.

Crosnoe, Robert. 2006. "The Connection between Academic Failure and Adolescent Drinking in Secondary School." Sociology of Education 79:44–60.

Crowder, Kyle, Scott J. South, and Erick Chavez. 2006. "Wealth, Race, and Inter-Neighborhood Migration." American Sociological Review 71:72–94.

Crutchfield, Robert D. 1989. "Labor Stratification and Violent Crime." Social Forces 68:489–512.

Cureton, Steven. 2000. "Justifiable Arrests or Discretionary Justice: Predictors of Racial Arrest Differentials." Journal of Black Studies 30:703–719.

Currie, Elliott. 1998. Crime and Punishment in America. New York: Henry Holt.

Curry, Theodore R. 1996. "Conservative Protestantism and the Perceived Wrongfulness of Crimes: A Research Note." Criminology 34:453–464.

Curtis, James E., Edward G. Grabb, and Douglas E. Baer. 1992. "Voluntary Association Membership in Fifteen Countries: A Comparative Analysis." American Sociological Review 57:139–152.

Curtis, James E., Douglas E. Baer, and Edward G. Grabb. 2001. "Nations of Joiners: Explaining Voluntary Association Membership in Democratic Societies." American Sociological Review 66:783–805.

DaCosta, Kimberly McClain. 2007. Making Multiracials: State, Family, and Market in the Redrawing of the Color Line. Palo Alto, CA: Stanford University Press.

Dailard, Cynthia. 2003. "Understanding 'Abstinence': Implications for Individuals, Programs and Policies." Guttmacher Report 6(5):4–6.

Dalby, Andrew. 2003. Language in Danger: The Loss of Linguistic Diversity and the Threat to Our Future. New York: Columbia University Press.

Daly, Martin, and Margo Wilson. 1983. Sex, Evolution, and Behavior (2nd ed.). Boston: Willard Grant.

Davidman, Lynn. 1991. Tradition in a Rootless World: Women Turn to Orthodox Judaism. Berkeley: University of California Press.

Davis, James, and Mark Stasson. 1988. "Small-Group Performance: Past and Future Research Trends." Advances in Group Processes 5:245–277.

Davis, Kingsley. 1973. "Introduction." In Kingsley Davis (ed.), Cities. New York: W. H. Freeman.

Davis, Kingsley, and Wilbert E. Moore. 1945. "Some Principles of Stratification." American Sociological Review 10:242–249.

Dawson, Lorne L. 2006. Comprehending Cults: The Sociology of New Religious Movements. New York: Oxford University Press.

Death Penalty Information Center. 2009a. "Facts about the Death Penalty." http://www.deathpenaltyinfo.org/documents/FactSheet.pdf. Accessed May 2009.

———. 2009b. "National Statistics on the Death Penalty and Race." http://www.deathpenaltyinfo.org/race-death-row-inmates-executed-1976#inmaterace. Accessed May 2009.

Demo, David H., and Martha J. Cox. 2000. "Families with Young Children: A Review of Research in the 1990s." Journal of Marriage and the Family 62:876–895.

DeNavas-Walt, Carmen, and Robert W. Cleveland. 2002. "Money Income in the United States: 2001." Current Population Reports, Issue 218. U.S. Census Bureau, Washington, D.C..

Deutsch, Claudia H. 2006. "Companies and Critics Try Collaboration." New York Times. May 17:D11.

Devine, Joel, Joseph Sheley, and M. Dwayne Smith. 1988. "Macroeconomic and Social Control Policy Influences in Crime Rate Changes, 1948–85." American Sociological Review 53:407–420.

Diamond, Jared M. 1997. Guns, Germs, and Steel: The Fates of Human Societies. New York: W.W. Norton.

——. 2005. Collapse: How Societies Choose to Fail or Succeed. New York: Viking.

DiMaggio, Paul, Eszter Hargittai, W. Russell Neuman, and John P. Robinson. 2001. "Social Implications of the Internet." Annual Review of Sociology 27:307–336.

Dolnick, Edward. 1993. "Deafness as Culture." The Atlantic Monthly, September: 37–53.

Domhoff, G. William. 2009. Who Rules America: Challenges to Corporate and Class Dominance (6th ed.). New York: McGraw-Hill.

Drew, Rob. 2005. "'Once More, with Irony': Karaoke and Social Class." Leisure Studies 24:371–383.

Dreze, Jean, and Amartya Sen. 1989. Hunger and Public Action. Oxford, England: Clarendon Press.

Durkheim, Emile. 1938. The Rules of Sociological Method. New York: Free Press. (Originally published 1895.)

——. 1951. Suicide: A Study in Sociology. New York: Free Press. (Originally published 1897.)

——. 1961. The Elementary Forms of the Religious Life. London: Allen & Unwin. (Originally published 1915.)

Dworkin, Shari. 2003. "Holding Back: Negotiating a Glass Ceiling on Women's Muscular Strength." In Rose Weitz (ed.), Politics of Women's Bodies: Sexuality, Appearance, and Behavior. New York: Oxford University Press.

Dye, Jane Lawler. 2005. "Fertility of American Women: June 2004." Current Population Reports P20–555. U.S. Bureau of the Census, Washington, D.C., 2005.

Eaton, William W., and Carles Muntaner. 1999. "Social Stratification and Mental Disorder." In Allan V. Horwitz and Teresa L. Scheid (eds.), A Handbook for the Study of Mental Health: Social Contexts, Theories, and Systems. Cambridge, UK: Cambridge University Press.

Eberhardt, Jennifer L., Paul G. Davies, Valerie J. Purdie-Vaughns, and Sheri Lynn Johnson. 2006. "Looking Deathworthy: Perceived Stereotypicality of Black Defendants Predicts Capital-Sentencing Outcomes." Psychological Science 17:383–386.

Eckholm, Erik. 2007. "Boys Cast Out by Polygamists Find Help." New York Times. September 9: http://www.nytimes.com/2007/09/09/us/09polygamy.html?_r=1&scp=1&sq=boys%20polygamy&st=cse.

——. 2008. "Abuses Are Found in Online Sales of Medication." New York Times. July 9: http://www.nytimes.com/2008/07/09/health/09drugs.html?scp=14&sq=internet%20health&st=cse.

Economic Policy Institute. 2009. The State of Working America. Washington, D.C.: Economic Policy Institute.

Edin, Kathryn, and Maria Kefalas. 2006. Promises I Can Keep: Why Poor Women Put Motherhood before Marriage. Berkeley: University of California Press.

Edwards, Tom. 2009. "Residential Mover Rate in U.S. Is Lowest since Census Bureau Began Tracking in 1948." http://www.census.gov/PressRelease/www/releases/archives/mobility_of_the_population013609.html. Accessed June 2009.

Ehrenreich, Barbara. 2001. Nickel and Dimed: On (Not) Getting By in America. New York: Metropolitan.

Eichenwald, Kurt. 2005. Conspiracy of Fools: A True Story. New York: Broadway Books.

Ellison, Christopher G., John P. Bartkowski, and Michelle L. Segal. 1996. "Conservative Protestantism and the Parental Use of Corporal Punishment." Social Forces 74:1003–1028.

Emergingchurch.info. Accessed June 2009.

Emerson, Michael O., and David Hartman. 2006. "The Rise of Religious Fundamentalism." Annual Review of Sociology 32:127–144.

Entman, Robert M., and Andrew Rojecki. 2000. The Black Image in the White Mind: Media and Race in America. Chicago: University of Chicago Press.

Entwisle, Doris E., and Karl L. Alexander. 1992. "Summer Setback: Race, Poverty, School Composition, and Mathematics Achievement in the First Two Years of School." American Sociological Review 57:72–84.

Envirowatch. 2006. "Environmental Contamination at Grassy Narrows." http://www.envirowatch.org/gnfnindex.htm. Accessed May 2006.

Erlanger, Steven. 2009. "Economic Crisis Pits Europe against Its Parts." New York Times. June 9:A1.

Evans, E. Margaret, Heidi Schweingruber, and Harold W. Stevenson. 2002. "Gender Differences in Interest and Knowledge Acquisition: The United States, Taiwan, and Japan." Sex Roles 47:153–167.

Evans, Sara. 2003. Tidal Wave: How Women Changed America at Century's End. New York: Free Press.

"Everybody In, Nobody Out." 2005. http://www.everybodyinnobodyout.org/DOCS/Polls.htm#GovHI. Accessed May 2006.

Fagan, Jeffery. 2006. "Death and Deterrence Redux: Science, Law and Causal Reasoning on Capital Punishment." Ohio State Journal of Criminal Law 4:255–320.

Farkas, George, and Kurt Beron. 2003. "The Detailed Age Trajectory of Oral Vocabulary Knowledge: Differences by Class and Race." Social Science Research 33:464–97.

Farkas, George, Robert P. Grobe, Daniel Sheehan, and Yuan Shuan. 1990. "Cultural Resources and School Success." American Sociological Review 55:127–142.

Feagin, Joe R., and Robert Parker. 1990. Building American Cities: The Urban Real Estate Game. Englewood Cliffs, NJ: Prentice-Hall.

Federal Bureau of Investigation. 2009. "Crime in the United States: 2007." http://www.fbi.gov/ucr. Accessed April 2009.

Federal Interagency Forum on Aging Related Statistics. 2008. Older Americans 2008: Key Indicators of Well-Being. Washington, D.C.: Federal Interagency Forum on Aging-Related Statistics, Washington, D.C.: U.S. Government Printing Office.

Federal Reserve. 2009a. "G19: Consumer Credit." http://www.federalreserve.gov/releases/g19/Current/. Accessed April 2009.

FedEx. 2006. "About FedEx: Hybrid Electric Vehicles." http://www.fedex.com/us/about/responsibility/environment/hybridelectricvehicle.html. Accessed May 2006.

Felson, Richard B. 1996. "Mass Media Effects on Violent Behavior." Annual Review of Sociology 22:103–128.

Ferguson, Ann. 2000. Bad Boys: Public Schools in the Making of Black Masculinities. Ann Arbor, MI: University of Michigan Press.

Ferriss, Abbott L. 2002. "Religion and the Quality of Life." Journal of Happiness Studies 3:199–215.

Fessenden, Ford. 2006. "Farther Afield: Americans Head Out beyond the Exurbs." New York Times. May 7:WK14.

Figueroa, Eric B., and Rose A. Woods. 2007. "Employment Outlook: 2006–2016. Monthly Labor Review November:53–85.

Filkins, Dexter. 2009. "Afghan Girls, Scarred by Acid, Defy Terror, Embracing School." New York Times. January 14:A1.

Fine, Gary Alan. 1996. Kitchens: The Culture of Restaurant Work. Berkeley: University of California Press.

Fisher, Allen P. 2003. "Still 'Not Quite as Good as Having Your Own'? Toward a Sociology of Adoption." Annual Review of Sociology 29:335–361.

Fishman, Charles. 2006. The Wal-Mart Effect: How the World's Most Powerful Company Really Works—and How It's Transforming the American Economy. New York: Penguin.

Flanagan, William G. 1993. Contemporary Urban Sociology. New York: Cambridge University Press.

Forbes Magazine. 2008. "The Forbes 400." Forbes.com. http://www.forbes.com/lists/2008/54/400list08_The-400-Richest-Americans_Rank.html. Accessed April 2009.

Foster, John L. 1990. "Bureaucratic Rigidity Revisited." Social Science Quarterly 71:223–238.

Fox, James Alan, and Marianne W. Zawitz. 2004. Homicide Trends in the United States. Washington, D.C.: Department of Justice.

Frankenberg, Erika, and Chungme Lee. 2002. Race in American Public Schools: Rapidly Resegregating School Districts. Cambridge, MA: Harvard University Civil Rights Project.

Freedman, Estelle B. 2002. No Turning Back: The History of Feminism and the Future of Women. New York: Ballantine.

Freeman, Sue J. M. 1990. Managing Lives: Corporate Women and Social Change. Amherst: University of Massachusetts Press.

Freud, Sigmund. 1925. The Standard Edition of the Complete Psychological Works of Sigmund Freud. Volume 19. London: Hogarth Press. (Republished in 1971.)

Freudenheim, Milt, and Robert Pear. 2006. "Health Hazard: Computers Spilling Your History." New York Times. December 3: http://www.nytimes.com/2006/12/03/business/yourmoney/03health.html?scp=4&sq=electronic%20medical%20records%20security%20risks&st=cse.

Friedlander, D., and B. S. Okun. 1996. "Fertility Transition in England and Wales: Continuity and Change." Health Transition Review Supplement 1–18.

Furstenberg Jr., Frank F., Sheela Kennedy, Vonnie C. McLoyd, Ruben G. Rumbaut, and Richard A. Settersten, Jr. 2004. "Growing up Is Harder to Do." Contexts 3:33–41.

Gallup.com. 2009. "Environment." http://www.gallup.com/poll/1615/Environment.aspx. Accessed June 2009.

Gamoran, Adam. 1992. "The Variable Effects of High School Tracking." American Sociological Review 57:812–828.

Gamson, William. 1990. The Strategy of Social Protest (2nd ed.). Belmont, CA: Wadsworth.

Garfinkel, H. 1967. Studies in Ethnomethodology. Englewood Cliffs, NJ: Prentice-Hall.

Garreau, Joel. 1991. Edge City: Life on the New Frontier. New York: Doubleday.

Gastil, John, Stephanie Burkhalter, and Laura W. Black. 2007. "Do Juries Deliberate? A Study of Deliberation, Individual Difference, and Group Member Satisfaction at a Municipal Courthouse." Small Group Research 38:337–359.

Gatto, John Taylor. 2002. Dumbing Us Down: The Hidden Curriculum of Compulsory Schooling (10th ed.). New York: New Society.

Gautier, Ann H., and Jan Hatzius. 1997. "Family Benefits and Fertility: An Econometric Analysis." Population Studies 51:295–306.

Gelbard, Alene, Carl Haub, and Mary M. Kent. 1999. "World Population beyond 6 Billion." Population Bulletin 54(1).

General Social Survey. 2009. University of California–Berkeley. http://sda.berkeley.edu.

Genocide Watch. 2009. "Countries at Risk of Genocide, 2008." http://www.genocidewatch.org. Accessed May 2009.

Geyer, Michael. 2002. "People's War: The German Debate about a Levee en Masse in October 1918." In Daniel Moran and Arthur Waldron (ed.), The People in Arms: Military Myth and National Mobilization since the French Revolution. Cambridge, UK: Cambridge University Press: 124–58.

Ghaziani, Amin, and Gary Alan Fine. 2008. "Infighting and Ideology: How Conflict Informs the Local Culture of the Chicago Dyke March." International Journal of Politics, Culture and Society 20:51–67.

Giddens, Anthony. 1984. The Constitution of Society. Cambridge, UK: Polity Press.

Gilligan, Carol. 1993. In a Different Voice: Psychological Theory and Women's Development. Harvard University Press.

Giridharadas, Anand. 2009. "A Pocket-Size Leveler in an Outsize Land." New York Times. May 10:WK3.

Gladwell, Malcolm. 2008. Outliers: The Story of Success. New York: Little, Brown and Company.

Glassner, Barry. 2004. "Narrative Techniques of Fear Mongering." Social Research 71:819–827.

Glick, Daniel. 2001. "Web-Exclusive Excerpt: 'Powder Burn.'" http://www.msnbc.com/news/512636.asp?cp1=1. Accessed June 2003.

Goddard, Stephen B. 1994. Getting There: The Epic Struggle between Road and Rail in the American Century. Chicago: University of Chicago Press.

Goffman, Erving. 1959. The Presentation of Self in Everyday Life. New York: Doubleday.

———. 1961a. Asylums: Essays on the Social Situation of Mental Patients and Other Inmates. New York: Doubleday.

———. 1961b. Encounters: Two Studies in the Sociology of Interaction. Indianapolis, IN: Bobbs-Merrill.

———. 1963. Behavior in Public Places: Notes on the Social Organization of Gatherings. New York: Free Press.

Goldin, Claudia. 1992. Understanding the Gender Gap: An Economic History of American Women. New York: Oxford University Press.

Gordon, Suzanne. 2005. Nursing against the Odds: How Health Care Cost Cutting, Media Stereotypes, and Medical Hubris Undermine Nurses and Patient Care. Ithaca, NY: Cornell University Press.

Gould, Arthur. 2001. "The Criminalization of Buying Sex: The Politics of Prostitution in Sweden." Journal of Social Policy 30:437–456.

Gouldner, Alvin. 1960. "The Norm of Reciprocity." American Sociological Review 25:161–178.

Graff, E. J. 2008. "The Lie We Love." Foreign Policy 169:58–66.

Granovetter, Mark. 1973. "The Strength of Weak Ties." American Journal of Sociology 78:1360–1380.

———. 1974. Getting a Job: A Study of Contacts and Careers. Cambridge, MA: Harvard University Press.

Grant, Don S., and J. R. Martinez-Ramiro. 1997. "Crime and Restructuring of the U.S. Economy: A Reconsideration of the Class Linkages." Social Forces 75:769–798.

Grant, Kathleen M., Stephanie Sinclair Kelley, Sangeeta Agrawal, Jane L.Meza, James R. Meyer, and Debra Romberger. J. 2007. "Methamphetamine Use in Rural Midwesterners." American Journal on Addictions 16:79–84.

Gray, Mike. 2000. Drug Crazy : How We Got into This Mess and How We Can Get Out. New York: Routledge.

Grekin, Emily R., Kenneth J. Sher, and Jennifer L. Krull. 2007. "College Spring Break and Alcohol Use: Effects of Spring Break Activity." Journal of Studies on Alcohol & Drugs. 68(5):681–8.

Grimes, Michael D. 1989. "Class and Attitudes toward Structural Inequalities: An Empirical Comparison of Key Variables in Neo and Post-Marxist Scholarship." Sociological Quarterly 30:441–463.

Grogan, Sarah. 2008. Body Image: Understanding Body Dissatisfaction in Men, Women and Children. New York: Routledge.

Guadagno, Rosanna E., and Robert B. Cialdini. 2007. "Gender Differences in Impression Management in Organizations: A Qualitative Review." Sex Roles 56(7–8):483–494.

Gullotta, Thomas P., Gerald R. Adams, and Carol A. Markstrom. 2000. The Adolescent Experience (4th ed.). San Diego: Academic.

Gurney, Joan N., and Kathleen, J. Tierney. 1982. "Relative Deprivation and Social Movements: A Critical Look at Twenty Years of Theory and Research." Sociological Quarterly 23:33–47.

Gustafsson, Siv S. and Frank P. Stafford. 2009. "Links between Early Childhood Programs and Maternal Employment in Three Countries." http://www.futureofchildren.org/information2826/information_show.htm?doc_id=77717. Accessed May 2009.

Hacker, Jacob S. 2006. The Great Risk Shift: The Assault on American Jobs, Families, Health Care, and Retirement—And How You Can Fight Back. New York: Oxford University Press.

Hafferty, Frederic W. 1991. Into the Valley: Death and the Socialization of Medical Students. New Haven, CT: Yale University Press.

Hagan, Frank. 2002. Introduction to Criminology. Belmont, CA: Wadsworth.

Hagan, John, and Wenona Rymond-Richmond. 2008. "The Collective Dynamics of Racial Dehumanization and Genocidal Victimization in Darfur." American Sociological Review 73:875–902.

Hagan, John, A. R. Gillis, and John Simpson. 1985. "The Class Structure of Gender and Delinquency: Toward a Power/Control Theory of Common Delinquent Behavior." American Journal of Sociology 90:1151–1178.

Halle, David. 1993. Inside Culture: Art and Class in the American Home. Chicago: University of Chicago Press.

Hallett, M.A. 2002. "Race, Crime, and For-Profit Imprisonment: Social Disorganization as Market Opportunity." Punishment & Society 4:369–393.

Hallinan, Maureen T. 1994. "School Differences in Tracking Effects on Achievement." Social Forces 72:799–820.

Hamill, Sean D. 2009. "Students Sue Prosecutor in Cellphone Photos Case." New York Times. March 26:A1+.

Hamilton, Brady E., Joyce A. Martin, and Stephanie J. Ventura. 2009. "Births: Preliminary Data for 2007." National Vital Statistics Reports 57(12):1–23.

Handel, Gerald, Spencer Cahill, and Frederick Elkin. 2007. Children and Society: The Sociology of Children and Childhood Socialization. Los Angeles: Roxbury Publishing.

Handel, Michael J. 2002. Sociology of Organizations: Classic, Contemporary, and Critical Readings. Thousand Oaks, CA: Sage.

Harlow, Harry F., and M. K. Harlow. 1966. "Learning to Live." Scientific American 1:244–272.

Harlow, Harry F., and Stephen J. Suomi. 1971. "Social Recovery by Isolation-Reared Monkeys." Proceedings of the National Academy of Sciences 68:1534–1538.

Harris, Irving B. 1996. Children in Jeopardy. New Haven, CT: Yale University Press.

Harris, Judith Rich. 1998. The Nurture Assumption: Why Children Turn Out the Way They Do. New York: Free Press.

Harry, Beth, and Janette Klingner. 2005. Why Are So Many Minority Students in Special Education? Understanding Race and Disability in Schools. New York: Teachers College Press.

Haub, Carl. 1993. "Tokyo Now Recognized as World's Largest City." Population Today 21 (March):1–2.

Haub, Carl. 1994. "Population Change in the Former Soviet Republics." Population Bulletin, Vol. 49. Washington, D.C.: Population Reference Bureau.

Haugen, Steven E. 2009. Measures of Labor Underutilization from the Current Population Survey. Working Paper 424. U.S. Bureau of Labor Statistics.

Hayes-Bautista, David, Paul Hsu, and Aide Perez. 2002. "The Browning of the Graying in America: Diversity in the Elderly Population and Policy Implications." Generations 26(3):15–24.

Haynie, Dana L., and D. Wayne Osgood. 2005. "Reconsidering Peers and Delinquency: How Do Peers Matter?" Social Forces 84:1109–1130.

Hechter, Michael. 1987. Principles of Group Solidarity. Berkeley: University of California Press.

Herd, Denise. 2005. "Changes in the Prevalence of Alcohol Use in Rap Song Lyrics, 1979–97." Addiction 100:1258–1269.

Herdt, Gilbert (ed.). 1994. Third Sex, Third Gender: Beyond Sexual Dimorphism in Culture and History. New York: Zone Books.

Herring, Cedric. 2003. Skin Deep: How Race and Complexion Matter in the "Color-Blind" Era. Champlain, IL: University of Illinois Press.

Hertz, Rosanna. 2006. Single by Chance, Mothers by Choice: How Women are Choosing Parenthood without Marriage and Creating the New American Family. New York: Oxford University Press.

Hesse-Biber, Sharlene, and Gregg Lee Carter. 2000. Working Women in America: Split Dreams. New York: Oxford University Press.

Heuveline, Patrick, and Matthew Weinshenker. 2008. "The International Child Poverty Gap: Does Demography Matter?" Demography 45:173–191

Hewitt, John, and Randall Stokes. 1975. "Disclaimers." American Sociological Review 40:1–11.

Higley, Stephen R. 1995. Privilege, Power, and Place: The Geography of the American Upper Class. Lanham, MD: Rowman & Littlefield.

Hillygus, D. Sunshine. 2005 "The Missing Link: Exploring the Relationship between Higher Education and Political Engagement." Political Behavior 27:25–47.

Hironaka, Ann. 2005. Neverending Wars: Weak States, the International Community, and the Perpetuation of Civil War. Cambridge, MA: Harvard University Press.

Hitlin, Steven, and Jane Allyn Piliavin. 2004. "Values: Reviving a Dormant Concept." Annual Review of Sociology 30:359–393.

Hochschild, Arlie R. 1985. The Managed Heart: The Commercialization of Human Feeling. Berkeley: University of California Press.

———. 1997. The Time Bind: When Work Becomes Home and Home Becomes Work. New York: Holt.

Hoffman, Bruce. 2006. Inside Terrorism. New York: Columbia University Press.

Hogan, Dennis P., David J. Eggebeen, and Clifford C. Clogg. 1993. "The Structure of Intergenerational Exchanges in American Families." American Journal of Sociology 98:1428–1458.

Holloway, Marguerite. 1994. "Trends in Women's Health: A Global View." Scientific American. August: 76–83.

Hondagneu-Sotelo, Pierrette. 2001. Domestica: Immigrant Workers Cleaning and Caring in the Shadows of Affluence. Berkeley: University of California Press.

Horwitz, Allan V. 2002. Creating Mental Illness. Chicago: University of Chicago Press.

Horwitz, Allan V., and Jerome C. Wakefield. 2007. The Loss of Sadness: How Psychiatry Transformed Normal Sorrow into Depressive Disorder. New York: Oxford University Press.

Hoyt, Homer. 1939. The Structure and Growth of Residential Neighborhoods in American Cities. Washington, D.C.: Federal Housing Administration.

Hull, Elizabeth A. 2005. The Disenfranchisement of Ex-Felons. Philadelphia: Temple University Press.

Hultin, Mia. 2003. "Some Take the Glass Escalator, Some Hit the Glass Ceiling? Career Consequences of Occupational Sex Segregation." Work and Occupations 30:30–61.

Human Rights Watch. 2004. "Women's Rights Division." http://www.hrw.org/women/. Accessed June 2004.

———. 2006. "Q & A: Crisis in Darfur." http://hrw.org/english/docs/2004/05/05/darfur8536_txt.htm

Huntford, Roald. 2000. Scott and Amundsen. London, UK: Abacus History.

Hurley, Dan. 2005. "Divorce Rate: It's Not as High as You Think." New York Times. April 19:D7.

Iceland, John, and Rima Wilkes. 2006. "Does Socioeconomic Status Matter? Race, Class, and Residential Segregation." Social Problems 53:248–273.

Innocence Project. 2009. http://www.innocenceproject.org. Accessed May 2009.

International Centre for Prison Studies. 2008. "World Prison Brief." http://www.prisonstudies.org/. Accessed April 2009.

International Telecommunications Union. World Telecommunication/ICT Indicators 2008. www.itu.int/ITU-D/ict/statistics. Accessed June 2009.

Isaacs, Julia B., Isabel V. Sawhill, and Ron Haskins. 2008. Getting Ahead or Losing Ground: Economic Mobility in America. Philadelphia: Pew Charitable Trusts.

Jackson, J. W. 2007. "Group Decision Making." Encyclopedia of Social Psychology. Newbury Park, CA: Sage.

http://sage-ereference.com/socialpsychology/Article_n238.html. Accessed May 2009.

Jacobs, David. 1988. "Corporate Economic Power and the State: A Longitudinal Assessment of Two Explanations." American Journal of Sociology 93:852–881.

Jacobs, David, and Ronald E. Helms. 1997. "Testing Coercive Explanations for Order: The Determinants of Law Enforcement Strength over Time." Social Forces 75:1361–1392.

Jacobs, Jerry. 1989. Revolving Doors: Sex Segregation and Women's Careers. Stanford, CA: Stanford University Press.

Jacobs, Jerry A., and Kathleen Gerson. 2004. The Time Divide: Work, Family, and Gender Inequality. Cambridge, MA: Harvard University Press.

Jacobson, Matthew Frye. 1998. Whiteness of a Different Color: European Immigrants and the Alchemy of Race. Cambridge, MA: Harvard University Press.

Jacquard, Roland. 2002. In the Name of Osama Bin Laden. Durham, NC: Duke University Press.

James, Angela D., David M. Grant, and Cynthia Cranford. 2000. "Moving Up, but How Far? African American Women and Economic Restructuring in Los Angeles, 1970–1990." Sociological Perspective 43:399–420.

James, Doris J., and Lauren E. Glaze. 2006. Mental Health Problems of Prison and Jail Inmates. CJ 213600. Washington, D.C.: U.S. Department of Justice, Bureau of Justice Statistics.

Janis, Irving. 1982. Groupthink: Psychological Studies of Policy Decisions and Fiascoes. Boston: Houghton Mifflin.

Jasper, James M., and Jane D. Poulsen. 1995. "Recruiting Strangers and Friends: Moral Shocks and Social Networks in Animal Rights and Anti-Nuclear Protests." Social Problems 42:493–512.

Jenkins, J. Craig, and Barbara Brent. 1989. "Social Protest, Hegemonic Competition, and Social Reform." American Sociological Review 54:891–909.

Jenkins, J. Craig, and Craig M. Eckert. 1986. "Channeling Black Insurgency: Elite Patronage and Professional Social Movement Organizations in the Development of the Black Movement." American Sociological Review 51:812–829.

Jensen, Leif, Diane K. McLaughlin, and Tim Slack. 2003. "Rural Poverty: The Persistent Challenge." In David L. Brown and Louis E. Swanson (eds.), Challenges for Rural American in the Twenty-First Century. University Park, PA: Pennsylvania State University Press: 118–131.

Johnson, Kenneth M. 2003. "Unpredictable Directions of Rural Population Growth and Migration." In David L. Brown and Louis E. Swanson (eds.), Challenges for Rural American in the Twenty-First Century. University Park, PA: Pennsylvania State University Press: 19–32.

Johnson, Michael P., and Kathleen J. Ferraro. 2000. "Research on Domestic Violence in the 1990s: Making Distinctions." Journal of Marriage and the Family 62:948–963.

Jones, Adam. 2006. Genocide: A Comprehensive Introduction. New York: Routledge.

Jones, Steve. 2003. Let the Games Begin: Gaming Technology and Entertainment among College Students. Washington, D.C.: Pew Internet and American Life Project.

Joseph, Brian D., Johanna DeStephano, Neil G. Jacobs, and Ilse Lehiste. 2003. When Languages Collide: Perspectives on Language Conflict, Language Competition, and Language Coexistence. Columbus, OH: Ohio State University Press.

Jurik, Nancy. 2004. "Imagining Justice: Challenging the Privatization of Public Life." Social Problems 51:1–15.

Jurik, Nancy C., Gray Cavender, and Julie Cowgill. 2009. "Resistance and Accommodation in a Post-Welfare Social Service Organization." Journal of Contemporary Ethnography 38:25–51.

Kaiser Commission on Medicaid and the Uninsured. 2008. The Uninsured: A Primer. Washington, D.C.: Kaiser Family Foundation.

Kaiser Family Foundation. 2005. "Generation M: Media in the Lives of 8 to 18 Year Olds." http://kff.org/entmedia/7251.cfm.

Kalish, Susan. 1994. "Culturally Sensitive Family Planning: Bangladesh Story Suggests It Can Reduce Family Size." Population Today 22(2):5.

Kalmijn, Matthijs. 1998. "Intermarriage and Homogamy: Causes, Patterns, and Trends." Annual Review of Sociology 24:395–421.

Kalmijn, Matthijs, and Gerbert Kraaykamp. 1996. "Race, Cultural Capital, and Schooling: An Analysis of Trends in the United States." Sociology of Education 69:22–34.

Kao, Grace, and Jennifer S. Thompson. 2003. "Racial and Ethnic Stratification in Educational Achievement and Attainment." Annual Review of Sociology 29:417–441.

Katz, Jack. 1988. The Seductions of Crime: Moral and Sensual Attractions of Doing Evil. New York: Basic Books.

Kaufman, Peter, and Kenneth A. Feldman. 2004. "Forming Identities in College: A Sociological Approach." Research in Higher Education 45:463–496.

Kemper, Peter. 1992. "Use of Formal and Informal Home Care by the Disabled Elderly." Health Services Research 27:421–451.

Kerbo, Harold R. 2005. Social Stratification and Inequality. New York: McGraw-Hill.

Kessler, Ronald C., Patricia Berglund, Olga Demler, Robert Jim, and Ellen E. Walters. 2005. "Lifetime Prevalence and Age-of-Onset Distributions of DSM IV Disorders in the National Comorbidity Survey Replication." Archives of General Psychiatry 62:593–602.

Kestnbaum, Meyer. 2009. "The Sociology of War and the Military." Annual Review of Sociology 35:235–54.

Khor, Martin. 2001. Rethinking Globalization: Critical Issues and Policy Choices. New York: Zed Books.

Kimball, Dan. 2003. The Emerging Church. Grand Rapids, MI: Zondervan.

Kimmel, Michael S. 2000. The Gendered Society. New York: Oxford University Press.

Kinzer, Stephen. 2003. All the Shah's Men: An American Coup and the Roots of Middle East Terror. New York: Wiley.

Kiple, Kenneth F. 1993. Cambridge World History of Human Disease. New York: Cambridge University Press.

Kirp, David L. 2007. The Sandbox Investment: The Preschool Movement and Kids-First Politics. Cambridge, MA: Harvard University Press.

Klandermas, Bert. 1984. "Mobilization and Participation: Social Psychological Expansion of Resource Mobilization Theory." American Sociological Review 49: 583–600.

Koblik, Steven. 1975. Sweden's Development from Poverty to Affluence 1750–1970. Minneapolis: University of Minnesota Press.

Kohut, Andrew, and Bruce Stokes. 2006. America against the World. New York: Henry Holt.

Kollock, Peter, Phillip Blumstein, and Pepper Schwartz. 1985. "Sex and Power in Conversation: Conversational Privileges and Duties." American Sociological Review 50:34–46.

Konig, René. 1968. "Auguste Comte." In David J. Sills (ed.), International Encyclopedia of the Social Sciences. Vol. 3. New York: Macmillan and Free Press.

Korpi, Walter. 1989. "Power, Politics, and State Autonomy in the Development of Social Citizenship." American Sociological Review 54:309–328.

Kosmin, Barry A., and Ariela Keysar. 2009. American Religious Identification Survey: Summary Report. Hartford, CT: Trinity College.

Kozol, Jonathan. 2005. The Shame of the Nation: The Restoration of Apartheid Schooling in America. New York: Crown.

Krakauer, Jon. 2003. Under the Banner of Heaven: A Story of Violent Faith. New York: Doubleday.

Kraska, Peter B., and Victor E. Kappeler. 1997. "Militarizing American Police: The Rise and Normalization of Paramilitary Units." Social Problems 44:1–18.

Krech, Shepard I. 1999. The Ecological Indians: Myth and History. New York: W. W. Norton.

Kreider, Rose M. 2005. "Number, Timing, and Duration of Marriages and Divorces: 2001." Current Population Reports, P70–97. U.S. Census Bureau, Washington, D.C.

Krysan, Maria. 2000. "Prejudice, Politics, and Public Opinion: Understanding the Sources of Racial Policy Attitudes." Annual Review of Sociology 26:135–168.

Ku, Leighton, Freya Sonenstein, Laura Lindberg, Carolyn H. Bradner, Scott Boggess, and Joseph Pleck. 1998. "Understanding Changes in Sexual Activity among Young Metropolitan Men: 1979–1995." Family Planning Perspectives 30(6):256–262.

Kunda, Gideon. 1993. Engineering Culture: Control and Commitment in a High Tech Corporation. Philadelphia: Temple University Press.

Lamanna, Mary Ann, and Agnes Riedmann. 2000. Marriages and Families: Making Choices in a Diverse Society (7th ed.). Belmont, CA: Wadsworth.

Lambert, Bruce, and Fernanda Santos. 2006. "'First' Suburbs Growing Older and Poorer, Report Warns." New York Times. February 16:B1.

Lamont, Michelle, and Marcel Fournier (eds.). 1992. Cultivating Differences: Symbolic Boundaries and the Making of Inequality. Chicago: University of Chicago Press.

Lane, Christopher. 2007. Shyness: How Normal Behavior Became a Sickness. New Haven, CT: Yale University Press.

Lane, Sandra D., and Donald A. Cibula. 2000. "Gender and Health." In Gary L. Albrecht, Ray Fitzpatrick, and Susan C. Scrimshaw (eds.), Handbook of Social Studies in Health and Medicine. Thousand Oaks, CA: Sage Publications: 136–153.

Langdon, Philip. 1994. A Better Place to Live: Reshaping the American Suburb. New York: Harper.

Lappé, Frances Moore, Joseph Collins, and Peter Rosset. 1998. World Hunger: Twelve Myths. New York: Grove.

Lareau, Annette. 2003. Unequal Childhoods: Class, Race, and Family Life. Berkeley: University of California Press.

Laumann, Edward O., John H. Gagnon, Robert T. Michael, and Stuart Michael. 1994. The Social Organization of Sexuality: Sexual Practices in the United States. Chicago: University of Chicago Press.

Lawrence, Bruce B. 1998. Shattering the Myth: Islam beyond Violence. Princeton, NJ: Princeton University Press.

Lawton, Julia. 2003. "Lay Experiences of Health and Illness: Past Research and Future Agendas." Sociology of Health and Illness 25:23–40.

Le, C. N. 2006. "Employment & Occupational Patterns." Asian-Nation: The Landscape of Asian America. http://www.asian-nation.org/index.shtml. Accessed June 2006.

Leahey, Erin, and Guang Guo. 2001. "Gender Differences in Mathematical Trajectories." Social Forces 80:713–732.

Leblanc, Lauraine. 1999. Pretty in Punk: Girls' Gender Resistance in a Boys' Subculture. New Brunswick, NJ: Rutgers University Press.

Lee, Christine M., Jennifer L. Maggs, and Lela A. Rankin. 2006 "Spring Break Trips as a Risk Factor for Heavy Alcohol Use among First-Year College Students." Journal of Studies on Alcohol. 67(6):911–916.

Lee, Richard B., and Richard Daly. 2005. Cambridge Encyclopedia of Hunters and Gatherers. Cambridge, UK: Cambridge University Press.

Lee, Yun-Suk, Barbara Schneider, and Linda J. Waite. 2003. "Children and Housework: Some Unanswered Questions." Sociological Studies of Children and Youth 9:105–125.

Lefkowitz, Bernard. 1997. Our Guys: The Glen Ridge Rape and the Secret Life of the Perfect Suburb. Berkeley: University of California Press.

Lemert, Edwin. 1981. "Issues in the Study of Deviance." Sociological Quarterly 22:285–305.

Lenhart, Amanda, Joseph Kahne, Ellen Middaugh, Alexandra Rankin Macgill, Chris Evans, and Jessica Vitak. 2008. Teens, Video Games, and Civics. Washington, D.C.: Pew Internet and American Life Project.

Lenski, Gerhard. 1966. Power and Privilege: A Theory of Social Stratification. New York: McGraw-Hill.

Leonhardt, David. 2004. "As Wealthy Fill Top Colleges, Concerns Grow over Fairness." New York Times. April 22:A1+.

Levine, John M. 2007. "Conformity." Encyclopedia of Social Psychology. Newbury Park, CA: Sage. http://sage-ereference.com/socialpsychology/Article_n99.html. Accessed May 2009.

Levy, Emanuel. 1990. "Stage, Sex, and Suffering: Images of Women in American Films." Empirical Studies of the Arts 8(1):53–76.

Lewin, Tamar. 2006. "At Colleges, Women Are Leaving Men in the Dust." New York Times. July 9:A11.

Lewin, Tamar, and Jennifer Medina. 2003. "To Cut Failure Rate, Schools Shed Students." New York Times. July 31:11.

Lewis, Oscar. 1969. "The Culture of Poverty." In Daniel P. Moynihan (ed.), On Understanding Poverty. New York: Basic Books.

Lichtenstein, Nelson. 2003. State of the Union: A Century of American Labor. Princeton, NJ: Princeton University Press.

Lichter, Daniel T., Felicia B. LeClere, and Diane K. McLaughlin. 1991. "Local Marriage Markets and the Marital Behavior of Black and White Women." American Journal of Sociology 96:843–867.

Lichter, Daniel T., Diane K. McLaughlin, George Kephart, and David J. Landry. 1992. "Race and the Retreat from Marriage: A Shortage of Marriageable Men?" American Sociological Review 57:781–799.

Link, Bruce G., Francis T. Cullen, James Frank, and John F. Wozniak. 1987. "The Social Rejection of Former Mental Patients: Understanding Why Labels Matter." American Journal of Sociology 92:1461–1500.

Link, Bruce G., Elmer L. Struening, Michael Rahav, Jo C. Phelan, and Larry Nuttbrock. 1997. "On Stigma and Its Consequences: Evidence from a Longitudinal Study of Men with Dual Diagnoses of Mental Illness and Substance Abuse." Journal of Health and Social Behavior 38:177–190.

Linn, James Weber, and Anne Firor Scott. 2000. Jane Addams: A Biography. Champlain, IL: University of Illinois Press.

Liska, Allen E., Mitchell B. Chamlin, and Mark Reed. 1985. "Testing the Economic Production and Conflict Models of Crime Control." Social Forces 64:119–138.

Lo, Clarence Y. H. 1982. "Countermovements and Conservative Movements in the Contemporary U.S." Annual Review of Sociology 8:10–34.

Loe, Meika. 2004. The Rise of Viagra: How the Little Blue Pill Changed Sex in America. New York: New York University Press.

Logan, John R., and Glenna D. Spitze. 1994. "Family Neighbors." American Journal of Sociology 100:453–476.

Lohr, Steve. 2006. "Outsourcing Is Climbing Skills Ladder." New York Times. February 16:C1+.

Lorber, Judith. 1994. Paradoxes of Gender. New Haven, CT: Yale University Press.

Love et al. 2003. "Child Care Quality Matters: How Conclusions May Vary with Context." Child Development 74:1021–1033.

Lowney, Kathleen S. 2003. "Wrestling with Criticism: The World Wrestling Federation's Ironic Campaign against the Parents Television Council." Symbolic Interaction 26:427–446.

Luker, Kristen. 1985. Abortion and the Politics of Motherhood. Berkeley: University of California Press.

———. 1996. Dubious Conceptions: The Politics of Teenage Pregnancy. Cambridge, MA: Harvard University Press.

Lundquist, Jennifer Hickes. 2008. "Ethnic and Gender Satisfaction in the Military: The Effect of a Meritocratic Institution." American Sociological Review 73: 477–496.

Luxembourg Income Study (LIS). "Key Figures." http://www.lisproject.org/keyfigures.htm. Accessed May 2009.

Lye, Diane N. 1996. "Adult Child-Parent Relationships." Annual Review of Sociology 22:79–102.

Lynn, Barry C. 2006. "Breaking the Chain: The Antitrust Case against Wal-Mart." Harpers 313(1874):29–36.

MacKenzie, Doris Layton, David B., Wilson, and Suzanne B. Kider. 2001. "Effects of Correctional Boot Camps on Offending." Annals of the American Academy of Political and Social Science 578:126–143.

Maher, Lisa, and Kathleen Daly. 1996. "Women in the Street-Level Drug Economy: Continuity or Change?" Criminology 34:465–491.

Mahler, Vincent A., and David K. Jesuit. "Fiscal Redistribution in Developed Countries: New Insight from the Luxembourg Income Study." Socio-Economic Review 4:483–511.

Manza, Jeff, and Christopher Uggen. 2006. Locked Out: Felon Disenfranchisement and American Democracy. New York: Oxford University Press.

Marcus, Eric. 2002. Making Gay History: The Half Century Fight for Lesbian and Gay Equal Rights. New York: Harper.

Marger, Martin. 2003. Race and Ethnic Relations (6th ed.). Belmont, CA: Wadsworth.

Marmot, Michael G. 2004. The Status Syndrome: How Your Social Standing Directly Affects Your Health and Life Expectancy. London: Bloomsbury.

Marsden, George M. 2006. Fundamentalism and American Culture. New York: Oxford University Press.

Martey, Rosa Mikeal, and Jennifer Stromer-Galley. 2007. "The Digital Dollhouse: Context and Social Norms in The Sims Online." Games and Culture 2:314–334.

Martin, Karin A. 1998. "Becoming a Gendered Body: Practices of Preschools." American Sociological Review 63:494–511.

Marx, Gary T., and Douglas McAdam. 1994. Collective Behavior and Social Movements: Process and Structure. Englewood Cliffs, NJ: Prentice Hall.

Marx, Karl, and Friedrich Engels. 1967. The Communist Manifesto. London: Penguin.

Massey, Douglas S. 2006. "The Wall That Keeps Illegal Workers In." New York Times. April 4:A23.

———. 2007. Categorically Unequal: The American Stratification System. New York: Russell Sage.

Maume, David. 1999. "Occupational Segregation and the Career Mobility of White Men and Women." Social Forces 77:1433–1459.

_____. 2004. "Wage Discrimination over the Life Course: A Comparison of Explanations." Social Problems 51:505–527.

McAdam, Doug. 1986. "Recruitment to High-Risk Activism." American Journal of Sociology 92:64–90.

McAdam, Doug, and Ronnelle Paulsen. 1993. "Specifying the Relationship between Social Ties and Activism." American Journal of Sociology 99:640–667.

McAdam, Doug, and David A. Snow. 1997. Social Movements: Readings on their Emergence, Mobilization, and Dynamics. Los Angeles: Roxbury Publishing Company.

McBrier, Debra Branch. 2003. "Gender and Career Dynamics within a Segmented Professional Labor Market: the Case of Law Academia." Social Forces 81:1201–1266.

McCarthy, Bill. 2002. "New Economics of Sociological Criminology." Annual Review of Sociology 28:417–442.

McClain, Dylan Loeb. 2005. "Richer Than Ever, but Watch Out for Missing Costs." New York Times. December 5:C6.

McDermott, Monica, and Frank L. Samson. 2005. "White Racial and Ethnic Identity in the United States." Annual Review of Sociology 31:245–261.

McDill, Edward L., Gary Natriello, and Aaron Pallas. 1986. "A Population at Risk: Potential Consequences of Tougher School Standards for School Dropouts." American Journal of Education 94:135–181.

McFarland, Daniel A., and Reuben J. Thomas. 2006. "Bowling Young: How Youth Voluntary Associations Influence Adult Political Participation." American Sociological Review 71:401–425.

McGranahan, David A. 2003. "How People Make a Living in Rural America." In David L. Brown and Louis E. Swanson (eds.), Challenges for Rural American in the Twenty-First Century. University Park, PA: Pennsylvania State University Press: 135–151.

McGuffey, C. Shawn, and B. Lindsay Rich. 1999. "Playing in the Gender Transgression Zone: Race, Class, and Hegemonic Masculinity in Middle Childhood." Gender and Society 13:608–627.

McKinlay, John B. 1994. "A Case for Refocusing Upstream: The Political Economy of Illness." In Peter Conrad and Rachelle Kern (eds.), The Sociology of Health and Illness. New York: St. Martin's: 509–530.

McKinlay, John B., and Sonja J. McKinlay. 1977. "The Questionable Effect of Medical Measures on the Decline of Mortality in the United States in the Twentieth Century." Milbank Memorial Fund Quarterly 55:405–428.

McLellan, David. 2006. Karl Marx: A Biography (4th ed.). London: Palgrave Macmillan.

McPherson, J. Miller, Pamela A. Popielarz, and Sonja Drobnic. 1992. "Social Networks and Organizational Dynamics." American Sociological Review 57:153–170.

McPherson, J. Miller, and Lynn Smith-Lovin. 2002. "Cohesion and Membership Duration: Linking Groups, Relations, and Individuals in an Ecology of Affiliation." Advances in Group Processes 19:1–36.

McPherson, J. Miller, Lynn Smith-Lovin, and Matthew E. Brashears. 2006. "Social Isolation in America: Changes in Core Discussion Networks over Two Decades." American Sociological Review 71:353–375.

Mead, George Herbert. 1934. Mind, Self, and Society: From the Standpoint of a Social Behaviorist (Charles W. Morris, ed.). Chicago: University of Chicago Press.

Mead, Lawrence M. 1986. Beyond Entitlement: The Social Obligations of Citizenship. New York: Free Press.

_____. 1992. The New Politics of Poverty: The Nonworking Poor in America. New York: Basic Books.

Mead, Sara. 2006. The Truth about Boys and Girls. Washington, D.C.: Education Sector.

Menjívar, Cecilia. 2000. Fragmented Ties. Berkeley, CA.: University of California Press.

Merton, Robert. 1957. Social Theory and Social Structure (2nd ed.). New York: Free Press.

Messerschmidt, James W. 1993. Masculinities and Crime: Critique and Reconceptualization of Theory. Lanhan, MD: Rowman & Littlefield.

Messner, Steven F. 1989. "Economic Discrimination and Societal Homicide Rates: Further Evidence on the Cost of Inequality." American Sociological Review 54:597–611.

Meyer, David S. 2004. "Protest and Political Opportunities." Annual Review of Sociology 30:125–145.

Meyer, David S., and Suzanne Staggenborg. 1996. "Movements, Countermovements, and the Structure of Political Opportunity." American Journal of Sociology 101:1628–1660.

Meyer, Pamela A., Timothy Pivetz, Timothy A. Dignam, David M. Homa, Jaime Schoonover, and Debra Brody. 2003. "Surveillance for Elevated Blood Lead Levels among Children—United States, 1997–2001." Morbidity and Mortality Weekly Report 2003:52(No. SS-10):1–21.

Meyerowitz, Joanne. 2002. How Sex Changed: A History of Transsexuality in the United States. Cambridge, MA: Harvard University Press.

Milgram, Stanley. 1974. Obedience to Authority: An Experimental View. New York: Harper & Row.

Milkie, Melissa A. 1999. "Social Comparisons, Reflected Appraisals, and Mass Media: The Impact of Pervasive Beauty Images on Black and White Girls' Self Concepts." Social Psychology Quarterly 62:190–210.

Mills, C. Wright. 1940. "Situated Actions and Vocabularies of Motives." American Sociological Review 5:904–913.

———. 1956. The Power Elite. New York: Oxford University Press.

———. 1959. The Sociological Imagination. Oxford, UK: Oxford University Press.

Minino, Arialdi M. 2002. "Deaths: Final Statistics for 2000." National Vital Statistics Report 50(15):1–120.

Mizruchi, Mark. 1989. "Similarity of Political Behavior among Large American Corporations." American Journal of Sociology 95:401–424.

Moe, Richard, and Carter Wilkie. 1997. Changing Places: Rebuilding Community in the Age of Sprawl. New York: Henry Holt.

Mokdad, Ali H., James S. Marks, Donna F. Stroup, and Julie L. Gerberding. 2004. "Actual Causes of Death in the United States, 2000." Journal of the American Medical Association 291:1238–1245.

Molm, Linda D. 2003. "Theoretical Comparisons of Forms of Exchange." Sociological Theory 21:1–17.

Molm, Linda D., and Karen S. Cook. 1995. "Social Exchange Theory." In Karen S. Cook, Gary A. Fine, and James S. House (eds.), Sociological Perspectives on Social Psychology. New York: Allyn & Bacon: 209–235.

Montemurro, Beth. 2006. Something Old, Something Bold: Bridal Showers and Bachelorette Parties. New Brunswick, NJ: Rutgers University.

Moody, Kim. 1997. Workers in a Lean World: Unions in the International Economy. London: Verso.

Moore, Solomon. 2009. "Missouri System Treats Juvenile Offenders with Lighter Hand." New York Times. March 27: A12+.

Morris, Aldon D., and Carol Mueller. 1992. Frontiers in Social Movement Theory. New Haven, CT: Yale University Press.

Morris, Martina, and Bruce Western. 1999. "Inequality in Earnings at the Close of the Twentieth Century." Annual Review of Sociology 25:623–657.

Morton, Lois Wright. 2003. In David L. Brown and Louis E. Swanson (eds.), Challenges for Rural America in the Twenty-First Century. University Park, PA: Pennsylvania State University Press: 290–304.

Murr, Andrew, and Tom Morganthau. 2001. "Burning Suburbia." Newsweek. January 15:32–33.

Murray, Charles A. 1984. Losing Ground: American Social Policy 1950–1980. New York: Basic Books.

Muslim West Facts Project. 2009. Muslim-Americans: A National Portrait. Gallup, Inc.

Mutran, Elizabeth, and Donald C. Reitzes. 1984. "Intergenerational Support Activities and Well-Being among the Elderly: A Convergence of Exchange and Symbolic Interaction Perspectives." American Sociological Review 49:117–130.

"NAFTA and Workers' Rights and Jobs." 2003. Public Citizen: Global Trade Watch. http://www.citizen.org/ trade/nafta/ jobs/. Accessed June 2003.

Nardi, Peter. 1992. Men's Friendships: Research on Men and Masculinities. Newbury Park, CA: Sage.

National Adoption Information Clearinghouse. "Voluntary Relinquishment for Adoption: Numbers and Trends." http://naic .acf.hhs.gov/pubs/s_place.cfm.

National Center for Education Statistics. 2009. Digest of Education Statistics. http://nces.ed.gov/programs/ digest/. Accessed July 2009.

National Center for Health Statistics. 2009. Health, United States, 2008. Hyattsville, MD.

National Coalition for the Homeless. 2008. "How Many People Experience Homelessness?" NCH Fact Sheet #2. http://www.nationalhomeless.org/publications/facts/ How_Many.pdf. Accessed June 2009.

National Gay and Lesbian Task Force. 2009. "State Nondiscrimination Laws in the U.S." http://www. thetaskforce.org/downloads/reports/issue_maps/ non_discrimination_7_09_color.pdf. Accessed August 2009.

National Highway Traffic Safety Administration. 2005. "Traffic Safety Facts: Motorcycle Helmet Use Laws." Washington, D.C.: U.S. Department of Transportation.

National Resource Defense Council 2008. "Bottled Water." http://www.nrdc.org/water/drinking/qbw.asp. Accessed May 2009.

Nelson, Alan R., Brian D. Smedley, and Adrienne Y. Stith. 2002. Unequal Treatment: Confronting Racial and Ethnic Disparities in Health Care. Washington, D.C.: Institute of Medicine, National Academy Press.

Newman, Katherine S. 1999a. Falling from Grace: Downward Mobility in the Age of Affluence (rev. ed.). Berkeley: University of California Press.

———. 1999b. No Shame in My Game: The Working Poor in the Inner City. New York: Knopf.

Newman, Katherine S., and Victor Tan Chen. 2008. The Missing Class: Portraits of the Near Poor in America. New York: Beacon Press.

Newman, Katherine S., and Rebekah Peeples Massengill. 2006. "The Texture of Hardship: Qualitative Sociology of

Poverty, 1995–2005." Annual Review of Sociology 32:1–24.

_____. 2006. "Separate and Unequal for Gypsies." *New York Times* March 11. http://query.nytimes.com/gst/fullpage.html?res=9E04EFDA1331F932A25750C0A96 09C8B63. Accessed May 2009.

_____. 2009. Poll: April 26–29, 2009. New York: New York Times.

Newton, Michael. 2004. Savage Girls and Wild Boys: A History of Feral Children. London: Picador.

NHSDA Report. 2003. Substance Use among American Indians or Alaska Natives. Washington D.C.: Substance Abuse and Mental Health Services Administration.

Nichols, Sharon L., and David C. Berliner. 2007. Collateral Damage: How High-Stakes Testing Corrupts America's Schools. Cambridge, MA: Harvard Education Press.

NIDA Research Report. 2006. "Methamphetamine Abuse and Addiction." NIH Publication Number 06-4210.

Norrish, Barbara R., and Thomas G. Rundall. 2001. "Hospital Restructuring and the Work of Registered Nurses." Milbank Quarterly 79:55–79.

NSDUH Report. 2006. "Underage Alcohol Use among Full-Time College Students." Issue 31.

Ogletree, Charles, Jr., and Austin Sarat. 2006. From Lynch Mobs to the Killing State: Race And the Death Penalty in America. New York: New York University Press.

Oldenburg, Ray. 1997. The Great Good Place: Cafés, Coffee Shops, Community Centers, Beauty Parlors, General Stores, Bars, Hangouts, and How They Get You Through the Day. New York: Marlowe.

Oleksyn, Veronika. 2006. "Birth of a Notion: Incentives Offered for Having More Kids." Seattle Times. July 22:A3.

Olsen, Gregg M. 1996. "Re-modeling Sweden: The Rise and Demise of the Compromise in a Global Economy." Social Problems 43:1–20.

Ontario Consultants on Religious Tolerance. 2006. "Religious Affiliation." http://www.religioustolerance.org/compuswrld.htm. Accessed June 2006.

_____. 2009. "Islam." http://www.religioustolerance.org/islam.htm. Accessed June 2009.

Orenstein, Peggy. 1994. School Girls: Young Women, Self-Esteem, and the Confidence Gap. New York: Doubleday.

Organization for Economic Cooperation and Development. 2009. Society at a Glance 2009—OECD Social Indicators. http://www.oecd.org/els/social/indicators/SAG.

Osgood, D. Wayne, Janet K. Wilson, Patrick M. O'Malley, Jerald G. Bachman, and Lloyd D. Johnston. 1996.

"Routine Activities and Individual Deviant Behavior." American Sociological Review 61:635–55.

"Out of Sight, Out of Mind." 2000. Economist. May 20:27–28.

Owens, Timothy J., Sheldon Stryker, and Norman Goodman. 2001. Extending Self-Esteem Research: Sociological and Psychological Currents. New York: Cambridge University Press.

Padden, Carol A., and Tom L. Humphries. 2006. Inside Deaf Culture. Cambridge, MA.: Harvard University Press.

Pager, Devah, and Hana Shepherd. 2008. "The Sociology of Discrimination: Racial Discrimination in Employment, Housing, Credit, and Consumer Markets." Annual Review of Sociology 34:181–209.

Paret, Peter. 1992. Understanding War: Essays on Clausewitz and the History of Military Power. Princeton, NJ: Princeton University Press.

Park, Kristin. 2005. "Choosing Childlessness: Weber's Typology of Action and Motives of the Voluntarily Childless." Sociological Inquiry 75:372–402.

Parreñas, Rhacel Salazar. 2000. "Migrant Filipina Domestic Workers and the International Division of Reproductive Labor." Gender and Society 14:560–581.

Parsons, Talcott. 1951. The Social System. New York: Free Press.

Pascoe, C. J. 2007. Dude, You're a Fag. Berkeley: University of California Press.

Passos, Nikos, and Robert Agnew (eds.). 1997. The Future of Anomie Theory. Boston: Northeastern University Press.

Paternoster, Raymond. 1989. "Absolute and Restrictive Deterrence in a Panel of Youth: Explaining the Onset, Persistence/Desistance, and Frequency of Delinquent Offending." Social Problems 36:289–309.

People for the American Way. 2008. "Memo on Election Day '08 Voting Issues." http://site.pfaw.org/site/PageServer?pagename=issues_vote_post_election_memo_08. Accessed May 2009.

Pérez-Peña, Richard. 2004. "Study Says 50% of Children Enter Shelters with Asthma." New York Times. March 2:A23.

Perrow, Charles. 1984. Normal Accidents: Living with High-Risk Technologies. New York: Basic Books.

Perrucci, Robert, and Earl Wysong. 2002. The New Class Society: Goodbye American Dream? (2nd ed.). Lanham, MD: Rowman & Littlefield.

Petersilia, Joan. 1999. "A Decade of Experimenting with Intermediate Sanctions: What Have We Learned?" Justice Research and Policy 1:9–23.

Peterson, Richard A., and Albert Simkus. 1992. "How Musical Tastes Mark Occupational Status Groups."

In Michelle Lamont and Marcel Fournier (eds.), Cultivating Differences: Symbolic Boundaries and the Making of Inequality. Chicago: University of Chicago Press: 152–186.

Petras, James, and Henry Veltmeyer. 2001. Globalization Unmasked: Imperialism in the 21st Century. New York: Zed Books.

Pew Research Center. 2003. Views of a Changing World 2003. Washington, D.C.: Pew Research Center.

———. Huge Racial Divide over Katrina and Its Consequences. Washington, D.C.: Pew Research Center.

———. 2007. "Global Attitudes Project: Spring 2007 Survey." http://pewglobal.org/reports/pdf/256topline.pdf. Accessed April 2009.

———. 2008. "Inside Obama's Sweeping Victory." http://pewresearch.org/pubs/1023/exit-poll-analysis-2008. Accessed June 2009.

Pew Research Center for the People and the Press. 2009. "Environment, Immigration, Health Care Slip Down The List: Economy, Jobs Trump All Other Policy Priorities in 2009." http://people-press.org/reports/pdf/485.pdf. Accessed June 2009.

Phillips, Kevin. 2008. Bad Money: Reckless Finance, Failed Politics, and the Global Crisis of American Capitalism. New York: Vintage.

Physicians for a National Health Program. 2009. "What Is Single Payer?" http://www.pnhp.org/facts/what_is_single_payer.php. Accessed May 2009.

Piaget, Jean. 1954. The Construction of Reality in the Child. New York: Basic Books.

Piliavin, Irving, Rosemary Gartner, Craig Thornton, and Ross Matsueda. 1986. "Crime, Deterrence, and Rational Choice." American Sociological Review 51:101–119.

Pitts-Taylor, Victoria. 2003. In the Flesh: The Cultural Politics of Body Modification. New York: Palgrave Macmillan.

Piven, Frances Fox, and Richard A. Cloward. 1988. Why Americans Don't Vote. New York: Pantheon.

Plummer, Gayle. 1985. "Haitian Migrants and Backyard Imperialism." Race and Class 26:35–43.

Population Reference Bureau. 2008. "World Population Data Sheet." Washington D.C.: Population Reference Bureau.

Porter, Eduardo. 2006. "Cost of Illegal Immigration May Be Less Than Meets the Eye." New York Times. April 16:B3.

Posner, Richard A. 2009. A Failure of Capitalism: The Crisis of '08 and the Descent into Depression. Cambridge, MA: Harvard University Press.

Poston, Dudley, Jr. 2000. "Social and Economic Development and the Fertility Transitions in Mainland China and Taiwan." Population and Development Review 26(Supplement):40–60.

Preston, Julia. 2007. "Survey Points to Tensions among Chief Minorities." New York Times. December 13: A1+.

———. 2009. "Mexico Data Say Migration to U.S. Has Plummeted." New York Times. May 15: A1+.

Project on Student Debt. 2008. Student Debt and the Class of 2007. Berkeley: Project on Student Debt.

Prüss-Üstün, Annette, Robert Bos, Fiona Gore, and Jamie Bartram. 2008. Safer Water, Better Health: Costs, Benefits and Sustainability of Interventions to Protect and Promote Health. Geneva, Switzerland: World Health Organization.

Public Citizen Water Privatization Overview. 2006. http://www.citizen.org/cmep/Water/general. Accessed April 2006.

Putnam, Robert D. 2000. Bowling Alone: The Collapse and Revival of American Community. New York: Simon & Schuster.

Qian, Zhenchao, and Daniel T. Lichter. 2007. "Social Boundaries and Marital Assimilation: Interpreting Trends in Racial and Ethnic Intermarriage." American Sociological Review 72:68–94.

Quadagno, Jill. 2005. One Nation Uninsured: Why the U.S. Has No National Health Insurance. New York: Oxford University Press.

Quart, Alissa. 2003. Branded: The Buying and Selling of Teenagers. Cambridge, MA: Perseus.

Quillian, Lincoln. 1996. "Group Threat and Regional Change in Attitudes toward African-Americans." American Journal of Sociology 102:816–860.

———. 2006. "New Approaches to Understanding Racial Prejudice and Discrimination." Annual Review of Sociology 32:299–328.

Quinney, Richard. 1980. Class, State, and Crime (2nd ed.). New York: Longman.

Rajagopal, Balakrishnan. 2003. International Law from Below: Development, Social Movements, and Third World Resistance. Cambridge: Cambridge University Press.

Raley, J. Kelly. 1996. "A Shortage of Marriageable Men? A Note on the Role of Cohabitation in Black-White Differences in Marriage Rates." American Sociological Review 61:973–983.

Reid, Lori L. 2002. "Occupational Segregation, Human Capital, and Motherhood: Black Women's Higher Exit Rates from Full-Time Employment." Gender and Society 16:728–747.

Reiman, Jeffrey. 2005. The Rich Get Richer and the Poor Get Prison: Ideology, Class, and Criminal Justice (8th ed.). Boston: Allyn & Bacon.

Renzulli?, Linda A., and Vincent J. Roscigno. 2007. "Charter Schools and the Public Good." Contexts 6(1):31–3.

Reskin, Barbara. 1989. "Women Taking 'Male' Jobs Because Men Leave Them." IlliniWeek. July 20:7.

Reverby, Susan. 1987. Ordered to Care: The Dilemma of American Nursing. New York: Cambridge University Press.

Ricento, Thomas, and Barbara Burnaby (eds.). 1998. Language and Politics in the United States and Canada: Myths and Realities. Mahwah, NJ: Erlbaum.

Rich, Paul. 1999. "American Voluntarism, Social Capital, and Political Culture." Annals of the American Academy of Political and Social Science 565:15–34.

Ridgeway, Cecilia L., and Lynn Smith-Lovin. 1999. "The Gender System and Interaction." Annual Review of Sociology 25:191–216.

Rieker, Patricia R., and Chloe E. Bird. 2000. "Sociological Explanations of Gender Differences in Mental and Physical Health." In Chloe E. Bird, Peter Conrad, and Allan Fremont (eds.), Handbook of Medical Sociology. New York: Prentice-Hall: 98–113.

Risman, Barbara J. 1998. Gender Vertigo: American Families in Transition. New Haven, CT: Yale University Press.

Ritzer, George. 1996. The McDonaldization of Society (rev. ed.). Thousand Oaks, CA: Pine Forge Press.

_____. Modern Sociological Theory. New York: McGraw-Hill.

Robert, Stephanie A., and James S. House. 2000. "Socioeconomic Inequalities in Health: Integrating, Individual-, Community-, and Societal-Level Theory and Research." In Gary L. Albrecht, Ray Fitzpatrick, and Susan C. Scrimshaw (eds.), Handbook of Social Studies in Health and Medicine. Thousand Oaks, CA: Sage: 115–135.

Robert Wood Johnson Foundation. 2006. "Sex and Intoxication More Common among Women on Spring Break, According to AMA Poll." http://www.rwjf.org/pr/product.jsp?id=21831. Accessed 2006.

Romaine, Suzanne. 2000. Language in Society: An Introduction to Sociolinguistics (2nd ed.). Oxford, UK: Oxford University Press.

Roschelle, Anne R., and Peter Kaufman. 2004. "Fitting In and Fighting Back: Stigma Management Strategies among Homeless Kids." Symbolic Interaction 27:23–46.

Rose, Peter. 1981. They and We: Racial and Ethnic Relations in the United States (3rd ed.). New York: Random House.

Rosen, Jeffrey. 2000. The Unwanted Gaze: The Destruction of Privacy in America. New York: Random House.

Ross, Catherine E., and John Mirowsky. 2002. "Family Relationships, Social Support and Subjective Life Expectancy." Journal of Health and Social Behavior 43:469–489.

Ross, Stephen L., and Margery A. Turner. 2005. "Housing Discrimination in Metropolitan America: Explaining Changes between 1989 and 2000." Social Problems 52:152–180.

Rothenberg, Paula S. (ed.) 2002. White Privilege: Essential Readings on the Other Side of Racism. New York: Worth Publishers.

Rothman, Barbara Katz. 2000. Recreating Motherhood: Ideology and Technology in a Patriarchal Society. New Brunswick, NJ: Rutgers University Press.

_____. 2005. Weaving a Family: Untangling Race and Adoption. New York: Beacon.

Rothman, David J. 1997. Beginnings Count: The Technological Imperative in American Health Care. New York: Oxford University Press.

Rothstein, Richard. 2004. Class and Schools: Using Social, Economic, and Educational Reform to Close the Black-White Achievement Gap. Washington, D.C.: Economic Policy Institute.

Rotow, Thomas. 2000. "A Time to Join, a Time to Quit: The Influence of Life Cycle Transitions on Voluntary Association Membership." Social Forces 78:1133–1161.

Rubin, Barry. 2002. Islamic Fundamentalism in Egyptian Politics (2nd ed.). New York: Macmillan.

Rutter et al., and the English and Romanian Adoptees (ERA) Study Team. 1999. "Quasi-Autistic Patterns Following Severe Early Global Privation." Journal of Child Psychology 40:537–549.

Saad, Linda. 2006. "Anti-Muslim Sentiments Fairly Commonplace." http://www.gallup.com/poll/24073/AntiMuslim-Sentiments-Fairly-Commonplace.aspx. Accessed May 2009.

_____. 2008. "Americans Evenly Divided on Morality of Homosexuality." http://www.gallup.com/poll/108115/Americans-Evenly-Divided-Morality-Homosexuality.aspx. Accessed May 2009.

Sacks, Peter. 2007. Tearing Down the Gates: Confronting the Class Divide in American Education. Berkeley: University of California Press.

Sadker, Myra, and David Sadker. 1994. Failing at Fairness: How Our Schools Cheat Girls. New York: Simon & Schuster.

Saliba, John A. 2003. Understanding New Religious Movements. Lanham, MD: AltaMira Press.

Sampson, Robert, Jeffrey Morenoff, and Felton Earls. 1999. "Beyond Social Capital: Spatial Dynamics of Collective Efficacy for Children." American Sociological Review 64:633–660.

Sampson, Robert, and Stephen W. Raudenbush. 1999. "Systematic Social Observation of Public Spaces: A New Look at Disorder in Urban Neighborhoods." American Journal of Sociology 105:603–651.

Sampson, Robert J., Jeffrey D. Morenoff, and Thomas Gannon-Rowley. 2002. "Assessing Neighborhood Effects: Social Processes and New Directions in Research." Annual Review of Sociology 28:443–478.

Sanday, Peggy Reeves. 1990. Fraternity Gang Rape: Sex, Brotherhood, and Privilege on Campus. New York: New York University Press.

Saporito, Salvatore. 2003. "Private Choices, Public Consequences: Magnet School Choice and Segregation by Race and Poverty." Social Problems 50:181–203.

Sassen, Saskia. 2001. The Global City: New York, London, Tokyo (2nd ed.). Princeton, N.J.: Princeton University Press.

———. 2006. Sociology of Globalization. New York: Norton.

Schneider, Mark, Paul Teske, and Melissa Marschall. 2000. Choosing Schools: Consumer Choice and the Quality of American Schools. Princeton, NJ: Princeton University Press.

Schor, Juliet B. 1998. The Overspent American: Upscaling, Downshifting, and the New Consumer. New York: Basic Books.

Schur, Edwin M. 1979. Interpreting Deviance: A Sociological Introduction. New York: Harper & Row.

Schwartz, Barry. 1983. "George Washington and the Whig Conception of Heroic Leadership." American Sociological Review 48:18–33.

Schwartz, Barry, Hazel Rose Markus, and Alana Conner Snibbe. 2006. "Is Freedom Just Another Word for Many Things to Buy?" New York Times Magazine. February 26:14–15.

Scott, Janny. 2006. "Cities Shed Middle Class, and Are Richer and Poorer for It." New York Times. July 23:A1+.

Scott, Marvin B., and Stafford M. Lyman. 1968. "Accounts." American Sociological Review 33:46–62.

Scott, W. Richard. 2004. "Reflections on a Half-Century of Organizational Sociology." Annual Review of Sociology 30:1–21.

Seale, Clive, Sue Ziebland, and Jonathan Charteris-Black. 2006. "Gender, Cancer Experience, and Internet Use: A Comparative Keyword Analysis of Interviews and Online Cancer Support Groups." Social Science & Medicine 62(10):2577–2590.

Seccombe, Karen, and Rebecca L. Warner. 2004. Marriage and Families: Relationships in Context. Belmont, CA: Wadsworth.

Sedlak, Andrea, and Diane D. Broadhurst. 1996. Third National Incidence Study of Child Abuse and Neglect. Washington, D.C.: U.S. Department of Health and Human Services.

Sedgh, Gilda, Stanley Henshaw, Susheela Singh, Elisabeth Åhman, and Iqbal H. Shah. 2007. "Induced Abortion: Estimated Rates and Trends Worldwide." The Lancet 370:1338–1345.

Selman, Peter. 2007. "Diaper Diaspora." Foreign Policy 158:32–33.

Semmerling, Tim Jon. 2006. "Evil" Arabs in American Popular Film: Orientalist Fear. Austin, TX: University of Texas Press.

Sen, Amartya. 1999. Development as Freedom. New York: Knopf.

Sengupta, Somini. 2008. "Crusader Sees Wealth as Cure for Caste Bias." New York Times. August 29: A1+.

Settersten, Richard A. Jr., Frank F. Furstenberg, Jr., and Rubén G. Rumbaut (eds.) 2006. On the Frontier of Adulthood: Theory, Research, and Public Policy. Chicago: University of Chicago Press.

Shackelford, Todd K. 2005. "An Evolutionary Psychological Perspective on Cultures of Honor." Evolutionary Psychology 3:381–391.

Shapiro, Laura. 1994. "A Tomato with a Body That Just Won't Quit." Newsweek. June 6:80–82.

Shapiro, Thomas M. 2004. The Hidden Cost of Being African American: How Wealth Perpetuates Inequality. New York: Oxford University Press.

Sherif, Muzafer. 1936. The Psychology of Social Norms. New York: Harper & Row.

Sherkat, Darren E., and Christopher G. Ellison. 1999. "Recent Developments and Current Controversies in the Sociology of Religion." Annual Review of Sociology 25:363–394.

Shierholz, Heidi. 2009. "Nearly Five Unemployed Workers for Every Available Job." http://www.epi.org/publications/entry/jolts_20090512/. Economic Policy Institute. Accessed June 2009.

Shkilnyk, Anastasia M. 1985. A Poison Stronger Than Love: The Destruction of an Ojibwa Community. New Haven, CT: Yale University Press.

Shover, Neal. 2006. Choosing White-Collar Crime. New York: Cambridge University Press.

Simons, Ronald, and Phyllis Gray. 1989. "Perceived Blocked Opportunity as an Explanation of Delinquency among Lower-Class Black Males." Journal of Research on Crime and Delinquency 26:90–101.

Simpson, Richard L. 1985. "Social Control of Occupations and Work." Annual Review of Sociology 11: 415–436.

Skenazy, Lenore. 2009. Free-Range Kids. freerangekids.wordpress.com. Accessed May 2009.

Skocpol, Theda. 1996. Boomerang: Clinton's Health Security Effort and the Turn against Government in U.S. Politics. New York: Norton.

Smaje, Chris. 2000. Natural Hierarchies: The Historical Sociology of Race and Caste. Malden, MA: Blackwell.

Small, Mario Luis, and Katherine Newman. 2002. "Urban Poverty after 'The Truly Disadvantaged': The Rediscovery of the Family, the Neighborhood, and Culture." Annual Review of Sociology 27:23–45.

Small, Meredith. 2001. Kids: How Biology and Culture Shape the Way We Raise Our Children. New York: Doubleday.

Smith, Aaron. 2009. The Internet's Role in Campaign 2008. Washington, D.C.: Pew Internet & American Life Project.

Smith, Christian. 1991. The Emergence of Liberation Theology: Radical Religion and Social Movement Theory. Chicago: University of Chicago Press.

Smith, Jane I. 1999. Islam in America. New York: Columbia University Press.

Smith, Kirsten P., and Nicholas A. Christakis. 2008. "Social Networks and Health." Annual Review of Sociology 34:405–29.

Smith, Ryan A. 1997. "Race, Income, and Authority at Work: A Cross-Temporal Analysis of Black and White Men (1972–1994)." Social Problems 44:19–37.

Smock, Pamela J. 2000. "Cohabitation in the United States: An Appraisal of Research Themes, Findings, and Implications." Annual Review of Sociology 26:1–20.

Smolensky, Eugene, and Jennifer Appleton Gootman (eds.). Working Families and Growing Kids: Caring for Children and Adolescents. Washington: National Academies Press.

Snow, David A., E. Burke Rochford, Jr., Steven K. Worden, and Robert D. Benford. 1986. "Frame Alignment Processes, Micromobilization, and Movement Participation." American Sociological Review 51:464–481.

Social Institutions & Gender Index. 2009. "Gender Equality and Social Institutions in Ghana" http://www.gender-index.org. Accessed June 2009.

Sohoni, Neera Kuckreja. 1994. "Where Are the Girls?" Ms. Magazine. July/August:96.

Sönmez, Sevil, Yorghos Apostolopoulos, Chong Ho Yu, Shiyi Yang, Anna Mattila, and Lucy C. Yu. 2006. "Binge Drinking and Casual Sex on Spring Break." Annals of Tourism Research 33(4):895–917.

Southwell, Priscilla Lewis, and Marcy Jean Everest. 1998. "The Electoral Consequences of Alienation: Nonvoting and Protest Voting in the 1992 Presidential Race." Social Science Journal 35:53–51.

Spring, Joel. 2004. Deculturalization and the Struggle for Equality: A Brief History of the Education of Dominated Cultures in the United States (4th ed.). New York: McGraw-Hill.

Stacey, Judith, and Timothy J. Biblarz. 2001. "(How) Does the Sexual Orientation of Parents Matter?" American Sociological Review 66:159–183.

Stalp, Marybeth C., M. Elise Radina, and Annette Lynch. 2008. "'We Do It Cuz It's Fun': Gendered Fun and Leisure for Midlife Women through Red Hat Society Membership." Sociological Perspectives 51: 325–348.

Stanton, Gregory H. 2009. "Stages of Genocide." http://www.genocidewatch.org/aboutgenocide/8stagesofgenocide.html. Accessed May 2009.

Stark, Rodney, and Roger Finke. 2000. Acts of Faith. Berkeley: University of California Press.

Starr, Paul. 1982. The Social Transformation of American Medicine. New York: Basic Books.

Steffensmeier, Darrell, and Emilie Allan. 1996. "Gender and Crime: Toward a Gendered Theory of Female Offending." Annual Review of Sociology 22:459–487.

Steffensmeier, Darrell J., Emilie Allan, Miles Harer, and Cathy Streifel. 1989. "Age and the Distribution of Crime." American Journal of Sociology 94:803–831.

Steinhauer, Jennifer. 2009. "To Cut Costs, States Relax Prison Policies." New York Times. March 25:A1+.

Stenner, Karen. 2005. The Authoritarian Dynamic. New York: Cambridge University Press.

Stern, Jessica. 2003. Terror in the Name of God: Why Religious Militants Kill. New York: HarperCollins.

Stiglitz, Joseph E. 2003. Globalization and Its Discontents. New York: Norton.

Stolte, John F., Gary Alan Fine, and Karen S. Cook. 2001. "Sociological Miniaturism: Seeing the Big through the Small in Social Psychology." Annual Review of Sociology 27:387–412.

Street, Marc D. 1997. "Groupthink: An Examination of Theoretical Issues, Implications, and Future Research Suggestions." Small Group Research 28:72–93.

Sulik, Gayle A., and Astrid Eich-Krohm. 2008. "No Longer a Patient: The Social Construction of a Medical Consumer Advances in Medical Sociology." Advances in Medical Sociology 10:33–28.

Sullivan, Deborah. 2001. Cosmetic Surgery: The Cutting Edge of Commercial Medicine in America. New Brunswick, NJ: Rutgers University Press.

Sullivan, Teresa A., Elizabeth Warren, and Jay Lawrence Westbrook. 2000. The Fragile Middle Class: Americans in Debt. New Haven, CT: Yale University Press.

Sutherland, Edwin H. 1961. White-Collar Crime. New York: Holt, Reinhart & Winston.

Swidler, Ann. 1986. "Culture in Action: Symbols and Strategies." American Sociological Review 51:273–286.

Szabo, Liz, and Julie Appleby. 2009. "21% of Americans Scramble to Pay Medical, Drug Bills." USA Today. March 11:1A+.

Takamura, Jeanette. 2002. "Social Policy Issues and Concerns in a Diverse Aging Society." Generations 26(3):33–38.

Tannen, Deborah. 1990. You Just Don't Understand. New York: Morrow.

———. 1994. Talking from 9 to 5: How Women's and Men's Conversational Styles Affect Who Gets Heard, Who Gets Credit, and What Gets Done at Work. New York: Morrow.

Teachman, Jay. 2002. "Stability across Cohorts in Divorce Risk Factors." Demography 39:331–351.

Teachman, Jay D., Lucky M. Tedrow, and Kyle D. Crowder. 2000. "The Changing Demography of America's Families." Journal of Marriage and the Family 62:1234–1246.

Tedeschi, James T., and Marc Riess. 1981. "Identities, the Phenomenal Self, and Laboratory Research." In J. T. Tedeschi (ed.), Impression Management Theory and Social Psychological Research. Orlando, FL: Academic.

Thoits, Peggy A., and Lyndi N. Hewitt. 2001. "Volunteer Work and Well Being." Journal of Health and Social Behavior 42:115–131.

Thomas, W. I., and Dorothy Thomas. 1928. The Child in America: Behavior Problems and Programs. New York: Knopf.

Thompson, Kevin. 1989. "Gender and Adolescent Drinking Problems: The Effects of Occupational Structure." Social Problems 36:30–47.

Thornberry, Terence P., and Margaret Farnworth. 1982. "Social Correlates of Criminal Involvement: Further Evidence on the Relationship between Social Status and Criminal Behavior." American Sociological Review 47:505–518.

Tilly, Charles. 1998. Durable Inequality. Berkeley: University of California Press.

———. 2004. Social Movements, 1768–2004. Boulder, CO: Paradigm Publishers.

Tjaden, Patricia, and Nancy Thoennes. 1998. "Prevalence, Incidence, and Consequences of Violence against Women: Findings from the National Violence against Women Survey." National Institute of Justice Research in Brief. November.

———. 2000. "Nature and Consequences of Intimate Partner Violence. Research Report 181867. Washington, D.C.: U.S. Department of Justice, National Institute of Justice.

Todd, Chuck, and Sheldon Gawiser. 2009. How Barack Obama Won: A State-by-State Guide to the Historic 2008 Presidential Election. New York: Vintage.

Tomaskovich-Devey, Donald, and Sheryl Skaggs. 2002. "Sex Segregation, Labor Process Organization, and Gender Earnings Inequality." American Journal of Sociology 108:102–128.

Troeltsch, Ernst. 1931. The Social Teaching of the Christian Churches. New York: Macmillan.

Troop, Don. 2007. "You're Never Gonna Believe This One." Chronicle of Higher Education. January 19:A4.

Trudgill, Peter. 2000. Sociolinguistics: An Introduction to Language and Society. New York: Penguin.

Turk, Austin. 2004. "Sociology of Terrorism." Annual Review of Sociology 30:271–286.

Turner, Jonathan, and Leonard Beeghley. 1981. The Emergence of Sociological Theory. Homewood, IL: Dorsey.

Turner, Jonathan, and David Musick. 1985. American Dilemmas. New York: Columbia University Press.

Turner, R. Jay, and William R. Avison. 2003. "Status Variations in Stress Exposure: Implications for the Interpretation of Research on Race, Socioeconomic Status, and Gender." Journal of Health and Social Behavior 44:488–505.

Turner, R. Jay, Blair Wheaton, and Donald A. Lloyd. 1995. "The Epidemiology of Social Stress." American Sociological Review 60:104–125.

Turtle Island Native Network News. "Spotlight on Aboriginal Rights." 2009. http://www.turtleisland.org/news/news-grassy.htm. Accessed April 2009.

Twenge, Jean, M. W. Keith Campbell, and Craig A Foster. 2003. "Parenthood and Marital Satisfaction: A Meta-Analytic Review." Journal of Marriage and Family 65:574–583.

Uehara, Edwina S. 1995. "Reciprocity Reconsidered: Gouldner's Moral Norm of Reciprocity and Social Support." Journal of Social and Personal Relationships 12:483–502.

Uggen, Christopher, and Jeff Manza. 2002. "Democratic Contraction? Political Consequences of Felon Disenfranchisement in the United States." American Sociological Review 67:777–803.

Umberson, Debra, Kristi Williams, Daniel A. Powers, Meichu D. Chen, and Anna M. Campbell. 2005. "As Good as It Gets? A Life Course Perspective on Marital Quality." Social Forces 84:487–506.

UNAIDS/WHO. 2007. AIDS Epidemic Update: 2007. Geneva, Switzerland: World Health Organization.

United Food and Commercial Workers. 2006. "Wal-Martization and Wages." http://www.ufcw.org. Accessed July 2006.

United Nations Development Programme. 2007. Human Development Report. New York, NY: Palgrave Macmillan.

United Nations High Commissioner for Refugees. 2009. Population of Concerns to UNHCR: 2008. New York: United Nations.

United Nations Population Fund. 2000. State of World Population 2000. New York: United Nations Population Fund.

_____. 2006. International Migration and Development: Report of the Secretary General. New York: United Nations.

_____. 2009. Migration: A World on the Move. http://www.unfpa.org/pds/migration.html. Accessed June 2009.

United Nations Statistics Division. 2009. "Composition of Macro Geographical (Continental) Regions, Geographical Sub-Regions, and Selected Economic and Other Groupings." http://unstats.un.org/unsd/methods/m49/m49regin.htm#developed. Accessed May 2009.

U.S. Attorney, Eastern District of Pennsylvania. 2009. "Pharmaceutical Company Eli Lilly to Pay Record $1.415 Billion for Off-Label Drug Marketing." http://www.usdoj.gov/usao/pae/News/Pr/2009/jan/lillyrelease.pdf. Accessed May 2009.

U.S. Bureau of the Census. 1975. "Historical Statistics of the United States: Colonial Times to 1970 (Bicentennial ed., Part 1)." Washington, D.C.: U.S. Government Printing Office.

_____. 2005. "Asset Ownership of Households: 2000." http://www.census.gov/hhes/www/wealth/1998_2000/wlth00-1.html. Accessed May 2009.

_____. 2006. "The Effects of Government Taxes and Transfers on Income and Poverty: 2004." http://www.census.gov/hhes/www/poverty/effect2004/effectofgovtandt2004.html. Accessed April 2009.

_____. 2008a. "Historical Income Tables—Households." http://www.census.gov/hhes/www/income/histinc/h01AR.html. Accessed May 2009.

_____. 2008b. "Income, Poverty, and Health Insurance Coverage in the United States: 2007." http://www.census.gov/hhes/www/poverty/poverty07.html. Accessed 2009.

_____. 2008c. Statistical Abstract of the United States: 2008. Washington, D.C.: U.S. Government Printing Office.

_____. 2009a. Statistical Abstract of the United States: 2009. Washington, D.C.: U.S. Government Printing Office.

_____. 2009b. American Fact Finder. http://factfinder.census.gov. Accessed April 2009.

_____. 2009c. "Poverty 2007 Tables." http://www.census.gov/hhes/www/poverty/poverty07. Accessed April 2009.

_____. 2009d. "America's Families and Living Arrangements: 2008." http://www.census.gov/population/www/socdemo/hh-fam/cps2008.html. Accessed May 2009.

_____. 2009e. International Data Base. http://www.census.gov/ipc/www/idb. Accessed June 2009.

U.S. Bureau of Labor Statistics. 2002. Occupational Outlook Quarterly. Winter 2001–02.

_____. 2005. "Highlight of Women's Earnings in 2004." Report 987. Washington D.C.: U.S. Government Printing Office.

_____. 2009. Occupational Outlook Handbook 2008–09. Washington D.C.: U.S. Government Printing Office.

U.S. Department of Education. 2009. Digest of Education Statistics 2008. Washington D.C.: U.S. Government Printing Office.

U.S. Department of Justice. 1995. Crime in the United States: Uniform Crime Reports, 1995. Bureau of Justice Statistics. Washington D.C.: U.S. Government Printing Office.

_____. 2009. Sourcebook of Criminal Justice Statistics. http://www.albany.edu/sourcebook/. Accessed April 2009.

U.S. Department of State. 2009. "Background Note: Sudan." http://www.state.gov/outofdate/bgn/s/115612.htm. Accessed June 2009.

U.S. General Accounting Office. 1996. "Death Penalty Sentencing: Research Indicates Pattern of Racial Disparities." In H. A. Bedau (ed.), The Death Penalty in America: Current Controversies. New York: Oxford University Press: 268–272.

United States Holocaust Memorial Museum. 2009. Holocaust Encyclopedia. http://www.ushmm.org/wlc. Accessed May 2009.

Useem, Bert, and Jack A. Goldstone. 2002. "Forging Social Order and Its Breakdown: Riot and Reform in U.S. Prisons." American Sociological Review 67: 499–525.

van de Walle, Etienne, and John Knodel. 1980. "Europe's Fertility Transition." Population Bulletin 34(6):1–43.

Van Gundy, Karen. 2006. "Substance Abuse in Rural and Small Town America." Reports on Rural America. 1(2):1–38.

van Vugt, Mark, and Mark Snyder (eds.). 2002. "Special Issue: Cooperation in Society: Fostering Community Action and Civic Participation." American Behavioral Scientist 45:769–782.

Vannini, Phillip. 2004. "The Meanings of a Star: Interpreting Music Fans' Reviews." Symbolic Interaction 27: 47–69.

Vares, Tiina, and Virginia Braun. 2006. "Spreading the Word, but What Word Is That? Viagra and Male Sexuality in Popular Culture." Sexualities 9:315–332.

Vaughan, Diane. 1996. The Challenger Launch Decision: Risky Technology, Culture, and Deviance at NASA. Chicago: University of Chicago Press.

Veblen, Thorstein. 1919. The Vested Interests and the State of the Industrial Arts. New York: Huebsch.

Venkatesh, Sudhir A. 2000. American Project: The Rise and Fall of a Modern Ghetto. Cambridge, MA.: Harvard University Press.

_____. 2008. Gang Leader for a Day: A Rogue Sociologist Takes to the Streets. New York: Penguin.

_____. 2009. Off the Books: The Underground Economy of the Urban Poor. Cambridge, MA: Harvard University Press.

Wade, Robert. 2001. "Making the World Development Report: Attacking Poverty." World Development 29:1435–1441.

_____. 2003. "What Strategies Are Viable for Developing Countries Today? The World Trade Organization and the Shrinking of 'Development Space.'" Review of International Political Economy 10:621–644.

Wagner, Dennis. 2006. "Meth Lays Siege to Indian Country." USA Today. March 30:3A.

Waite, Linda J., and Evelyn L. Lehrer. 2003. "The Benefits from Marriage and Religion in the United States: A Comparative Analysis." Population and Development Review 29:255–275.

Wald, Kenneth D. 1987. Religion and Politics in the United States. New York: St. Martin's.

Walker, Edward T. 2008. "Contingent Pathways from Joiner to Activists: The Indirect Effect of Participation in Voluntary Associations on Civic Engagement." Sociological Forum 23:116–143.

Walker, Iain, and Heather J. Smith. 2002. Relative Deprivation: Specification, Development, and Interpretation. New York: Cambridge University Press.

Wallerstein, Immanuel. 2004. World-Systems Analysis: An Introduction. Durham, NC: Duke University Press.

Waltman, Jerold. 2000. The Politics of the Minimum Wage. Urbana, IL: University of Illinois Press.

Washburn, Jennifer. 2006. University, Inc.: The Corporate Corruption of American Higher Education. New York: Basic Books.

Watamura, Sarah E., Bonny Donzella, Jan Alwin, and Megan R. Gunnar. 2003. "Morning-to-Afternoon Increases in Cortisol Concentrations for Infants and Toddlers at Child Care: Age Differences and Behavioral Correlates." Child Development 74:1006–1020.

Waters, Mary C., and Tomas R. Jimenez. 2005. "Assessing Immigrant Assimilation: New Empirical and Theoretical Challenges." Annual Review of Sociology 31:105–125.

Watkins, Kevin. 2006. Human Development Report 2006. New York: Palgrave Macmillan.

Weber, Max. 1958. The Protestant Ethic and the Spirit of Capitalism. (Talcott Parsons, trans.) New York: Scribners. (Originally published 1904–1905.)

_____. 1970a. "Bureaucracy." In H. H. Gerth and C. Wright Mills (trans.), From Max Weber: Essays in Sociology. New York: Oxford University Press. (Originally published 1910.)

_____. b "Religion." In H. H. Gerth and C. Wright Mills (trans.), From Max Weber: Essays in Sociology. New York: Oxford University Press. (Originally published 1910.)

Weil, Frederick. 1989. "The Sources and Structure of Legitimation in Western Democracies." American Sociological Review 54:682–706.

Weisberg, Robert. 2005. "The Death Penalty Meets Social Science: Deterrence and Jury Behavior Under New Scrutiny." Annual Review of Law and Social Science 1:151–170.

Weitz, Rose. 2001. "Women and Their Hair: Seeking Power through Resistance and Accommodation." Gender & Society 15:667–686.

_____. 2004. Rapunzel's Daughters: What Women's Hair Tells Us about Women's Lives. New York: Farrar, Straus, and Giroux.

_____. 2010. The Sociology of Health, Illness, and Health Care: A Critical Approach (5th ed.). Belmont, CA: Wadsworth.

Weitzer, Ronald. 2007. "The Social Construction of Sex Trafficking: Ideology and Institutionalization of a Moral Crusade." Politics and Society 35:447–475.

Wellman, Barry (ed.). 1999. Networks in the Global Village: Life in Contemporary Communities. Boulder, CO: Westview.

Wellman, Barry, Janet Salaff, Dimitrina Dimitrova, Laura Garton, and Milena Gulia. 1996. "Computer Networks as Social Networks: Collaborative Work, Telework, and Virtual Community." Annual Review of Sociology 22:213–238.

Wellman, Barry, and Scot Wortley. 1990. "Different Strokes from Different Folks: Community Ties and Social Support." American Journal of Sociology 96:558–588.

Welsh, Sandy. 1998. "Gender and Sexual Harassment." Annual Review of Sociology 25:169–190.

West, Candace, and Don H. Zimmerman. 1987. "Doing Gender." Gender and Society 1:125–151.

Whorf, Benjamin L. 1956. Language, Thought, and Reality. Cambridge, MA: MIT Press.

Wilkins, Amy C. 2008. "Wannabes, Goths, and Christians: The Boundaries of Sex, Style, and Status." Chicago: University of Chicago Press.

Wilkinson, Richard G. 1996. Unhealthy Societies: The Afflictions of Inequality. London: Routledge.

———. 2005. The Impact of Inequality. London: New Press.

Williams, Christine L. 1992. "The Glass Escalator: Hidden Advantages for Men in the 'Female' Professions." Social Problems 39:253–267.

———. 2006. Inside Toyland: Working, Shopping, and Social Inequality. Berkeley: University of California Press.

Williams, David R. 1998. "African-American Health: The Role of the Social Environment." Journal of Urban Health: Bulletin of the New York Academy of Medicine 75:300–321.

Williams, David R., and Pamela Braboy Jackson. 2005. "Social Sources of Racial Disparities in Health." Health Affairs 24:325–35.

Williams, Kirk, and Susan Drake. 1980. "Social Structure, Crime, and Criminalization: An Empirical Examination of the Conflict Perspective." Sociological Quarterly 21:563–575.

Williams, Marian R., and Jefferson E. Holcolm. 2001. "Racial Disparity and Death Sentences in Ohio." Journal of Criminal Justice 29:207–218.

Williams, Rhys H. 1995. "Constructing the Public Good: Social Movements and Cultural Resources." Social Problems 42:124–144.

Wilson, Edward O. 1978. "Introduction: What Is Sociobiology?" In Michael S. Gregory, Anita Silvers, and Diane Sutch (eds.), Sociobiology and Human Nature. San Francisco: Jossey-Bass.

Wilson, James Q. 1992. "Crime, Race, and Values." Society 30:90–93.

Wilson, William J. 1978. The Declining Significance of Race. Chicago: University of Chicago Press.

———. 1987. The Truly Disadvantaged. Chicago: University of Chicago Press.

———. 1996. When Work Disappears: The World of the New Urban Poor. New York: Knopf.

———. 2009. More Than Just Race: Being Black and Poor in the Inner City. New York: Norton.

Winders, Bill. 1999. "The Roller Coaster of Class Conflict: Class Segments, Mass Mobilization, and Voter Turnout in the U.S., 1840–1996." Social Forces 77:833–860.

Winerip, Michael. 2003. "Rigidity in Florida and its Consequences." New York Times. July 23:A15.

Winkler, Anne E., Timothy D. McBride, and Courtney Andrews. 2005. "Wives Who Outearn Their Husbands: A Transitory or Persistent Phenomenon for Couples?" Demography 42:523–53.

Wirth, Louis. 1938. "Urbanism as a Way of Life." American Journal of Sociology 44(1):1–24.

Wolf, Martin. 2005. Why Globalization Works (2nd ed.). New Haven, CT: Yale University Press.

World Health Organization. 2008a. "Fact Sheet No. 241: Female Genital Mutilation." Geneva, Switzerland: World Health Organization.

———. 2008b. "WHO Report on the Global Tobacco Epidemic, 2008." Geneva, Switzerland: World Health Organization.

———. 2009. "Maternal Mortality." http://www.who.int/making_pregnancy_safer/topics/maternal_mortality/en/index.html. Accessed May 2009.

Wright, Erik O. 1985. Classes. London: Verso.

Wrong, Dennis. 1961. "The Oversocialized Conception of Man in Modern Sociology." American Sociological Review 26(April):183–193.

———. 1979. Power. New York: Harper & Row.

Wykes, Maggie, and Barrie Gunter. 2005. The Media and Body Image: If Looks Could Kill. Thousand Oaks, CA: Sage.

Zhou, Min, and James V. Gatewood. 2000. "Mapping the Terrain: Asian American Diversity and the Challenges of the Twenty-First Century." Asian American Policy Review 9:5–29.

Zhu, Wei Xing, Li Lu, and Therese Hesketh. 2009. "China's Excess Males, Sex Selective Abortion, and One Child Policy: Analysis of Data from 2005 National Intercensus Survey." British Medical Journal 338: 920–923.

Zweigenhaft, Richard L., and G. William Domhoff. 1998. Diversity in the Power Elite: Have Women and Minorities Reached the Top? New Haven, CT: Yale University Press.

Photo Credits

This page constitutes an extension of the copyright page. We have made every effort to trace the ownership of all copyrighted material and to secure permission from copyright holders. In the event of any question arising as to the use of any material, we will be pleased to make the necessary corrections in future printings. Thanks are due to the following authors, publishers, and agents for permission to use the material indicated.

Chapter 1: p. 1, Angela Hampton/Bubbles Photolibrary/Alamy; p. 4, AP Images; p. 6, © Brown Brothers; p. 6, © Brown Brothers; p. 7, © Brown Brothers; p. 8, © Bettmann/Corbis; p. 9, © Brown Brothers; p. 9, © Ed Kashi; p. 10, © Corbis; p. 10, © Bettmann Archive/Corbis; p. 12, AP Images; p. 14, © Jeff Greenberg/PhotoEdit; p. 15, © John Van Hasselt/Corbis; p. 23, © David Young Wolff/PhotoEdit; p. 25, © Scott Houston/Sygma/Corbis; p. 26, Reuters/Scott Olson.

Chapter 2: p. 31, Jon Arnold Images Ltd/Alamy; p. 33, © Andre Perlstein/Sygma/Corbis; p. 36, AP Images; p. 38, © Nathan Benn/Corbis; p. 40, © David Young Wolff/PhotoEdit; p. 44, Bob Daemmrich/The Image Works. Reproduced by permission; p. 45, © David Austen/Woodfin Camp & Associated; p. 47, 2009/Jupiterimages; p. 51, AP Images.

Chapter 3: p. 55, David Sacks; p. 56, © Martin Rogers/Stock Boston Inc.; p. 57, Cynthia Johnson/Getty Images News/Getty Images; p. 60, altrendo images/Altrendo/Getty Images; p. 64, Foto Begsteiger/Alexa BenteWoodyStock/Alamy; p. 66, Alyson Aliano/Riser/Getty Images; p. 70, © Sean Sprague/The Image Works. Reproduced by permission; p. 71, © O'Brien Productions/Corbis; p. 72, © Owen Franken/Terra/Corbis; p. 76.

Chapter 4: Zoran Milich/Masterfile; p. 80, Chip Somodevilla/Getty Images News/Getty Images; p. 82, © Catherine Karnow/Woodfin Camp & Associates; p. 84, © Anthony Bannister; Gallo Images/Terra/Corbis; p. 88, MIKE CASSESE/Reuters/Landov; p. 90, AP Images; p. 91, ©iStockphoto.com/Yuri Arcurs; p. 93, © Jose Luis Pelaez, Inc./Surf/Corbis; p. 94, Ladi Kirn/Alamy; p. 96, © Patrick Ward/Stock Boston Inc.; p. 97, Janine Wiedel/Photolibrary/Alamy.

Chapter 5: p. 100, Image copyright Monkey Business Images, 2009. Used under license from Shutterstock.com.; p. 103, © M. Greenlar/The Image Works. Reproduced by permission.; p. 105, Ghislain & Marie David de Lossy/Taxi/Getty Images.; p. 107, © Bob Daemmrich/Stock, Boston Inc.; p. 110, Milgram, Stanley, Pps Manuscripts & Archives, Yale University; p. 114, Image copyright Monkey Business Images, 2009. Used under license from Shutterstock.com; p. 118, © 2009 Jupiterimages; p. 119, © Mark Peterson/Corbis News/Corbis; p. 122, MATTHIAS SCHRADER/dpa/Landov.

Chapter 6: p. 126, Image copyright Lisa F. Young, 2009. Used under license from Shutterstock.com.; p. 129, Janine Wiedel Photolibrary/Alamy; p. 132, bobhdeering/Alamy; p. 134, © Alonv Reininger/Contact Press Images; p. 135, Stockbyte/White/PhotoLibrary; p. 140, iStockphoto.com/Stephanie Phillips; p. 144, © Joel Gordon; p. 148, © A. Ramey/Woodfin Camp & Associates.

Chapter 7: p. 151, ©2009/Jupiterimages; p. 153, brianindia/Alamy; p. 157, © Paul Barton/Surf/Corbis; p. 157, © Stephanie Maze/Woodfin Camp & Associates; p. 160, ©Michael Dwyer/Stock Boston Inc.; p. 162, Copyright © Spencer Grant/Photo Edit; p. 167, AP Images; p. 168, Joe Raedle/Getty

Images; p. 171, © Gerd Ludwig/Woodfin Camp & Associates; p. 177, © Les Stone/The Image Works. Reproduced by permission.; p. 180, AP Images.

Chapter 8: p. 184, Mario Tama/Getty Images; p. 186, Austrophoto/F1online digitale Bildagentur GmbH/Alamy; p. 189, Image copyright Monkey Business Images, 2009. Used under license from Shutterstock.com; p. 191, AP Images; p. 193, © Ted Spiegel/Encyclopedia/Corbis; p. 196, © Sandy Felsenthal/Encyclopedia/Corbis; p. 201, Image copyright Andresr, 2009. Used under license from Shutterstock.com; p. 204, © Ed Kashi/Corbis News/Corbis.

Chapter 9: p. 209, Brand X Pictures/Jupiterimages; p. 214, © Tony Freeman/PhotoEdit; p. 221, © Tom & Dee Ann McCarthy/Comet/Corbis; p. 223, © A. Ramey/Stock, Boston Inc.; p. 225, ©2009/Jupiterimages; p. 227, Ian Waldie/Getty Images.

Chapter 10: p. 232, IMAGEMORE Co., Ltd./Getty Images; p. 234, ©2009/Jupiterimages; p. 236, ©Bob Daemmrich/The Image Works. Reproduced by permission.; p. 237, Image copyright Konstantin Sutyagin , 2009. Used under license from Shutterstock.com; p. 239, AP Images; p. 243, © Mark Peterson/Corbis; p. 245, © Joel Gordon; p. 249, Francis Hogan/Electronic Publishing Services Inc., N.Y.C.; p. 252, ©2009/Jupiterimages.

Chapter 11: p. 257, ©2009/Jupiterimages; p. 262, AP Images; p. 263, © Gerd Ludwig/Woodfin Camp & Associates; p. 266, ©2009/Jupiter Images.; p. 268, ©2009/Jupiterimages; p. 269, Image copyright Erwin Wodicka, 2009. Used under license from Shutterstock.com; p. 271, © Peter Beck/Corbis; p. 274, © Big Cheese Photo LLC/Alamy; p. 277, © Janine Wiedel Photolibrary/Alamy; p. 280, © Richard Hutchings/Photo Researchers, Inc.

Chapter 12: p. 283, John F Clarke/PhotoLibrary; p. 285, © Charles Gupton/Stock, Boston Inc.; p. 287, ©Bob Daemmrich/The Image Works. Reproduced by permission.; p. 289, ©Laurence Gough/Fotolia; p. 293, ©iStockphoto.com/Joseph C. Justice Jr.; p. 297, AP Images; p. 301, Martin Thomas Photography/Alamy; p. 304, Bill Lyons/Alamy; p. 309, Henry Francis du Pont Winterthur Museum.

Chapter 13: p. 312, DAVID NOBLE PHOTOGRAPHY/Alamy; p. 314, AP Images; p. 317, 1994 Tom Muscionico/Contact Press Images; p. 318, REUTERS/Mariana Bazo/Landov; p. 322, Photo by Marc Serota/Getty Images; p. 324, KYNDELL HARKNESS/MCT/Landov; p. 325, Joel Stettenheim/Corbis; p. 328, © Joe Rodriguez/Black Star Publishing/Picture Quest; p. 329, © Mira/Alamy; p. 331, AP Images; p. 335, © Seth Resnick; p. 337, Enigma/Alamy.

Chapter 14: p. 344, Jeremy Richards/Shutterstock; p. 350, Jenny Matthews/Alamy; p. 352, © Claudia Kunin/Surf/Corbis; p. 354, © Harold Castro; p. 355, Photo by Tim Graham/Getty Images; p. 357, AP Images; p. 361, © Irwin Thompson/Dallas Morning News/Corbis; p. 364, Photo by Bentley Archive/Popperfoto/Getty Images; p. 366, Peter Menzel ; p. 368, Alan Chandler/Photo Library; p. 371, AP Images; p. 372, Jeff Morgan tourism and leisure/Alamy.

Chapter 15: p. 375, Ellen McKnight/Alamy; p. 377, © Neal Preston/Encyclopedia/Corbis; p. 380, Visions LLC/Photo Library; p. 383, Alex Macnaughton/Photo Library; p. 385, © Sylvia Johnson/Woodfin Camp & Associates; p. 386, AP Images; p. 391, ©iStockphoto.com/© Alberto L. Pomares G.; p. 393, © Chris Pancewicz/Alamy; p. 394, EGimages/Alamy.

Name Index

Subject Index